T0399234

DEEP SHALE OIL AND GAS

DEEP SHALE OIL AND GAS

JAMES G. SPEIGHT, PhD, DSc
CD&W Inc., Laramie, WY, United States

AMSTERDAM • BOSTON • HEIDELBERG • LONDON
NEW YORK • OXFORD • PARIS • SAN DIEGO
SAN FRANCISCO • SINGAPORE • SYDNEY • TOKYO
Gulf Professional Publishing is an imprint of Elsevier

Gulf Professional Publishing is an imprint of Elsevier
50 Hampshire Street, 5th Floor, Cambridge, MA 02139, United States
The Boulevard, Langford Lane, Kidlington, Oxford, OX5 1GB, United Kingdom

Notices
Knowledge and best practice in this field are constantly changing. As new research and experience broaden our understanding, changes in research methods, professional practices, or medical treatment may become necessary.

Practitioners and researchers must always rely on their own experience and knowledge in evaluating and using any information, methods, compounds, or experiments described herein. In using such information or methods they should be mindful of their own safety and the safety of others, including parties for whom they have a professional responsibility.

To the fullest extent of the law, neither the Publisher nor the authors, contributors, or editors, assume any liability for any injury and/or damage to persons or property as a matter of products liability, negligence or otherwise, or from any use or operation of any methods, products, instructions, or ideas contained in the material herein.

Library of Congress Cataloging-in-Publication Data
A catalog record for this book is available from the Library of Congress

British Library Cataloguing-in-Publication Data
A catalogue record for this book is available from the British Library

ISBN: 978-0-12-803097-4

For information on all Gulf Professional publications visit our website at https://www.elsevier.com/

Publisher: Joe Hayton
Senior Acquisition Editor: Katie Hammon
Senior Editorial Project Manager: Kattie Washington
Production Project Manager: Kiruthika Govindaraju
Cover Designer: Matthew Limbert

Typeset by SPi Global, India

CONTENTS

About the Author *vii*
Preface *ix*

1. Gas and Oil in Tight Formations **1**
1 Introduction 1
2 Definitions 2
3 Tight Gas and Tight Oil 16
4 Origin and Reservoirs 29
5 Oil Shale and Shale Oil 39
6 Tight Oil, Tight Gas, and Energy Security 51
7 Resources and Reserves 53
References 56

2. Reservoirs and Reservoir Fluids **63**
1 Introduction 63
2 Sediments 68
3 Reservoirs and Reservoir Evaluation 75
4 Tight Formations 91
5 Core Analyses for Tight Reservoirs 110
References 115

3. Gas and Oil Resources in Tight Formations **121**
1 Introduction 121
2 United States 127
3 Canada 158
4 Other Countries 164
5 The Future of Resources in Tight Formations 167
References 169

4. Development and Production **175**
1 Introduction 175
2 Tight Reservoirs and Conventional Reservoirs 178
3 Well Drilling and Completion 182
4 Production Trends 200
References 206

5. Hydraulic Fracturing **209**
1 Introduction 209
2 Reservoir Evaluation 214
3 The Fracturing Process 224
4 Hydraulic Fracturing in Tight Reservoirs 247
References 256

6. Fluids Management **261**
1 Introduction 261
2 Water Requirements, Use, and Sources 264
3 Water Disposal 283
4 Waste Fluids 286
5 Water Management and Disposal 293
6 Fluids Analysis 300
References 305

7. Properties Processing of Gas From Tight Formations **307**
1 Introduction 307
2 Tight Gas and Gas Composition 310
3 Gas Processing 316
4 Tight Gas Properties and Processing 326
5 Tight Gas Processing 332
References 346

8. Properties and Processing of Crude Oil From Tight Formations **349**
1 Introduction 349
2 Tight Oil Properties 353
3 Transportation and Handling 370
4 Behavior in Refinery Processes 374
5 Mitigating Refinery Impact 389
References 394

9. Environmental Impact **397**
1 Introduction 397
2 Environmental Regulations 402
3 Environmental Impact 410
4 Remediation Requirements and Outlook 423
References 428

Conversion Factors *431*
Glossary *433*
Index *465*

ABOUT THE AUTHOR

Dr. James G. Speight CChem., FRSC, FCIC, FACS earned his BSc and PhD degrees from the University of Manchester, England. He also holds a DSC in The Geological Sciences (VINIGRI, St. Petersburg, Russia) and a PhD in Petroleum Engineering, Dubna International University, Moscow, Russia. Dr. Speight is the author of more than 60 books in petroleum science, petroleum engineering, and environmental sciences. Formerly the CEO of the Western Research Institute (now an independent consultant), he has served as adjunct professor in the Department of Chemical and Fuels Engineering at the University of Utah and in the Departments of Chemistry and Chemical and Petroleum Engineering at the University of Wyoming. In addition, he has also been a visiting professor in Chemical Engineering at the following universities: the University of Missouri-Columbia, the Technical University of Denmark, and the University of Trinidad and Tobago.

Dr. Speight was elected to the Russian Academy of Sciences in 1996 and awarded the Gold Medal of Honor that same year for outstanding contributions to the field of petroleum sciences. He has also received the Scientists without Borders Medal of Honor of the Russian Academy of Sciences. In 2001, the Academy also awarded Dr. Speight the Einstein Medal for outstanding contributions and service in the field of Geological Sciences

PREFACE

Natural gas and crude oil production from hydrocarbon rich deep shale formations, known (incorrectly) as *shale oil* and *shale gas*, is one of the most quickly expanding trends in onshore domestic oil and gas exploration. Vast new natural gas and oil resources are being discovered every year across North America—drilling activity is at a 25-year high and supplies are rapidly growing. One major source of these "new resources" comes from the development of deep shale formations, typically located many thousands of feet below the surface of the Earth in tight, low-permeability formations. Recent technological advancements in horizontal drilling and hydraulic fracturing are unlocking an abundance of deep shale natural gas and oil in the United States. However, the issues of oil and gas recovery relate to the differences in shale plays.

Natural gas and crude oil from shale formations are considered an unconventional source as the gas may be attached to or adsorbed onto organic matter. The gas is contained in difficult-to-produce reservoirs—shale is a rock that can hold huge amounts of gas, not only in the zones between the particles, but within the particles themselves as they are organic-like sponges. Evaluation of the shale gas potential of sedimentary basins has now become an important area of development internationally and is of great national interest as shale gas potential evaluation will have direct positive impact on the energy security of many countries which have sizeable resources in sedimentary basins. However, it will be appreciated that the reserves estimations are not static and are changing annually based upon new discoveries and improvement in drilling and recovery techniques.

Three decades ago natural gas and crude oil from shale formations were of limited importance, but due to issues related to price and availability at times of natural disasters, concerns grew that natural gas prices would continue to escalate. Furthermore, natural gas and crude oil from tight shale formations have not only changed the energy distribution in the United States as a result of this new-found popularity, but development of these resources also brings change to the environmental and socioeconomic landscape, particularly in those areas where gas development is new. With these changes have come questions about the nature of shale development, the potential environmental impacts, and the ability of the current regulatory structure to deal with these issues.

Increasing global significance of shale gas plays has led to the need for deeper understanding of shale behavior. Increased understanding of gas shale reservoirs will provide a better decision regarding the development of these resources. To find these reserves may be easy, but the technology to produce gas is very expensive. This is due to the fact that the hole drilled straight through the gas bearing rock has very little exposure for the gas to escape. Hydraulic fracturing is the only way to increase this exposure for successful gas production rate.

This book presents (1) shale gas and oil potential, (2) the evolution of shale plays, (3) exploitation through discovery, drilling, and reservoir evaluation, (4) production, (5) the efficiency of these completion operations, and (6) impact on the environment. A descriptive Glossary is also included.

The objective of this book is to present an introduction to shale gas resources, as well as offer an understanding of the geomechanical properties of shale, the need for hydraulic fracturing, and an indication of shale gas processing. The book also introduces the reader to issues regarding the nature of shale gas development, the potential environmental impacts, and the ability of the current regulatory structure to deal with these issues. The book also serves to introduce scientists, engineers, managers, regulators, and policymakers to objective sources of information upon which to make decisions about meeting and managing the challenges that may accompany development of these resources.

Dr. James G. Speight
Laramie, Wyoming

CHAPTER ONE

Gas and Oil in Tight Formations

1 INTRODUCTION

The generic terms *crude oil* (also referred to as *petroleum*) and *natural gas* apply to the mixture of liquids and gases, respectively, commonly associated with petroliferous (petroleum-producing, petroleum-containing) geologic formations (Speight, 2014a) and which has been extended to gases and liquids from the recently developed deep shale formations (Speight, 2013b).

Crude oil and natural gas are the most important raw materials consumed in modern society—they provide raw materials for the ubiquitous plastics and other products as well as fuels for energy, industry, heating, and transportation. From a chemical standpoint, natural gas and crude oil are a mixture of hydrocarbon compounds and nonhydrocarbon compounds with crude oil being much more complex than natural gas (Mokhatab et al., 2006; Speight, 2007, 2012a, 2014a). The fuels that are derived from these two natural products supply more than half of the world's total supply of energy. Gas (for gas burners and for the manufacture of petrochemicals), gasoline, kerosene, and diesel oil provide fuel for automobiles, tractors, trucks, aircraft, and ships (Speight, 2014a). In addition, fuel oil and natural gas are used to heat homes and commercial buildings, as well as to generate electricity.

Crude oil and natural gas are carbon-based resources. Therefore, the geochemical carbon cycle is also of interest to fossil fuel usage in terms of petroleum formation, use, and the buildup of atmospheric carbon dioxide. Thus, the more efficient use of natural gas and crude oil is of paramount importance and the technology involved in processing both feedstocks will supply the industrialized nations of the world for (at least) the next 50 years until suitable alternative forms of energy are readily available (Boyle, 1996; Ramage, 1997; Speight, 2011a,b,c).

In this context, it is relevant to introduce the term *peak oil* which refers to the maximum rate of oil production, after which the rate of production of this natural resources (in fact the rate of production of any natural resource) enters a terminal decline (Hubbert, 1956, 1962). Peak oil production usually

Deep Shale Oil and Gas
http://dx.doi.org/10.1016/B978-0-12-803097-4.00001-2

occurs after approximately half of the recoverable oil in an oil reserve has been produced (i.e., extracted). Peaking means that the rate of world oil production cannot increase and that oil production will thereafter decrease with time; even if the demand for oil remains the same or increases. Following from this, the term *peak energy* is the point in time after which energy production declines and the production of energy from various energy sources is in decline. In fact, most oil-producing countries—including Indonesia, the United Kingdom, Norway, and the United States—have passed the peak crude oil production apex several years or decades ago. Their production declines have been offset by discoveries and production growth elsewhere in the world and the so-called *peak energy precipice* will be delayed by such discoveries as well as by further development or crude oil and natural gas resources that are held in tight formations and tight shale formations (Islam and Speight, 2016).

However, before progressing any further, a series of definitions are used to explain the terminology used in this book.

2 DEFINITIONS

The *definitions* of crude oil and natural gas have been varied and diverse and are the product of many years of the growth of the crude oil and natural gas processing industries. Of the many forms of the definitions that have been used not all have survived but the more common, as illustrated here, are used in this book. Also included for comparison are sources of gas—such as gas hydrates and other sources of gas—that will present an indication of future sources of gaseous energy.

2.1 Crude Oil

The term *crude oil* (and the equivalent term *petroleum*) covers a wide assortment of naturally occurring hydrocarbon-type liquids consisting of mixtures of hydrocarbons and nonhydrocarbon compounds containing variable amounts of sulfur, nitrogen, and oxygen as well as heavy metals such as nickel and vanadium, which may vary widely in volatility, specific gravity, and viscosity along with varying physical properties such as API gravity and sulfur content (Tables 1.1 and 1.2) as well as being accompanied by variations in color that ranges from colorless to black—the lower API gravity crude oils are darker than the higher API gravity crude oils (Speight, 2012a, 2014a; US EIA, 2014). The metal-containing constituents, notably those compounds consisting of derivatives of vanadium and nickel with, on

Table 1.1 Selected Crude Oils Showing the Differences in API Gravity and Sulfur Content

Country	Crude Oil	API	Sulfur (%*w*/*w*)
Abu Dhabi (UAE)	Abu Al Bu Khoosh	31.6	2.00
Abu Dhabi (UAE)	Murban	40.5	0.78
Angola	Cabinda	31.7	0.17
Angola	Palanca	40.1	0.11
Australia	Barrow Island	37.3	0.05
Australia	Griffin	55.0	0.03
Brazil	Garoupa	30.0	0.68
Brazil	Sergipano Platforma	38.4	0.19
Brunei	Champion Export	23.9	0.12
Brunei	Seria	40.5	0.06
Cameroon	Lokele	20.7	0.46
Cameroon	Kole Marine	32.6	0.33
Canada (Alberta)	Wainwright-Kinsella	23.1	2.58
Canada (Alberta)	Rainbow	40.7	0.50
China	Shengli	24.2	1.00
China	Nanhai Light	40.6	0.06
Dubai (UAE)	Fateh	31.1	2.00
Dubai (UAE)	Margham Light	50.3	0.04
Egypt	Ras Gharib	21.5	3.64
Egypt	Gulf of Suez	31.9	1.52
Gabon	Gamba	31.4	0.09
Gabon	Rabi-Kounga	33.5	0.07
Indonesia	Bima	21.1	0.25
Indonesia	Kakap	51.5	0.05
Iran	Aboozar (Ardeshir)	26.9	2.48
Iran	Rostam	35.9	1.55
Iraq	Basrah Heavy	24.7	3.50
Iraq	Basrah Light	33.7	1.95
Libya	Buri	26.2	1.76
Libya	Bu Attifel	43.3	0.04
Malaysia	Bintulu	28.1	0.08
Malaysia	Dulang	39.0	0.12
Mexico	Maya	22.2	3.30
Mexico	Olmeca	39.8	0.80
Nigeria	Bonny Medium	25.2	0.23
Nigeria	Brass River	42.8	0.06
North Sea (Norway)	Emerald	22.0	0.75
North Sea (UK)	Innes	45.7	0.13
Qatar	Qatar Marine	36.0	1.42
Qatar	Dukhan (Qatar Land)	40.9	1.27
Saudi Arabia	Arab Heavy (Safaniya)	27.4	2.80

Continued

Table 1.1 Selected Crude Oils Showing the Differences in API Gravity and Sulfur Content—cont'd

Country	Crude Oil	API	Sulfur (%*w/w*)
Saudi Arabia	Arab Extra Light (Berri)	37.2	1.15
USA (California)	Huntington Beach	20.7	1.38
USA (Michigan)	Lakehead Sweet	47.0	0.31
Venezuela	Leona	24.4	1.51
Venezuela	Oficina	33.3	0.78

Table 1.2 API Gravity and Sulfur Content of Selected Heavy Oils

Country	Crude Oil	API	Sulfur (%*w/w*)
Canada (Alberta)	Athabasca	8.0	4.8
Canada (Alberta)	Cold Lake	13.2	4.11
Canada (Alberta)	Lloydminster	16.0	2.60
Canada (Alberta)	Wabasca	19.6	3.90
Chad	Bolobo	16.8	0.14
Chad	Kome	18.5	0.20
China	Qinhuangdao	16.0	0.26
China	Zhao Dong	18.4	0.25
Colombia	Castilla	13.3	0.22
Colombia	Chichimene	19.8	1.12
Ecuador	Ecuador Heavy	18.2	2.23
Ecuador	Napo	19.2	1.98
USA (California)	Midway Sunset	11.0	1.55
USA (California)	Wilmington	18.6	1.59
Venezuela	Boscan	10.1	5.50
Venezuela	Tremblador	19.0	0.80

occasion iron and copper, usually occur in the more viscous crude oils in amounts up to several thousand parts per million and can have serious consequences for the equipment and catalysts used in processing of these feedstocks (Speight and Ozum, 2002; Parkash, 2003; Hsu and Robinson, 2006; Gary et al., 2007; Speight, 2014a). The presence of iron and copper has been subject to much speculation insofar as it is not clear if these two metals are naturally occurring in the crude oil or whether they are absorbed by the crude oil during recovery and transportation in metal pipelines.

Crude oil exists in reservoirs that consist of more porous and permeable sediments, such as *sandstone* and *siltstone*. A series of reservoirs within a common rock structure or a series of reservoirs in separate but neighboring formations is commonly referred to as an *oil field*. A group of fields often found in a single geologic environment is known as a *sedimentary basin* or *province*. In the underground locale, crude oil is much more fluid (mobile) than it is on the surface and is generally much more mobile under reservoir conditions because the elevated temperatures (the *geothermal gradient*) in subterranean formations decrease the viscosity of the oil. Although the geothermal gradient varies from place to place, the increase in temperature with depth below the surface is generally recognized to be on the order of 25–30°C/km (15°F/1000 ft or 120°C/1000 ft, i.e., 0.015°C per foot of depth or 0.012°C per foot of depth).

The major components of conventional crude oil are *hydrocarbons* with *nonhydrocarbon compounds (compounds contain nitrogen, oxygen, sulfur, and metals) in the minority and* all of which display a substantial variation in molecular structure. The simplest hydrocarbons are *paraffin derivatives* which extend from methane (the simplest hydrocarbon, CH_4) to the liquids that are refined into gasoline, diesel fuel, and fuel oil to the highly crystalline wax and the more complex multi-ring constituents (C_{60+}); the latter constituents (i.e., the more complex multi-ring constituents) are typically nonvolatile and are found in the residuum from atmospheric distillation or vacuum distillation.

The *nonhydrocarbon constituents* of crude oil include organic derivatives of nitrogen, oxygen, sulfur, and metals (predominantly nickel and vanadium) and are often referred to as polar aromatics, which include the resin and asphaltene constituents (Fig. 1.1) (Speight, 2014a, 2015a). In the case of heavy oils and tar sand bitumen, there is a lesser amount of hydrocarbon constituents (with very little of the material boiling below 200°C (390°F) volatile constituents) in favor of increasing amounts of nonhydrocarbon constituents (low-volatile and nonvolatile constituents) (Speight, 2014a, 2016c). While most of the higher boiling constituents are removed during refining by conversion of hydrocarbon products, the low-volatile and nonvolatile constituents greatly influence the choice and effectiveness of recovery processes and whether or not fracturing is to be considered as a recovery process enhancement (Speight, 2014a, 2016a,c).

In terms of crude oil recovery, geologic techniques can give valuable information about the existence of rock formations as well as crude oil properties, drilling is the only sure way to ascertain the presence of crude oil in the formation and the suitability of the crude oil for recovery. With modern rotary equipment, wells can be drilled to depths of >30,000 ft (9000 m). Once oil is found, it may be recovered (brought to the surface) by the reservoir pressure created by the presence of natural gas or water within the reservoir. Crude oil can also be brought to the surface by injecting water

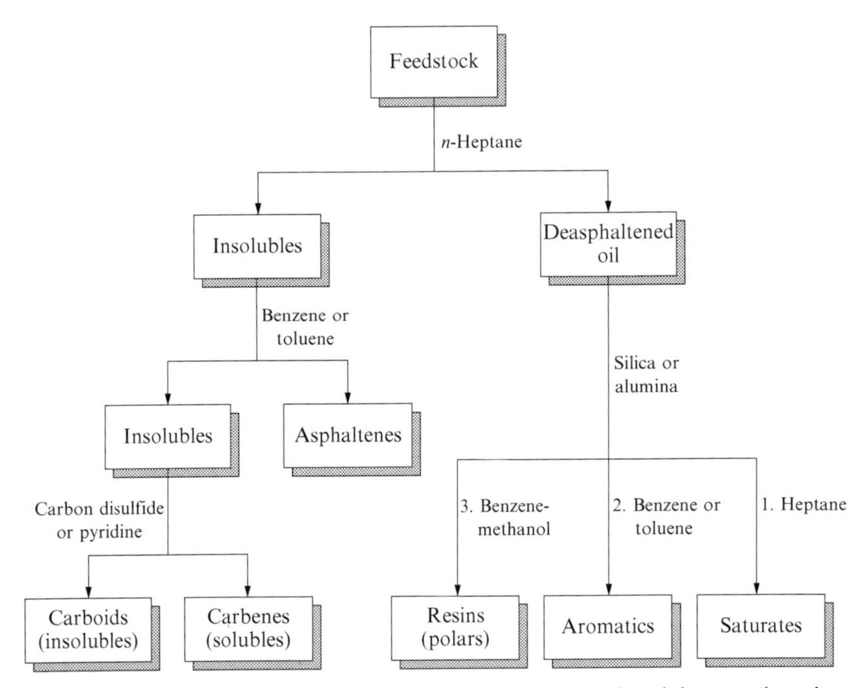

Fig. 1.1 Typical fractionating sequence for conventional crude oil, heavy oil, and tar sand bitumen.

or steam into the reservoir to raise the pressure artificially or by injecting such substances as carbon dioxide, polymers, and solvents to reduce crude oil viscosity. Thermal recovery methods are frequently used to enhance the production of heavy crude oils, especially when extraction of the heavy oil is impeded by viscous resistance to flow at reservoir temperatures (Speight, 2014a, 2016a,b).

Crude oil reserves are conveniently divided into (1) conventional resources and (2) unconventional resources. However, in a more general sense, unconventional resources also include heavy oil which is not a subject of this text and crude oil from shale formations and from tight formations. The latter type of resources (and crude oil from shale formations and from tight formations) are the subject of this book.

2.2 Natural Gas

The generic term *natural gas* applies to gases commonly associated with petroliferous (crude oil-producing, crude oil-containing) geologic formations.

Natural gas is found in deep underground rock formations and is also found in coal seams (coalbed methane) and generally contains high proportions of methane (CH_4) and some of the higher molecular weight higher paraffins (C_nH_{2n+2}) generally containing up to six or eight carbon atoms may also be present in small quantities (Table 1.3). The hydrocarbon constituents of natural gas are combustible, but nonflammable nonhydrocarbon components such as carbon dioxide, nitrogen, and helium are often present in the minority and are regarded as contaminants that are removed during gas processing (gas cleaning, gas refining) (Mokhatab et al., 2006; Speight, 2007, 2014a).

The hydrocarbon constituents of natural gas are combustible, but nonflammable nonhydrocarbon components such as carbon dioxide, nitrogen, and helium are often present in the minority and are regarded as contaminants. In addition to the natural gas found in petroleum reservoirs, there are also those reservoirs in which natural gas may be the sole occupant. In such cases the liquids that occur with the gas are gas condensate and such reservoirs may also be referred to as *gas-condensate reservoirs*. Differences in natural gas composition occur between different reservoirs, and two wells in the same field may also yield gaseous products that are different in composition (Mokhatab et al., 2006; Speight, 2007, 2014a).

In addition to the natural gas that is found in crude oil reservoirs, there are also those reservoirs in which natural gas may be the sole occupant. The principal constituent of this type of natural gas is methane, but other hydrocarbons, such as ethane, propane, and butane, may (but not always) also be present in somewhat lesser amount than natural gas from a crude oil reservoir. Carbon dioxide is also a common constituent of natural gas. Trace

Table 1.3 Constituents of Natural Gas

Name	Formula	Vol. (%)
Methane	CH_4	>85
Ethane	C_2H_6	3–8
Propane	C_3H_8	1–5
Butane	C_4H_{10}	1–2
Pentane$^+$	C_5H_{12}	1–5
Carbon dioxide	CO_2	1–2
Hydrogen sulfide	H_2S	1–2
Nitrogen	N_2	1–5
Helium	He	<0.5

Pentane$^+$: pentane and higher molecular weight hydrocarbons, including benzene and toluene (Speight, 2014a,b,c).

amounts of rare gases, such as helium, may also occur, and certain natural gas reservoirs are a source of these rare gases. Just as crude oil can vary in composition, so can natural gas. Differences in natural gas composition occur between different reservoirs, and two wells in the same field may also yield gaseous products that are different in composition (Speight, 1990, 2014a, 2016a).

Like crude oil resources, natural gas resources are typically divided into two categories: (1) conventional gas and unconventional gas (Mokhatab et al., 2006; Speight, 2007, 2014a). However, in a more general sense, unconventional resources also include coalbed methane (which is not a subject of this text but which is included for comparison) and natural gas from shale formations and from tight formations. The latter type of resources (natural gas from shale formations and from tight formations) are the subject of this book. Conventional gas is typically found in reservoirs with a permeability greater than 1 milliDarcy (>1 mD) and can be extracted by means of traditional recovery methods. In contrast, unconventional gas is found in reservoirs with relatively low permeability (<1 mD) and hence cannot be extracted by conventional methods.

In addition, there are several general definitions that have been applied to natural gas. Thus, *lean* gas is gas in which methane is the major constituent. *Wet* gas contains considerable amounts of the higher molecular weight hydrocarbons and these paraffins are present in the gas, in fact >0.1 gal/1000 ft^3. On the other hand, *dry* natural gas indicates that there is <0.1 gallon (1 gallon (US) = 264.2 m^3) of gasoline vapor (higher molecular weight paraffins) per 1000 ft^3 (1 ft^3 = 0.028 m^3). Other terms of interest are sour gas, residua gas, and casinghead gas (casing head gas). *Sour* gas contains hydrogen sulfide whereas *sweet* gas contains very little, if any, hydrogen sulfide. *Residue gas* is natural gas from which the higher molecular weight hydrocarbons have been extracted and *casinghead gas* is derived from petroleum but is separated at the separation facility at the well-head.

There are several general definitions that have been applied to natural gas. Thus, *associated natural gas* or *dissolved natural gas* occurs either as free gas or as gas in solution in the crude oil. Gas that occurs as a solution in the crude oil is *dissolved* gas whereas the gas that exists in contact with the crude oil (*gas cap*) is *associated* gas. Such gas typically has higher proportions of the higher boiling hydrocarbons that commonly occur as low-boiling crude oil constituents (Table 1.4). *Lean* gas is gas in which methane is the major constituent while *wet* gas contains considerable amounts of the higher molecular weight hydrocarbons. *Sour* gas contains hydrogen sulfide whereas

Table 1.4 Composition of Associated Natural Gas From a Petroleum Well

Category	Component	Amount (%)
Paraffinic	Methane (CH_4)	70–98
	Ethane (C_2H_5)	1–10
	Propane (C_3H_8)	Trace–5
	Butane (C_4H_{10})	Trace–2
	Pentane (C_5H_{12})	Trace–1
	Hexane (C_6H_{14})	Trace–0.5
	Heptane and higher (C_7^+)	None–trace
Cyclic	Cyclopropane (C_3H_6)	Traces
	Cyclohexane (C_6H_{12})	Traces
Aromatic	Benzene (C_6H_6), others	Traces
Nonhydrocarbon	Nitrogen (N_2)	Trace–15
	Carbon dioxide (CO_2)	Trace–1
	Hydrogen sulfide (H_2S)	Trace occasionally
	Helium (He)	Trace–5
	Other sulfur and nitrogen compounds	Trace occasionally
	Water (H_2O)	Trace–5

sweet gas contains very little, if any, hydrogen sulfide. To further define the terms *dry* and *wet* in quantitative measures, the term *dry* natural gas indicates that there is <0.1 gallon (1 gallon (US) = 264.2 m^3) of gasoline vapor (higher molecular weight paraffins) per 1000 ft^3 (1 ft^3 = 0.028 m^3). The term *wet natural gas* indicates that there are such paraffins present in the gas, in fact >0.1 gal/1000 ft^3.

Residue gas is natural gas from which the higher molecular weight hydrocarbons have been extracted and *casing head gas* is derived from crude oil but is separated at the separation facility at the well-head. The use of term *residue* as it is applied to natural gas is in direct contrast to the residue left in the distillation tower during crude oil refining—this residue is the high boiling fraction of the crude oil from which the lower boiling constituents have been removed as part of the distillation process.

Other components such as carbon dioxide (CO_2), hydrogen sulfide (H_2S), mercaptans (thiols, RSH), as well as trace amounts of other constituents may also be present. Thus, there is no single composition of components which might be termed *typical* natural gas. Methane and ethane often constitute the bulk of the combustible components; carbon dioxide (CO_2) and nitrogen (N_2) are the major noncombustible (inert) components. Thus, sour gas is natural gas that occurs mixed with higher levels of sulfur compounds (such as hydrogen sulfide, H_2S, and mercaptans or thiols, RSH) and which constitutes a corrosive gas (Speight, 2014c). The sour gas requires additional processing for purification (Mokhatab et al., 2006; Speight, 2007).

Natural gas condensate (*gas condensate, natural gasoline*) is a low-density low-viscosity mixture of hydrocarbon liquids that may be present as gaseous components under reservoir conditions and which occur in the raw natural gas produced from natural wells. The constituents of *condensate* separate from the untreated (raw) gas if the temperature is reduced to below the hydrocarbon dew point temperature of the raw gas. Briefly, the dew point is the temperature to which a given volume of gas must be cooled, at constant barometric pressure, for vapor to condense into liquid. Thus, the dew point is the saturation point.

On a worldwide scale, there are many gas-condensate reservoirs and each has its own unique gas-condensate composition. However, in general, gas condensate has a specific gravity on the order of ranging from 0.5 to 0.8 and is composed of hydrocarbons such as propane, butane, pentane, hexane, heptane and even octane, nonane, and decane in some cases. In addition, the gas condensate may contain additional impurities such as hydrogen sulfide, thiols (mercaptans, RSH), carbon dioxide, cyclohexane (C_6H_{12}), and low-molecular weight aromatics such as benzene (C_6H_6), toluene (C_6H5CH_3), ethylbenzene ($C_6H_5CH_2CH_3$), and xylenes ($H_3CC_6H_4CH_3$) (Mokhatab et al., 2006; Speight, 2007, 2014a).

When condensation occurs in the reservoir, the phenomenon known as condensate blockage can halt flow of the liquids to the wellbore. Hydraulic fracturing is the most common mitigating technology in siliciclastic reservoirs (reservoirs composed of clastic rocks), and acidizing is used in carbonate reservoirs (Speight, 2016a). Briefly, clastic rocks are composed of fragments, or clasts, of preexisting minerals and rock. A clast is a fragment of geological detritus, chunks, and smaller grains of rock broken off other rocks by physical weathering. Geologists use the term *clastic* with reference to sedimentary rocks as well as to particles in sediment transport whether in suspension or as bed load, and in sedimentary deposits.

In addition, production can be improved with less drawdown in the formation. For some gas-condensate fields, a lower drawdown means single-phase production above the dew point pressure can be extended for a longer time. However, hydraulic fracturing does not generate a permanent conduit past a condensate saturation buildup area. Once the pressure drops below the dew point, saturation will increase around the fracture, just as it did around the wellbore. Horizontal or inclined wells are also being used to increase contact area within formations.

2.3 Coalbed Methane

Natural gas is often located in the same reservoir as with crude oil, but it can also be found trapped in gas reservoirs and within coal seams. The occurrence of methane in coal seams is not a new discovery and methane (called *firedamp* by the miners because of its explosive nature) was known to coal miners for at least 150 years (or more) before it was *rediscovered* and developed as coalbed methane (Speight, 2013b). The gas occurs in the pores and cracks in the coal seam and is held there by underground water pressure. To extract the gas, a well is drilled into the coal seam and the water is pumped out (*dewatering*) which allows the gas to be released from the coal and brought to the surface.

Coalbed methane (sometime referred to as *coalmine methane*) is a generic term for the methane found in most coal seams. To the purist, coalmine methane is the fraction of coalbed methane that is released during the mining operation. Thus, in practice, the terms coalbed methane and coalmine methane usually refer to different sources of gas—both forms of gas, whatever the name, are equally dangerous to the miners.

Coalbed methane is a gas formed as part of the geological process of coal generation and is contained in varying quantities within all coal. coalbed methane is exceptionally pure compared to conventional natural gas, containing only very small proportions of higher molecular weight hydrocarbons such as ethane and butane and other gases (such as hydrogen sulfide and carbon dioxide). Coalbed gas is over 90% methane and, subject to gas composition, may be suitable for introduction into a commercial pipeline with little or no treatment (Rice, 1993; Levine, 1993; Mokhatab et al., 2006; Speight, 2007, 2013a). Methane within coalbeds is not structurally trapped by overlying geologic strata, as in the geologic environments typical of conventional gas deposits (Speight, 2007, 2013a, 2014a). Only a small amount (on the order of 5–10% *v/v*) of the coalbed methane is present as free gas within the joints and cleats of coalbeds. Most of the coalbed methane is contained within the coal itself (adsorbed to the sides of the small pores in the coal).

As the coal forms, large quantities of methane-rich gas are produced and subsequently adsorbed onto (and within) the coal matrix. Because of its many natural cracks and fissures, as well as the porous nature, coal in the seam has a large internal surface area and can store much more gas than a conventional natural gas reservoir of similar rock volume. If a seam is disturbed, either during mining or by drilling into it before mining, methane is released from the surface of the coal. This methane then leaks into any open spaces such as fractures in the coal seam (known as *cleats*). In these cleats, the coalmine methane mixes with nitrogen and carbon dioxide

(CO_2). Boreholes or wells can be drilled into the seams to recover the methane. Large amounts of coal are found at shallow depths, where wells to recover the gas are relatively easy to drill at a relatively low cost. At greater depths, increased pressure may have closed the cleats, or minerals may have filled the cleats over time, lowering permeability and making it more difficult for the gas to move through the coal seam. coalbed methane has been a hazard since mining began. To reduce any danger to coal miners, most effort is addressed at minimizing the presence of coalbed in the mine, predominantly by venting it to the atmosphere. Only during the past two decades has significant effort been devoted to recovering the methane as an energy resource. Another source of methane from a working mine is the methane mixed with ventilation air, the so-called ventilation air methane (VAM). In the mine, ventilation air is circulated in sufficient quantity to dilute the methane to low concentrations for safety reasons. VAM is often too low in concentration to be of commercial value.

In coalbeds (coal seams), methane (the primary component of natural gas) is generally adsorbed to the coal rather than contained in the pore space or structurally trapped in the formation. Pumping the injected and native water out of the coalbeds after fracturing serves to depressurize the coal, thereby allowing the methane to desorb and flow into the well and to the surface. Methane has traditionally posed a hazard to underground coal miners, as the highly flammable gas is released during mining activities. Otherwise inaccessible coal seams can also be tapped to collect this gas, known as coalbed methane, by employing similar well-drilling and hydraulic-fracturing techniques as are used in shale gas extraction.

The primary (or natural) permeability of coal is very low, typically ranging from 0.1 to 30 mD and, because coal is a very weak (low modulus) material and cannot take much stress without fracturing, coal is almost always highly fractured and cleated. The resulting network of fractures commonly gives coalbeds a high secondary permeability (despite coal's typically low primary permeability). Groundwater, hydraulic-fracturing fluids, and methane gas can more easily flow through the network of fractures. Because hydraulic fracturing generally enlarges preexisting fractures in addition to creating new fractures, this network of natural fractures is very important to the extraction of methane from the coal.

The gas from coal seams can be extracted by using technologies that are similar to those used to produce conventional gas, such as using wellbores. However, complexity arises from the fact that the coal seams are generally of low permeability and tend to have a lower flow rate (or permeability) than conventional gas systems, gas is only sourced from close to the well and as such a higher density of wells is required to develop a coalbed methane resource as an unconventional resource (such as tight gas) than a conventional gas resource.

Technologies such as horizontal and multilateral drilling with hydraulic fracturing are sometimes used to create longer, more open channels that enhance well productivity but not all coal seam gas wells require application of this technique. Water present in the coal seam, either naturally occurring or introduced during the fracturing operation, is usually removed to reduce the pressure sufficiently to allow the gas to be released, which leads to additional operational requirements, increased investment, and environmental concerns. There are some horizontally drilled coalbed methane wells and some that receive hydraulic-fracturing treatments. However, some coalbed methane reservoirs are also underground sources of drinking water, and as such, there are restrictions on hydraulic-fracturing operations. The coalbed methane wells are mostly shallow, as the coal matrix does not have the strength to maintain porosity under the pressure of significant overburden thickness.

2.4 Gas Hydrates

Methane hydrates (also called *methane clathrates*) is a resource in which a large amount of methane is trapped within a crystal structure of water, forming a solid similar to ice (Kvenvolden, 1995; Buffett and Archer, 2004; Gao et al., 2005; Gao, 2008; USGS, 2011). Natural gas hydrates are solids that form from a combination of water and one or more hydrocarbon or non-hydrocarbon gases. In physical appearance, gas hydrates resemble packed snow or ice. In a gas hydrate, the gas molecules (such as methane, hence the name *methane hydrates*) are trapped within a cage-like crystal structure composed of water molecules. Gas hydrates are stable only under specific conditions of pressure and temperature. Under the appropriate pressure, they can exist at temperatures significantly above the freezing point of water. The maximum temperature at which gas hydrate can exist depends on pressure and gas composition. For example, methane plus water at 600 psia forms hydrate at 5°C (41°F), while at the same pressure, methane with 1% *v/v* propane forms a gas hydrate at 9.4°C (49°F). Hydrate stability can also be influenced by other factors, such as salinity (Edmonds et al., 1996).

Per unit volume, gas hydrates contain a high amount of gas. For example, 1 cubic yard of hydrate disassociates at atmospheric temperature and pressure to form approximately 160 cubic yards of natural gas plus 0.8 cubic yards of water (Kvenvolden, 1993). The natural gas component of gas hydrates is typically dominated by methane, but other natural gas components (e.g., ethane, propane, carbon dioxide) can also be incorporated into a hydrate. The origin of the methane in a hydrate can be either thermogenic gas or biogenic gas. Bacterial gas formed during early digenesis of organic matter can become part of a gas hydrate in continental shelf sediment. Similarly, thermogenic gas leaking to the surface from a deep thermogenic gas accumulation can form a gas hydrate in the same continental shelf sediment (Boswell and Collett, 2011; Collett, 2002; Gupta, 2004; Demirbas, 2010a,b,c; Chong et al., 2016).

Generally, methane clathrates are common constituents of the shallow marine geosphere, and they occur both in deep sedimentary structures and form outcrops on the ocean floor. Methane hydrates are believed to form by migration of gas from depth along geological faults, followed by precipitation, or crystallization, on contact of the rising gas stream with cold seawater. When drilling in petroleum-bearing and gas-bearing formations submerged in deep water, the reservoir gas may flow into the wellbore and form gas hydrates owing to the low temperatures and high pressures found during deep water drilling. The gas hydrates may then flow upward with drilling mud or other discharged fluids. When the hydrates rise, the pressure in the annulus decreases and the hydrates dissociate into gas and water. The rapid gas expansion ejects fluid from the well, reducing the pressure further, which leads to more hydrate dissociation (sometimes explosive dissociation) and further fluid ejection.

Estimates of the global inventory of methane clathrate exceed 10^{19} g of carbon (MacDonald, 1990a,b; Gornitz and Fung, 1994), which is comparable to estimates of potentially recoverable coal, oil, and natural gas. The proximity of this methane reservoir to the seafloor has motivated speculations about a release of methane in response to climate change (MacDonald, 1990b). Increases in temperature or decreases in pressure (through changes in sea level) tend to dissociate clathrate, releasing methane into the near-surface environment. Release of methane from clathrate has been invoked to explain abrupt increases in the atmospheric concentration of methane during the last glacial cycle.

2.5 Other Sources of Gas

Biogenic gas (predominantly methane) is produced by certain types of bacteria (methanogens) during the process of breaking down organic matter in an oxygen-free environment. Livestock manure, food waste, and sewage are all potential sources of biogenic gas, or biogas, which is usually considered a form of renewable energy. Small-scale biogas production is a well-established technology in parts of the developing world, particularly Asia, where farmers collect animal manure in vats and capture the methane given off while it decays.

Landfills offer another under-utilized source of biogas. When municipal waste is buried in a landfill, bacteria break down the organic material contained in garbage such as newspapers, cardboard, and food waste, producing gases such as carbon dioxide and methane. Rather than allowing these gases to go into the atmosphere, where they contribute to global warming, landfill gas facilities can capture them, separate the methane, and combust it to generate electricity, heat, or both.

2.6 Wax

Naturally occurring wax, often referred to as *mineral wax*, occurs as a yellow to dark brown, solid substance that is composed largely of paraffins (Wollrab and Streibl, 1969). Fusion points vary from as low as 37°C (99°F) with a boiling range in excess of >370°C (698°F). They are usually found associated with considerable mineral matter, as a filling in veins and fissures or as an interstitial material in porous rocks. The similarity in character of these native products is substantiated by the fact that, with minor exceptions where local names have prevailed, the original term *ozokerite* (*ozocerite*) has served without notable ambiguity for mineral wax deposits.

Ozokerite (*ozocerite*), from the Greek meaning *odoriferous wax*, is a naturally occurring hydrocarbon material composed chiefly of solid paraffins and cycloparaffins (i.e., hydrocarbons) (Wollrab and Streibl, 1969). Ozocerite usually occurs as stringers and veins that fill rock fractures in tectonically disturbed areas. It is predominantly paraffinic material (containing up to 90% nonaromatic hydrocarbons) with a high content (40–50%) of normal or slightly branched paraffins as well as cyclic paraffin derivatives. Ozocerite contains approximately 85% carbon, 14% hydrogen, and 0.3% each of sulfur and nitrogen and is, therefore, predominantly a mixture of pure hydrocarbons; any nonhydrocarbon constituents are in the minority. Ozocerite is soluble in solvents that are commonly employed for dissolution of petroleum derivatives, for example, toluene, benzene, carbon disulfide, chloroform, and ethyl ether.

While naturally occurring paraffin wax is of particular interest in some conventional crude oil reservoirs, it occurs in shale as part of the crude oil and becomes of interest when the crude oil from shale formations is blended with other paraffinic liquids that can lead to deposition of the wax during transportation and refining from which fouling can occur (Speight, 2014a, 2015b).

3 TIGHT GAS AND TIGHT OIL

The terms *tight oil* and *tight gas* refer to crude oil (primarily light sweet crude oil) and natural gas, respectively, that are contained in formations such as shale or tight sandstone, where the low permeability of the formation makes it difficult for producers to extract the crude oil or natural gas except by unconventional techniques such as horizontal drilling and hydraulic fracturing. The terms *unconventional oil* and *unconventional gas* are umbrella terms for crude oil and natural gas that are produced by methods that do not meet the criteria for conventional production. Thus, the terms tight oil and tight gas refer to natural gas trapped in organic-rich rocks dominated by shale while tight gas refers to natural gas trapped in sandstone or limestone formations that exhibit very low permeability and such formations may also contain condensate. Given the low permeability of these reservoirs, the gas must be developed via special drilling and production techniques including fracture stimulation (hydraulic fracturing) in order to be produced commercially (Gordon, 2012).

Unlike conventional mineral formations containing natural gas and crude oil reserves, shale and other tight formations have low permeability, which naturally limits the flow of natural gas and crude oil. In such formations, the natural gas and crude oil are held in largely unconnected pores and natural fractures. Hydraulic fracturing is the method commonly used to connect these pores and allow the gas to flow. The process of producing natural gas and crude oil from tight deposits involves many steps in addition to hydraulic fracturing, all of which involve potential environmental impacts (Chapters 4, 5 and 9) (Speight, 2016b). Hydraulic fracturing is often misused as an umbrella term to include all of the steps involved in gas and oil production from shale formations and tight formations. These steps include road and well-pad construction, drilling the well, casing, perforating, hydraulic fracturing, completion, production, abandonment, and reclamation (Chapter 5). Thus, shale formations and other tight formations are impermeable formations that restrict migration of gases and fluids. Economic production of natural gas and crude oil from such formations was, until recently, unfeasible mainly related to the very low to ultralow permeability of the formation, a parameter that determines the connectivity and flow between pores where natural gas and crude oil reside (Chapter 2).

Tight sandstone formations and shale formations are heterogeneous and vary widely over relatively short distances. Thus, even in a single horizontal drill hole, the amount of gas or oil recovered may vary, as may recovery within a field or even between adjacent wells. This makes evaluation of tight plays (a *play* is a group of fields sharing geological similarities where the reservoir and the trap control the distribution of oil and gas). Because of the variability of the reservoirs—even reservoirs within a play—is different, decisions regarding the profitability of wells on a particular lease are difficult. Furthermore, the production of crude oil from tight formations requires that at least 15–20% *v/v* of the reservoir pore space is occupied by natural gas to provide the necessary reservoir energy to drive the oil toward the borehole; tight reservoirs which contain only oil cannot be economically produced (US EIA, 2013).

In tight shale reservoirs and other tight reservoirs, there are areas known as *sweet spots* which are preferential targets for drilling and releasing the gas and oil. In these areas, the permeability of the formation is significantly higher than the typical permeability of the majority of the formations. The occurrence of a sweet spot and the higher permeability may often result from open natural fractures, formed in the reservoir by natural stresses, which results in the creation of a dense pattern of fractures. Such fractures may have reclosed, filled in with other materials, or may still be open. However, a well that can be connected through hydraulic fracturing to open natural fracture systems can have a significant flow potential.

The development of deep shale oil and gas resources is typically found thousands of feet below the surface of the Earth in tight, low-permeability shale formations. Until recently the vast quantities of natural gas in these formations were thought to be unrecoverable. Through the use of hydraulic fracturing, combined with recently improved horizontal drilling techniques, extraordinary amounts of natural gas and crude oil are produced from deep shale formations across the United States.

3.1 Tight Gas

In the context of this book, the focus is on *shale gas* and, when necessary, reference will also be made to *tight gas*. In respect of the low permeability of these reservoirs, the gas must be developed via special techniques including stimulation by hydraulic fracturing (or *fraccing*, *fracking*) in order to be produced commercially.

Conventional gas typically is found in reservoirs with permeability >1 mD and can be extracted via traditional techniques (Fig. 1.2). A large proportion of the gas produced globally to date is conventional, and is relatively easy and inexpensive to extract. In contrast, unconventional gas is found in reservoirs with relatively low permeability (<1 mD) (Fig. 1.2) and hence cannot be extracted via conventional methods. However, there are several types of unconventional gas resources that are currently under production but the three most common types are (1) shale gas, (2) tight gas, and (3) coalbed methane although methane hydrates are often included with these gases under the general umbrella of *unconventional gas* but while included here for comparison and not the subject of this text.

Generally, shale gas is a natural gas contained in predominantly fine, low-permeable sedimentary rocks, in consolidated clay-sized particles, at the scale of nanometers. Gas shale formations are organic-rich formations that are both source rock and reservoir. The expected value of permeability to gas flow is in the range of micro- to nanodarcy. The gas retained in such deposits is in the form of adsorbed material on rock, trapped in pore spaces and as an interbedding material with shales. Although the shale gas is usually very clean, it is hard to recover from deposits because of the structural complexity and low hydrodynamic conductivity of shales.

Shale gas is part of a continuum of unconventional gas that progresses from tight gas sand formations, tight gas shale formations to coalbed methane in which horizontal drilling and fracture stimulation technology can enhance the natural fractures and recover gas from rocks with low permeability. Gas can be found in the pores and fractures of shales and also bound to the matrix, by a process known as adsorption, where the gas molecules adhere to the surfaces within the shale. During enhanced fracture stimulation drilling technology, fluid is pumped into the ground to make the reservoir more permeable, then the fractures are propped open by small particles, and can enable the released gas to flow at commercial rates. By

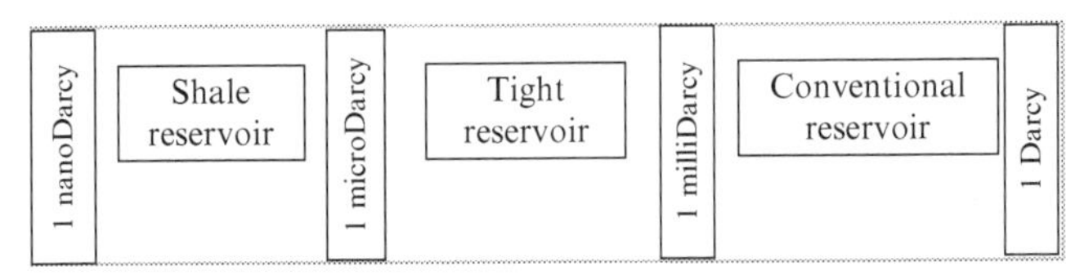

Fig. 1.2 Representation of the differences in permeability of shale reservoirs, tight reservoirs, and conventional reservoirs.

drilling multilateral horizontal wells followed by hydraulic fracturing (Chapter 5), a greater rock volume can be accessed.

More specifically, shale gas is natural gas that is produced from a type of sedimentary rock derived from clastic sources often including mudstones or siltstones, which is known as shale. Clastic sedimentary rocks are composed of fragments (clasts) of preexisting rocks that have been eroded, transported, deposited, and lithified (hardened) into new rocks. Shales contain organic material which was lain down along with the rock fragments. In areas where conventional resource plays are located, shales can be found in the underlying rock strata and can be the source of the hydrocarbons that have migrated upwards into the reservoir rock. Furthermore, a tight gas reservoir is commonly defined as is a rock with matrix porosity of 10% or less and permeability of 0.1 mD or less, exclusive of fracture permeability.

Shale gas resource plays differ from conventional gas plays in that the shale acts as both the source for the gas and also the zone (also known as the reservoir) in which the gas is trapped. The very low permeability of the rock causes the rock to trap the gas and prevent it from migrating toward the surface. The gas can be held in natural fractures or pore spaces, or can be adsorbed onto organic material. With the advancement of drilling and completion technology, this gas can be successfully exploited and extracted commercially as has been proven in various basins in North America.

Aside from permeability, the key properties of shales, when considering gas potential, are total organic carbon (TOC) and thermal maturity. The total organic content is the total amount of organic material (kerogen) present in the rock, expressed as a percentage by weight. Generally, the higher the total organic content, the better the potential for hydrocarbon generation. The thermal maturity of the rock is a measure of the degree to which organic matter contained in the rock has been heated over time and potentially converted into liquid and/or gaseous hydrocarbons. Thermal maturity is measured using vitrinite reflectance (R_o).

Because of the special techniques required for extraction, shale gas can be more expensive than conventional gas to extract. On the other hand, the in-place gas resource can be very large given the significant lateral extent and thickness of many shale formations. However, only a small portion of the total world resources of shale gas is theoretically producible and even less likely to be producible in a commercially viable manner. Therefore, a key determinant of the success of a shale play is whether, and how much, gas can be recovered to surface and at what cost.

3.2 Tight Oil

In addition, oil from tight sandstone and from shale formations (*tight oil*) is another type of crude oil which varies from a gas condensate-type liquid to a highly volatile liquid (Table 1.5) (McCain, 1990; Dandekar, 2013; Speight, 2014a, 2016b; Terry and Rogers, 2014). Tight oil refers to the oil preserved in tight sandstone or tight carbonate rocks with low matrix permeability—in these reservoirs, the individual wells generally have no natural productivity or their natural productivity is lower than the lower limit of industrial oil flow, but industrial oil production can be obtained under certain economic conditions and technical measures. Such measures include acid fracturing, multistage fracturing, horizontal wells, and multilateral wells.

The term *light tight oil* is also used to describe oil from shale reservoirs and tight reservoirs because the crude oil produced from these formations is light crude oil. The term *light crude oil* refers to low-density petroleum that flows freely at room temperature and these light oils have a higher proportion of light hydrocarbon fractions resulting in higher API gravities (between 37 and 42 degrees) (Speight, 2014a). However, the crude oil contained in shale reservoirs and in tight reservoirs will not flow to the wellbore without assistance from advanced drilling (such as horizontal drilling) and fracturing (hydraulic fracturing) techniques. There has been a tendency to refer to this oil as *shale oil*. This terminology is incorrect insofar as it is confusing and the use of such terminology should be discouraged as illogical since shale oil has (for decades, even centuries) been the name given to the distillate produced from oil shale by thermal decomposition (Scouten, 1990; Speight, 2012b, 2014a, 2016b). There has been the recent (and logical) suggestion that shale oil can be referred to as *kerogen oil* (IEA, 2013).

Table 1.5 Typical Properties of Fluids Occurring in Shale Formations and in Tight Formations

Constituents (%v/v)	Dry Gas	Wet Gas	Condensate	Volatile Oil[a]
CO_2	0.1	1.4	2.4	1.8
N_2	2.1	0.3	0.3	0.2
C_1	86.1	92.5	73.2	57.6
C_2	5.9	3.2	7.8	7.4
C_3	3.6	1.0	3.6	4.2
Butane derivatives (C_4)	1.7	0.5	2.2	2.8
Pentane derivatives (C_5)	0.5	0.2	1.3	1.5
Hexane derivatives (C_{6+})		0.1	1.1	1.9
Heptane derivatives (C_{7+})		0.8	8.2	22.6

[a]Representative of crude oil from tight formations and tight shale formations.

The challenges associated with the production of crude oil from shale formations are a function of the compositional complexity and the varied geological formations where they are found. These oils are light, but they often contain high proportions of waxy constituents and, for the most part, reside in oil-wet formations. These phenomena create some of the predominant difficulties associated with crude oil extraction from the shale formations and include (1) scale formation, (2) salt deposition, (3) paraffin wax deposits, (4) destabilized asphaltene constituents, (5) equipment corrosion, and (6) bacteria growth. Thus, multicomponent chemical additives are added to the stimulation fluid to control these problems.

While crude oil from tight shale formations is characterized by a low content of asphaltene constituents and low-sulfur content, there can be a significant proportion of wax constituents in the oil. These constituents may exhibit a broad distribution of the molecular weight. For example, paraffin carbon chains of C_{10}–C_{60} have been found and some tight crude oil may even have hydrocarbon carbon chains (wax) up to C_{72}. While this may be a relief from recovery of high-asphaltene heavy oils, the joy is short-lived and the deposition of waxy constituents can cause as many problems as asphaltene incompatibility. To control deposition and plugging in formations due to paraffin, a variety of wax dispersants are available for use. In upstream applications, the paraffin wax dispersants are applied as part of multifunctional additive packages where, for convenience, asphaltene stability and corrosion control can also be addressed simultaneously.

Scale deposits of calcite, carbonates, and silicates must also be controlled during production or plugging problems arise. A wide range of scale additives is available. These additives can be highly effective when selected appropriately. Depending on the nature of the well and the operational conditions, a specific chemistry is recommended or blends of products are used to address scale deposition.

The most notable tight oil plays in North America include the Bakken shale, the Niobrara formation, the Barnett shale, the Eagle Ford shale, and the Miocene Monterey play of California's San Joaquin Basin in the United States, and the Cardium play in Alberta. In many of these tight formations, the existence of large quantities of oil has been known for decades and efforts to commercially produce those resources have occurred sporadically with typically disappointing results. However, starting in the mid-2000s, advancements in well-drilling and stimulation technologies combined with high oil prices have turned tight oil resources into one of the most actively explored and produced targets in North America.

Furthermore, of the tight oil plays, perhaps the best understood is the Bakken which straddles the border between Canada and the United States in North Dakota, Montana, and Saskatchewan. Much of what is known about the exploitation of tight oil resources comes from industry experiences in the Bakken and the predictions of future tight oil resource development described in this study are largely based on that knowledge. The Bakken tight oil play historically includes three zones, or members, within the Bakken Formation. The upper and lower members of the Bakken are organic-rich shales which serve as oil source rocks, while the rocks of the middle member may be siltstone formations, sandstone formations, or carbonate formations that are also typically characterized by low permeability and high oil content. Since 2008 the Three Forks Formation, another tight oil-rich formation which directly underlies the lower Bakken shale, has also yielded highly productive oil wells. Drilling, completion, and stimulation strategies for wells in the Three Forks Formation are similar to those in the Bakken and the light, sweet crude oil that is produced from both plays has been geochemically determined to be essentially identical. Generally, the Three Forks Formation is considered to be part of the Bakken play, though the authors of published works will sometimes refer to it as the Bakken-Three Forks play.

Other known tight formations (on a worldwide basis) include the R'Mah Formation in Syria, the Sargelu Formation in the northern Persian Gulf region, the Athel Formation in Oman, the Bazhenov formation and Achimov Formation in West Siberia, Russia, the Coober Pedy in Australia, the Chicontepex formation in Mexico, and the Vaca Muerta field in Argentina (US EIA, 2011, 2013). However, tight oil formations are heterogeneous and vary widely over relatively short distances. Thus, even in a single horizontal drill hole, the amount of oil recovered may vary as may recovery within a field or even between adjacent wells. This makes evaluation of *shale plays* and decisions regarding the profitability of wells on a particular lease difficult and a tight reservoir which contains only crude oil (without natural gas as the pressurizing agent) cannot be economically produced (US EIA, 2011, 2013).

Success in extracting crude oil and natural gas from shale reservoirs depends largely on the hydraulic-fracturing process (Speight, 2016b) that requires an understanding of the mechanical properties of the subject and confining formations. In hydraulic-fracturing design, Young's modulus is a criterion used to determine the most-appropriate fracturing fluid and other design considerations. Young's modulus provides an indication of the fracture conductivity that can be expected under the width and embedment considerations. Without adequate fracture conductivity, production from the hydraulic fracture will be minimal or nonexistent (Akrad et al., 2011; Speight, 2016b).

Typical of the crude oil from tight formations (*tight oil—tight light oil and tight shale oil* have been suggested as alternate terms) is the Bakken crude oil which is a light highly volatile crude oil. Briefly, Bakken crude oil is a light sweet (low-sulfur) crude oil that has a relatively high proportion of volatile constituents. The production of the oil also yields a significant amount of volatile gases (including propane and butane) and low-boiling liquids (such as pentane and natural gasoline), which are often referred to collectively as (low boiling or light) naphtha. By definition, natural gasoline (sometime also referred to as *gas condensate*) is a mixture of low-boiling liquid hydrocarbons isolate from crude oil and natural gas wells suitable for blending with light naphtha and/or refinery gasoline (Mokhatab et al., 2006; Speight, 2007, 2014a). Because of the presence of low-boiling hydrocarbons, low-boiling naphtha (*light naphtha*) can become extremely explosive, even at relatively low ambient temperatures. Some of these gases may be burned off (flared) at the field well-head, but others remain in the liquid products extracted from the well (Speight, 2014a).

Bakken crude oil is considered to be a low-sulfur (*sweet*) crude oil and there have been increasing observations of elevated levels of hydrogen sulfide (H_2S) in the oil. Hydrogen sulfide is a toxic, highly flammable, corrosive, explosive gas (hydrogen sulfide) and there have been increasing observations of elevated levels of hydrogen sulfide in Bakken oil. Thus, the liquids stream produced from the Bakken Formation will include the crude oil, the low-boiling liquids, and gases that were not flared, along with the materials and byproducts of the hydraulic-fracturing process. These products are then mechanically separated into three streams: (1) produced salt water, often referred to as brine, (2) gases, and (3) crude oil liquids, which include condensates, natural gas liquids, and light oil. Depending on the effectiveness and appropriate calibration of the separation equipment which is controlled by the oil producers, varying quantities of gases remain dissolved and/or mixed in the liquids, and the whole is then transported from the separation equipment to the well-pad storage tanks, where emissions of volatile hydrocarbons have been detected as emanating from the oil.

Oil from tight shale formation is characterized by low-asphaltene content, low-sulfur content, and a significant molecular weight distribution of the paraffinic wax content (Speight, 2014a, 2015a). Paraffin carbon chains of C_{10}–C_{60} have been found, with some shale oils containing carbon chains up to C_{72}. To control deposition and plugging in formations due to paraffins, the dispersants are commonly used. In upstream applications, these paraffin dispersants are applied as part of multifunctional additive packages where asphaltene stability and corrosion control are also addressed simultaneously (Speight, 2014a,b,c, 2015a,b). In addition, scale deposits of calcite ($CaCO_3$), other carbonate minerals (minerals containing the carbonate ion, CO_3^{2-}), and silicate minerals (minerals classified on the basis of the structure of the silicate group, which contains different ratios of silicon and oxygen) must be controlled during production or plugging problems arise. A wide range of scale additives is available which can be highly effective when selected appropriately. Depending on the nature of the well and the operational conditions, a specific chemistry is recommended or blends of products are used to address scale deposition.

Another challenge encountered with oil from tight shale formations—many of which have been identified but undeveloped—is the general lack (until recently) of transportation infrastructure. Rapid distribution of the crude oil to the refineries is necessary to maintain consistent refinery throughput—a necessary aspect of refinery design. Some pipelines are in use, and additional pipelines are being (and need to be) constructed to provide consistent supply of the oil to the refinery. During the interim, barges and railcars are being used, along with a significant expansion in trucking to bring the various crude oil to the refinery. For example, with development of suitable transportation infrastructure, production of Eagle Ford tight oil is estimated to increase by a substantial amount to approximately 2,000,000 bpd by 2017. Similar expansion in crude oil production is estimated for Bakken and other identified (and perhaps as yet unidentified and, if identified, undeveloped) tight formations.

While the basic approach toward developing a tight oil play is expected to be similar from area to area, the application of specific strategies, especially with respect to well completion and stimulation techniques, will almost certainly differ from play to play, and often even within a given play. The differences depend on the geology (which can be very heterogeneous, even within a play) and reflect the evolution of technologies over time with increased experience and availability.

Finally, the properties of tight oil are highly variable. Density and other properties can show wide variation, even within the same field. The Bakken crude is light and sweet with an API of 42 degrees and a sulfur content of

0.19% *w/w*. Similarly, Eagle Ford is a light sweet feed, with a sulfur content of approximately 0.1% *w/w* and with published API gravity between 40 and 62 degrees API.

In terms of refining, although tight oil is considered sweet (low-sulfur content) and amenable to refinery options, this is not always the case. Hydrogen sulfide gas, which is flammable and poisonous, comes out of the ground with the crude oil and must be monitored at the drilling site as well as during transportation. Amine-based hydrogen sulfide scavengers are added to the crude oil prior to transport to refineries. However, mixing during transportation due to movement, along with a change in temperature that raises the vapor pressure of the oil, can cause the release of entrained hydrogen sulfide during offloading, thereby creating a safety hazard. For example, such crude that is loaded on railcars in winter and then transported to a warmer climate becomes hazardous clue to the higher vapor pressure. The shippers and receivers of the oil should be aware of such risks.

Paraffin waxes are present in tight oil and remain on the walls of railcars, tank walls, and piping. The waxes are also known to foul the preheat sections of crude heat exchangers (before they are removed in the crude desalter). Paraffin waxes that stick to piping and vessel walls can trap amines against the walls which can create localized corrosion (Speight, 2014c). *Filterable solids* also contribute to fouling in the crude preheat exchangers and a tight crude can contain over seven times more filterable solids than a traditional crude oil. To mitigate filter plugging, the filters at the entrance of the refinery require automated monitoring because they need to capture large volumes of solids. In addition, wetting agents are added to the desalter to help capture excess solids in the water, rather than allowing the undesired solids to travel further downstream into the process.

In many refineries, *blending two or more crude oils* as the refinery feedstock is now standard operating procedure which allows the refiner to achieve the right balance of feedstock qualities. However, the blending of the different crude oils may cause problems if the crude oils being mixed are incompatible (Speight, 2014a). When crude oils are incompatible, there is increased deposition of the asphaltene constituents (Speight, 2014a) which accelerates fouling in the heat exchanger train downstream of the crude desalter (Speight, 2014c). Accelerated fouling increases the amount of energy that must be supplied by the crude fired heater, which limits throughput when the fired heater reaches its maximum duty and may also necessitate an earlier shutdown for cleaning.

Mixing stable crude oil blends with asphaltic and paraffinic oils creates the potential for precipitating the unstable asphaltene constituents—a high content of paraffinic naphtha in tight oils also creates favorable conditions for precipitation of the asphaltene constituents (Speight, 2014a,c). It should be noted that the ratio of crude oils in a blend may have an impact on crude incompatibility. For example, a low amount of tight oil in a blend may not cause accelerated fouling whereas a blend containing a higher amount of tight oil may bring about accelerated fouling. The key is the separation of any constituents that can cause fouling (Speight, 2014c).

Shale gas (also called *tight gas*) is a description for a field in which natural gas accumulation is locked in tiny bubble-like pockets within layered low-permeability sedimentary rock such as shale. The terms *shale gas* and *tight gas* are often used interchangeably but there are differences—while shale gas is trapped in rock, tight gas describes natural gas that is dispersed within low-porosity silt or sand areas that create a tight-fitting environment for the gas. Typically, tight gas refers to natural gas that has migrated into a reservoir rock with high porosity but low permeability. These types of reservoirs are not usually associated with oil and commonly require horizontal drilling and hydraulic fracturing to increase well output to cost-effective levels. In general, the same drilling and completion technology that is effective with shale gas can also be used to access and extract tight gas. Shell uses proven technology in responsible ways to access this needed resource.

Tight gas is the fastest growing natural gas resource in the United States and worldwide as a result of several recent developments (Nehring, 2008). Advances in horizontal drilling technology allow a single well to pass through larger volumes of a shale gas reservoir and, thus, produce more gas. The development of hydraulic-fracturing technology has also improved access to shale gas deposits. This process requires injecting large volumes of water mixed with sand and fluid chemicals into the well at high pressure to fracture the rock, increasing permeability and production rates.

To extract tight gas, a production well is drilled vertically until it reaches the shale formation, at which point the wellbore turns to follow the shale horizontally. As drilling proceeds, the portion of the well within the shale is lined with steel tubing (casing). After drilling is completed, small explosive charges are detonated to create holes in the casing at intervals where hydraulic fracturing is to occur. In a hydraulic-fracturing operation, the fracturing fluid is pumped in at a carefully controlled pressure to fracture the rock out to several hundred feet from the well. Sand mixed with the fracturing fluid acts to prop these cracks open when the fluids are subsequently pumped out.

After fracturing, gas will flow into the wellbore and up to the surface, where it is collected for processing and sales.

Shale gas is natural gas produced from shale formations that typically function as both the reservoir and source rocks for the natural gas. In terms of chemical makeup, shale gas is typically a dry gas composed primarily of methane (60–95% *v/v*), but some formations do produce wet gas. The Antrim and New Albany plays have typically produced water and gas. Gas shale formations are organic-rich shale formations that were previously regarded only as source rocks and seals for gas accumulating in the strata near sandstone and carbonate reservoirs of traditional onshore gas development.

Thus, by definition, *shale gas* is the hydrocarbon gas present in organic-rich, fine-grained, sedimentary rocks (shale and associated lithofacies). The gas is generated and stored in situ in gas shale as both adsorbed gas (on organic matter) and free gas (in fractures or pores). As such, shale containing gas is a self-sourced reservoir. Low-permeable shale requires extensive fractures (natural or induced) to produce commercial quantities of gas.

Shale is a very fine-grained sedimentary rock, which is easily breakable into thin, parallel layers. It is a very soft rock, but it does not disintegrate when it becomes wet. The shale formations can contain natural gas, usually when two thick, black shale deposits *sandwich* a thinner area of shale. Because of some of the properties of the shale deposits, the extraction of natural gas from shale formations is more difficult and perhaps more expensive than that of conventional natural gas. Shale basins are scattered across the United States.

There are several types of unconventional gas resources that are currently produced: (1) deep natural gas—natural gas that exists in deposits very far underground, beyond "conventional" drilling depths, typically 15,000 ft or more, (2) shale gas—natural gas that occurs in low-permeability shale formations, (3) tight natural gas—natural gas that occurs in low-permeability formations, (4) geopressurized zones—natural underground formations that are under unusually high pressure for their depth, (5) Coalbed methane—natural gas that occurs in conjunction with coal seams, and (6) methane hydrates—natural gas that occurs at low-temperature and high-pressure regions such as the sea bed and is made up of a lattice of frozen water, which forms a *cage* around the methane.

Coalbed methane is produced from wells drilled into coal seams which act as source and reservoir to the produced gas (Speight, 2013b). These wells often produce water in the initial production phase, as well as natural gas. Economic coalbed methane reservoirs are normally shallow, as the coal matrix tends to have insufficient strength to maintain porosity at depth. On the other hand, shale gas is obtained from ultra-low permeability shale formations that may also be the source rock for other gas reservoirs. The natural gas volumes can be stored in fracture porosity, within the micropores of the shale itself, or adsorbed onto the shale.

To prevent the fractures from closing when the pressure is reduced several tons of sand or other *proppant* is pumped down the well and into the pressurized portion of the hole. When the fracturing occurs millions of sand grains are forced into the fractures. If enough sand grains are trapped in the fracture, it will be propped partially open when the pressure is reduced. This provides an improved permeability for the flow of gas to the well.

It has been estimated that there is on the order of 750 trillion cubic feet (Tcf, 1×10^{12} ft^3) of technically recoverable shale gas resources in the United States and represents a large and very important share of the United States recoverable resource base and in addition, by 2035, approximately 46% of the natural gas supply of the United States will come from shale gas (EIA, 2011).

Tight gas is a form of unconventional natural gas that is contained in a very low-permeability formation underground—usually hard rock or a sandstone or limestone formation that is unusually impermeable and nonporous (tight sand). In a conventional natural gas deposit, once drilled, the gas can usually be extracted quite readily and easily (Speight, 2007). Like shale gas reservoirs, tight gas reservoirs are generally defined as having low permeability (in many cases <0.1 mD, Law and Spencer, 1993). Tight gas makes up a significant portion of the natural gas resource base—>21% *v/v* of the total recoverable natural gas in the United States is in tight formations and represents an extremely important portion of natural gas resources (GAO, 2012).

In tight gas sands (low-porosity sandstones and carbonate reservoirs), gas is produced through wells and the gas arose from a source outside the reservoir and migrates into the reservoir over geological time. Some *tight gas reservoirs* have also been found to be sourced by underlying coal and shale formation source rocks, as appears to be the case in the *basin-centered gas accumulations*.

However, extracting gas from a tight formation requires more severe extraction methods—several such methods do exist that allow natural gas to be extracted, including hydraulic fracturing and acidizing. It has been projected that shale formations and tight formations containing natural gas and crude oil with a permeability as low as 1 nD may be economically productive with optimized spacing and completion of staged fractures to maximize yield with respect to cost (McKoy and Sams, 2007). In any case, with all unconventional natural gas and crude oil reserves, the economic incentive must be there to encourage companies to extract this gas and oil instead of more easily obtainable, conventional natural gas and crude oil.

4 ORIGIN AND RESERVOIRS

Unlike oil shale projects (Scouten, 1990; Lee, 1991; Lee et al., 2007; Speight, 2008, 2012b), the producible portions of deep shale oil and natural gas formations exist many thousands of feet below the surface—typically at depths ranging from 5000 to 12,000 ft underground. Tight formations scattered throughout North America have the potential to produce not only gas (*tight gas*) but also crude oil (*tight oil*) (Law and Spencer, 1993; US EIA, 2011, 2013; Speight, 2013a; Islam, 2014). Such formations might be composted of shale sediments or sandstone sediments. In a conventional sandstone reservoir, the pores are interconnected so gas and oil can flow easily from the rock to a wellbore. In tight sandstones, the pores are smaller and are poorly connected by very narrow capillaries which results in low permeability. Tight gas and tight oil occur in sandstone sediments that have an effective permeability of <1 mD (Fig. 1.2) and in addition, the tight oil is light oil with a high volatility (Tables 1.5 and 1.6). In addition, application of typical fractionating techniques (Fig. 1.1) to oil from tight formations would show very little, if any, resin and asphaltene constituents. The majority of the crude oil is typically low-boiling paraffin constituents and aromatic constituents.

One of the newest terms in the crude oil lexicon is the arbitrarily-named (even erroneously named) *shale oil* which is used to describe crude oil that is produced from tight shale formations. The use of this term should be discontinued and when used it should not be confused with the older term *shale oil*, which is crude oil that is produced by the thermal treatment of oil shale and the ensuing decomposition of the kerogen contained within the shale (Scouten, 1990; Speight, 2012b) and does not fall under the umbrella definition of tight oil. Oil shale represents one of the largest unconventional hydrocarbon deposits in the world with an estimated eight trillion barrels (8×10^{12} bbls) of oil in place. Approximately six trillion barrels of oil in place is located in the United States including the richest and most concentrated deposits found in the Green River Formation in Colorado, Utah, and Wyoming. Documented efforts to develop oil shale to produce shale oil in the

Table 1.6 Simplified Differentiation between Conventional Crude Oil and Crude Oil from Shale Formations

Conventional crude oil
Medium-to-high API gravity
Low-to-medium sulfur content
Mobile in the reservoir
High-permeability reservoir
Primary recovery
Secondary recovery
May use tertiary recovery when reservoir energy is depleted
Tight oil
High API gravity
Low-sulfur content
Immobile in the reservoir
Low-to-zero permeability reservoir
Primary, secondary, and tertiary methods of recovery ineffective
Horizontal drilling into reservoir
Fracturing (typically multifracturing) to release reservoir fluids

United States go back to approximately 1900, even earlier in Scotland (Scouten, 1990; Lee, 1991; Lee et al., 2007; Speight, 2008, 2012b). These prior efforts have produced a wealth of knowledge regarding the geological description as well as technical options and challenges for development. Thus far, however, none of these efforts have produced a commercially viable business in the United States. There needs to be economically viable, socially acceptable, and environmentally responsible development solutions.

Recently, the introduction of the term *shale oil* to define crude oil from tight shale formations is the latest term to add confusion to the system of nomenclature of crude oil-heavy oil-bitumen materials. The term has been used without any consideration of the original term shale oil produced by the thermal decomposition of kerogen in oil shale. It is not quite analogous, but is certainly similarly confusing, to the term *black oil* that has been used to define crude oil by color rather than by any meaningful properties or recovery behavior (Speight, 2014a, 2015a).

Briefly, the characteristic feature of oil shale is the presence of kerogen (organic matter) from which oil is obtained by thermal treatment in the absence of oxygen (Scouten, 1990; Lee, 1991; Lee et al., 2007; Speight, 2008, 2012b). Shale oil is the main product of oil shale thermal treatment or processing (sometimes also called retorting). Utilization of oil shale is technically and economically feasible only if can at least 20% of its kerogen be converted into oil by thermal processing. If oil yield is less, one can but talk about some kind of low-kerogen rock and not true oil shale (Scouten, 1990; Lee, 1991; Lee et al., 2007; Speight, 2008, 2012b).

Generally, unconventional tight oil and natural gas are found at considerable depths in sedimentary rock formations that are characterized by very low permeability. While some of the tight oil plays produce oil directly from shales, tight oil resources are also produced from low-permeability siltstone formations, sandstone formations, and carbonate formations that occur in close association with a shale source rock. It is important to note that in the context of this text, the term tight oil does not include resources that are commonly known as "oil shales" which refers to oil or kerogen-rich shale formations that are either heated in situ and produced or if surface accessible mined and heated (Scouten, 1990; Lee, 1991; Lee et al., 2007; Speight, 2008, 2012b).

4.1 Origin

In the same manner as conventional natural gas and crude oil, natural gas in shale formations and in tight formations has, essentially, formed from the remains of plants, animals, and microorganisms that lived millions of years ago. Though there are different theories on the origins of fossil fuels, the most widely accepted is that they are formed when organic matter (such as the remains of a plant or animal) is buried and compressed (even heated but the actual temperature of the maturation process remains unknown and, at best, is only speculative) for geological long time (millions of years).

More specifically, natural gas and crude oil in shale formations and in tight formations are generated in two different ways: (1) as a thermogenic product that is generated thermally from the organic matter in the matrix and (2) as a biogenic product—an example is the Antrim shale gas field in Michigan in which the gas has been generated from microbes in areas of fresh water recharge (Shurr and Ridgley, 2002; Martini et al., 1998, 2003, 2004).

By way of explanation, the origin of natural gas and crude oil is an important aspect of evaluating shale reservoir. For example, thermogenic systems often produce natural gas liquids with the methane, which can add value to production, whereas biogenic systems typically generate methane only. Thermogenic systems can also lead to the generation of carbon dioxide as an impurity in the natural gas, which must be removed during the gas processing operations (Chapter 7). Also, reservoirs classed as having thermogenic origins tend to flow at high rates but are normally exploited through the extensive use of horizontal drilling and are therefore more expensive to develop than biogenic plays, which flow at lower rates and are exploited through shallow, closely spaced vertical wells instead.

Thus, the thermogenic product is associated with mature organic matter that has been subjected to relatively high temperature and pressure in order to generate hydrocarbons. Moreover, all other factors being equal the more mature organic matter should generate higher gas-in-place resources than less mature organic matter (Schettler and Parmely, 1990; Martini et al., 1998). However, generation of the gas and oil within individual shale formations and tight formations may differ significantly. Better knowledge is needed, for example, on basin modeling, petrophysical characterization, or gas flow in shales for an improved understanding of unconventional reservoirs.

The biogenic product can be associated with either mature or immature organic matter and can add substantially to shale gas reserves. For example, the San Juan Basin coalbed methane gas field is a mixture of both gases and has generated much of its gas from biogenic processes (Scott et al., 1994). Likewise, gas from the Antrim Shale Formation in the Michigan Basin is largely biogenic gas that has been generated in the last 10,000–20,000 years (Martini et al., 1998, 2003, 2004) and has produced >2.4 Tcf as of 2006. A mixture of gases is suggested for the New Albany Shale formation in the Illinois Basin (Wipf and Party, 2006) and is certainly possible in Alberta shale.

Tight gas resources and tight oil resources differ from conventional natural gas resources insofar as the shale acts as both the source for the gas and oil, and also the zone (the reservoir) in which the gas and oil are trapped. The very low permeability of the rock causes the rock to trap the gas or oil and prevent it from migrating toward the surface. The gas and oil can be held in natural fractures or pore spaces or can be adsorbed onto organic material. With the advancement of drilling and completion technology, this gas can be successfully exploited and extracted commercially as has been proven in various basins in North America.

Kerogen is often cited as the precursor to oil and gas but, however, the role played by kerogen in the natural gas and crude oil maturation process is not fully understood (Tissot and Welte, 1978; Durand, 1980; Hunt, 1996; Scouten, 1990; Speight, 2014a). What obviously needs to be addressed more fully in terms of kerogen participation in petroleum generation is the potential to produce petroleum constituents from kerogen by low-temperature processes rather than by processes that involve the use of temperatures in excess of 250°C (>480°F) (Burnham and McConaghy, 2006; Speight, 2014a).

If such geochemical studies are to be pursued, a thorough investigation is needed to determine the potential for such high temperatures being present during the main phase, or even various phases, of petroleum generation on order to give stronger indications that kerogen is a precursor to petroleum (Speight, 2014a).

4.2 Shale Formations

By way of definition and in the context of this book, a shale play is a defined geographic area containing an organic-rich fine-grained sedimentary rock that underwent physical and chemical compaction during diagenesis to produce the following characteristics: (1) clay to silt sized particles, (2) high percentage of silica, and sometimes carbonate minerals, (3) thermally mature, (4) hydrocarbon-filled porosity—on the order of 6–14%, (5) low permeability—on the order of <0.1 mD, (6) large areal distribution, and (7) fracture stimulation required for economic production.

In the current context, natural gas and crude oil from shale formation are natural gas and crude oil that are produced from a type of sedimentary rock derived from clastic sources often including mudstone or siltstone, which is known as shale. Clastic sedimentary rocks are composed of fragments (clasts) of preexisting rocks that have been eroded, transported, deposited, and lithified (hardened) into new rocks. Shale deposits typically contain organic material which was lain down along with the rock fragments.

Shale is a sedimentary rock that was once deposited as mud (clay and silt) and is generally a combination of clay minerals, silica minerals (e.g., quartz), carbonate minerals (calcite or dolomite), and organic material. While shale formations are generally considered to be rich in clay minerals, the proportions of the constituents are more likely to be highly variable. Shale formations may also exist in thin beds or laminae of sandstone (SiO_2), limestone ($CaCO_3$), or dolostone, a sedimentary carbonate rock that contains a high proportion of the mineral dolomite ($CaCO_3 \cdot MgCO_3$). The mud—in the form of microscopic mineral particles—was deposited in deep, quiet (calm) water such as in large lakes or deep seas and oceans. The organic matter in the mud was algae, plant matter, or plankton that died and sank to the sea floor or lake bed before being buried.

A more technical definition of a *shale formation* is a fissile, terrigenous sedimentary rock in which particles are mostly of silt and clay size in which the term *fissile* refers to the rock's ability to split into thin sheets along bedding while *terrigenous* refers to the sediment's origins, that it is the product of weathering of rocks (Blatt and Tracy, 2000). In addition, a *bed* is layer of sediment thicker than 1 cm, whereas a lamina (plural: laminae) is a layer of sediment, typically thinner than 1 cm (Blatt and Tracy, 2000). Also, the pore spaces in shale, through which the natural gas must move if the gas is to flow into any well, are on the order of 1000 times smaller than the pores in a conventional sandstone reservoir (Bowker, 2007). The gaps that connect pores (the pore throats) are smaller still, only 20 times larger than a single methane molecule. Therefore, a shale formation has very low permeability. However, fractures, (natural fractures) which act as conduits for the movements for natural gas, may naturally exist in the shale and increase the permeability of the formation.

A shale reservoir originated as a formation that is an organic-rich, and fine-grained sediment that contains natural gas or crude oil (Bustin, 2006; Bustin et al., 2008). However, the term *shale* is used very loosely and does not describe the lithology of the reservoir. Lithological variations in American shale gas reservoirs indicate that natural gas is retained in the reservoir not only in shale but also a wide spectrum of lithology and texture from mudstone (i.e., nonfissile shale) to siltstone and fine-grained sandstone, any of which may be of siliceous or carbonate composition. For example, in many basins, much of what is described as shale is often siltstone, or encompasses multiple rock types, such as siltstone or sandstone laminations interbedded with shale laminations or beds. The presence of multiple rock types in organic-rich shale formations implies that there are multiple gas and oil storage mechanisms, as gas or oil constituents may be adsorbed on organic matter and stored as free gas in micropores and macropores.

Laminations serve a dual purpose because they both store free gas and oil and transmit gas and oil desorbed from organic matter in shale to the wellbore. The determination of the permeability and porosity of the laminations, and the linking of these laminations via a hydraulic fracture to the wellbore, are key requirements for efficient development. Additionally, solute or solution gas may be held in micropores and nanopores of bitumen (Bustin, 2006) and may be an additional source of gas, although traditionally this is thought to be a minor component. Free gas may be a more dominant source of production than desorbed gas or solute gas in a shale gas reservoir. Determining the percentage of free gas versus solute gas versus desorbed gas is important for resource and reserve evaluation and is a significant issue in gas production and reserve calculations, as desorbed gas diffuses at a lower pressure than free gas.

In areas where conventional resources are located, shale can be found in the underlying rock strata and can be the source of the hydrocarbons that have migrated upwards into the reservoir rock. Over time, as the rock matures, hydrocarbons are produced from the kerogen. These may then migrate, as either a liquid (petroleum) or a gas (natural gas), through existing fissures and fractures in the rock until they reach the earth's surface or until they become trapped by strata of impermeable rock. Porous areas beneath these traps collect the hydrocarbons in a conventional reservoir, frequently of sandstone.

Shale gas reservoirs generally allow recovery of less gas (from <5% to 20% *v/v*) relative to conventional gas reservoirs (approximately 50–90% *v/v*) (Faraj et al., 2004), although the naturally well-fractured Antrim Shale may have a recovery factor as high as 50–60% *v/v*. More recently, there have been suggestions that the Haynesville shale in Louisiana may have a recovery factor as high as 30% (Durham, 2008). To increase the recovery factor, innovation in drilling and completion technology is paramount in low-permeability shale reservoirs. In the initial state of pool development, permeability "sweet spots" are often sought because they result in higher rates of daily production and increased recovery of gas compared to less permeable shale.

But these sweet spots are small, relative to the size of unconventional pools, so horizontal drilling and new completion techniques (such as *staged fracs* and *simultaneous fracs*) (Cramer, 2008) were developed to improve economics both inside and outside of the reservoir sweet spots. The result is a significant increase in economically producible reserves and a substantial extension of the area of economically producible gas. However, well-fractured shale that typically contains an abundance of mature organic matter and is deep or under high pressure will yield a high initial flow rate. For example, horizontal wells in the Barnett with a high initial reservoir pressure can yield an initial flow rate of a few million cubic feet per day after induced fracturing. However, after the first year gas flow may be dominated by the rate of diffusion from the matrix to the induced fractures (Bustin et al., 2008).

Aside from permeability, the key properties of shale, when considering gas and oil potential, are (1) total organic content and (2) thermal maturity. The total organic content is the total amount of organic material present in the rock, expressed as a percentage by weight. Generally, the higher the total organic content, the better the potential for hydrocarbon generation. The *thermal maturity* of the rock is a measure of the degree to which organic matter contained in the rock has been heated over time and potentially converted into liquid and/or gaseous hydrocarbons.

The gas and oil storage properties of shale are quite different to conventional reservoirs. In addition to having gas or oil present in the matrix system of pores similar to that found in conventional reservoir rocks, shale also has gas bound or adsorbed to the surface of organic materials in the shale. The relative contributions and combinations of free gas from matrix porosity and from desorption of adsorbed gas is a key determinant of the production profile of the well.

The amount and distribution of gas or oil within the shale is determined by, amongst other things, the initial reservoir pressure, the petrophysical properties of the rock, and its adsorption characteristics. During production there are three main processes at play. Initial gas production is dominated by depletion of gas from the fracture network. This form of production declines rapidly due to limited storage capacity. After the initial decline rate stabilizes, the depletion of gas stored in the matrix becomes the primary process involved in production. The amount of gas or oil held in the matrix is dependent on the particular properties of the shale reservoir which can be hard to estimate. Secondary to this depletion process is desorption whereby adsorbed gas is released from the rock as pressure in the reservoir declines. The rate of gas production via the desorption process depends on there being a significant drop in reservoir pressure. Pressure changes typically advance through the rock very slowly due to low permeability. Tight well spacing can therefore be required to lower the reservoir pressure enough to cause significant amounts of adsorbed gas to be desorbed.

The ultimate recovery (Chapter 3) of the gas or oil in place surrounding a particular shale well can be in the order of 28–40% of the original volume in place whereas the recovery per conventional well may be as high as 60–80% *v/v*. The development of shale gas plays, therefore, differs significantly from the development of conventional resources. With a conventional reservoir, each well is capable of draining oil or gas over a relatively large area (dependent on reservoir properties). As such, only a few wells (normally vertical) are required to produce commercial volumes from the field. With shale gas projects, a large number of relatively closely spaced wells are required to produce large enough volumes to make the plays economic. As a result, many wells must be drilled in a shale play to drain the reservoir sufficiently—in the Barnett shale resource in the United States, the drilling density can exceed one well per 60 acres.

TOC is a fundamental attribute of tight gas and tight oil formations and is a measure of organic richness. The content of total organic carbon together with the thickness of organic shale and organic maturity are key indicators that can be used to determine the economic viability of a shale gas play. There is no unique combination or minimum amount of these factors that determines economic viability. The factors are highly variable between shale of different ages and can vary, in fact, within a single deposit or stratum of shale over short distances.

Induced fracturing may occur many times during the productive life of a shale gas reservoir (Walser and Pursell, 2007). Shale, in particular, exhibits permeability lower than coalbed methane or tight gas and, because of this, forms the source and seal of many conventional oil and gas pools. Hence, not all shale is capable of sustaining an economic rate of production. In this respect, permeability of the shale matrix is the most important parameter influencing sustainable shale gas production (Bennett et al., 1991a,b; Davies et al., 1991; Davies and Vessell, 2002; Gingras et al., 2004; Pemberton and Gingras, 2005; Bustin et al., 2008).

To sustain yearly production, gas must diffuse from the low-permeability matrix to induced or natural fractures. Generally, higher matrix permeability results in a higher rate of diffusion to fractures and a higher rate of flow to the wellbore (Bustin et al., 2008). Furthermore, more fractured shale (i.e., shorter fracture spacing), given sufficient matrix permeability, should result in higher production rates (Bustin et al., 2008), a greater recovery of hydrocarbons, and a larger drainage area (Cramer, 2008; Walser and Pursell, 2007). Additionally, microfractures within shale matrix may be important for economic production; however, these microfractures are not easily determined in situ in a reservoir (Tinker and Potter, 2007), and only further research and analysis will determine their role in shale gas production.

An additional factor to consider is shale thickness. The substantial thickness of shale is one of the primary reasons, along with a large surface area of fine-grained sediment and organic matter for adsorption of gas, that shale resource evaluations yield such high values. Therefore, a general rule is that thicker shale is a better target. Shale targets such as the Bakken oil play in the Williston Basin (itself a hybrid conventional-unconventional play), however, are <50 m thick in many areas and are yielding apparently economic rates of flow. The required thickness to economically develop a shale gas target may decrease as drilling and completion techniques improve, as porosity and permeability detection techniques progress in unconventional targets and, perhaps, as the price of gas increases. Such a situation would add a substantial amount of resources and reserves to the province.

4.3 Tight Formations

The term *tight formation* refers to a formation consisting of extraordinarily impermeable, hard rock. Tight formations are relatively low-permeability, non-shale, sedimentary formations that can contain oil and gas.

A *tight reservoir* (*tight sands*) is a low-permeability sandstone reservoir that produce primarily dry natural gas. A tight gas reservoir is one that cannot be produced at economic flow rates or recover economic volumes of gas unless the well is stimulated by a large hydraulic fracture treatment and/or produced using horizontal wellbores. This definition also applies to coalbed methane and tight carbonate reservoirs—shale gas reservoirs are also included by some observers (but not in this text). Typically, tight formations which formed under marine conditions contain less clay and are more brittle, and thus more suitable for hydraulic fracturing than formations formed in fresh water which may contain more clay. The formations become more brittle with an increase in quartz content (SiO_2) and carbonate content (such as calcium carbonate, $CaCO_3$, or dolomite, $CaCO_3 \cdot MgCO_3$).

By way of explanation, in a conventional sandstone reservoir the pores are interconnected so that natural gas and crude oil can flow easily through the reservoir and to the production well. However, in tight sandstone formations, the pores are smaller and are poorly connected (if at all) by very narrow capillaries which results in low permeability and immobility of the natural gas and crude oil. Such sediments typically have an effective permeability of <1 mD (Fig. 1.2) and, in the case of a crude oil reservoir, the oil is a highly volatile light sweet crude oil (Tables 1.5 and 1.6). In addition, application of typical fractionating techniques (Fig. 1.1) to the crude oil from tight formations shows the relative absence (compared to conventional crude oil) of resin constituents and asphaltene constituents. The majority of the crude oil is typically low-boiling paraffin constituents (including high-molecular weight waxy constituents) and aromatic constituents.

4.4 Geopressurized Zones

Geopressurized zones are natural underground formations that are under high pressure that is not always expected from the depth of the zone. These areas are formed by layers of clay minerals that are deposited and compacted on top of (or above) more porous, absorbent material such as sandstone or silt. Water and natural gas that are present in this clay formation are forced out by the rapid compression and enter into the more porous sandstone or silt deposits. The natural gas or crude oil is deposited in this sandstone or silt under very high pressure (*geopressure*). Geopressurized zones are typically located at great depths—on the order of usually 10,000–25,000 ft below the surface of the earth. The combination of the above factors makes the extraction of natural gas or crude oil located in geopressurized zones quite complicated.

Most of the geopressurized natural gas in the United States is located in the Gulf Coast region. Although the amount of natural gas in these geopressurized zones is uncertain, it is estimated to be on the order of 5000–49,000 trillion cubic feet ($5000–49,000 \times 10^{12}$ ft^3) or 5–49 quadrillion cubic feet, $5–40 \times 10^{15}$ ft^3. Thus, geopressurized zones offer an incredible opportunity for increasing the natural gas supply of the United States.

5 OIL SHALE AND SHALE OIL

References was made earlier to the erratic and incorrect terminology by which tight oil has been referred to as *shale oil*. This terminology is also confusing and the use of such terminology should be discouraged as illogical since shale oil has been (for decades) the name given to the distillate produced from oil shale by thermal decomposition (Lee, 1990; Scouten, 1990; Speight, 2012b, 2014a, 2016b). Therefore, it is appropriate at this point to clarify the names oil shale and shale oil.

Oil shale is found at shallow depths, from surface outcrops to 3000 ft deep. It is a very low-permeability sedimentary rock that contains a large proportion of kerogen, a mixture of solid organic compounds. In kerogen-containing shale reservoirs (IEA, 2013), conversion of the kerogen into liquid oil (shale oil, kerogen oil) has not taken place because the high temperatures required have not been experienced. Oil shale consists of minerals of variable composition mixed with organic matter commonly occurring finely dispersed in the matrix or in thin laminae (Eseme et al., 2007). To allow an appreciation of their complex behavior especially at high temperature, it is useful to consider an oil shale as a three-phase material: (1) minerals, (2) kerogen, and (3) pore fill and the bulk mechanical properties strongly depend on the volume fractions of these phases. Most studies on mechanical properties of oil shales are accompanied by an indication of organic matter content. Commonly, the parameter reported is the oil yield given in gallons per ton (GPT, 1gallon = 4.2 L) determined by the standardized Fischer assay technique (heating 100 g of crushed rock to 500°C, 930°F).

Oil shale is a prolific synthetic fuel resources on earth and copious deposits of oil shale are found in a number of countries, including Australia, Brazil, China, Estonia, Israel, Jordan, Mongolia, and the United States. Preliminary geologic surveys and evidence from oil shale outcrops indicate that Mongolia may also have oil shale resources of a size and quality that are commercially viable. The United States is recognized as having the largest oil shale deposits in the world, the richest of which are located in the Green River Basin, an overlapping area of Colorado, Utah, and Wyoming. Estimates of the total resource that could be conceivably recovered exceeds 2 trillion barrels (2×10^{12} bbls) of thermally produced shale oil.

In the context of this book—natural gas and crude oil from tight formations, including shale formations—a comparison between shale oil and oil from tight shales should end. With the evolution of tight oil production, there has been confusion regarding the difference between tight oil and shale oil. Often, the terms are used incorrectly and often interchangeably, thereby adding further confusion to their distinction. The two resources are vastly different and markedly different in composition and methods of production.

By definition, oil shale is shale that contains a carbonaceous immobile material (kerogen) which, by the application of heat, can be converted into gas and synthetic crude oil—only applied heat will produce shale oil (a synthetic crude oil) from oil shale. Furthermore, the term synthetic crude oil is typically applied to oil that is produced by thermal means and which contains constituents that are not indigenous to the oil or to the shale formation. There is no requirement for elaborate horizontal drilling or hydraulic fracturing of the shale deposit to allow flow paths through which the oil and gas will be produced. On the other hand, tight oil is a conventional crude oil that is created naturally through the maturation process but is trapped in shale deposits. Shale gas is similarly produced from precursors that were trapped within the shale formations.

5.1 Oil Shale

The term *oil shale* describes an organic-rich rock from which little carbonaceous material can be removed by extraction (with common crude oil-based solvents) but which produces variable quantities of distillate (*shale oil*) when raised to temperatures in excess of 350°C (660°F). Thus, oil shale is assessed by the ability of the mineral to produce shale oil in terms of gallons per ton (g/t) by means of a test method (Fischer assay) in which the oil shale is heated to 500°C (930°F).

Oil shale represents a large and mostly untapped hydrocarbon resource. Like tar sand (*oil sand* in Canada) and coal, oil shale is considered unconventional because oil cannot be produced directly from the resource by sinking a well and pumping. Oil has to be produced thermally from the shale. The organic material contained in the shale is called *kerogen*, a solid material intimately bound within the mineral matrix (Baughman, 1978; Allred, 1982; Scouten, 1990; Lee, 1991; Speight, 2008, 2013b, 2014a).

Oil shale is distributed widely throughout the world with known deposits in every continent. Oil shale ranging from Cambrian to Tertiary in age occurs in many parts of the world (Table 1.1). Deposits range from small occurrences of little or no economic value to those of enormous size that occupy thousands of square miles and contain many billions of barrels of potentially extractable shale oil. However, crude oil-based crude oil is cheaper to produce today than shale oil because of the additional costs of mining and extracting the energy from oil shale. Because of these higher costs, only a few deposits of oil shale are currently being exploited in China, Brazil, and Estonia. However, with the continuing decline of crude oil supplies, accompanied by increasing costs of crude oil-based products, oil shale presents opportunities for supplying some of the fossil energy needs of the world in the future (Culbertson and Pitman, 1973; Bartis et al., 2005; Andrews, 2006).

Oil shale is not generally regarded as true shale by geologists nor does it contain appreciable quantities of free oil (Scouten, 1990; Speight, 2008). The fracture resistance of all oil shales varies with the organic content of the individual lamina and fractures preferentially initiate and propagate along the leaner horizontal laminas of the depositional bed.

Oil shale was deposited in a wide variety of environments including freshwater to saline ponds and lakes, epicontinental marine basins, and related subtidal shelves as well as shallow ponds or lakes associated with coal-forming peat in limnic and coastal swamp depositional environments. This give rise to a variety of different oil shale types (Hutton, 1987, 1991) and it is not surprising, therefore, that oil shale exhibits a wide range in organic and mineral composition (Scouten, 1990; Mason, 2006; Ots, 2014; Wang et al., 2009). Most oil shale contains organic matter derived from varied types of marine and lacustrine algae, with some debris of land plants, depending upon the depositional environment and sediment sources.

Organic matter in oil shale is a complex moisture and is derived from the carbon-containing remains of algae, spores, pollen, plant cuticle and corky fragments of herbaceous and woody plants, plant resins, pant waxes, and other cellular remains of lacustrine, marine, and land plants (Scouten, 1990; Dyni, 2003, 2006). These materials are composed chiefly of carbon, hydrogen, oxygen, nitrogen, and sulfur. Generally, the organic matter is unstructured and is best described as amorphous (*bituminite*)—the origin of which has not been conclusively identified but is theorized to be a mixture of degraded algal or bacterial remains. Other carbon-containing materials such as phosphate and carbonate minerals may also be present which, although of organic origin, are excluded from the definition of organic matter in oil shale and are considered to be part of the mineral matrix of the oil shale.

Finally, much of the work performed on oil shale has referenced the oil shale from the Green River formation in the western United States. Thus, unless otherwise stated, the shale referenced in the following text is the Green River shale.

5.1.1 General Properties

Oil shale is typically a fine-grained sedimentary rock containing relatively large amounts of organic matter (*kerogen*) from which significant amounts of shale oil and combustible gas can be extracted by thermal deposition with ensuing distillation from the reaction zone. However, oil shale does not contain any oil—this must be produced by a process in which the kerogen is thermally decomposed (cracked) to produce the liquid product (shale oil). The mineral matter (shale) consists of fine-grained silicate and carbonate minerals. The ratio of kerogen-to-shale for commercial grades of oil shale is typically in the range 0.75:5 to 1.5:5—as a comparison, for coal the organic matter-to-mineral matter ratio in coal is usually greater than 4.75:5 (Speight, 2013b).

In the United States there are two principal oil shale types, the shale from the Green River Formation in Colorado, Utah, and Wyoming, and the Devonian-Mississippian black shale of the East and Midwest (Table 1.3) (Baughman, 1978). The Green River shale is considerably richer, occurs in thicker seams, and has received the most attention for synthetic fuel.

The common property of these two types of oil shale is the presence of the ill-defined kerogen. The chemical composition of the kerogen has been the subject of many studies (Scouten, 1990) but whether or not the data are indicative of the true nature of the kerogen is extremely speculative. Based on solubility/insolubility in various solvents (Koel et al., 2001) it is, however, a reasonable premise (remembering that regional and local variations in the flora that were the precursors to kerogen) led to differences in kerogen composition and properties of kerogen from different shale samples—similar to the varying in quality, composition, and properties of crude oil from different reservoirs (Speight, 2014a).

The organic matter is derived from the varied types of marine and lacustrine algae, with some debris of land plants, is largely dependent on the depositional environment and sediment sources. Bacterial processes were probably important during the deposition and early diagenesis of most oil shale deposits—these processes could produce significant quantities of biogenic methane, carbon dioxide, hydrogen sulfide, and ammonia. These gases in turn could react with dissolved ions in the sediment waters to form authigenic minerals (minerals generated where they were found or observed) such as calcite ($CaCO_3$), dolomite ($CaCO_3 \cdot MgCO_3$), pyrite (FeS_2), and even such rare authigenic minerals as buddingtonite (ammonium feldspar—$NH_4AlSi_3O_8 \cdot 0.5H_2O$).

5.1.2 Mineral Constituents

Oil shale has often been termed as (incorrectly and for various illogical reasons) high-mineral coal. Nothing could be further from the truth than this misleading terminology. Coal and oil shale are fraught with considerable differences (Speight, 2008, 2013b) and such terminology should be frowned upon.

In terms of mineral and elemental content, oil shale differs from coal in several distinct ways. Oil shale typically contains much larger amounts of inert mineral matter (60–90%) than coal, which has been defined as containing <40% mineral matter (Speight, 2013b). The organic matter of oil shale, which is the source of liquid and gaseous hydrocarbons, typically has a higher hydrogen and lower oxygen content than that of lignite and bituminous coal.

The mineral component of some oil shale deposits is composed of carbonates including calcite ($CaCO_3$), dolomite ($CaCO_3 \cdot MgCO_3$), siderite ($FeCO_3$), nahcolite ($NaHCO_3$), dawsonite [$NaAl(OH)_2CO_3$], with lesser amounts of aluminosilicates—such as alum [$KAl(SO_4)_2 \cdot 12H_2O$]—and sulfur, ammonium sulfate, vanadium, zinc, copper, and uranium, which add byproduct value (Beard et al., 1974). For other deposits, the reverse is true—silicates including quartz (SiO_2), feldspar [$xAl(AlSi)_3O_8$, where x

can be sodium (Na), and/or calcium (Ca), and/or potassium (K)], and clay minerals are dominant and carbonates are a minor component.

Briefly, clay minerals are the characteristic minerals of the earths near-surface environments. They form in soils and sediments, and by diagenetic and hydrothermal alteration of rocks. Water is essential for clay mineral formation and most clay minerals are described as hydrous aluminosilicates. Structurally, the clay minerals are composed of planes of cations, arranged in sheets, which may be tetrahedral-coordinated or octahedrally coordinated with oxygen, which in turn are arranged into layers often described as 2:1 if they involve units composed of two tetrahedral and one octahedral sheet or 1:1 if they involve units of alternating tetrahedral and octahedral sheets. Additionally some 2:1 clay minerals have interlayers sites between successive 2:1 units which may be occupied by interlayer cations, which are often hydrated. The planar structure of clay minerals gives rise to characteristic platy habit of many and to perfect cleavage, as seen for example in larger hand specimens of mica minerals.

Many oil-shale deposits contain small, but ubiquitous, amounts of sulfides including pyrite (FeS_2) and marcasite (FeS_2, but which physically and crystallographically distinct from pyrite), indicating that the sediments probably accumulated in dysaerobic (a depositional environment with 0.1–1.0 mL of dissolved oxygen per liter of water) to anoxic waters that prevented the destruction of the organic matter by burrowing organisms and oxidation.

Green River oil shale contains abundant carbonate minerals including dolomite, nahcolite, and dawsonite. The latter two minerals have potential byproduct value for their soda ash and alumina content, respectively. The oil shale deposits of the eastern United States are low in carbonate content but contain notable quantities of metals, including uranium, vanadium, molybdenum, and others which could add significant byproduct value to these deposits. There is the potential for low emissions due to the inherent presence of carbonate minerals. Calcium carbonate present in oil shale ash binds sulfur dioxide and it is not necessary to add limestone for desulfurization:

$$CaCO_3 \rightarrow CaO + CO_2$$

$$2CaO + SO_2 + O_2 \rightarrow CaSO_4$$

Illite (a layered aluminosilicate, $[(K,H_3O)(Al,Mg,Fe)_2(Si,Al)_4O_{10}(OH)_2, (H_2O)]$) is ever-present in Green River oil shale—it is generally associated with other clay minerals but frequently occurs as the only clay mineral found in the oil shale (Tank, 1972). Smectite (a group of clay minerals that includes montmorillonite, which tends to swell when exposed to water) is present in all three members of the Green River Formation, but its presence frequently shows an inverse relationship to both analcime (a white, grey, or colorless tectosilicate mineral which consists of hydrated sodium aluminum silicate, $NaAlSi_2O_6 \cdot H_2O$) and loughlinite (a silicate of magnesium, $Na_2Mg_3Si_6O_{16} \cdot 8H_2O$). Chlorite (a group of mostly monoclinic but also triclinic or orthorhombic micaceous phyllosilicate minerals) occurs only in the silty and sandy beds of the Tipton Shale Member. The distribution of random mixed-layer structures and amorphous material is irregular. Several independent lines of evidence favor an in situ origin for many of the clay minerals. Apparently the geochemical conditions favoring the accumulation of the oil shale also favored in situ generation of illite.

Finally, precious metals and uranium are contained in good amounts in oil shale of the Eastern United States. It may not be in the near future to recover these mineral resources, since a commercially favorable recovery process has not yet been developed. However, there are many patents on recovery of alumina from Dawsonite-bearing beds [$NaAl(CO_3)(OH)_2$] by leaching, precipitation, and calcination.

5.1.3 Grade

The grade of oil shale has been determined by many different methods with the results expressed in a variety of units (Scouten, 1990; Dyni, 2003, 2006). For example, the heating value is useful for determining the quality of an oil shale that is burned directly in a power plant to produce electricity. Although the heating value of a given oil shale is a useful and fundamental property of the rock, it does not provide information on the amounts of shale oil (which does not fall under the definition of *tight oil*) or combustible gas that would be yielded by retorting (destructive distillation). Alternatively, the grade of oil shale can be determined by measuring the yield of distillable oil produced from a shale sample in a laboratory retort (Scouten, 1990). This is perhaps the most common type of analysis that has been, and still is, used to evaluate an oil-shale resource—however the end result of the evaluation depends upon the source of the sample and whether or not the sample is representative of the deposit.

The method commonly used in the United States is the *modified Fischer assay* test method (ASTM D3904). Some laboratories have further modified the Fischer assay method to better evaluate different types of oil shale and different methods of oil-shale processing. The standard Fischer assay test method (ASTM D3904, now withdrawn but still used in many laboratories) consists of heating a 100-g sample crushed to −8 mesh (2.38-mm) screen in a small aluminum retort to 500°C (930°F) at a rate of 12°C (21.6°F) per minute and held at that temperature for 40 min. The distilled vapors of oil, gas, and water are passed through a condenser cooled with ice water into a graduated centrifuge tube. The oil and water are then separated by centrifuging. The quantities reported are the weight percentages of shale oil (and its specific gravity), water, shale residue, and (by difference) gas plus losses.

Another method for characterizing organic richness of oil shale is a pyrolysis test developed by the Institut Français du Pétrole for analyzing source rocks (Allix et al., 2011). The Rock-Eval test heats a 50–100-mg sample through several temperature stages to determine the amounts of hydrocarbon and carbon dioxide generated. The results can be interpreted for kerogen type and potential for oil and gas generation. The method is faster than the Fischer assay and requires less sample material (Kalkreuth and Macauley, 1987).

5.1.4 Porosity

The porosity (void fraction) is a measure of the void spaces in a material such as a reservoir rock, and is the volume of void space over the total volume and is expressed as a fractional number between 0 and 1, or as a percentage between 0 and 100.

The porosity of porous material can be measured in a number of different ways, depending on what specific pores are looked at and how the void volumes are measured. They include (1) interparticle porosity, (2) intraparticle porosity, (3) internal porosity, (4) porosity by liquid penetration, (5) porosity by saturation, (6) porosity by liquid absorption, (7) superficial porosity, (8) total open porosity, (9) bed porosity—the bed void fraction, and (10) packing porosity.

Except for the two low-yield oil shale samples, naturally occurring porosities in the raw oil shales are almost negligible and they do not afford access to gases (Table 1.4). Porosity may exist to some degree in the oil shale formation where fractures, faults, or other structural defects occurred. It is also believed that a good portion of pores is either blind or very inaccessible. Cracking and fractures, or other structural defects often create new pores and also break up some of the blind pores—closed or blind pores are normally not accessible by mercury porosimetry even at high pressures. Due to the severity of mercury poisoning, the instrument based on pressurized mercury penetration through pores is no longer used.

5.1.5 Permeability

The permeability of raw oil shale is essentially zero, because the pores are filled with a nondisplaceable organic material. In general, oil shale constitutes a highly impervious system. Thus, one of the major challenges of any in situ retorting project is in the creation of a suitable degree of permeability in the formation. This is why an appropriate rubbelization technique is essential in the success of an in situ pyrolysis project.

Of practical interest is the dependency of porosity or permeability on temperature and organic contents. Upon heating to 510°C (950°F), an obvious increase in oil shale porosity is noticed. These porosities, which vary from 3% to 6% *v*/*v* of the initial bulk oil shale volume, represented essentially the volumes occupied by the organic matter before the retorting treatment. Therefore, the oil shale porosity increases as the extent of pyrolysis reaction proceeds.

5.1.6 Compressive Strength

Raw oil shale has high compressive strengths both perpendicular and parallel to the bedding plane (Eseme et al., 2007). After heating, the inorganic matrices of low-yield Fischer assay oil shale retain high compressive strength in both perpendicular and parallel planes. This indicates that a high degree of inorganic cementation exists between the mineral particles comprising each lamina and between adjacent laminae. With an increase in organic matter of oil shale the compressive strength of the respective organic-free mineral matrices decreases, and it becomes very low in those rich oil shales.

5.1.7 Thermal Conductivity

Measurements of thermal conductivity of oil shale show that blocks of oil shale are anisotropic about the bedding plane and thermal conductivity as a function of temperature, oil shale assay and direction of heat flow, parallel to the bedding plane (parallel to the earth's surface for a flat oil shale bed), was slightly higher than the thermal conductivity perpendicular to the bedding plane. As layers of material were laid to form the oil shale bed over a long period of geological years, the resulting continuous strata have slightly higher resistance to heat flow perpendicular to the strata than parallel to the strata (Table 1.5).

The thermal conductivity of oil shale is, in general, only weakly dependent on the temperature. However, extreme caution needs to be exercised in the interpretation of results at temperatures close to the decomposition temperature of the shale organic matter. This is due to the fact that the kerogen decomposition reaction (or, pyrolysis reaction) is endothermic in nature and as such the temperature transients can be confounded between the true rate of heat conduction and the rate of heat of reaction.

5.1.8 Thermal Decomposition

High-yield oil shale sustains combustion hence the older Native American name *the rock that burns* but in the absence of air (oxygen) three carbonaceous end products result when oil shale is thermally decomposed. Distillable oil is produced as are noncombustible gases and a carbonaceous (high-carbon) deposit remains on the rock on (the surface or in the pores) as char—a coke-like residue. The relative proportions of oil, gas, and char vary with the pyrolysis temperature and to some extent with the organic content of the raw shale. All three products are contaminated with nonhydrocarbon compounds and the amounts of the contaminants also vary with the pyrolysis temperature (Bozak and Garcia, 1976; Scouten, 1990).

At temperatures on the order of 500–520°C (930–970°F), oil shale produces shale oil while the mineral matter of the oil shale is not decomposed. The yield and quality of the products depend on a number of factors, whose impact has been identified and quantified for some of the deposits, notably the US Green River deposits and the Estonian deposits (Brendow, 2003, 2009). A major factor is that oil shale ranges widely in organic content and oil yield. Commercial grades of oil shale, as determined by the yield of shale oil, range from about 25 to 50 gallons per ton of rock (typically using the Fischer assay method).

One simple aspect of the thermal decomposition of oil shale kerogen is the relationships of the organic hydrogen and nitrogen contents, and Fischer assay oil yields. Stoichiometry suggests that kerogen with a higher organic hydrogen-to-carbon atomic ratio can yield more oil per weight of carbon than kerogen that is relatively hydrogen-poor (Scouten, 1990). However, the hydrogen-to-carbon atomic ratio is not the only important factor. South African kerogen with an atomic hydrogen-to-carbon ratio of 1.35 has a lower oil yield than Brazilian kerogen with an atomic hydrogen-to-carbon ratio of 1.57. In general, the oil shale containing kerogen that is converted efficiently to oil contains relatively low levels of nitrogen (Scouten, 1990). Furthermore, variation of product distribution with time in the reaction zone can cause a change in product distribution (Hubbard and Robinson, 1950).

During retorting, kerogen decomposes into three organic fractions: (1) shale oil, (2) gas, and (3) carbonaceous residue. Oil shale decomposition begins at relatively low retort temperatures (300°C, 572°F) but proceeds more rapidly and more completely at higher temperatures (Scouten, 1990). The highest rate of kerogen decomposition occurs at retort temperatures of 480–520°C (895–970°F). In general, the yield of shale oil decreases, the yield of gas increases, and the aromaticity of the oil increases with increasing decomposition temperature (Dinneen, 1976; Scouten, 1990).

However, there is an upper limit on optimal retorting temperature as the mineral content of the shale may decompose if the temperature is too high. For example, the predominant mineral component of Estonian kukersite shales is calcium carbonate, a compound that dissociates at high temperatures (600–750°C, 1112–1382°F for dolomite, and 600–900°C, 1112–1652°F for calcite). Thus carbon must be anticipated as a product of oil shale decomposition process, which will dilute the off-gases (adding to emissions issues) produced from the retorting process. The gases and vapors leaving the retort are cooled to condense the reaction products, including oils and water.

The active devolatilization of oil shale begins at about 350–400°C (660–750°F), with the peak rate of oil evolution at about 425°C (800°F), and devolatilization essentially completes in the range of 470–500°C (890–930°F) (Hubbard and Robinson, 1950; Shih and Sohn, 1980). At temperatures near 500°C (930°F), the mineral matter, consisting mainly of calcium/magnesium and calcium carbonates, begins to decompose yielding carbon dioxide as the principal product. The properties of crude shale oil are dependent on the retorting temperature, but more importantly on the temperature-time history because of the secondary reactions accompanying the evolution of the liquid and gaseous products. The produced shale oil is dark brown, odoriferous, and tending to waxy oil.

Kinetic studies (Scouten, 1990) indicate that below 500°C (930°F) the kerogen (organic matter) decomposes into an extractable product (*bitumen*) with subsequent decomposition into oil, gas, and carbon residue. The actual kinetic picture is influenced by the longer time required to heat the organic material which is dispersed throughout the mineral matrix and to the increased resistance to the outward diffusion of the products by the matrix which does not decompose. From the practical standpoint of oil shale retorting, the rate of oil production is the important aspect of kerogen decomposition.

Contrary to other oil shales, obtaining high oil yields of distillable oil from kukersite needs specific conditions of processing. It can be explained by the fact that on thermal processing of kukersite, its elevated moisture percentage and the predominance of calcium carbonate in its mineral part result in high values of specific heat consumption in the process (Yefimov and Purre, 1993). Also shale is rich in organic matter and must pass the temperatures of thermobitumen formation and coking at a relatively high speed to avoid caking and secondary pyrolysis of oil.

5.2 Kerogen and Shale Oil

The name *kerogen* is also generally used for organic matter in sedimentary rocks that is insoluble in common organic and inorganic solvents. Thus, the term *kerogen* is used throughout this text to mean the carbonaceous material that occurs in sedimentary rocks, carbonaceous shale, and oil shale. This carbonaceous material is, for the most part, insoluble in common organic solvents. A soluble fraction, *bitumen*, coexists with the kerogen. The bitumen is not to be confused with the material found in tar sand deposits (Speight, 2008, 2009, 2014a). However, like many naturally occurring organic materials, kerogen does yield a hydrocarbonaceous oil when heated to temperatures sufficiently high (typically > 300°C, 570°F) to cause thermal decomposition with simultaneous removal of distillate.

Kerogen is the naturally occurring, solid, insoluble organic matter that occurs in source rocks and can yield oil upon heating. Typical organic constituents of kerogen are algae and woody plant material. Kerogen has a high-molecular weight and is generally insoluble in typical organic solvents (Speight, 2009, 2014a) and has been conveniently divided into four types: (1) Type I, which consists mainly algal and amorphous constituents, (2) Type II, which consists of mixed terrestrial and marine source material, (3) Type III, which consists of woody terrestrial source material, and (4) Type IV, which consists mostly of decomposed organic matter in the form of polycyclic aromatic hydrocarbons and has a low (<0.5) hydrogen-to-carbon (H/C) atomic ratio.

Kerogen is a solid, waxy, organic substance that forms when pressure and heat from the Earth act on the remains of plants and animals. Given geological time, it has been proposed that kerogen converts to various liquid and gaseous *hydrocarbons* at a depth of approximately 4.5 miles or more (approximately 7 km) and a temperature between 50°C and 100°C (122°F and 212°F) (USGS, 1995), which has been assigned to the presence of the *thermal gradient*.

Briefly, the *geothermal gradient* is the variation of temperature with depth in subterranean formations of the Earth (in this chapter). Although the geothermal gradient varies from place to place, it is generally on the order of 22° F per 1000 ft of depth or 12°C per 1000 ft of depth, that is, 0.022°F per foot of depth or 0.012°C per foot of depth. This would require a depth on the order of 25,000 to attain temperatures of 300°C (570°F). However, the thermal evolution of kerogen is unknown and the role of kerogen in crude oil formation is at best, highly speculative.

The precise structure of kerogen is unknown and the precise role of kerogen-rock interactions in determining the properties of oil shale is also unknown. In addition, the precise role played by kerogen in the natural gas and crude oil maturation process is not fully understood (Tissot and Welte, 1978; Durand, 1980; Hunt, 1996; Scouten, 1990; Speight, 2014a). What obviously needs to be addressed more fully in terms of kerogen participation in crude oil generation is the potential to produce crude oil constituents from kerogen by low-temperature processes rather than by processes that involve the use of temperatures in excess of 250°C (>480°F) (Burnham and McConaghy, 2006; Speight, 2014a).

6 TIGHT OIL, TIGHT GAS, AND ENERGY SECURITY

Energy security is the continuous and uninterrupted availability of energy, to a specific country or region. The security of energy supply conducts a crucial role in decisions that are related to the formulation of energy policy strategies. The economies of many countries are depended by the energy imports in the notion that their balance of payments is affected by the magnitude of the vulnerability that the countries have in crude oil and natural gas (Speight, 2011b).

Energy security has been an on-again-off-again political issue in the United States since the first Arab oil embargo in 1973. Since that time, the speeches of various Presidents and the Congress of the United States have continued to call for an end to the dependence on foreign oil and gas by the United States. The congressional rhetoric of energy security and energy independence continues but meaningful suggestions of how to address this issue remain few and far between.

The energy literature and numerous statements by officials of oil-and-gas-producing and oil-and-gas-consuming countries indicate that the concept of energy security is elusive. Definitions of energy security range from uninterrupted oil supplies to the physical security of energy facilities to support for biofuels and renewable energy resources. Historically, experts and politicians referred to *security of oil supplies* as *energy security*. Only recently policy makers started to include natural gas supplies in the portfolio of energy definitions.

The security aspects of natural gas are similar, but not identical, to those of oil. Compared with oil imports, natural gas imports play a smaller role in most importing countries—mainly because it is less costly to transport liquid crude oil and petroleum products than natural gas. Natural gas is transported by pipeline over long distances because of the pressurization costs of transmission; the need to finance the cost of these pipelines encourages long-term contracts that dampen price volatility.

The past decade has yielded substantial change in the natural gas industry. Specifically, there has been rapid development of technology allowing the recovery of natural gas from shale formations. Since 2000, rapid growth in the production of natural gas from shale formations in North America has dramatically altered the global natural gas market landscape. Indeed, the emergence of shale gas is perhaps the most intriguing development in global energy markets in recent memory.

Beginning with the Barnett shale in northeast Texas, the application of innovative new techniques involving the use of horizontal drilling with hydraulic fracturing has resulted in the rapid growth in production of natural gas from shale. Knowledge of the shale gas resource is not new as geologists have long known about the existence of shale formations, and accessing those resources was long held in the geology community to be an issue of technology and cost. In the past decade, innovations have yielded substantial cost reductions, making shale gas production a commercial reality. In fact, shale gas production in the United States has increased from virtually nothing in 2000 to over 10 billion cubic feet per day (bcfd, 1×10^9 ft^3 per day) in 2010, and it is expected to more than quadruple by 2040, reaching 50% or more of total US natural gas production by the decade starting in 2030.

Natural gas—if not disadvantaged by government policies that protect competing fuels, such as coal—stands to play a very important role in the U.S. energy mix for decades to come. Rising shale gas production has already delivered large beneficial impacts to the United States. Shale gas resources are generally located in close proximity to end-use markets where natural gas is utilized to fuel industry, generate electricity, and heat homes. This offers both security of supply and economic benefits (Medlock et al., 2011).

The *Energy Independence and Security Act of 2007* (originally named the *Clean Energy Act of 2007*) is an Act of Congress concerning the energy policy of the United States. The stated purpose of the act is "to move the United States toward greater energy independence and energy security, to increase the production of clean renewable fuels, to protect consumers, to increase the efficiency of products, buildings, and vehicles, to promote research on and deploy greenhouse gas capture and storage options, and to improve the energy performance of the Federal Government, and for other purposes."

The bill originally sought to cut subsidies to the petroleum industry in order to promote petroleum independence and different forms of alternative energy. These tax changes were ultimately dropped after opposition in the Senate, and the final bill focused on automobile fuel economy, development of biofuels, and energy efficiency in public buildings and lighting. It was, and still is, felt by many observers that there should have been greater recognition of the role that natural gas can play in energy security. In fact, viewed from the perspective of the energy-importing countries as a whole, diversification in oil supplies has remained constant over the last decade while diversification in natural gas supplies has steadily increased. Given the increasing importance of natural gas in world energy use, this is an indicator of an increase in overall energy security (Cohen et al., 2011).

However, natural gas is an attractive fuel, and its attraction is growing because of its clean burning characteristics, compared to oil or coal, and because of its price advantage, on an energy equivalent basis, compared to oil. Accordingly, analysts predict significant future growth in natural gas consumption worldwide and growth in the trade of natural gas. Significant investments are being made to meet this future demand by bringing the so-called *stranded gas* (including *shale gas*) to market.

Current trends suggest that natural gas will gradually become a global commodity with a single world market, just like oil, adjusted for transportation differences. The outcome of a global gas market is inevitable; once this occurs, the tendency will be toward a world price of natural gas, as with oil today, and the prices of oil and gas each will reach a global equivalence based on energy content (Deutch, 2010).

7 RESOURCES AND RESERVES

Throughout this book there is frequent reference to resources and reserves and it is appropriate to present explanation of these terms and any related terms (Fig. 1.3). Thus, terms such as (1) original oil or gas in

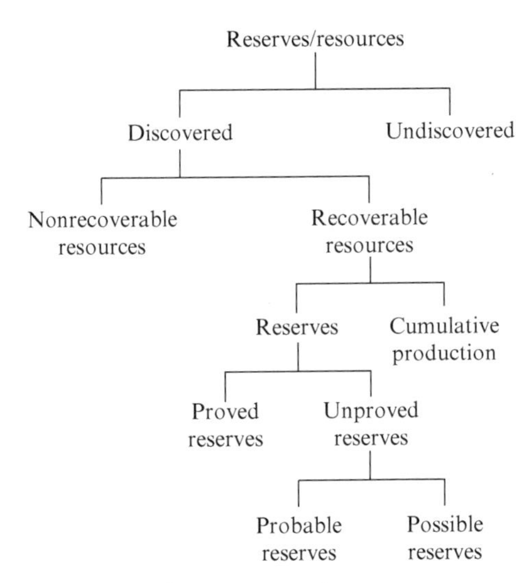

Fig. 1.3 Simplified subdivision of resources and reserves.

place, OOIP or OGIP, (2) ultimately recoverable resources, URR, (3) technically recoverable resources, TRR, (4) estimate ultimate recovery, EUR, (5) economically recoverable resources, ERR, (6) proved reserves, (7) probable reserves, and (8) possible reserves.

The *original oil or gas in place* (OOIP or OGIP) is the total volume of the resource that is estimated to be present in a given field, play, or region. However, it is not possible to recover all (100%) of the volume in place and the amount that is recoverable is referred to by a *recovery factor*. This factor is a key factor in the estimation of natural gas or crude oil availability and can vary significantly depending on geological conditions, recovery technology employed, and the economic environment (such as cost of recovery, gas or oil price).

The *ultimately recoverable resources* (URR) is the sum of all natural gas or crude oil that is expected to be produced from a field or region over the lifetime of the production site. This estimate includes not only (1) the gas or oil already produced, (2) the gas or oil resources already discovered, but also (3) the natural gas and crude oil which is not currently producible either technically or economically but is expected to be so in the future as recovery technology evolves, and (4) undiscovered gas or oil which is expected to be discovered in the future. This definition is very dependent upon the assumption employed in the estimation, such as future gas and oil prices, future developments in recovery technology, and future discoveries of gas or oil. The estimate for URR is closely related to *estimated ultimate recovery* (EUR) which is commonly used to refer to the production from a single well but for all other purposes is synonymous with the definition of URR.

The *technically recoverable resources* (TRR) is the natural gas or crude oil that can be recovered (produced) using currently available recover technology, but this definition tends to exclude any economic aspect of the recovery operations. However, there is some ambiguity as to whether this classification includes undiscovered gas or oil but the majority of evidence suggests that undiscovered gas oil should be included. Another definition, *remaining technically recoverable resources* (RTRR), can be used to exclude the cumulative production.

The *economically recoverable resources* (ERR) defines the technically *and* economically producible gas given current technical and economic conditions. As such this definition is highly sensitive to changes in economic conditions and it is difficult to defend the basis for any assumptions on the economic producibility of undiscovered gas and gas resources.

In terms of definitions of reserves, there are the probabilistic definitions: (1) proved reserves, (2) probable reserves, and (3) possible reserves.

The term *proved reserves* (*proven reserves*) is typically limited to those quantities that are commercial under current economic conditions, while probable and possible reserves may be based on future economic conditions. In general, quantities should not be classified as reserves unless there is an expectation that the accumulation will be developed and placed on production within a reasonable timeframe. On the other hand, the *probable reserves* are those reserves of natural gas or crude oil that are nearly certain but about which a slight doubt exists while *possible reserves* are those reserves of petroleum with an even greater degree of uncertainty about recovery but about which there is some information. An additional term *potential reserves* is also used on occasion; these reserves are based upon geological information about the types of sediments where such resources are likely to occur and they are considered to represent an educated guess. The term *inferred reserves* is also commonly used in addition to, or in place of, *potential reserves*. Inferred reserves are regarded as of a higher degree of accuracy than potential reserves, and the term is applied to those reserves that are estimated using an improved understanding of reservoir frameworks. The term also usually includes those reserves that can be recovered by further development of recovery technologies.

Proved reserves is sometimes referred to as P90 and represents an estimate with a 90% probability of being exceeded. *2P* or *proved and probable reserves* is sometimes referred to as P50 and represents an estimate with a 50% chance of being exceeded (the median estimate). *3P or proved, probable,* and *possible reserves* is sometimes referred to as P10 and represents an estimate with a 10% chance of being exceeded.

Caution is advised when using these definitions and care must be taken to define each term carefully with the explanation of any assumptions so that anyone following the definitions can know precisely what is expected from the definition. In summary, the use of resource and reserve definitions is inconsistent with variations between company definitions and government definitions. Thus, when reporting reserve and resource data, it is essential to be explicit about resource definitions.

Finally, a *stranded resource* (*stranded reserve*) is a resource (reserve) that is not economical to recover and transport to an existing market. The resource (reserve) may be too remote from a market, making construction of a pipeline prohibitively expensive or the resource (reserve) may be in a region where demand for gas or oil is saturated and the cost of exporting gas beyond this region is excessive. Such a resource (reserve) is likely to be developed in the future when existing sources begin to deplete.

REFERENCES

Akrad, O., Miskimins, J., Prasad, M., 2011. The effects of fracturing fluids on shale rock-mechanical properties and proppant embedment. In: Proceedings of the SPE Annual Technical Conference and Exhibition, Denver, Colorado, October 30–November 2. Society of Petroleum Engineers, Richardson, TX. Paper no. SPE 146658.

Allix, P., Burnham, A., Fowler, T., Herron, M., Kleinberg, R., Symington, B., 2011. Coaxing oil from shale. Oilfield Rev. 22 (4 (Winter 2010/2011)), 4–5.

Allred, V.D. (Ed.), 1982. Oil Shale Processing Technology. Center for Professional Advancement, East Brunswick, NJ.

Andrews, A., 2006. Oil shale: history, incentives, and policy. Specialist, industrial engineering and infrastructure policy resources, science, and industry division. In: Congressional Research Service. The Library of Congress, Washington, DC.

Bartis, J.T., LaTourrette, T., Dixon, L., 2005. Oil Shale Development in the United States: Prospects and Policy Issues. Prepared for the National Energy Technology of the United States Department of Energy. Rand Corporation, Santa Monica, CA.

Baughman, G.L., 1978. Synthetic Fuels Data Handbook, second ed. Cameron Engineers, Inc., Denver, CO.

Beard, T.M., Tait, D.B., Smith, J.W., 1974. Nahcolite and dawsonite resources in the Green River Formation, Piceance Creek Basin, Colorado. In: Guidebook to the Energy Resources of the Piceance Creek Basin, 25th Field Conference. Rocky Mountain Association of Geologists, Denver, CO, pp. 101–109.

Bennett, R.H., Bryant, W.R., Hulbert, M.H. (Eds.), 1991a. Microstructure of Fine-Grained Sediments: From Mud to Shale. Springer-Verlag, New York.

Bennett, R.H., O'Brien, N.R., Hulbert, M.H., 1991b. Determinants of clay and shale microfabric signatures: processes and mechanisms. In: Bennett, R.H., Bryant, W.R.,

Hulbert, M.H. (Eds.), Microstructure of Fine-Grained Sediments: From Mud to Shale. Springer-Verlag, New York, pp. 5–32.

Blatt, H., Tracy, R.J., 2000. Petrology: Igneous, Sedimentary, and Metamorphic. W.H. Freeman and Company, New York.

Boswell, R., Collett, T.S., 2011. Current perspectives on gas hydrate resources. Energy Environ. Sci. 4, 1206–1215.

Bowker, K.A., 2007. Development of the Barnett Shale play, Fort Worth Basin. W. Tex. Geol. Soc. Bull. 42 (6), 4–11. http://www.searchanddiscovery.com/documents/2007/07023bowker/index.htm?q=%2Btext%3Agas (accessed 05.01.16.).

Boyle, G. (Ed.), 1996. Renewable Energy: Power for a Sustainable Future. Oxford University Press, Oxford.

Bozak, R.E., Garcia Jr., M., 1976. Chemistry in the oil shales. J. Chem. Educ. 53 (3), 154–155.

Brendow, K., 2003. Global oil shale issues and perspectives. Oil Shale 20 (1), 81–92.

Brendow, K., 2009. Oil shale—a local asset under global constraint. Oil Shale 26 (3), 357–372.

Buffett, B., Archer, D., 2004. Global inventory of methane clathrate: sensitivity to changes in the deep ocean. Earth Planet. Sci. Lett. 227 (3–4), 185.

Burnham, A.K., McConaghy, J.R., 2006. Comparison of the acceptability of various oil shale processes. In: Proceedings of the AICHE 2006 Spring National Meeting, Orlando, FL, March 23, 2006 through March 27.

Bustin, R.M., 2006. Geology report: where are the high-potential regions expected to be in Canada and the U.S.? Capturing opportunities in Canadian shale gas. In: 2nd Annual Shale Gas Conference, January 31–February 1. The Canadian Institute, Calgary.

Bustin, A.M.M., Bustin, R.M., Cui, X., 2008. Importance of fabric on the production of gas shales. In: Proceedings of the Unconventional Gas Conference, Keystone, Colorado, February 10–12. SPE paper no. 114167.

Chong, Z.R., Yang, S.H.B., Babu, P., Linga, P., Li, X.S., 2016. Review of natural gas hydrates as an energy resource: prospects and challenges. Appl. Energy 162, 1633–1652.

Cohen, G., Joutz, F., Loungani, P., 2011. Measuring energy security: trends in the diversification of oil and natural gas supplies. IMF working paper WP/11/39, International Monetary Fund, Washington, DC.

Collett, T.S., 2002. Energy resource potential of natural gas hydrates. AAPG Bull. 86, 1971–1992.

Cramer, D.D., 2008. Stimulating unconventional reservoirs: lessons learned, successful practices, areas for improvement. In: Proceedings of the Unconventional Gas Conference, Keystone, Colorado, February 10–12, 2008. SPE paper no. 114172.

Culbertson, W.C., Pitman, J.K., 1973. Oil Shale in United States mineral resources. Paper no. 820, United States Geological Survey, Washington, DC.

Dandekar, A.Y., 2013. Petroleum Reservoir Rock and Fluid Properties, second ed. CRC Press, Taylor & Francis Group, Boca Raton, FL.

Davies, D.K., Vessell, R.K., 2002. Gas production from non-fractured shale. In: Scott, E.D., Bouma, A.H. (Eds.), Depositional Processes and Characteristics of Siltstones, Mudstones and Shale.Society of Sedimentary Geology, GCAGS Siltstone Symposium 2002. GCAGS (Gulf Coast Association of Geological Societies) Transactions, vol. 52, pp. 177–202.

Davies, D.K., Bryant, W.R., Vessell, R.K., Burkett, P.J., 1991. Porosities, permeabilities, and microfabrics of Devonian shales. In: Bennett, R.H., Bryant, W.R., Hulbert, M.H. (Eds.), Microstructure of Fine-Grained Sediments: From Mud to Shale. Springer-Verlag, New York, pp. 109–119.

Demirbas, A., 2010a. Methane from gas hydrates in the Black Sea. Energy Sources, Part A 32, 165–171.

Demirbas, A., 2010b. Methane hydrates as potential energy resource: part 1—importance, resource and recovery facilities. Energy Convers. Manage. 51, 1547–1561.

Demirbas, A., 2010c. Methane hydrates as potential energy resource: part 2—methane production processes from gas hydrates. Energy Convers. Manage. 51, 1562–1571.

Deutch, J., 2010. Oil and Gas Energy Security Issues. Resource for the Future. National Energy Policy Institute, Washington, DC.

Dinneen, G.U., 1976. Retorting technology of oil shale. In: Yen, T.F., Chilingar, G.V. (Eds.), Oil Shale. Elsevier Science Publishing Company, Amsterdam, Netherlands, pp. 181–198.

Durand, B., 1980. Kerogen: Insoluble Organic Matter from Sedimentary Rocks. Editions Technip, Paris, France.

Durham, L., 2008, July. Louisiana Play a Company Maker? AAPG Explorer, 18. 20, 36.

Dyni, J.R., 2003. Geology and resources of some world oil-shale deposits. Oil Shale 20 (3), 193–252.

Dyni, J.R., 2006. Geology and resources of some world oil shale deposits. Report of investigations 2005-5295, United States Geological Survey, Reston, VA.

Edmonds, B., Moorwood, R., Szczepanski, R., 1996. A practical model for the effect of salinity on gas hydrate formation. Paper no. 35569, Society of Petroleum Engineers, Richardson, TX.

EIA, 2011. Review of Emerging Resources: US Shale Gas and Shale Oil Plays. Energy Information Administration, United States Department of Energy, Washington, DC.

Eseme, E., Urai, J.L., Krooss, B.M., Littke, R., 2007. Review of the mechanical properties of oil shales: implications for exploitation and basin modelling. Oil Shale 24 (2), 159–174.

Faraj, B., Williams, H., Addison, G., McKinstry, B., 2004. Gas potential of selected shale formations in the western Canadian sedimentary basin. GasTIPS 10 (1), 21–25.

Gao, S., 2008. Investigation of interactions between gas hydrates and several other flow assurance elements. Energy Fuel 22 (5), 3150–3153.

GAO, 2012. Information on Shale resources, development, and environmental and public health risks. Report no. GAO-12-732. Report to Congressional Requesters, In: United States Government Accountability Office, Washington, DC.

Gao, S., House, W., Chapman, W.G., 2005. NMR MRI study of gas hydrate mechanisms. J. Phys. Chem. B 109 (41), 19090–19093.

Gary, J.G., Handwerk, G.E., Kaiser, M.J., 2007. Petroleum Refining: Technology and Economics, fifth ed. CRC Press, Taylor & Francis Group, Boca Raton, FL.

Gingras, M.K., Mendoza, C.A., Pemberton, S.G., 2004. Fossilized worm burrows influence the resource quality of porous media. AAPG Bull. 88 (7), 875–883.

Gordon, D., 2012. Understanding unconventional oil. The Carnegie papers, The Carnegie Endowment for International Peace, Washington, DC.

Gornitz, V., Fung, I., 1994. Potential distribution of methane hydrate in the world's oceans. Glob. Biogeochem. Cycles 8, 335–347.

Gupta, A.K., 2004. Marine gas hydrates: their economic and environmental importance. Curr. Sci. 86, 1198–1199.

Hsu, C.S., Robinson, P.R. (Eds.), 2006. Practical Advances in Petroleum Processing Volume 1 and Volume 2. Springer Science, New York.

Hubbard, A.B., Robinson, W.E., 1950. A thermal decomposition study of Colorado Oil Shale. Report of investigations no. 4744, United States Bureau of Mines, Washington, DC.

Hubbert, M.K., 1956. Nuclear Energy and the Fossil Fuels—Drilling and Production Practice. American Petroleum Institute, Washington, DC.

Hubbert, M.K., 1962. Energy resources. Report to the Committee on Natural Resources, National Academy of Sciences, Washington, DC.

Hunt, J.M., 1996. Petroleum Geochemistry and Geology, second ed. W.H. Freeman, San Francisco.
Hutton, A.C., 1987. Petrographic classification of oil shales. Int. J. Coal Geol. 8, 203–231.
Hutton, A.C., 1991. Classification, organic petrography and geochemistry of oil shale. In: Proceedings of the 1990 Eastern Oil Shale Symposium. Institute for Mining and Minerals Research, University of Kentucky, Lexington, KY, pp. 163–172.
IEA, 2013. Resources to Reserves 2013: Oil, Gas and Coal Technologies for the Energy Markets of the Future. OECD Publishing, International Energy Agency, Paris, France.
Islam, M.R., 2014. Unconventional Gas Reservoirs. Elsevier, Amsterdam, Netherlands.
Islam, M.R., Speight, J.G., 2016. Peak Energy—Myth or Reality? Scrivener Publishing, Beverly, MA.
Kalkreuth, W.D., Macauley, George, 1987. Organic petrology and geochemical (Rock-Eval) studies on oil shales and coals from the Pictou and Antigonish areas, Nova Scotia, Canada. Can. Pet. Geol. Bull. 35, 263–295.
Koel, M., Ljovin, S., Hollis, K., Rubin, J., 2001. Using neoteric solvents in oil shale studies. Pure Appl. Chem. 73 (1), 153–159.
Kvenvolden, K.A., 1993. Gas hydrates as a potential energy resource—a review of their methane content. In: Howell, D.G. (Ed.), The Future of Energy Gases. United States Geological Survey, Washington, DC, pp. 555–561. Professional paper no. 1570.
Kvenvolden, K., 1995. A review of the geochemistry of methane in natural gas hydrate. Org. Geochem. 23 (11–12), 997–1008.
Law, B.E., Spencer, C.W., 1993. Gas in tight reservoirs—an emerging major source of energy. In: Howell, D.G. (Ed.), The Future of Energy Gases. United States Geological Survey, Reston, VA, pp. 233–252. Professional paper no. 157.
Lee, S., 1991. Oil Shale Technology. CRC Press, Taylor & Francis Group, Boca Raton, FL.
Lee, S., Speight, J.G., Loyalka, S.K., 2007. Handbook of Alternative Fuel Technologies. CRC-Taylor & Francis Group, Boca Raton, FL.
Levine, J.R., 1993. Coalification: the evolution of coal as a source rock and reservoir rock for oil and gas. Am. Assoc. Pet. Geol., Stud. Geol. 38, 39–77.
MacDonald, G.J., 1990a. The future of methane as an energy resource. Annu. Rev. Energy 15, 53–83.
MacDonald, G.J., 1990b. Role of methane clathrates in past and future climates. Climate Change 16, 247–281.
Martini, A.M., Walter, L.M., Budai, J.M., Ku, T.C.W., Kaiser, C.J., Schoell, M., 1998. Genetic and temporal relations between formation waters and biogenic methane: upper Devonian Antrim Shale, Michigan Basin, USA. Geochim. Cosmochim. Acta 62 (10), 1699–1720.
Martini, A.M., Walter, L.M., Ku, T.C.W., Budai, J.M., McIntosh, J.C., Schoell, M., 2003. Microbial production and modification of gases in sedimentary basins: a geochemical case study from a Devonian shale gas play, Michigan basin. AAPG Bull. 87 (8), 1355–1375.
Martini, A.M., Nüsslein, K., Petsch, S.T., 2004. Enhancing microbial gas from unconventional reservoirs: geochemical and microbiological characterization of methane-rich fractured black shales. Final report. Subcontract no. R-520, GRI-05/0023, Research Partnership to Secure Energy for America, Washington, DC.
Mason, G.M., 2006. Fractional differentiation of silicate minerals during oil shale processing: a tool for prediction of retort temperatures. In: Proceedings of the 26th Oil Shale Symposium, October 16–19. Colorado School of Mines, Golden, CO.
McCain Jr., W.D., 1990. Petroleum Fluids, second ed. PennWell Publishing Corp., Tulsa, OK.
McKoy, M.L., Sams, W.N., 2007. Tight Gas Reservoir Simulation: Modeling Discrete Irregular Strata-Bound Fracture Networks and Network Flow, Including Dynamic

Recharge From the Matrix. Contract No. DE-AC21-95MC31346, Federal Energy Technology Center, United States Department of Energy, Morgantown, WV.
Medlock III, K.B., Jaffe, A.M., Hartley, P.R., 2011. Shale Gas and US National Security. James A. Baker III Institute for Public Policy, Rice University, TX.
Mokhatab, S., Poe, W.A., Speight, J.G., 2006. Handbook of Natural Gas Transmission and Processing. Elsevier, Amsterdam, Netherlands.
Nehring, R., 2008. Growing and indispensable: the contribution of production from tight-gas sands to U.S. gas production. In: Cumella, S.P., Shanley, K.W., Camp, W.K. (Eds.), Understanding, Exploring, and Developing Tight-Gas Sands. 2005 Vail Hedberg Conference: AAPG Hedberg Series, vol. 3, pp. 5–12 (accessed 15.01.15.).
Ots, A., 2014. Estonian oil shale properties and utilization in power plants. Energetika 53 (2), 8–18.
Parkash, S., 2003. Refining Processes Handbook. Gulf Professional Publishing, Elsevier, Amsterdam, Netherlands.
Pemberton, G.S., Gingras, M.K., 2005. Classification and characterization of biogenically enhanced permeability. AAPG Bull. 89, 1493–1517.
Ramage, J., 1997. Energy: A Guidebook. Oxford University Press, Oxford.
Rice, D.D., 1993. Composition and origins of coalbed gas. Am. Assoc. Pet. Geol., Stud. Geol. 38, 159–184.
Schettler, P.D., Parmely, C.R., 1990. The measurement of gas desorption isotherms for Devonian shale. Gas Shales Technol. Rev. 7 (1), 4–9.
Scott, A.R., Kaiser, W.R., Ayers, W.B., 1994. Thermogenic and secondary biogenic gases, San Juan Basin, Colorado and New Mexico: implications for coalbed gas productivity. AAPG Bull. 78 (8), 1186–1209.
Scouten, C.S., 1990. Oil shale. In: Speight, J.G. (Ed.), Fuel Science and Technology Handbook. Marcel Dekker Inc., New York, pp. 795–1053 (Chapters 25–31).
Shih, S.M., Sohn, H.Y., 1980. Non-isothermal determination of the intrinsic kinetics of oil generation from oil shale. Ind. Eng. Chem. Process. Des. Dev. 19, 420–426.
Shurr, G.W., Ridgley, J.R., 2002. Unconventional shallow gas biogenic systems. AAPG Bull. 86 (11), 1939–1969.
Speight, J.G. (Ed.), 1990. Fuel Science and Technology Handbook. Marcel Dekker, New York.
Speight, J.G., 2007. Natural Gas: A Basic Handbook. GPC Books, Gulf Publishing Company, Houston, TX.
Speight, J.G., 2008. Synthetic Fuels Handbook: Properties, Processes, and Performance. McGraw-Hill, New York.
Speight, J.G., 2009. Enhanced Recovery Methods for Heavy Oil and Tar Sands. Gulf Publishing Company, Houston, TX.
Speight, J.G., 2011a. The Refinery of the Future. Gulf Professional Publishing, Elsevier, Oxford.
Speight, J.G., 2011b. An Introduction to Petroleum Technology, Economics, and Politics. Scrivener Publishing, Salem, MA.
Speight, J.G. (Ed.), 2011c. The Biofuels Handbook. Royal Society of Chemistry, London.
Speight, J.G., 2012a. Crude Oil Assay Database. Knovel, New York. Online version available at: http://www.knovel.com/web/portal/browse/display?_EXT_KNOVEL_DISPLAY_bookid=5485&VerticalID=0.
Speight, J.G., 2012b. Shale Oil Production Processes. Gulf Professional Publishing, Elsevier, Oxford.
Speight, J.G., 2013a. Shale Gas Production Processes. Gulf Professional Publishing, Elsevier, Oxford.
Speight, J.G., 2013b. The Chemistry and Technology of Coal, third ed. CRC Press, Taylor & Francis Group, Boca Raton, FL.

Speight, J.G., 2014a. The Chemistry and Technology of Petroleum, fifth ed. CRC Press, Taylor & Francis Group, Boca Raton, FL.
Speight, J.G., 2014b. High Acid Crudes. Gulf Professional Publishing, Elsevier, Oxford.
Speight, J.G., 2014c. Oil and Gas Corrosion Prevention. Gulf Professional Publishing, Elsevier, Oxford.
Speight, J.G., 2015a. Handbook of Petroleum Product Analysis, second ed. John Wiley & Sons Inc., Hoboken, NJ.
Speight, J.G., 2015b. Fouling in Refineries. Gulf Professional Publishing, Elsevier, Oxford.
Speight, J.G., 2016a. Introduction to Enhanced Recovery Methods for Heavy Oil and Tar Sands, second ed. Gulf Publishing Company, Taylor & Francis Group, Waltham, MA.
Speight, J.G., 2016b. Handbook of Hydraulic Fracturing. John Wiley & Sons Inc., Hoboken, NJ.
Speight, J.G., 2016c. Introduction to Enhanced Recovery Methods for Heavy Oil and Tar Sands, second ed. Gulf Professional Publishing, Elsevier, Oxford.
Speight, J.G., Ozum, B., 2002. Petroleum Refining Processes. Marcel Dekker Inc., New York.
Tank, R.W., 1972. Clay minerals of the Green River Formation (Eocene) of Wyoming. Clay Miner. 9, 297.
Terry, R.E., Rogers, J.B., 2014. Applied Petroleum Reservoir Engineering, third ed. Prentice Hall, Upper Saddle River, NJ.
Tinker, S.W., Potter, E.C., 2007. Unconventional gas research and technology needs. In: Proceedings of the Society of Petroleum Engineers R&D Conference: Unlocking the Molecules, San Antonio, TX, April 26–27.
Tissot, B., Welte, D.H., 1978. Petroleum Formation and Occurrence. Springer-Verlag, New York.
US EIA, 2011. Review of Emerging Resources. US Shale Gas and Shale Oil Plays. Energy Information Administration, United States Department of Energy, Washington, DC.
US EIA, 2013. Technically Recoverable Shale Oil and Shale Gas Resources: An Assessment of 137 Shale Formations in 41 Countries Outside the United States. Energy Information Administration, United States Department of Energy, Washington, DC.
US EIA, 2014. Crude Oils and Different Quality Characteristics. Energy Information Administration, United States Department of Energy, Washington, DC. http://www.eia.gov/todayinenergy/detail.cfm?id=7110.
USGS, 1995. United States Geological Survey. Dictionary of Mining and Mineral-Related Terms, second ed. Bureau of Mines & American Geological Institute, US Bureau of Mines, US Department of the Interior, Washington, DC. Special Publication SP 96-1.
USGS, 2011. U.S. geological survey gas hydrates project: database of Worldwide gas hydrates. http://woodshole.er.usgs.gov/project-pages/hydrates/database.html (accessed 07.08.14.).
Walser, D.W., Pursell, D.A., 2007. Making mature shale gas plays commercial: process and natural parameters. In: Proceedings of the Society of Petroleum Engineers, Eastern Regional Meeting, Lexington, October 17–19. SPE paper no. 110127.
Wang, D.-M., Xu, Y.-M., He, D.-M., Guan, J., Zhang, O.-M., 2009. Investigation of mineral composition of oil shale. Asia-Pac. J. Chem. Eng. 4, 691–697.
Wipf, R.A., Party, J.M., 2006. Shale plays—a US overview. In: AAPG Energy Minerals Division Southwest Section Annual Meeting.
Wollrab, V., Streibl, M., 1969. Earth waxes, peat, montan wax, and other organic brown coal constituents. In: Eglinton, G., Murphy, M.T.J. (Eds.), Organic Geochemistry. Springer-Verlag, New York, p. 576.
Yefimov, V., Purre, T., 1993. Characteristics of kukersite oil shale, some regularities and features of its retorting. Oil Shale 10 (4), 313–319.

CHAPTER TWO

Reservoirs and Reservoir Fluids

1 INTRODUCTION

The term *reservoir fluid* is used in this text to collectively include any fluid that exists in a reservoir, which includes gases, liquids, and solids—water may also be included in this terminology and is an important aspect of the reservoir fluid category. Nevertheless, water notwithstanding, the specific types of fluids of interest are (1) natural gas, (2) crude oil, which includes gas condensate and paraffin wax, (3) heavy oil, and (4) tar sand bitumen. However, the focus of this text is predominantly on the first three categories of fluids. More specifically, the water that occurs in natural gas and crude oil reservoirs is usually a *brine* which consists of dissolved sodium chloride (NaCl) as well as salts (minerals) which include calcium (Ca), magnesium (Mg), sulfate (SO_4), bicarbonate (HCO_3), iodide (I), and bromide (Br). Under reservoir conditions, the brine that is sharing pore space with hydrocarbons always contains a limited amount of solution gas (predominantly methane) but increasing salinity decreases gas in solution. Reservoir brines exhibit only slight shrinkage (<5%) when produced to the surface.

In addition, paraffin wax when occurring naturally (and not as part of crude oil) may also be classed as a reservoir fluid. The pure material is a white or colorless soft solid that consists of a mixture of hydrocarbons containing between 20 and 40 carbon atoms (Gruse and Stevens, 1960; Wollrab and Streibl, 1969; Musser and Kilpatrick, 1998; Huang et al., 2003; Speight, 2014). Paraffin wax is solid at room temperature and begins to melt above approximately 37°C (99°F) with a boiling range in excess of >370°C (698°F). While naturally occurring paraffin wax is of particular interest in some conventional crude oil reservoirs, it occurs in shale as part of the crude oil (with perhaps not the same molecular range as in conventional crude oil) and becomes of interest when the crude oil from shale formations is blended with other paraffinic liquids (such as paraffinic naphtha) that can lead to deposition of the wax during transportation and refining from which fouling can occur (Speight, 2014, 2015).

Deep Shale Oil and Gas
http://dx.doi.org/10.1016/B978-0-12-803097-4.00002-4

The fluids, particularly natural gas and crude oil as well as heavy oil, which exist in a reservoir (and which vary widely in properties) (Tables 2.1–2.3) must be determined very early after the discovery of the reservoir. Fluid type is a critical consideration in the decisions that must be made about producing the fluids. Furthermore, fluid properties play a key role in the design and optimization of injection/production strategies and surface facilities for efficient reservoir management and longevity. Inaccurate fluid characterization will lead to uncertainty in the amount of the resource that is in place as well as predictions of recovery efficiency. Prior to production (Chapter 4), determination of the fluid properties may only represent laboratory (hence, ex situ) properties but once production commences, variations in fluid composition because of pressure changes and flow throughout the reservoir will become apparent from which the in-place properties can be assessed accurately as well as a measure of reservoir longevity can be assessed.

Moreover, reservoir fluids vary greatly in composition—in some fields, the fluid is in the gaseous state and in others it is in the liquid state but

Table 2.1 Constituents of Natural Gas

Name	Formula	Vol. (%)
Methane	CH_4	85+
Ethane	C_2H_6	4
Propane	C_3H_8	1–5
Butane	C_4H_{10}	1–2
Pentane$^+$	C_5H_{12}	1–5
Carbon dioxide	CO_2	1–2
Hydrogen sulfide	H_2S	1–2
Nitrogen	N_2	1–5
Helium	He	<0.5

Pentane$^+$: pentane and higher molecular weight hydrocarbons up to approximately C_{10}, including benzene and toluene.

Table 2.2 Selected Crude Oils Showing the Differences in API Gravity and Sulfur Content

Country	Crude Oil	API	Sulfur (%*w/w*)
Abu Dhabi (UAE)	Abu Al Bu Khoosh	31.6	2.00
Abu Dhabi (UAE)	Murban	40.5	0.78
Angola	Cabinda	31.7	0.17
Angola	Palanca	40.1	0.11

Table 2.2 Selected Crude Oils Showing the Differences in API Gravity and Sulfur Content—cont'd

Country	Crude Oil	API	Sulfur (%*w*/*w*)
Australia	Barrow Island	37.3	0.05
Australia	Griffin	55.0	0.03
Brazil	Garoupa	30.0	0.68
Brazil	Sergipano Platforma	38.4	0.19
Brunei	Champion Export	23.9	0.12
Brunei	Seria	40.5	0.06
Cameroon	Lokele	20.7	0.46
Cameroon	Kole Marine	32.6	0.33
Canada (Alberta)	Wainwright-Kinsella	23.1	2.58
Canada (Alberta)	Rainbow	40.7	0.50
China	Shengli	24.2	1.00
China	Nanhai Light	40.6	0.06
Dubai (UAE)	Fateh	31.1	2.00
Dubai (UAE)	Margham Light	50.3	0.04
Egypt	Ras Gharib	21.5	3.64
Egypt	Gulf of Suez	31.9	1.52
Gabon	Gamba	31.4	0.09
Gabon	Rabi-Kounga	33.5	0.07
Indonesia	Bima	21.1	0.25
Indonesia	Kakap	51.5	0.05
Iran	Aboozar (Ardeshir)	26.9	2.48
Iran	Rostam	35.9	1.55
Iraq	Basrah Heavy	24.7	3.50
Iraq	Basrah Light	33.7	1.95
Libya	Buri	26.2	1.76
Libya	Bu Attifel	43.3	0.04
Malaysia	Bintulu	28.1	0.08
Malaysia	Dulang	39.0	0.12
Mexico	Maya	22.2	3.30
Mexico	Olmeca	39.8	0.80
Nigeria	Bonny Medium	25.2	0.23
Nigeria	Brass River	42.8	0.06
North Sea (Norway)	Emerald	22.0	0.75
North Sea (UK)	Innes	45.7	0.13
Qatar	Qatar Marine	36.0	1.42
Qatar	Dukhan (Qatar Land)	40.9	1.27
Saudi Arabia	Arab Heavy (Safaniya)	27.4	2.80
Saudi Arabia	Arab Extra Light (Berri)	37.2	1.15
USA (California)	Huntington Beach	20.7	1.38
USA (Michigan)	Lakehead Sweet	47.0	0.31
Venezeula	Leona	24.4	1.51
Venezuela	Oficina	33.3	0.78

Table 2.3 API Gravity and Sulfur Content of Selected Heavy Oils

Country	Crude Oil	API	Sulfur (%*w/w*)
Canada (Alberta)	Athabasca	8.0	4.8
Canada (Alberta)	Cold Lake	13.2	4.11
Canada (Alberta)	Lloydminster	16.0	2.60
Canada (Alberta)	Wabasca	19.6	3.90
Chad	Bolobo	16.8	0.14
Chad	Kome	18.5	0.20
China	Qinhuangdao	16.0	0.26
China	Zhao Dong	18.4	0.25
Colombia	Castilla	13.3	0.22
Colombia	Chichimene	19.8	1.12
Ecuador	Ecuador Heavy	18.2	2.23
Ecuador	Napo	19.2	1.98
USA (California)	Midway Sunset	11.0	1.55
USA (California)	Wilmington	18.6	1.59
Venezuela	Boscan	10.1	5.50
Venezuela	Tremblador	19.0	0.80

generally gases and liquids frequently coexist in a reservoir—in some reservoirs (or deposits) solids may exist as a wax or as a *tar mat* (Wilhelms and Larter, 1994a,b; Zhang and Zhang, 1999; Speight, 2014). The rocks which contain these reservoir fluids also vary considerably in composition and can influence the physical properties and the flow properties. Other factors, such as producing area, height of the fluid column, natural fracturing, or faulting, and water production also serve to distinguish one reservoir from another and which also affect the choice of the production method.

In fact, in keeping with understanding the nature of the reservoir fluids, understanding the elastic properties of reservoir rocks is crucial for exploration and successful production of natural gas and crude oil from tight shale and tight formation reservoirs. In the case of the static and dynamic elastic properties of shale from Barnett, Haynesville, Eagle Ford, and Fort St. John shale formations, the matter is not so straightforward since the elastic properties of these rocks vary significantly between reservoirs (and even within a reservoir) due to the wide variety of minerals that form the reservoir rock as well as the microstructures exhibited by these shale reservoirs and tight reservoirs. For example, the static (Young's modulus) and dynamic (P- and S-wave moduli) elastic parameters generally decrease monotonically with the content of the clay minerals plus any kerogen. However, the elastic properties of the shale formations are strongly anisotropic (the properties are not identical in all directions) since the degree of anisotropy correlates with the organic content of the shale as well as the minerals that constitute the amount and type of clay minerals that constitute the shale (Tables 2.4 and 2.5) (Hillier, 2003; Bergaya et al., 2011). This is not (and should not be) uprising considering the complex and varying composition of clay minerals (Sone and Zoback, 2013a,b).

More generally, the production of natural gas and crude oil occurs from two classes of rock: (1) source rock and (2) reservoir rock, although it is generally believed that crude oils have, at some time during the history of the formation of the crude oil, there has been migration of the crude oil (or a precursor to the crude oil) from the source rock to the reservoir rock (Speight, 2014) this differentiating between the original and final maturation state of the crude oil. The same rationale can be applied to natural gas. Typically, source rocks are sedimentary rocks in which natural gas and crude oil commences formation from organic debris. After forming in the source

Table 2.4 General Types of Clay Minerals

Group	Minerals in Group
Kaolin	Kaolinite
	Dickite
	Halloysite
	Nacrite (polymorphs of $Al_2Si_2O_5(OH)_4$)
Smectite	Montmorillite
	Nontronite
	Beidellite
	Saponite
Illite	Illite
	Clay-micas
Chlorite	Considerable chemical variation throughout this group

Table 2.5 Chemical Formulas of Clay Minerals

Group	Layer Type	Layer Charge	Chemical Formula
Kaolinite	1:1	<0.01	$[Si_4]Al_4O_{10}(OH)_8 \cdot nH_2O$ ($n=0$ or 4)
Illite	2:1	1.4–2.0	$M_x[Si_{6.8}Al_{1.2}]Al_3Fe.025Mg_{0.7}5O20(OH)_4$
Vermiculite	2:1	1.2–1.8	$M_x[Si_7Al]AlFe.05Mg0.5O_20(OH)_4$
Smectite	2:1	0.5–1.2	$M_x[Si_8]Al_{3.2}Fe_{0.2}Mg_{0.6}O_20(OH)_4$
Chlorite	2:1:1	Variable	$(Al(OH)_{2.55})_4[Si_{6.8}Al0_{1.2}\}Al_{3.4}Mg_{0.6})20$ $(OH)_4$

rock, the protopetroleum and any formed hydrocarbons as well as any potential hydrocarbon-forming constituents, which can vary from simple structures such as methane (Table 2.1) to more complex structures, such as those constituents of conventional crude oil heavy oil (as determined from the variation in properties), can migrate to the reservoir rock after which further maturation processes can take place (Speight, 2014).

Geologic formations that contain natural gas and crude oil include clastic or detrital rocks (pertaining to rock or rocks composed of fragments of older rocks or minerals), chemical rocks (formed by chemical precipitation of minerals), and organic rocks (formed by biological debris from shells, plant material, and skeletons). The three most common sedimentary rock types encountered in oil and gas fields are (1) shale, (2) sandstone, and (3) carbonate. Classifying these rock types primarily depends on characteristics such as grain size and composition, porosity (pore space within and between grains), and cementitious character (the manner in which the rock grains are held together), each of which can influence oil and gas production (Bustin et al., 2008). Historically, the majority of the crude oil and natural gas produced in the United States were withdrawn from carbonate and sandstone reservoirs. However, over the past decade, the production of natural gas and crude oil from shale formations and other tight rock formations has increased dramatically.

2 SEDIMENTS

Sediments (sedimentary rocks) are types of rock that are formed by the deposition of material within bodies of water and sedimentation is the process in which mineral matter and/or organic particles (detritus) settle and accumulate or which causes mineral matter to precipitate from a solution. In most cases, before being deposited, the sediment was formed by weathering and erosion in a source area, and then transported to the place of deposition by natural forces such as water, wind, ice, mass movement, or glaciers.

The sedimentary rock cover of the continents of the crust of the Earth is extensive, but the total contribution of sedimentary rocks is estimated to be only 8% of the total volume of the crust. Sedimentary rocks are deposited in layers (strata) and form a bedding structure and can provide information about the subsurface leading to discovery and development of natural resources, such as (in the context of this book) crude oil, natural gas, and coal seams as sources of coalbed methane.

Table 2.6 The Geologic Timescale[a]

Era	Period	Epoch	Duration (10^6 years)	Years Ago (10^6 years)
Cenozoic	Quaternary	Holocene	10,000 to present	
		Pleistocene	2	.01
	Tertiary	Pliocene	11	2
		Miocene	12	13
		Oligocene	11	25
		Eocene	22	36
		Paleocene	71	58
Mesozoic	Cretaceous		71	65
	Jurassic		54	136
	Triassic		35	190
Paleozoic	Permian		55	225
	Carboniferous		65	280
	Devonian		60	345
	Silurian		20	405
	Ordovician		75	425
	Cambrian		100	500
Precambrian			3380	600

[a]The numbers are approximate (±5%) due to variability of the data in literature sources; nevertheless, the numbers do give an indication of the extent of geologic time.

The geologic age of any sediment is an important determinant of the potential of the sediment to contain crude oil and natural gas. While many rocks of different ages produce oil and natural gas, the areas of prolific production include formations that are from several different geologic periods: (1) the Devonian period, approximately 405–345 million years ago; (2) the Carboniferous period, approximately 345–280 million years ago; (3) the Permian, approximately 280–225 million years ago; and (4) the Cretaceous period, approximately 136–71 million years ago (Table 2.6). During these periods, organic-rich materials accumulated with the sediments and, over geologic time (millions of years), chemical changes (induced by pressure from the overlying sediments and any resulting heat from increasing pressure) changed the original organic detritus thereby (eventually) producing natural gas and crude oil.

2.1 Rock Types

Sandstone is the second most abundant clastic sedimentary rock and is the most commonly encountered reservoir rock for natural gas and crude oil and sandstone formations are created by larger sediment particles, and are typically deposited in river channels, deltas, and shallow sea environments. A clast is a fragment of geological detritus, chunks and smaller grains of rock broken off other rocks by physical weathering and clastic rocks are composed of fragments (clasts) of preexisting minerals and rock. The term clastic is used with reference to sedimentary rocks as well as to particles in sediment transport whether in suspension or as bed load, and in sedimentary deposits (Marshak, 2012). The predominant clastic sedimentary rocks are (1) *conglomerate*, in which the grains are predominately rounded and on the order of 64 to >256 mm in size, (2) breccia, in which the angular grains are on the order of 2–64 mm in size, and (3) sandstone in which the grains range from 2 to 1/16 mm (c.f., shale formations are composed of particles <1/16 mm in size).

Conglomerate formations are the least abundant sediment type and are typically consolidated gravel deposits with variable amounts of sand and mud between the particles (sometimes referred to as *pebbles*). Conglomerates accumulate in stream channels, along the margins of mountain ranges, and on beaches and are composed largely of angular pebbles (*breccias*) and some (*tillites*) are formed in glacial deposits. On the other hand, *sandstone formations* are composed essentially of cemented sand and comprise approximately one-third of all sedimentary rocks. The most abundant mineral in sandstone is quartz (SiO_2), along with lesser amounts of calcite ($CaCO_3$), gypsum ($CaSO_4 \cdot 2H_2O$), and various iron compounds. These formations tend to be more porous than shale formations and consequently make excellent reservoir rocks—as long as impermeable basement rocks and cap rocks (such as shale formation) are present. The third most abundant formations, *carbonate formations*, are created by the accumulation of shells and skeletal remains of water-dwelling organisms in marine environments.

Chemical and organic sedimentary rocks are the other main group of sediments (besides clastic sediments) and are formed by weathered material in solution precipitating from water or as biochemical rocks made of dead marine organisms and special conditions (such as high temperature, high evaporation, and high organic activity) are required for the formation of these rocks. Some chemical sediment is deposited directly from the water in which the material is dissolved—for example, upon evaporation of seawater. Such deposits are generally referred to as *inorganic chemical sediments*. Chemical sediments that have been deposited by or with the assistance of plants or animals are classed as *organic sediments* or *biochemical sediments*.

Sedimentary rocks formed from sediments created by inorganic processes include (1) limestone, (2) dolomite, and (3) evaporites. *Limestone* ($CaCO_3$, calcite) is precipitated by organisms usually to form a shell or other skeletal structure. Accumulation of these skeletal remains results in the most common type of chemical sediment, limestone, which may also form by inorganic precipitation as well as by organic activity. *Dolomite* (magnesium limestone, $CaCO_3 \cdot MgCO_3$) occurs in the same settings as limestone and is formed when some of the calcium in limestone is replaced by magnesium. *Evaporite minerals* are sedimentary rocks (true chemical sediments) that are derived from minerals precipitated from seawater. There are two types of evaporite deposits: marine, which can also be described as ocean deposits, and nonmarine, which are found in standing bodies of water such as lakes. Rock salt, which is composed of halite (NaCl), and rock gypsum ($CaSO_4 \cdot 2H_2O$) are the most common types of evaporite minerals. High evaporation rates cause concentration of solids to increase due to water loss by evaporation.

Biochemical sedimentary rocks consist of sediments formed from the remains or secretions of organisms and are formed by the accumulation and subsequent consolidation of sediments into various types of rock. They include *fossiliferous limestone*, *coquina* (limestone composed of shells and coarse shell fragments), *chalk* (porous, fine-textured variety of limestone composed of calcareous shells), *lignite* (brown coal), and *bituminous* (soft) *coal*.

2.2 Characteristics

Sedimentary rocks possess definite physical characteristics and display certain features that make them readily distinguishable from igneous rocks (rocks formed through the cooling and solidification of magma or lava) or metamorphic rocks (rocks have been modified by heat, pressure, and chemical processes, usually while buried deep below surface of the Earth). Some of the most important sedimentary characteristics include the following: (1) stratification, (2) cross-bedding, (3) graded bedding, (4) texture, (5) ripple marks, (6) mud cracks, (7) concretions, (8) fossils, and (9) color.

Stratification, probably the most characteristic feature of sedimentary rocks, is their tendency of the rocks to occur in *beds* (*strata*), which are formed when geological agents such as wind, water, or ice gradually deposit sediment. *Cross-bedding* (*cross-stratification*) occurs to sets of beds that are inclined relative to one another. The beds are inclined in the direction that the wind or water was moving at the time of deposition and nay boundaries between sets of cross-beds usually represent an erosional surface. The cross-beds are common in beach deposits, sand dunes, and river-deposited sediment and enable determinations to be made about the origin and formation of ancient sediments. *Graded bedding* occurs as a result of a reduction in velocity (typically in a stream bed) and (1) larger or denser particles are deposited followed by (2) deposition of smaller particles. This results in the bedding showing a decrease in grain size from the bottom of the bed to the top of the bed (fine sediment particles at the top of the bed and coarse sediment particles at the bottom of the bed).

Fossils are the remains of once-living organisms that have been preserved in the crust of the Earth. Because life has evolved, fossils give clues to the relative age of the sediment and can be important indicators of past climates. In addition, to such information, the totality of fossils, both discovered and undiscovered, and their placement in *fossiliferous* (fossil-containing) rock formations and sedimentary layers (strata) is known as the *fossil record.*

Finally, the minerals in some sediments impart color to the sediment. However, color is one of the first noticed and most obvious characteristics of a rock, but it is also one of the most difficult to interpret. With the exception of gray and black, which mostly results from partially decayed organic matter, most rock colors are the result of iron staining. Ferric iron (Fe^{3+}) produces red, purple, and yellow colors—minerals such as hematite (iron oxide, Fe_2O_3, also spelled *haematite*) and limonite (an iron ore consisting of a mixture of hydrated iron(III) oxide-hydroxides in varying composition) produce a pink or red color. Ferrous iron (Fe^{2+}) produces greenish colors in sediments. Typically, a red color can be interpreted as a well-oxygenated environment, such as river channels, some flood plains, and very shallow seawater. Green colors mean an environment low in, or lacking, oxygen, often associated with marine environments. Dark gray to black colors mean anoxic conditions, which may mean deep water, but could also be a swamp environment. The conclusion is that environmental interpretations can only be made in relation to the other evidence present with the rock. When and how these colors originate in sedimentary rocks is still subject to speculation and, thus, the meaning of the color as it related to the environment in which the sediment was formed must viewed with caution and any conclusion must be confirmed with independent evidence.

2.3 Composition

A sediment is composed of three basic components: (1) grains, (2) matrix, and (3) cementitious materials. Grains (sometimes referred to as *framework grains*) refer to the larger, solid components in the sediment which form the basic small-scale units of sandstone reservoirs. The original grain composition is controlled by the composition of the sediment source (origin and history) as well as the physical and chemical processes under which the sediments are created and transported to the geologic basin. Often referred to as detrital grains, the grain composition of most sandstone reservoirs consists primarily of quartz, feldspars, and rock fragments (Berg, 1986).

Following deposition and burial, the framework grains (typically referred to as *authigenic grains*) are often altered by the physical effects of compaction as well as various chemical processes (diagenesis). The original grain composition governs the type and severity of diagenesis (Rushing et al., 2008). For example, some minerals are more brittle and may be more susceptible to compaction and/or failure during burial and the associated increase in stresses. Other minerals may be more reactive to natural fluids within the pores and may be altered (sometimes significantly) by adverse chemical reactions.

The matrix component in a sediment refers to the finer materials deposited between the larger grains and typically includes both clay minerals and shale minerals. Clay minerals may also be classified as either detrital or authigenic: (1) detrital clay minerals originate either from the sediment source material during deposition or may form from biogenic processes shortly after deposition, (2) authigenic clay minerals are formed by a chemical process, either by precipitation from formation fluids or regeneration of detrital clays. Clay regeneration refers to processes in which clay minerals develop by alteration of precursor clays (Wilson, 1982; Rushing et al., 2008). The principal clay minerals observed in sandstone reservoirs are kaolinite, smectite, illite, and chlorite.

Clay minerals vary widely in the structure or morphology of both the individual and aggregate particles and the presence of these minerals in the pore of sediments and can significantly reduce both permeability and primary porosity (Neasham, 1977a,b; Wilson and Pittman, 1977). This potential (and real) effect of clay minerals reinforces the importance of a comprehensive pore-scale program to identify clay type, origin, and the factors controlling its occurrence.

In terms of tight sand formations, a major component common for many tight gas sands is the grain cement (cementitious material) which typically refers to any mineral that forms during diagenesis and is precipitated after deposition of both grains and matrix components (Berg, 1986). As the name implies, the cement binds the minerals together in a competent mass in the rock and fills the pore system, thus reducing both permeability and porosity. The most common cement compositions are silica minerals and carbonate minerals. Silica minerals are precipitated as overgrowths or layers on quartz grains. Silica overgrowth cements may form soon after deposition but often continue to develop with increased pressure and temperature during burial. Carbonate cements are often precipitated early after deposition and tend to fill pore spaces between framework grains. Authigenic clay minerals may also act as cements by helping to bind rock particles together.

2.4 Texture

Sediment texture is another important aspect of sandstone reservoirs which includes grain size, sorting, packing, shape, and grain orientation since it not only affects properties of the sediment at deposition but also can impact the rate, magnitude, and severity of diagenesis. *Sediment texture* refers to the size, shape, and arrangement of materials that is derived from processes of weathering, transportation, deposition, and diagenesis.

The texture in sediment and sedimentary rocks is dependent on the processes that occur during each stage of formation which also includes (1) the nature of the source materials, (2) the nature of wind and water currents present, (3) the distance that materials were transported as well as the timer spent in the transportation process, (4) any biological activity, and (5) exposure to various chemical environments.

Grain size and distribution, sorting, shape and packing also govern the type and magnitude of the original porosity present following sediment deposition but before significant diagenesis has occurred. Generally, clean coarse-grained materials will have larger, better connected pores, while small-grained sands will have smaller and less well connected pores. Depending on the type and morphology, the presence of smaller matrix materials (i.e., clay minerals and shale minerals) in clean coarse-grained sands will tend to reduce both permeability and primary porosity. Other textural traits include grain shape and orientation. Grain shape is usually expressed as sphericity (a measure of the deviation of a grain from a spherical shape) and roundness (a measure of the roundness of the grain edges) (Berg, 1986). On the other hand, grain orientation refers to the preferred direction of the grain's long axes.

2.5 Structure

Sediment structure (including identification of bed geometry, bedding planes, contacts between beds, and bedding plane orientation) is an important element of the depositional process since the type of structure may help in identifying the original depositional environment. Understanding sedimentary structure is also an important component in optimizing field development activities since bed geometry and dimensions may impact both vertical and lateral continuity which would, in turn, dictate well spacing and the type of wellbore architecture. For example, significant vertical heterogeneity may determine how effectively horizontal wellbores will recover the natural gas or the crude oil.

Other aspects of sediment structure, insofar as the outcome is differences to the surface structure (appearance) of the sediment, are (1) ripple marks, (2) mud cracks. *Ripple marks* in a sediment are characteristic of deposition of the inorganic materials in shallow water and are caused by forces such as wave-related forces or wind-related forces that leave ripples of sand as typified by the marks seen on beach sand or on the bottom of a shallow stream. Thus, ripples of this type can provide information about the conditions of deposition when the sediment was originally deposited. *Mud cracks* provide other signs that are a guide to the environment in which the sediment was formed. Mud cracks result from the drying out of wet sediment on the bottom of dried-up lakes, ponds, or stream beds. Mud cracks may be many-sided (polygonal) shapes that present a honeycomb-type appearance on the surface of the beach or stream sand. If these marks are preserved in sedimentary rocks, there is the suggestion that the original material was subjected to alternating periods of flooding and drying. *Concretions* are spherical or flattened masses of rock enclosed in some (but not all) shale formations or in some limestone formations and which are generally harder than the rock enclosing them. Concretions are typically an indication that the softer environmentally prone rock was eroded away leaving the harder concretions intact.

Thus, understanding the depositional history of the reservoir is important in terms of predicting the long-term production behavior.

3 RESERVOIRS AND RESERVOIR EVALUATION

Production of crude oil and natural gas occurs in two classes of rock: source rocks and reservoir rocks. Source rocks are sedimentary rocks in which hydrocarbons (organic chemical compounds of hydrogen and carbon) form. Reservoir rocks are both porous, meaning that there are open spaces, or voids, within the rock, and permeable, meaning fluids are able to flow within them. After forming in the source rock, hydrocarbons, which can vary from simple structures, like methane (a constituent of natural gas), to very complex structures, like bitumen (contained in formation such as tar sand formations), can migrate to the reservoir rock.

Historically, nearly all hydrocarbons produced domestically were withdrawn from carbonate and sandstone reservoirs. However, over the past decade, production from shale and other tight rock formations, spurred by advances in exploration and production technology, has grown dramatically. Geologic formations that contain oil and gas include clastic or detrital rocks (formed from pieces of preexisting rocks or minerals), chemical rocks (formed by chemical precipitation of minerals), and organic rocks (formed by biological debris from shells, plant material, and skeletons). The three most common sedimentary rock types encountered in oil and gas fields are shales, sandstones, and carbonates. Classifying these rock types primarily depends on characteristics such as grain size and composition, porosity (pore space within and between grains), and cement (a chemically formed material that holds the grains together), each of which can influence oil and gas production.

The critical elements of a petroleum system are (1) the source rock, which is the rock containing the organic precursors which were converted into petroleum reservoir fluid, (2) the migration path, which is the path taken by the crude oil—or immature crude oil that is not fully matured—from the source rock to the reservoir, (3) the reservoir, which is a rock formation—such as sandstone, limestone, or dolomite—that has sufficient porosity to store the fluid and sufficient permeability for fluid mobility, (4) the seal, which is impermeable basement rock and cap rock that prevent the escape of the petroleum. For the purpose of this test, the critical part of the crude oil and/or natural gas system is the reservoir.

Crude oil (conventional crude or heavy oil) cannot be retained as an accumulation unless there is a trap, and this requires the boundary between the cap rock or other sealing agent but the exact form of the boundary varies widely. The simplest forms are the flat-lying convex lens, the anticline, and the dome, each of which has a convex upper surface. Many oil and gas accumulations are trapped in anticlines or domes, structures that are generally more easily detected than some other types of traps, such as fault traps and salt dome traps (Hunt, 1996; Dandekar, 2013; Speight, 2014). Thus, reservoir evaluation is an important aspect of oil and gas production. A reservoir is a subsurface porous permeable rock body or formation that has been created by the sequential steps of deposition, conversion, migration, and entrapment and has the ability to store fluids, such as natural gas, crude oil, and water. As such, each reservoir will exhibit individual properties that are specific to that reservoir (site-specific properties). Indeed, within a reservoir, these properties may even change with longitudinal extent and with vertical height of the reservoir.

Typically, reservoir rocks exhibit porosity—a measure of the openings in a rock in which crude oil and natural gas can exist. Another characteristic of reservoir rock is that it must be permeable—the pores of the rock must be interconnected thereby allowing so that crude oil and/or natural gas mobility within the reservoir and thence flow to a production well. A reservoir with high porosity but low permeable is a general indication of immobility of the gas and/or oil within the reservoir. In such a case, variations in gas composition and crude oil composition from different locations within the reservoir might be expected. Thus, reservoirs that are to be developed for crude oil and/or natural gas production are characteristically large and extensive in volume with a good fluid-holding capacity (high porosity) and also have the capability to transmit fluids once penetrated by geological disturbances (such as earthquakes) or anthropological disturbances, such as drilling a well into the reservoir.

As a result, effective resource exploitation requires a comprehensive reservoir description and characterization program to quantify gas in place and to identify those reservoir properties which control production. In terms of reservoir evaluation, reservoirs are generally evaluated on the basis of (1) rock types, (2) structural types, (3) heterogeneity, and (4) porosity and permeability, which are obtained through core analysis.

3.1 Rock Types

Fundamental to the process of reservoir evaluation is the identification and comparison of three different rock types: (1) *depositional rock types*, (2) *petrographic rock types*, and (3) *hydraulic rock types* (Rushing et al., 2008).

Depositional rock types are rock types that are derived from *core-based descriptions* of genetic units which are collections of rocks grouped according to similarities in composition, texture, sedimentary structure, and stratigraphic sequence as influenced by the environment at the time of deposition. These rock types also represent original large-scale rock properties present at deposition. The *original* rock properties will vary depending on many factors, including the depositional environments, sediment source and depositional flow regimes, sand grain size and distribution, and the type and volume of clay deposited. Thus, depositional rock types are an aid in defining the geological architecture and to describe large-scale reservoir compartments. Mapping the distribution of depositional rock types should also define the extent of the reservoir as well as the natural gas and/or crude oil in place.

Petrographic rock types are also described within the context of the geological framework, but the rock type criteria are based on pore-scale, microscopic imaging of the *current* pore structure, as well as the rock texture and composition, clay mineralogy, and diagenesis. *Hydraulic rock types* are rock types that are also defined at the pore scale and are the rock types that quantify the physical flow and storage properties of the rock relative to the native fluid(s) as controlled by the dimensions, geometry, and distribution of the current pore and pore throat structure (Rushing et al., 2008). The hydraulic rock type classification provides a physical measure of the rock flow and storage properties at current conditions. The size, geometry, and distribution of pore throats, as determined by capillary pressure measurements, control the magnitude of porosity and permeability for a given rock.

In addition, all three rock types should be similar if the rocks have been subjected to little or no diagenesis (Rushing et al., 2008). For example, the permeability-porosity relationships for depositional rock types would be expected to be applicable to petrographic and hydraulic rock types. However, as diagenetic effects increase in severity and occurrence, the original rock texture and composition, pore geometry, and physical rock properties are modified. Under these conditions, we would expect to see no or very poor correlations among the permeability-porosity relationships derived for each of the different rock types (Rushing et al., 2008).

Each rock type represents different physical and chemical processes affecting rock properties during the depositional and paragenetic cycles. Since most tight gas sands have been subjected to postdepositional diagenesis, a comparison of all three rock types will allow us to assess the impact of diagenesis on rock properties. If diagenesis is minor, the depositional environment (and depositional rock types) as well as the expected rock properties derived from those depositional conditions will be good predictors of rock quality. However, if the reservoir rock has been subjected to significant diagenesis, the original rock properties present at deposition will be quite different than the current properties. More specifically, use of the depositional environment and the associated rock types (in isolation) to guide field development activities may result in ineffective exploitation.

A natural gas and/or a crude oil reservoir is a subsurface collection (sometime referred to as a pool which leaves open the possibility of misinterpretation) of hydrocarbons and hydrocarbonaceous derivatives in porous or fractured rock formations. However, natural gas and crude oil reservoirs are typically classified as (1) conventional reservoirs and (2) unconventional reservoirs. In the former type of reservoir—the conventional reservoir—the natural gas and crude oil are trapped by an underlying rock formation (the basement rock) and by an overlying rock formation (the cap rock) with lower permeability than the reservoir rock. In the second type of reservoir—the unconventional reservoir—the reservoir rock typically has high porosity and low permeability in which the natural gas and crude oil are trapped in place without the need for a cap rock or basement rock.

Thus, in a conventional gas and oil reservoirs (Speight, 2007, 2014; GAO, 2012), oil and gas are fairly mobile and easily move through the permeable formation because of buoyancy (they are lighter than the water in the same formation and therefore rise) until they are trapped against an impermeable rock (i.e., a seal) that prevents further movement. This leads to localized pools of oil and gas while the rest of the formation is filled with water. However, both biogenic and thermogenic shale gas remain where they were first generated and can be found in three forms: (1) free gas in the pore spaces and fractures; (2) adsorbed gas, where the gas is electrically stuck to the organic matter and clay; and (3) a small amount of dissolved gas that is dissolved in the organic matter. In such reservoirs, typically an impermeable shale formation is either the basement rock or the cap rock of a sandstone formation thereby preventing any fluids within the sandstone from escaping or migrating to other formations. When a significant amount of organic matter has been deposited with the sediments, the shale rock can contain organic solid material (kerogen). The properties and composition of shale place it in the category of sedimentary rocks known as *mudstones*. Shale is distinguished from other mudstones because it is laminated and fissile—the shale is composed of many thin layers and readily splits into thin pieces along the laminations.

The evaluation of any reservoir, including reservoirs formed from shale formations and tight formations, should always begin with a thorough understanding of the geologic characteristics of the formation. The important geologic parameters for a trend or basin are (1) the structural and tectonic regime, (2) the regional thermal gradients, and (3) the regional pressure gradients, (4) the depositional system, (5) the genetic facies, (6) textural maturity, (7) mineralogy, (8) diagenetic processes, (9) reservoir dimensions, and (10) the presence of natural fractures, all of which can affect drilling, evaluation, completion, and stimulation. Without understanding the above-listed factors can lead to guesswork in determining reservoir behavior, performance, and longevity.

One of the most difficult parameters to evaluate in tight gas reservoirs is the drainage area size and shape of a typical well. In tight reservoirs, months or years of production are normally required before the pressure transients are affected by reservoir boundaries or well-to-well interference. As such, it may be necessary to estimate the drainage area size and shape for a typical well in order to estimate reserves. Knowledge of the depositional system and the effects of diagenesis on the rock are needed to estimate the drainage area size and shape for a specific well. In blanket, tight gas reservoirs, the average drainage area of a well largely depends on the number of wells drilled, the size of the fracture treatments pumped on the wells, and the time frame being considered. In lenticular or compartmentalized tight gas reservoirs, the average drainage area is likely a function of the average sand-lens size or compartment size and may not be a strong function of the size of the fracture treatment.

A main factor controlling the continuity of the reservoir is the depositional system. Generally, reservoir drainage per well is small in continental deposits and larger in marine deposits. Fluvial systems tend to be more lenticular whereas barrier-strand-plain systems tend to be more blanket and continuous. To date, most of the more successful tight gas plays are those in which the formation is a thick, continuous, marine deposit.

Thus, an understanding of the geology of a reservoir is essential to reservoir development, oil and gas production, and management, including reservoir longevity and environmental management. Furthermore, reservoir evaluation includes both the external geology of the reservoir (the forces responsible for the formation of the reservoir) and the internal geology of the reservoir (the nature of the rocks that constitute the reservoir). These aspects are even more important when hydraulic fracturing methodology is to be applied to the reservoir. In addition, the efficient extraction of crude oil and natural gas requires that the reservoir be visualized in three dimensions which can only be adequately provided through a variety of scientific and geological studies (Solano et al., 2013).

An important geologic aspect of the reservoir is the external geometry of the reservoir, defined by seals that inhibit the further migration of the natural gas and crude oil. Migration will cease, and a hydrocarbon reservoir will form, only where hydrocarbons encounter a trap, which are composed of a suitable gas-holding or oil-holding rock with the following types of seals: (1) top, (2) lateral, and (3) bottom seals. In addition, the geometry of traps can be (1) structural, (2) sedimentary, and (3) diagenetic (Hunt, 1996; Dandekar, 2013; Speight, 2014).

Another important geologic aspect of the reservoir is the internal architecture which involves the lateral distribution of depositional textures that is related to depositional environments, and the vertical stacking of textures is described by stratigraphy, which is the geological study of the following aspects of rock strata: (1) form, (2) arrangement, (3) geographic distribution, and (4) chronologic succession. Diagenesis, which refers to changes that happen to the sediment after deposition, can also control the lateral continuity and vertical stacking of reservoir rock types. This phenomenon is an important aspect of carbonate reservoirs, in which the conversion of limestone to dolostone and the dissolution of carbonate have a large effect on internal reservoir architecture (Tucker and Wright, 1990; Blatt and Tracy, 1996).

Briefly, dolostone or dolomite rock is a sedimentary carbonate-rich rock that contains a high proportion of dolomite ($CaCO_3 \cdot MgCO_3$), which has also been referred to as *magnesian limestone*. Most dolostone is formed by magnesium replacement of calcium in limestone ($CaCO_3$) prior to lithification—the process in which sediments compact under pressure, expel connate fluids, and gradually become solid rock. Dolostone is resistant to erosion and can act as an oil and natural gas reservoir.

3.2 Structural Types

Reservoirs are created by structural deformation of the geological strata and there are three basic forms of a structural trap in petroleum geology: (1) anticline trap, (2) fault trap, and (3) salt dome trap (Hunt, 1996; Dandekar, 2013; Speight, 2014). The *anticline trap* is a typical *structural trap* that is produced by compressional folding, by uplift, and by drape over older tectonically created features. An anticline is an example of rocks which were previously flat but have been bent into an arch. The rocks have been folded or bucked into the form of a dome and hydrocarbons accumulate in the hinge area of an anticline. The *fault trap* is formed by the movement of permeable and impermeable layers of rock along a fault line. The permeable reservoir rock faults such that it is now adjacent to an impermeable rock, preventing hydrocarbons from further migration. In some cases, there can be an impermeable substance smeared along the fault line (such as clay) that also acts to prevent migration. Another form of trap is the *stratigraphic trap* that is formed when other geologic formations seal a reservoir or when the permeability changes through a change in lithology, that is, a change due to the presence of rock with characteristics different to those of the reservoir rock. *Salt domes* are formed by flow of salt or removal of salt deposits.

Reservoir rocks tend to show greater variations in permeability than in porosity and in addition, these two properties, as measured on core samples from reservoir rocks, are not always identical with the values indicated for the bulk rock in the underground formation because of the non-representative nature of many core sample. Generally, porosity is on the order of 5–30%; while permeability falls between 0.005 Darcy (5 mD) and several Darcys (several thousand milliDarcys) (Kovscek, 2002).

3.3 Heterogeneity

In addition to understanding the petrophysics of the reservoir, oil recovery requires an understanding of displacement and flow through porous media—however, flow through porous media is complicated (Dawe, 2004; Maxwell and Norton, 2012). Within the reservoir there can be displacements and miscible and/or immiscible flow, with one, two, or sometimes three mobile phases (oil, gas, and water) (Grattoni and Dawe, 2003). Furthermore, heterogeneity in the form of layers, lenses, cross-beds, and quadrants can have a significant effect on fluid displacement patterns.

Low-permeability crude oil and natural gas reservoirs exhibit a high degree of bodily heterogeneity encompassing different scales within the hosting geological formation. Local variations of porosity, permeability, and pore geometry are variably affected by the compositional nature of the sediments and the depositional environment in which they formed, as well as the evolving diagenetic and tectonic history of the reservoir rocks.

Physically, natural gas and crude oil reservoirs are not the homogeneous porous media that is often envisaged on paper and used in calculations using data from laboratory simulations. Heterogeneity means that a specific property of interest will vary vertically and longitudinally within the reservoir (Dawe, 2004) much like the coal in a seam that varies in composition from one part of the seam to another (Speight, 2013). For example, well log and core analysis reports show that all reservoirs are heterogeneous with rock properties (such as porosity and pore saturation) varying within the reservoir. In addition, permeability heterogeneity causes variations in the fluid movements compared to a homogeneous system (Dawe, 2004). Furthermore, reservoir heterogeneity can arise from variations in permeability or variations in wettability. In fact, wettability of the reservoir rock by the crude oil (particularly adsorption of the polar constituents in heavy oil) can have significant effects on crude oil recovery (Anderson, 1986; Caruana and Dawe, 1996a,b; Dawe, 2004).

The *wettability* of reservoirs rocks refers to the tendency of the fluid (e.g., crude oil) to spread on or adhere to a solid surface in the presence of other immiscible fluids and is determined by complex *interface boundary conditions* acting within pore space of sedimentary rocks. The term *oil wet* refers to reservoir rock that is preferentially in contact with crude oil, which occupies the small pores and contacts the majority of the rock surface. Conversely, the term *water wet* refers to reservoir rock that is preferentially in contact with water. The minerals present in reservoir rocks are generally known as being intrinsically *hydrophilic* (i.e., preferentially water-wet) or *oleophilic* (i.e., preferentially oil-wet).

3.4 Porosity and Permeability

Porosity and permeability are related properties of any rock or loose sediment. Most oil and gas has been produced from sandstones which usually have high porosity and high permeability. Porosity and permeability are necessary to make a productive oil or gas well and are the result of both depositional and diagenetic factors (Alreshedan and Kantzas, 2015). Although a relationship between porosity and permeability is often difficult to interpret (Speight, 2014), there are trends that show a very general relationship between the two properties (Fig. 2.1).

In conventional reservoirs, hydrocarbons move easily through the formation until they are trapped against an impermeable rock that prevents further flow downward (basement rock) or upward (cap rock). This leads to localized pools of oil and natural gas that can be accessed with a vertical well

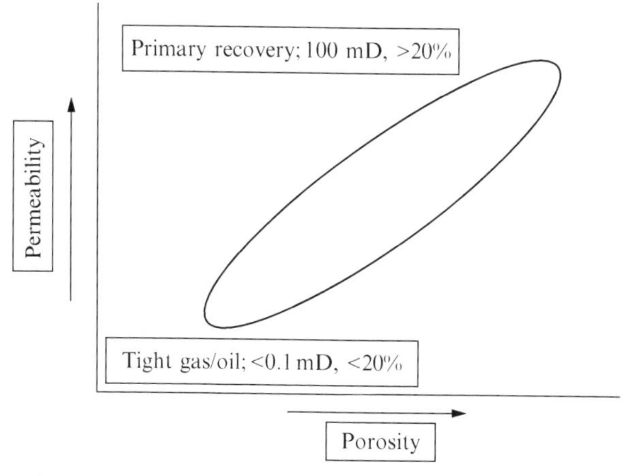

Fig. 2.1 General trends in the relationship between porosity and permeability.

drilled directly into the reservoir. In reservoirs that consist of tight formations, the natural gas and crude oil are often found within the same rocks they were generated in, trapped in their pore spaces, natural fractures, and within the organic matter itself (inside its pores and stuck to pore walls). Otherwise, hydrocarbons in tight reservoirs can be found in the pore spaces and natural fractures of any tight rocks into which they migrated. Shale formations, tight sandstone formations, and tight carbonate formations, and the resources contained therein, tend to be widely distributed over extensive areas rather than concentrated in specific locations.

In conventional reservoirs, pore space (pore volume, porosity) can vary from fairly large, visible openings to microscopic pores, and generally comprises <30% (*v*/*v*) of the reservoir rock volume. In tight reservoirs, porosity is commonly <10% (*v*/*v*) of the reservoir rock. However, regardless of the total porosity volume, if these pores are not efficiently connected one to the other (to give permeability), natural gas and crude oil cannot migrate. Thus, the higher the permeability, the greater the amount of fluid that can flow through the rock. Conventional reservoirs may have a permeability in the range of tens to hundreds of milliDarcys. Tight reservoirs usually have permeability from 0.1 to 0.001 mD, and shale reservoirs are even less permeable—in the 0.001–0.0001 mD range (Fig. 2.2). In tight reservoirs, the typical permeability is usually too small to allow commercial production unless unconventional completion techniques (horizontal drilling and hydraulic fracturing) are used.

The characterization of porosity and permeability is of fundamental importance for the proper evaluation of a reservoir. At the microscopic scale, porosity and permeability are highly dependent on the geometry of the pores and pore throats within volumetrically finite homogeneous systems. These microscopic, locally homogeneous domains are usually found as layered sediments and/or clusters which confer different degrees of heterogeneity to the reservoir (Radlinski et al., 2004). Thus, porosity is the proportion of void space to the total volume of rock and is a measure of the ability of the rock

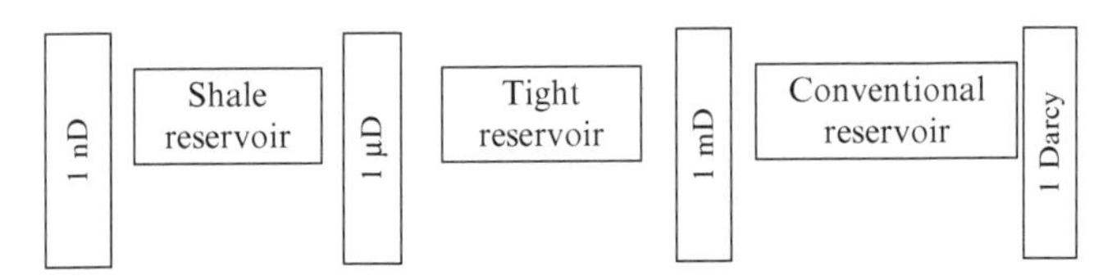

Fig. 2.2 Representation of differences in permeability of shale reservoirs, tight reservoirs, and conventional reservoirs.

to hold fluids (including gas). Mathematically, porosity is the open space in a rock divided by the total rock volume and is normally expressed as a percent of the total rock which is taken up by pore space. For example, sandstone may have 8% porosity—the other 92% is solid space-filling rock. In newly deposited sand formations and poorly consolidated sandstone formations, grain size correlates well with pore size and hence is a primary control on permeability.

In 1978, the U.S. government defined reservoirs with a permeability of <0.1 mD as *tight reservoirs*. However, several studies since have shown that permeability is not, in fact, the only aspect that can identify tight gas reservoirs (Shanley et al., 2004; Aguilera and Harding, 2008). Several researches characterized tight gas reservoirs by high capillary entry pressures, high irreducible water saturations, low to moderate porosity, and low permeability (Thomas and Ward, 1972; Dutton, 1993; Byrnes, 1997, 2003) as well as pore geometry (Soeder and Randolph, 1987) which was subdivided into three categories: (1) grain supported pores, (2) slot and solution pores, and (3) matrix-supported grains and emphasized that the most common pore structure of tight gas reservoirs consists of secondary solution pores connected to each other through narrow tortuous slots (or fractures) known as a dual porosity model. These slots have an important effect on permeability, as well as on gas flow through tight porous media, but they may not have significant contribution to porosity (Soeder and Randolph, 1987; Aguilera, 2008; Byrnes et al., 2009). Thus, the pore topology/structure of unconventional reservoirs is not only heterogeneous but also unlike conventional reservoirs. It has been found in coal, shale, and tight gas reservoirs that the liquid permeability is lower than measured gas permeability (Bloomfield and Williams, 1995; Byrnes, 1997; Cui et al., 2009; Cluff and Byrnes, 2010; Ziarani and Aguilera, 2012; Ghanizadeh et al., 2014). Throat radius is the accounted parameter for this variation (Bloomfield and Williams, 1995; Mehmani and Prodanović, 2014). The Knudsen number (*Kn*) is being used as a dimensionless correction factor to quantify the amount of different gas flow regimes (slip flow, transition, and free molecular) as a function of pore throat radius (Knudsen, 1909; Javadpour et al., 2007).

Moreover, pore structure controls both the natural gas in place and the crude oil in place and long-term deliverability in tight formations. This is controlled by mineralogy, which is dependent on depositional environment and postdepositional diagenetic processes. Understanding the means by which the mineral composition controls rock fabric, natural gas in place, crude oil in place, and the mechanical properties of the formation is critical to accurately assessing its potential. Pore throat size controls permeability of the rock, affecting the gas flow rate and porosity controls the distribution of the natural gas and crude oil in the formation. In formations with smaller pores, a larger amount of surface area is available for the constituents of the natural gas or crude oil to be adsorbed, and a larger fraction of the gas or oil may be present in the adsorbed state than in the free state.

Laboratory measurements for tight porous media are costly and time-consuming. Also, studying the effect of a certain parameter in such complex pore structures is difficult due to the complexity associated with the experimental setup. As an alternative approach, pore network modeling can be used to construct physically sound models of a real porous media. Pore network modeling gives a reasonable prediction of fluid flow properties at pore scale and offers the flexibility of studying macroscopic properties relationship with pore structure and geometry.

In pore network modeling, the complex pore structure in a rock is represented by a network of pore bodies (void spaces) and pore throats (narrow paths that connect pore bodies) with simplified geometries. When this is successfully established, then single and multiphase flow calculations can be performed (Okabe and Blunt, 2005). Detailed physics and productive capabilities for pore-scale modeling of multiphase flow at the pore scale have been reviewed in several studies (Blunt, 2001; Blunt et al., 2002). At the early stages of network modeling, capillary pressure and relative permeability curves of drainage using two-dimensional (2D) regular lattice networks where the radii were randomly assigned were predicted (Fatt, 1956) and later illustrated (Chatzis and Dullien, 1977) that 3D pore network models represent the real porous media more realistically than 2D pore network models. Following this early work, extensive studies on the importance of topology, pore bodies and throats size distribution and their spatial correlations were performed (Chatzis and Dullien, 1977; Jerauld and Salter, 1990; Grattoni and Dawe, 1994). However, most of these studies were based on regular lattice networks which are limited in reflecting the real topology and geometry of a rock. The capabilities of network modeling were enormously improved and have been applied to make many successful predictions of single and multiphase flow and transport properties including two- and three-phase relative permeability and capillary pressure of conventional formations (Øren et al., 1998; Patzek, 2001; Øren and Bakke, 2002, 2003; Valvatne and Blunt, 2003; Piri and Blunt, 2005a,b).

In complex sandstones, it is recommended to create first a 3D image-based representation of the pore space that should capture the statistics of the real rock. This can be generated using a direct imaging technique such as micro-CT scanning (Okabe and Blunt, 2005) or by various process/object-based modeling approaches (Bakke and Ören, 1997; Ören et al., 1998; Ören and Bakke, 2002, 2003). Subsequently, using various image-based network extraction techniques (Al-Kharusi and Blunt, 2007; Dong and Blunt, 2009), an equivalent pore network is then extracted from the 3D image to estimate the single and multiphase fluid flow properties.

Permeability and formation factor are physical properties of the rock. Permeability is defined by Darcy's law and formation factor is defined by Archie's law. Permeability plays a critical role in determining the potential of hydrocarbon flow in a porous medium. Further, formation factor gives an indication of the availability of pathways for transport. Both properties are sensitive to pore and throat size, connectivity and geometries.

Permeability values for tight formations usually fall into a range from the submicroDarcy ($<10^{-3}$ mD) up to the one hundred of milliDarcys. Permeability is usually measured either with core plugs or full-diameter core samples using unconfined, unsteady-state techniques. For the measurements of ultra-low permeability samples, crushed rock pressure decay techniques are used instead.

The porosity of rock samples is traditionally calculated from helium pycnometer measurements—as used for petroleum coke and other solids (ASTM D2638)—from which an accurate value of grain density is obtained. The relatively small size of helium molecules ensures that even subnanometer sized pores and pore throats are probed. In addition, the low adsorptive capacity of these molecules reduces the errors that might be introduced due to the absorption processes during the measurements. However, helium may be accessible to finer pores than crude oil and natural gas constituents in shale formations thereby overestimating accessible porosity (Cui et al., 2009).

On the other hand, permeability is the ability of fluid to move through the pores and is a measure of the ease with which fluids (including gas) pass through a rock. Thus, it is extremely important to know the values of formation permeability in every rock layer. The values of permeability control everything from gas flow rate to fracture fluid leak-off. It is impossible to optimize the location of the perforations, the length of the hydraulic fracture, the conductivity of the hydraulic fracture, and the well spacing, if one does not know the values of formation permeability in every rock layer. In addition, one must know the formation permeability to forecast gas reserves and to analyze postfracture pressure buildup tests. To determine the values of formation permeability, one can use data from logs, cores, production tests, and prefracture pressure buildup tests or injection falloff tests (Ahmed et al., 1991).

Permeability in petroleum-producing rocks is usually expressed in milliDarcys and most oil and gas reservoirs have permeability in the range up to several hundred milliDarcys (Fig. 2.2). The extremely low permeability of shale formations determines the effectiveness of basement rock and cap rock seals for many conventional reservoirs. However, grain-size combinations are the key in determining these seal and leakage characteristics. Although well-compacted clay-rich formations are generally considered as good seals, certain combinations with larger grains, such as silt grains, can reduce the seal quality. Clay-poor formations are generally considered to be poor seals, but recent studies have shown that certain combinations of large and smaller grains can contribute to improving their seal characteristics. Therefore, it is important to understand the effect of texture on seal quality and develop field methods to detect these characteristics. This understanding is also essential for predicting and detecting possible leakage mechanisms in rocks that are thought to be good seals.

Accumulations of petroleum and natural gas can only occur if all of the essential elements (source rock, reservoir rock, seal rock, and overburden rock) and processes (generation-migration-accumulation-trap formation) have operated adequately and in the proper time space framework (Magoon and Dow, 1994; Speight, 2014). Absence or inadequacy of even one of the elements or processes eliminates any chance of economic success. Thus, reservoir parameters (reservoir size, porosity, and permeability) are among the geologic controls that have to be included in the consideration of risk factors for reservoir development (Berg, 1970; Ahmed et al., 1991; Rose, 1992; White, 1993; Ramm and Bjørlykke, 1994; Yao and Holditch, 1996). The quantification and predictability of three major causes of anomalously high porosity are (1) grain coats and grain rims, (2) early emplacement of hydrocarbons, and (3) shallow development of fluid overpressure.

Grain coats are the result of authigenic processes and form subsequent to burial by growth outward from framework grain surfaces, except at points of grain-to-grain contact (Wilson and Pittman, 1977). Grain coats include clay minerals and microcrystalline quartz. Grain coats and grain rims retard quartz cementation and concomitant porosity and/or permeability reduction by blocking potential nucleation sites for quartz overgrowths on detrital-quartz seed grains. The effectiveness of grain coats or grain rims in preserving porosity is a function of the thermal history, grain size, and the abundance of quartz grains (Walderhaug, 1996; Bonnell et al., 1998). Grain coats and grain rims have no effect on porosity where the primary control of reservoir quality is the occurrence of cements such as carbonate minerals, sulfate minerals, or zeolites (Pittman et al., 1992).

Porosity and permeability generally decrease with increasing depth (thermal exposure and effective pressure); however, a significant number of deep (approximately 13,000 ft) sandstone reservoirs worldwide are characterized by anomalously high porosity and permeability (Bloch et al., 2002). Anomalous porosity and permeability can be defined as being statistically higher than the porosity and permeability values occurring in typical sandstone reservoirs of a given lithology (composition and texture), age, and burial/temperature history.

In tight gas (shale) reservoirs, areas where the reservoir quality and completion are high (*sweet spots*) are confined to areas that have high clay rim coverages but relatively low volume of clay minerals. By contrast, interstratified clay-free and coarse-grained nonreservoir rock is tightly cemented by quartz overgrowth (Wescott, 1983; Weimer and Sonnenberg, 1994). Furthermore, drilling success in any reservoir is dependent on finding the most prospective areas, or the *sweet spots*, and aligning the wellbore for maximum borehole exposure to these zones. In shale reservoirs, this means placing the well in the zones most conducive to fracturing. This requires a thorough understanding of the shale gas reservoir characteristics. Aiming for a middle-of-the-road operation is rarely a successful strategy—shale formations can have significant variance in thickness and composition.

The thickness of a tight formation is one of the primary reasons, along with a large surface area of fine-grained sediment and organic matter for adsorption of gas, that resource evaluations in tight formations yield such high values for total organics carbon content and potential producibility of the natural gas or crude oil. A general rule-of-thumb is that a thicker formation is a more attractive (producible and economic) shale. Shale targets such as the Bakken formation (Williston Basin), however, are <150 ft thick in many areas but are yielding economic rates of gas flow and recovery. The required thickness to economically develop a tight formation may decrease as drilling and completion techniques improve, as porosity and permeability detection techniques progress in unconventional targets and, perhaps, as the price of gas increases.

In summary, the success or failure of a hydraulic fracture treatment will depend on the quality of the candidate well selected for the treatment.

Evaluation and selection of a suitable candidate reservoir for stimulation is a move in the depiction of success, while choosing a poor candidate normally results in failure. To select the best candidate for stimulation, the design engineer must consider many variables of which the most critical parameters for hydraulic fracturing are (1) formation permeability, (2) the in situ stress distribution, (3) viscosity of the reservoir fluid, (4) reservoir pressure, (5) reservoir depth, (6) the condition of the wellbore, and (7) prior stimulation of, or damage to, the reservoir.

The best candidate wells for hydraulic fracturing treatments have a substantial volume of oil and gas in place and need to increase the productivity index. The characteristics of such reservoirs include (1) a thick pay zone, (2) medium to high pressure, (3) in situ stress barriers to minimize vertical height growth, and (4) either a low-permeability zone or a zone that has been damaged. On the other hands, reservoirs that are poor candidates for hydraulic fracturing are those with little oil or gas in place because of thinness (lack of thickness or depth) as well as low reservoir pressure and small areal extent. Reservoirs with extremely low permeability may not produce enough hydrocarbons to pay all the drilling and completion costs, even if successfully stimulated; thus, such reservoirs would not be good candidates for stimulation.

3.5 Reservoir Morphology

Finally, reservoir morphology defines the sand-body dimensions, geometry, orientation, heterogeneity, and continuity as developed by depositional and postdepositional processes. Both sand quality and quantity are controlled by primary and secondary depositional environments and processes. Quantification of the morphology helps define the reservoir architecture and compartments, and ultimately, to determine the original reservoir volume or "container." For example, the gas-in-place volumes and producing characteristics for a "blanket" sand will be much different than for a reservoir characterized by lenticular sands. Reservoir morphology will also affect the optimum well spacing for field development. Depositional environment and postdepositional diagenesis both have a significant bearing on morphology, including reservoir compartmentalization and heterogeneity. Reservoir compartments refer to intervals or sections of the sand deposits that are mostly or completely isolated (i.e., not in pressure communication) from other parts of the reservoir. Compartments may be created by significant changes in the depositional environment or by postdepositional processes (such as diagenesis and/or tectonic activity creating sand pinch-outs, and no-flow barriers). Reservoir heterogeneities, which are typically manifested by lateral and vertical variability in permeability and porosity within the same sand body, are mostly caused by postdepositional diagenesis. Most diagenetic processes do not cause completely isolated reservoir compartments—but such processes may yield complex and/or poor quality flow paths, which may result in low productivity for a given reservoir system.

4 TIGHT FORMATIONS

Geologic age is an important determinant of hydrocarbon potential, beyond the characteristics of source and reservoir rocks. Identifying fossils, other chemical markers, and correlating rocks across different formations allows earth scientists to determine the age of the rock and to understand the processes that influenced the sediments and organic material over time. While many rocks of different ages produce oil and natural gas, domestic areas of prolific production include formations from several different geologic periods: (1) Devonian period—416–359 million years ago, (2) Carboniferous period—359–299 million years ago, (3) Permian period—299–251 million years ago, and (4) Cretaceous period—145–65 million years ago. During these periods, organic-rich materials accumulated and, over time, heat and pressure chemically altered originally organic chemicals into natural gas and oil.

Unconventional natural gas resources and crude oil and crude oil resources found in shale formations and in tight formations (Ma et al., 2016; Moore et al., 2016) comprise a significant percentage of the North American natural gas and crude oil resource base and these systems represent an important source for future reserve growth and production. Similar to conventional natural gas and crude oil systems, unconventional reservoirs are characterized by complex geological and petrophysical systems as well as reservoir heterogeneity. In fact, low-permeability oil reservoirs exhibit a high degree of heterogeneity. Local variations of porosity, permeability, and pore geometry are variably affected by the compositional nature of the sediments and the depositional environment in which they formed, as well as the evolving diagenetic and tectonic history of the reservoir rocks (Solano et al., 2013). In addition, and unlike conventional reservoirs, unconventional reservoirs typically exhibit storage and flow characteristics which are uniquely tied to depositional and diagenetic processes.

The behavior and properties of each reservoir are symptomatic of the rock types that constitute the reservoir and each reservoir must be assessed differently according to the geochemical characteristics (Jarvie et al., 2011). For example, each rock type represents different physical and chemical processes affecting rock properties during the depositional and paragenetic cycles (equal equilibrium sequences of mineral phases). Since most tight gas sands have been subjected to postdepositional diagenesis, a comparison of the rock types allows an assessment of the impact of diagenesis on rock properties. If diagenesis is minor, the depositional environment (and depositional rock types) as well as the expected rock properties derived from those depositional conditions will be good predictors of rock quality. However, if the reservoir rock has been subjected to significant diagenesis the original rock properties present at deposition will be quite different than the current properties. Thus, it might be anticipated that use of the depositional environment and the associated rock types may not always result in effective exploitation of a formation or reservoir (Rushing et al., 2008).

Generally, a tight reservoir is a layered system and, in a clastic depositional system, the layers are composed of (1) shale, (2) mudstone, (3) and siltstone while in a carbonate system the layers are composed predominantly of (1) limestone, $CaCO_3$, (2) dolomite, $CaCO_3 \cdot MgCO_3$, (3), possibly halite, NaCl, or anhydrite, $CaSO_4$, and some shale. Thus to optimize the development of a tight reservoir, a multidisciplinary team consisting of (minimally) geoscientists, petrophysicists, and engineers must fully characterize all the layers of rock above, within, and below the pay zones in the reservoir.

4.1 Shale Formations

Shale is formed by the accumulation of very small sediments deposited in deep water, at the bottoms of rivers, lakes, and oceans. Shales are the most abundant clastic sedimentary rock, and because of their potential for a high organic content, shales are considered to be the primary source rocks for hydrocarbons. *Sandstone* is created by larger sediment, deposited in deserts, river channels, deltas, and shallow sea environments. These rocks tend to be more porous than shales and consequently make excellent reservoir rocks. Sandstone is the second most abundant clastic sedimentary rock and is the most commonly encountered reservoir rock in hydrocarbon production. *Carbonate* is created by the accumulation of shells and skeletal remains of water-dwelling organisms in marine environments. The third most abundant sedimentary rock, carbonate (e.g., limestone) rocks, is also very good reservoirs and is commonly encountered during hydrocarbon production.

On the other hand, a *shale reservoir* (*shale play*) is similar on a worldwide basis insofar as organic-rich, gas-prone shale is generally difficult to *discover*. Tight shale formations are heterogeneous and vary widely over relatively short distances. Thus, even in a single horizontal drill hole, the amount recovered may vary, as may recovery within a field or even between adjacent wells. This makes evaluation of plays and decisions regarding the profitability of wells on a particular lease difficult. However, the lack of a strict definition for shale causes an additional degree of difficulty for resource evaluation. Such a broad spectrum of lithology appears to form a transition with other resources, such as natural gas and crude oil, where the difference between the tight gas reservoir and gas shale may be the higher amount of sandstone (in the tight gas reservoir) and the tight gas reservoir may actually contain no organic matter. The properties and composition of shale place it in the category of sedimentary rocks known as *mudstones*. Shale is distinguished from other mudstones because it is laminated and fissile—the shale is composed of many thin layers and readily splits into thin pieces along the laminations.

Shale formations are a worldwide occurrence (Ma et al., 2016; Moore et al., 2016). Shale is a geological rock formation that is rich in clay, typically derived from fine sediments, deposited in fairly quiet environments at the bottom of seas or lakes, having then been buried over the course of millions of years. Shale formations can serve as pressure barriers in basins, as top seals, and as reservoirs in shale gas plays. More technically, shale is a fissile, terrigenous sedimentary rock in which particles are mostly of silt and clay size (Blatt and Tracy, 1996). In this definition, *fissile* refers to the ability of the shale to split into thin sheets along bedding and *terrigenous* refers to the origin of the sediment. In many basins, the fluid pressure of the aqueous system becomes significantly elevated, leading to the formation of a hydrofracture, and fluid bleed-off. However, the occurrence of a natural hydrofracture is an unlikely process in the circumstances that exist in most basins.

Shale is composed mainly of clay-size mineral grains, which are usually clay minerals such as illite, kaolinite, and smectite. Shale usually contains other clay-size mineral particles such as quartz, chert, and feldspar. Other constituents can include organic particles, carbonate minerals, iron oxide minerals, sulfide minerals and heavy mineral grains and the presence of such minerals in shale is determined by the environment under which the shale constituents were.

Shale is formed by the accumulation of very small sediments deposited in deep water, at the bottoms of rivers, lakes, and oceans. Shale is, by definition, a sedimentary rock that is predominantly composed of consolidated clay-sized particles but is separated from the more generally recognized sedimentary rocks that typically constitute a natural gas or petroleum reservoir. Shale formations are deposited as muds in low-energy environments such as tidal flats and deep water basins where the fine-grained clay particles fall out of suspension in the quiet waters. During the deposition process, in addition to the fine-grained material that eventually forms the sediment, there can also be the deposition and accumulation of organic matter in the form of algae, plant, and animal derived organic debris that is eventually converted to natural gas and crude oil (Davis, 1992).

Shale formations and silt formations are the most abundant sedimentary rocks in the Earth's crust. In petroleum geology, organic shale formations are source rocks as well as seal rocks that trap oil and gas (Speight, 2014). In reservoir engineering, shale formations are flow barriers. In drilling, the bit often encounters greater shale volumes than reservoir sands. In seismic exploration, shale formations interfacing with other rocks often form good seismic reflectors. As a result, seismic and petrophysical properties of shale formations and the relationships among these properties are important for both exploration and reservoir management.

Shale formations consist of consolidated clay and other fine particles (mud) that have hardened into rock. These formations are the most abundant of all sedimentary rocks, comprising approximately two-thirds of the sedimentary rocks. Typically, these formations are fine-grained and thinly bedded and readily split along dividing (bedding) planes. Shale is classified or typed by composition; for example, shale containing large amounts of clay is referred to as *argillaceous shale*, and shale containing appreciable amounts of sand is known as *arenaceous shale*. Shale with a high content of organic matter (*carbonaceous shale*) is typically black in color. Shale that contains large amounts of lime (*calcareous shale*) is used in the manufacture of Portland cement. Another type of shale, *oil shale*, is currently of great interest worldwide because of the supply and demand and increasing cost of crude oil. Oil shale contains kerogen (Chapter 1), a fossilized insoluble organic material that is converted into synthetic crude oil from which a variety of petroleum-type products can be produced.

The very fine-grained sheet-like clay mineral grains and any laminated layers of sediment result in a rock with permeability that is limited horizontally and even more limited when measured from samples taken from different vertical height within the formation. Such formations provide excellent cap rocks and basement rocks for conventional natural gas and crude oil (sandstone) reservoirs (Speight, 2014). However, in the current context, the low permeability of shale formations results in natural gas and crude oil that has originated in the shale remains trapped in the formation cannot move within (or migrate from) the rock except (in some instances) over geologic time (millions of years). Shale formations are the most abundant clastic sedimentary rock and, because of their potential for a high organic content, such formations are considered to be the primary source rocks for the formation of natural gas and crude oil.

Indeed, the variety of rock types observed in organic-rich shale formation confirms the implication that a range of different types of shale reservoirs exist. Each reservoir may have distinct geochemical and geological characteristics that may require equally unique methods of drilling, completion, production, and resource and reserve evaluation (Cramer, 2008)—leading to further necessary considerations when the shale gas had to be processed (Chapter 4). Additionally, it must not be forgotten that a shale formation is often a seal or cap rock for a conventional (sandstone) petroleum or natural gas reservoir and that not all shale are necessarily reservoir rocks (Speight, 2007, 2014).

There is also the possibility (only assiduous geological studies will tell) of hybrid shale formations, where the originally deposited mud was rich in sand. These foreign minerals (sand and silt) result in a natural higher permeability for the shale formation and result in greater susceptibility of the shale to hydraulic fracturing.

More specific to this text, four general types of shale formation (shale plays) have been defined as being predominant in the shale formations investigated and developed and have been given the simple designations: Type 1, Type 2, Type 3, and Type 4, without any order of preference but more on the basis of composition and behavior. The *Type 1 shale* is a fractured organic mudstone with high carbonate content—an example is the Barnett Shale—and primary typically involves a mix of gas released from fractures and micropores through gas desorption from the organic material and clay minerals. On the other hand, the *Type 2 shale* has laminated sands embedded in organic-rich shale—an example is the Bakken Formation, which is primarily an oil resource play. The *Type 3 shale* is an organic-rich black shale, such as the Marcellus Shale, and production occurs through gas desorption—the gas has been observed to carry with it gas condensate. Finally, the *Type 4 shale* (such as the Niobrara Shale) is a combination of the other three types of shale formations and production is through desorption, matrix structures, and fractures.

A significant factor associated with tight gas reservoirs is the low productivity, which is emphasized in the case of gas-condensate fluids, which exhibit complex phase behavior and complex flow behavior due to the appearance of condensate banking in the near-well region. This behavior differs essentially in their behavior from conventional gas reservoirs, especially for low permeability high yield condensate systems, which have more severe condensate banking problems. Thus, it is necessary to have a thorough understanding of the manner by which the condensate accumulation influences the productivity. Knowledge of the composition configuration in the liquid phase is very important to optimize the producing strategy for tight formations, to reduce the impact of condensate banking, and to improve the ultimate gas recovery.

A key factor that controls the gas-condensate well deliverability is the relative permeability, which is influenced directly by the condensate accumulation which not only reduces both the gas and liquid relative permeability but also changes the phase composition of the reservoir fluid, hence changes the phase diagram of reservoir fluid and varies the fluid properties (Wheaton and Zhang, 2000; Pedersen and Christiansen, 2006). In addition, different producing strategies may impact the composition configuration of the condensate for both the flowing phase and the static phase and the amount of the liquid trapped in the reservoir, which in turn may influence the well productivity and hence the ultimate gas and liquid recovery from the reservoir. Changing the manner in which the well is brought into flowing condition can affect the liquid dropout composition and can therefore change the degree of productivity loss.

As might be expected (even predicted), shale plays that cover large subterranean areas show considerable variations in mineralogical composition, formation depth, and thermal maturity. An example of a large-area play is the Marcellus Shale, which covers portions of the states of Pennsylvania, New York, Ohio, West Virginia, Virginia, Tennessee, and Maryland. Examples of plays confined to smaller areas include the Eagle Ford Shale, which covers a portion of south Texas, the Barnett Shale in central Texas and Haynesville Shale, which underlies parts of Arkansas,

Louisiana, and Texas. It must also be realized that the definitions given above are for convenience based on current investigations and it is likely (perhaps, even expected) that a wider variety of shale types exists and will be explored and developed in the future as technology improves.

On a more physical note, typical shale formations can be anywhere from 20 ft to a mile or so thick and extend over very wide geographic areas, a gas shale reservoir is often referred to as a *resource play*, where natural gas resources are widely distributed over extensive areas (perhaps several fields) rather than concentrated in a specific location. The volume of natural gas contained within a resource play increases as the thickness and areal extent of the deposit grows. Individual gas shale formations may have a billion cubic feet (1×10^9 ft^3) or even a trillion cubic feet (1×10^{12} ft^3) of gas in place spread over hundreds to thousands of square miles. The difficulty lies in extracting even a small fraction of that gas.

Shale formations exhibit a wide range of mechanical properties and significant anisotropy reflecting their wide range of material composition and fabric anisotropy (Sone, 2012). The elastic properties of these shale rocks are successfully described by tracking the relative amount of soft components (clay and solid organic materials) in the rock and also acknowledging the anisotropic distribution of the soft components. Gas shale formations also possess relatively stronger degree of anisotropy compared to other organic-rich shale formations, possibly due to the fact that these rocks come from peak-maturity source rocks. The deformational properties are influenced by the amount of soft components in the rock and exhibited mechanical anisotropy.

The pore spaces in shale, through which the natural gas must move if the gas is to flow into any well, are as much as one thousand times smaller than pores in conventional sandstone reservoirs. The gaps that connect pores (the pore throats) are smaller still, only 20 times larger than a single methane molecule. Therefore, shale has very low permeability but natural or induced fractures, which act as conduits for the movement for natural gas, will increase the permeability of the shale.

Shale comes in two general varieties based on organic content: (1) dark or (2) light. Dark colored or black shale formations are organic-rich, whereas the lighter colored shale formations are organic-lean. Organic-rich shale formations were deposited under conditions of little or no oxygen in the water, which preserved the organic material from decay. The organic matter was mostly plant debris that had accumulated with the sediment. Black organic shale formations are the source rock for many of the oil and natural gas deposits of the world. These black shale formations obtain their black color from tiny particles of organic matter that were deposited with the mud from which the shale formed. As the mud was buried and warmed within the earth some of the organic material was transformed into oil and natural gas.

A black color in sedimentary rocks almost always indicates the presence of organic materials. Just 1% or 2% organic materials can impart a dark gray or black color to the rock. In addition, this black color almost always implies that the shale formed from sediment deposited in an oxygen-deficient environment. Any oxygen that entered the environment quickly reacted with the decaying organic debris. If a large amount of oxygen was present, the organic debris would all have decayed. An oxygen-poor environment also provides the proper conditions for the formation of sulfide minerals such as pyrite, another important mineral found in black shale sediments or formations.

The presence of organic debris in black shale formations makes them the candidates for oil and gas generation. If the organic material is preserved and properly heated after burial oil and natural gas might be produced. The Barnett Shale, Marcellus Shale, Haynesville Shale, Fayetteville Shale, and other gas producing rocks are all dark gray or black shale formations that yield natural gas.

The oil and natural gas migrated out of the shale and upwards through the sediment mass because of their low density. The oil and gas were often trapped within the pore spaces of an overlying rock unit such as a sandstone formation. These types of oil and gas deposits are known as *conventional reservoirs* because the fluids can easily flow through the pores of the rock and into the extraction well.

Shale formations are ubiquitous in sedimentary basins: they typically form about 80% of what a well will drill through. As a result, the main organic-rich shale formations have already been identified in most regions of the world. Their depths vary from near surface to several thousand feet underground, while their thickness varies from a tens of feet to several hundred feet. Often, enough is known about the geological history to infer which shale formations are likely to contain gas (or oil, or a mixture of both). In that sense there may appear to be no real need for a major exploration effort and expense required for shale gas. However, the amount of gas present and particularly the amount of gas that can be recovered technically and economically cannot be known until a number of wells have been drilled and tested.

Each shale formation has different geological characteristics that affect the way gas can be produced, the technologies needed, and the economics of production. Different parts of the (generally large) shale deposits will also have different characteristics: *small sweet spots* or *core areas* may provide much better production than the remainder of the formation, often because of the presence of natural fractures that enhance permeability (Hunter and Young, 1953).

The amount of natural gas liquids (NGLs—hydrocarbons having a higher molecular weight than methane, such as propane, butane, pentane, hexane, heptane, and even octane, commonly associated with natural gas production) present in the gas can also vary considerably, with important implications for the economics of production. While most dry gas plays in the United States are probably uneconomic at the current low natural gas prices, plays with significant liquid content can be produced for the value of the liquids only (the market value of NGLs is correlated with oil prices rather than gas prices), making gas an essentially free byproduct.

The Barnett Shale of Texas was the first major natural gas field developed in a shale reservoir rock. Producing gas from the Barnett Shale was a challenge because the pore spaces in shale are so tiny that the gas has difficulty moving through the shale and into the well. Drillers discovered that the permeability of the shale could be increased by pumping water down the well under pressure that was high enough to fracture the shale. These fractures liberated some of the gas from the pore spaces and allowed that gas to flow to the well (hydraulic fracturing, hydrofracing).

Horizontal drilling and hydraulic fracturing revolutionized drilling technology and paved the way for developing several giant natural gas fields. These include the Marcellus Shale in the Appalachians, the Haynesville Shale in Louisiana, and the Fayetteville Shale in Arkansas. These enormous shale reservoirs hold enough natural gas to serve all of the United States' needs for 20 years or more. Hydraulic properties are characteristics of a rock such as permeability and porosity that reflect its ability to hold and transmit fluids such as water, oil, or natural gas. In this respect, shale has a very small particle size so the interstitial spaces are very small. In fact, they are so small that oil, natural gas, and water have difficulty moving through the rock. Shale can therefore serve as a cap rock for oil and natural gas traps and it also is an aquiclude that blocks or limits the flow of underground water.

Although the interstitial spaces in a shale formation are very small they can take up a significant volume of the rock. This allows the shale to hold significant amounts of water, gas, or oil but not be able to effectively transmit them because of the low permeability. The oil and gas industry overcomes these limitations of shale by using horizontal drilling and hydraulic fracturing to create artificial porosity and permeability within the rock.

Some of the clay minerals that occur in shale have the ability to absorb or adsorb large amounts of water, natural gas, ions, or other substances. This property of shale can enable it to selectively and tenaciously hold or freely release fluids or ions.

Thus, this shale gas resource can be considered a technology-driven resource as achieving gas production out of otherwise unproductive rock requires technology-intensive processes. Maximizing gas recovery requires far more wells than would be the case in conventional natural gas operations. Furthermore, horizontal wells with horizontal legs up to 1 mile or more in length are widely used to access the reservoir to the greatest extent possible.

Multistage hydraulic fracturing (Chapter 5), where the shale is cracked under high pressures at several places along the horizontal section of the well, is used to create conduits through which gas can flow. Microseismic imaging allows operators to visualize where this fracture growth is occurring in the reservoir. However, as a technology-driven resource, the rate of development of shale gas may become limited by the availability of required resources, such as fresh water, fracture proppant, or drilling rigs capable of drilling wells several two miles or more in length.

As in shale formations, which are believed to be the source rocks in which oil and gas form during geological time, crude oil and gas are contained in the pore space of the formation—which can include sandstone formations, siltstone formations, and carbonate formations. While a conventional formation containing crude oil and/or natural gas can be relatively easily drilled and extracted from the ground, tight gas and tight oil (natural gas and crude oil in tight formations) requires more effort to extract from a tight reservoir. In such formations, the pores in the formation in which the gas is trapped are either irregularly distributed or interconnection of the pores is poor, which adversely affects permeability. Without secondary production methods, gas and/or oil from a tight formation would flow at very slow rates, making production uneconomical.

While vertical wells may be easier and less expensive to drill, they are not the most conducive to developing tight formations. In a tight formation, it is important to expose as much of the reservoir as possible, making horizontal and directional drilling a necessity. Here, the well can run along the formation, opening up more opportunities for the natural gas and/or tight crude oil to enter the wellbore. A more common technique for developing reserves in tight formations includes drilling more wells which enhances the ability of the oil and gas to leave the formation and enter the wellbore. This can be achieved through drilling several myriad directional wells (the number of wells is formation specific) from one location, which lessens the environmental footprint of the drilling operation. After seismic data has illuminated the best well locations, and the wells have been drilled, production stimulation (through both fracturing and acidizing) is employed on tight reservoirs to promote a greater rate of flow.

Fracturing involves breaking apart the rocks in the formation (Chapter 5). After the well has been drilled and completed, hydraulic fracturing is achieved by pumping the well full of fracturing fluids under high pressure to cause rock fracturing in the reservoir and improve permeability. Additionally, acidizing the well is employed to improve permeability and production rates of tight gas formations, which involves pumping the well with acids that dissolve the limestone, dolomite, and calcite cement between the sediment grains of the reservoir rocks. This form of production stimulation helps to reinvigorate permeability by reestablishing the natural fissures that were present in the formation before compaction and cementation.

Typically, North American crude oil and natural gas formations that require fracturing are located 1 mile or more below the water table and below many layers of impermeable rock. These thousands of feet of rock overlying the tight gas formations, combined with the low permeability of the tight gas formations themselves, ensure that the natural gas and crude oil remain contained within the target formation and also help prevent migration of any hydraulic fracturing fluids that may be pumped into such formations. Drilling, casing, and cementing procedures must be designed to at least meet (or even exceed) regulatory requirements to protect groundwater by isolating the well from any groundwater supplies.

The upper portions of the well, where the wellbore passes through the water table, should be reinforced to prevent either gas or oil (and any fracturing fluids) from escaping into the surrounding ground. Wells are lined with steel pipes and sealed in place with cement from the surface to below the level of drinking water supplies, typically to a depth of 1000 ft or more. These barriers help to contain the fracturing fluid and, along with the depth at which fracturing takes place, prevent the fluid from mingling with drinking water close to the surface. During and after hydraulic fracturing, wells are monitored with pressure sensors to check that they are firmly sealed which helps to make production as efficient as possible and protects the environment (Chapters 5 and 9).

The manner in which gas and oil are trapped within tight formations and shale formations requires advanced technology to access these resources. Horizontal and directional drilling techniques are used to access a large underground area from a single well pad and when followed by hydraulic fracturing technology stimulates the release of the encapsulated oil and gas to flow into the wellbore. One of the most difficult parameters to evaluate in tight reservoirs is the drainage area size and shape of a typical well.

Finally, because of the properties of the shale formation (above) such as (1) the presence of *hard minerals* and (2) the *internal pressure* of the shale, it may be possible to isolate sections along the horizontal portion of the well, segments of the borehole for one-at-a-time fracking (multistage fraccing). By monitoring the process at the surface and in neighboring wells, it can be determined how far, how extensively, and in what directions the shale has cracked from the induced pressure. As has been discovered because of the natural gas and crude oil contents of shale formations, these formations can be re-fractured after production has declined, which may (1) allow the well to access to larger areas of the reservoir that may have been missed during the initial hydraulic fracturing or (2) to reopen fractures that may have closed due to the decrease in pressure as the reservoir was drained. Even with hydraulic fracturing, wells drilled into low-permeability reservoirs have difficulty communicating far into the formation. As a result, additional wells must be drilled to access as much gas as possible, typically three or four, but up to eight, horizontal wells per section.

4.2 Sandstone and Carbonate Formations

The term *tight formation* refers to a formation consisting of extraordinarily impermeable, hard rock. Tight formations are relatively low permeability, nonshale, sedimentary formations that can contain oil and gas. When a significant amount of organic matter has been deposited with the sediments, the shale rock can contain organic solid material (kerogen). Reservoirs with estimated in situ permeability of 0.1 mD or less were recognized in late 1970s and early 1980s as *tight reservoirs* (Moslow, 1993). Since that time, however, and for all practical purposes, a tight reservoir is generally recognized as any low-permeability formation in which special well completion techniques (such as horizontal drilling and hydraulic fracturing) are required to stimulate production (Chapter 5).

The low permeability (and porosity) associated with tight sands are attributed to a large distribution of small to very small pores and/or a complex system of pore throats connecting those pores. Furthermore, both small pores and pore throat systems can result from several processes such as (1) deposition of fine- to very fine-grained sediments, (2) the presence of various types of dispersed shale minerals and clay minerals in the pores, and/or (3) postdepositional diagenesis that alter the original pore structure (Rushing et al., 2008). Therefore, successful exploitation of a tight gas sand reservoir requires a basic understanding of the rock pore structure and properties as well as the processes affecting those properties.

Geologically, tight reservoirs are reservoirs having a low permeability and are often associated with conventional reservoirs, which could be sandstone, siltstone, limestone, dolomite, sandy carbonate minerals, and shale minerals having significant thickness. Tight reservoir sands are continuous and stacked sedimentary layers charged with hydrocarbons. Many tight gas sand formations (and shale formations) are naturally fractured and/or layered. Development of tightness and different geological complexities in sandstone reservoirs are due to different geological events. It is due to loss of porosity through diagenesis, occurrence of most porosity in secondary pore spaces. In comparison to the typical sandstone formation, tight sand formations have lesser void space and void connectivity.

Development of tight reservoirs includes two factors: (1) the provenance, mineralogy, grain size, its sorting, flow regime, sedimentary depositional environment, and the lithification and (2) diagenesis which involves compaction, cementation, and dissolution followed by tectonics and development of fractures. Regional and local tectonics plays a very important role in the evaluation of the tight sand reservoirs. Pressure and thermal gradient are affected by the tectonics and are also an important aspect of the evolution of these types of reservoirs. The most common tight sand formations generally consist of highly altered primary porosity, with authigenic quartz growth, coupled with secondary pore developments.

As expected for any complex reservoir system, successful exploitation of a tight reservoir requires basic understanding of the rock pore structure and properties as well as processes affecting those properties. Core analysis is direct evidence of the reservoir composition and mineral relationships and will help in understanding nature of facies, the depositional environment, sand texture, digenetic alteration, reservoir morphology, sand distribution and its orientation. The different sedimentary structures control tightness of the reservoir as well. The thin section analysis of the rock infers detail mineralogy, texture, sorting of grains, matrix and cementing material, and the type of porosity.

Low-permeability formations occur in almost all sedimentary basins worldwide. In North America, the vast majority of tight gas reservoirs can be grouped into two main geological categories: (1) Devonian shales from eastern United States and Canada, and (2) low-permeability sandstones from throughout the United States and from the Western Canada Sedimentary basin. It has been estimated that in the United States alone, tight sandstone formations are likely to have recoverable reserves ranging from 100 to 400 tcf, and Devonian shales have recoverable reserves of up to 100 tcf (Rushing et al., 2008). The successful exploitation of tight gas resources in the future will depend in large part on advancements made in the proper geological evaluation of low-permeability reservoirs.

A *tight reservoir* (*tight sands*) is a low-permeability sandstone reservoir—in some sedimentary basins tight reservoirs consist of mainly sandstone, silty sandstone, siltstone, argillaceous limestone, and dolomites. A tight reservoir is one that cannot be produced at economic flow rates or recover economic volumes of gas unless the well is stimulated by a large hydraulic fracture treatment and/or produced using horizontal wellbores. This definition also applies to coalbed methane and tight carbonate reservoirs—shale gas reservoirs are also included by some observers (but not in this text). Typically, tight formations which formed under marine conditions contain less clay and are more brittle, and thus more suitable for hydraulic fracturing than formations formed in fresh water which may contain more clay. The formations become more brittle with an increase in quartz content (SiO_2) and carbonate content (such as calcium carbonate, $CaCO_3$, or dolomite, $CaCO_3 \cdot MgCO_3$).

Thus, a tight reservoir is an umbrella term often used to refer to low-permeability reservoirs that produce mainly dry natural gas and volatile crude oil. Many of the low-permeability reservoirs that have been developed in the past are sandstone, but significant quantities of gas are also produced from low-permeability carbonate formations, shale formations, and coal seams (which is not included in this text). In tight reservoirs, the expected value of permeability is <1 mD (Fig. 2.2) whereas in shallow, thin, low-pressure reservoirs a permeability of several milliDarcys might be required to produce natural gas or crude oil at economic flow rates, even after successful fracture treatment. In this context, perhaps the best definition of a tight reservoir is a reservoir from which gas or liquids cannot be produced at economic flow rates nor recover economic volumes of natural gas unless the reservoir is stimulated by hydraulic fracture treatment or produced by use of a horizontal wellbore or multilateral wellbores. In fact, further to this definition, and remembering the role of economics in reservoir development, a typical tight reservoir does not exist—the reservoir can involve a range of physical parameters: (1) deep or shallow, (2) high pressure or low pressure, (3) high temperature or low temperature, (4) blanket or lenticular, (5) homogeneous or heterogeneous, (6) naturally fractured—which is unlikely—or require serious fracturing by hydraulic means, and (7) composed of a single layer or multiple layers. Therefore, the optimum drilling, completion, and stimulation methods for each well in a specific reservoir are functions of reservoir characteristics.

In tight reservoirs, the typical drainage area of a well largely depends on (1) the number of wells drilled, (2) the size of the fracture treatments, and (3) the time frame being considered. In lenticular or compartmentalized tight reservoirs, the drainage area is usually a function of the sand-lens size or compartment size, and may not be a strong function of the size of the fracture treatment. A main factor controlling the continuity of the reservoir is the depositional system. Generally, reservoir drainage per well is small in continental deposits and larger in marine deposits. Fluvial systems tend to be more lenticular whereas barrier-strand plain systems tend to be more continuous.

Also, in tight reservoirs *diagenesis* (any postdepositional process causing changes in the initial rock properties) is very important aspect of reservoir geology since it is the principal cause of both low permeability and low porosity. The diagenesis process may be either a physical process or a chemical process, or a combination of several different types of processes. The initial digenetic process is directly attributable to the prevailing local depositional environment as well as the sediment composition. Subsequent diagenesis is typically more widespread, often crossing multiple facies boundaries as a result of regional fluid migration patterns (Stonecipher and May, 1990). Thus, diagenesis is frequently caused by interactions between the sediment minerals and pore fluids at the elevated pressure and temperature conditions prevalent in the reservoir. The primary diagenetic processes commonly observed in tight sands are (1) mechanical compaction, (2) chemical compaction, (3) cementation, (4) mineral dissolution, (5) mineral leaching, and (6) clay genesis.

Mechanical compaction is caused by grain rearrangement, ductile and plastic rock deformation, and fracturing/shearing of brittle materials. This form of compaction may be mitigated by high pore pressure which tends to reduce stresses transferred to the grain materials. Chemical compaction refers to changes in grain size and geometry caused by physical and chemical reactions enhanced by pressure conditions, such as mineral solution. Generally, both mechanical and chemical compaction will reduce both permeability and porosity—permeability is reduced when the pore throats are partially or completed closed and porosity is reduced by a reduction in the primary pore volume. Cementation is a chemical process in which minerals are precipitated from pore fluids and bind with existing grains and rock fragments. The most common cement compositions in tight sands are composed of silica minerals and carbonate minerals. Silica minerals are precipitated as overgrowths or layers on quartz (SiO_2) grains which may form soon after deposition but often continue to develop with increased pressure and temperature during burial. Carbonate cements are often precipitated early after deposition and tend to fill pore spaces between minerals. Authigenic clay minerals (i.e., clay minerals that formed during sedimentation and were not transported from elsewhere—allogenic minerals—by water or wind) may also act as cementitious materials by helping to bind rock particles together. Most cements tend to reduce both permeability and porosity. However, the presence of authigenic grain coats and rims can retard quartz cementation and the associated reduction in permeability and porosity by blocking potential nucleation sites for quartz overgrowths on detrital-quartz grains (Bloch et al., 2002).

Another type of chemical diagenesis is mineral dissolution. For example, quartz (silica, SiO_2) can become soluble by the application of pressure (pressure solution, which can only occur at higher temperatures) caused by stress concentrations at grain contacts which results in silica dissolution, diffusion, transport of silica for reprecipitation in adjacent pores, and an associated loss in porosity. Another type of mineral dissolution is mineral leaching which often results in an increase in primary porosity and/or the creation of secondary porosity. A common source of secondary porosity creation is dissolution of carbonate cements which are often precipitated early after deposition and tend to fill pore spaces between framework grains.

Clay genesis refers to authigenic clay minerals created or generated after deposition. Such minerals that are found in tight sands include chlorites, mixed-layer smectite/illite clay minerals, and illite clay minerals. Authigenic chlorite minerals typically develop under iron-rich conditions and commonly occur as pore linings (or coatings). Since these clay minerals often do not completely cover the detrital grain surfaces, quartz overgrowths may develop on many grains, thus reducing the original primary porosity. Smectite clay minerals have been observed in sandstones that contain significant amounts of volcanic rock fragments. Illite minerals may also form from kaolinite and can develop either through precursor detrital or authigenic clays. Illite crystals can occur either as fibrous, sheet-like or plate structures—the fibers tend to break easily and accumulate in pore throats, causing a reduction or loss of permeability. Illite sheets and plates may also reduce permeability by blocking pore throats.

Both reservoir pressure and temperature affect the type, magnitude, and severity of diagenesis. Moreover, increasing temperatures increases the solubility of minerals and causes the pore waters to become saturated, thereby increasing precipitation and formation of cementitious materials.

Finally, rocks with interlaminated shale and siltstone is a shale gas target (e.g., Lewis Shale, New Mexico; Colorado Group, Alberta) that may require new techniques for detection in well logs, as well as new completion and drilling techniques. The silt laminations are too thin to be detected on well logs and to allow an accurate determination of how many laminations are in a given interval. Also, well logs are unable to accurately determine the percentage of porosity in shale or the laminations, the degree of water saturation in a reservoir, or the relative degree of permeability in each lamination. Laminations both store gas (free gas) and are pathways of transport for diffusion of gas from shale to the wellbore (Beaton et al., 2009). The laminations are also particularly difficult completion targets. Normally, induced fractures are meant to extend laterally rather than vertically in a reservoir, yet the laminations may span tens of hundreds of feet vertically. Therefore, a horizontal fracture may miss many productive shale and silt laminations. Induced fracturing techniques may have to be altered, or new techniques developed for this type of shale gas reservoir.

4.3 Development and Production

As a part of the development and production activities in a tight reservoir, one of the most difficult parameters to evaluate is the drainage area size and shape of a typical well. Knowledge of the depositional system and the effects of diagenesis on the rock are needed to estimate the drainage area size and shape for a specific well. In continuous-type (blanket-type) tight reservoirs, the average drainage area of a well largely depends on the number of wells drilled, the size of the fracture treatments pumped on the wells, and the time frame being considered. In lenticular-type or compartmentalized tight gas reservoirs, the average drainage area is likely a function of the average sand-lens size or compartment size and may not be a strong function of the size of the fracture treatment.

A main factor controlling the continuity of the reservoir is the depositional system. For example, reservoir drainage per well is small in continental deposits and larger in marine deposits while in fluvial systems the drainage area tends to be more lenticular. Tight reservoirs that have been more successfully developed, such as the Vicksburg in south Texas, the Cotton Valley Taylor in east Texas, the Mesa Verde in the San Juan Basin, and the Frontier in the Green River Basin, are tight sandstones that originated as marine deposits, which tend to be more continuous (blanket reservoirs). In fact, most of the more successful tight reservoirs are those in which the formation is a thick, continuous, marine deposit. However, there are other formations, such as the Travis Peak formation in east Texas, the Abo formation in the Permian Basin, and the Mesa Verde formation in parts of the Rocky Mountains, that are fluvial systems and tend to be highly lenticular. The Wilcox Lobo formation in south Texas is highly compartmentalized because of faulting. In lenticular reservoirs (compartmentalized reservoirs) the drainage area is controlled by the geology and must be estimated by the geologist or engineer.

Whether the reservoir is a tight sandstone reservoir or a shale reservoir, development of the reservoir (Chapter 4) requires a vertical well drilled and completed and must be successfully stimulated to produce at commercial gas flow rates and produce commercial gas volumes. Typically, hydraulic fracturing is required to produce natural gas and crude oil economically. In some naturally fractured tight reservoirs, horizontal wells and/or multilateral wells can be used to provide the stimulation required for commerciality. Moreover, to optimize the development of a tight reservoir, the number of wells drilled must be optimized along with the necessary drilling and completion procedures for each well. Often, more data and more engineering manpower are required to understand and develop tight gas reservoirs that are required for higher permeability, conventional reservoirs. This requires that more wells (or smaller well spacing) must be drilled into a tight reservoir to recover a large percentage of the original gas in place or original oil in place when compared to a conventional reservoir.

In all cases, a thorough understanding of the fundamental geochemical and geological attributes of tight formations and shale formations is essential for resource assessment, development, and environmental stewardship. Four very pertinent properties that define the characteristics of a reservoir are (1) the maturity of the organic matter, (2) the type of gas generated and stored in the reservoir—biogenic gas or thermogenic gas, (3) the total organic carbon content of the strata, and (4) the permeability of the reservoir. Only when these characteristics are known can the reservoir be successfully developed and exploited.

Thus, tight reservoirs can be subdivided into three rock types—(1) *depositional*, (2) *petrographic*, and (3) *hydraulic* (Newsham and Rushing, 2001; Rushing and Newsham, 2001; Rushing et al., 2008). Each rock type represents different physical and chemical processes affecting the rock properties during both depositional and paragenetic cycles. We define the tight gas sand rock types as follows.

Production from tight reservoirs usually requires enhancement due to problems associated with very low permeability (Akanji and Matthai, 2010; Akanji et al., 2013). The collection of the fundamental knowledge database needed to fully understand the key mechanisms affecting flow behavior in tight formation is still sparse. In this paper, we applied a new technique of measuring flow properties in porous media to characterize flow behavior in core samples of tight carbonate formations.

Moreover, the permeability of the samples is higher for the core samples obtained vertically through the parent rock. Tight formations are considered to be reservoirs with an absolute permeability of generally <10 mD and can range down to the microDarcy range, (10^6 Darcy) in many situations. These reservoirs could potentially serve as media for the storage of commercial accumulation of hydrocarbon. However, production and ultimate recovery are usually uneconomical due to a number of factors which include poor reservoir quality, unfavorable initial saturation condition, formation damage caused by drilling and completion operations, hydraulic or acid fracturing, kill or work-over treatments, and other production-related problems.

Many tight reservoirs are extremely complex producing oil or gas from multiple layers with low permeability that often require enhancement by natural fractures. Despite the marginal economics and low productivity from these reservoirs, the soaring demands for energy have necessitated a devotion of technologies to optimize recovery. Description of the pore geometry of reservoirs plays a major role in understanding the degree of pore interconnectivity, pore shapes and sizes, capillary trapping potentials, and flow behavior.

5 CORE ANALYSES FOR TIGHT RESERVOIRS

Obtaining and analyzing cores is crucial to the proper understanding of any layered, complex reservoir system. To obtain the data needed to understand the fluid flow properties, the mechanical properties, and the depositional environment of a specific reservoir requires that cores be cut, handled correctly, and tested in the laboratory using modern and sophisticated laboratory methods. Of primary importance is measuring the rock properties under restored reservoir conditions. The effect of net overburden pressure must be reproduced in the laboratory to obtain the most accurate quantitative information from the cores.

To provide all the data needed to characterize the reservoir and depositional system, a core should be cut in the pay interval and in the layers of rock above and below the pay interval. Core from the shales and mudstones above and below the pay interval help the geologist determine the environment of deposition. Knowing more about the deposition allows the reservoir engineer to better estimate the morphology and size of the gas-bearing reservoir layers. Also, mechanical property tests can be run on the shales to determine estimates of Poisson's ratio and Young's modulus. Additional tests can be run to measure the shale density and the sonic travel time in the shale to assist in the analyses of the density and sonic-log data. In fact, the rock quality governs the hydrocarbon storage properties whereas completion quality depends on the elastic properties. Successful hydraulic fracturing of low-permeability reservoirs requires identification of reservoir sections along the wellbore with good reservoir and completion qualities. Elastic properties govern in situ stress field, stress concentration around wellbore, and failure properties of the units, which lead to hydraulic fracture geometry and propagation behavior. Geomechanical modeling and characterization is one of the key components to get the desired fracturing results (Guha et al., 2013).

5.1 Handling and Testing Cores

After cutting the cores in the field, it is important to handle the core properly: (1) the core should not be hammered out of the barrel. It should be pumped out, (2) once the core is laid out on the pipe racks, it should be wiped with rags to remove the mud (do not wash with water), then described as quickly as possible, (3) bedding features, natural fractures, and lithology should be described foot by foot, (4) permanent markers should be used to label the depth of the core and clearly mark the up direction on the core, (5) as quickly as feasible, the core should be wrapped in heat shrinking plastic, then sealed in paraffin for the trip to the core analysis laboratory, and (6) precautions should be taken to minimize alteration of the core properties while retrieving and describing the core in the field.

Once in the laboratory, the core is unwrapped and slabbed, and plugs are cut for testing. Normally, a core plug should be cut every foot in the core, trying to properly sample all the rock—not just the cleaner pay zones. Routine core analyses can be run on these core plugs. Once the routine core analyses are completed, additional core plugs are cut for special core analyses. Sometimes samples of whole core are used for testing. Both the routine and the special core analyses are required to calibrate the open-hole logging data and to prepare the data sets required to design the optimum completion. The core plugs must also be treated with care. For example, if a core plug from a shale-containing sand is placed in a standard oven, it is likely that the clays in the pores will be altered as they dry out. A more accurate core analysis can be achieved if the core plugs are dried in a humidity-controlled oven in which the free water is evaporated, but the treatment should not be so severe as to affect the bound clay water.

5.2 Routine Core Analyses

Routine core analyses should be run on core plugs cut every foot along the core. Routine core analyses should consist of measurements of (1) grain density, (2) porosity and permeability to air, both unstressed and stressed, (3) cation exchange capacity, and (4) fluid saturations analysis. In addition, each core plug should be described in detail to understand the lithology and grain size and to note any natural fractures and other details that could be of importance to the geologist, petrophysicist, or engineer.

The porosity is used to determine values of gas in place and to develop correlations with permeability. The grain density should be used to determine how to correlate the density log values and to validate any calculation of lithology from log data. The cation exchange capacity can be used to determine how much electric current can be transmitted by the rock rather than the fluid in the pore space. The cation exchange capacity must be measured in the laboratory, using samples of rock, and is a function of the amount and type of clay in the rock. Saturation analysis measures the amount of water, oil, and gas in the core plugs in the laboratory. Saturation analysis can be misleading in rocks that are cored with water-based mud because of mud filtrate invasion during the coring process and problems that occur with core retrieval and handling prior to running the laboratory tests. However, the values of water saturation from the core analysis of cores cut with an oil-based mud can be used to calibrate the log data and to estimate values of gas in place in the reservoir.

The measurements of porosity and permeability are a function of the net stress applied to the rock when the measurements are taken. For low porosity rock, it is very important to take measurements at different values of net stress to fully understand how the reservoir will behave as the gas is produced and the reservoir pressure declines. In fact, a challenge in the development of resources from low-permeability formations is the ability to accurately evaluate rock properties (such as permeability, porosity, and capillary pressure). This information is necessary to quantify the resource potential of the formation and also to predict production behavior. However, due to the complex structure of the pore network, simple relationships relating permeability to porosity are not representative since the low permeability of tight formations renders the standard steady-state techniques, which are applicable in evaluating conventional natural gas and crude oil reservoirs, difficult to implement in terms of producing reliable and meaningful data.

Furthermore, the measurements of permeability and porosity of conventional, high permeability reservoirs are typically performed at low pressure in the laboratory and are not representative of in situ conditions which means that the effect of the overburden stress is largely ignored. As stress is increased, the high aspect ratio of the pore structures dominating conductive pathways are compressed and ultimately closed off, restricting fluid flow and increasing tortuosity of flow pathways. Thus, after the values of porosity and permeability are measured in the laboratory, the values should be correlated to the conditions in the tight formation (Thomas and Ward, 1972; Jones and Owens, 1980; Soeder and Randolph, 1987; Guha et al., 2013). In addition, however, it should be remembered that these estimates are from routine core analyses, which means the core has been tested dry with no water in the core. If similar measurements are made at connate water saturation, the permeability in the core is further reduced, maybe by a factor of 2 or even an order of magnitude in some cases. As such, in tight gas reservoirs, it is often found that in situ permeability to gas is 10–100 times lower than gas permeability measured at ambient conditions on dry core plugs cut from whole core. If cores come from a percussion sidewall device, the core plugs are typically altered, and the values of permeability under unstressed conditions can be even more optimistic.

5.3 Specialized Analyses

To fully understand the properties of tight gas formations, special core analyses must be run on selected core plugs to measure values of gas permeability versus water saturation, resistivity index, formation factor, capillary pressure, acoustic velocity, and the rock mechanical properties. The values of resistivity index and formation factor are used to better analyze the porosity and resistivity logs. The acoustic velocity can be used to better estimate porosity and to determine how to estimate the mechanical properties of the rock from log data. The mechanical properties are measured and correlated to log measurements and lithology. The capillary pressure measurements and the gas permeability versus water saturation relative permeability measurements are required to properly simulate fluid flow in the reservoir and to design hydraulic fracture treatments.

It is important to choose the correct core samples for conducting the special core analyses. Special core analysis tests are expensive and require weeks or months of special laboratory measurements. As such, the core samples must be chosen with care to provide the optimum data for designing the well completion and the well stimulation treatment and forecasting future gas recovery. A good way to select the core samples for special core analysis testing is to (1) form a team of geologists, engineers, and petrophysicists, (2) lay out the core, (3) have the routine core analysis and log analysis available, (4) determine how many rock types or lithology types that are contained in the core are important to the completion and stimulation process, and (5) pick three to six locations for each rock type or lithology where core plugs are cut for testing.

In addition to the composition of the shale (i.e., shale type), the depth of the formation is a major parameter since depth also influences thermal maturity, bottomhole temperatures, formation pressure, and overall formation behavior. For example, the Haynesville play is relatively deep which is reflected in high-pressure and high-temperature conditions whereas the Marcellus play is relatively shallow and is only mildly overpressured in the most productive areas. Pore pressure can also be used as an indicator of reservoir quality—higher pressures typically reflect a high degree of gas generation and storage. Furthermore, in order to better understand the effect of hydraulic stimulation, it is necessary to understand the stress state in conjunction with the elastic and strength properties of the rock mass. Such a study may commence with determination of the brittleness index (BI) of the rocks.

The BI, a measure of the ability of the rock to crack or fracture, also assists shale play developers when they evaluate a well and field and is primarily related to shale mineralogy and rock strength. A high BI is typically associated with high quartz content or high carbonate (dolomite) content, as illustrated by the Barnett shale play and the Woodford shale play. On the other hand, the BI decreases with increasing clay and organic matter, as is in the Marcellus shale play. Thus, the BI serves as a guide for placement of perforations, isolation points, and fracture stages (Wylie, 2012).

However, the BI should be used according to the relevant definition (Herwanger et al., 2015) within which the brittleness of rocks is characterized from (1) the elastic properties, (2) the petrophysical properties, and (3) the strength properties. The definition that is predominant in the geophysical literature relates to a high brittleness for rocks that exhibit a high Young's modulus E and low Poisson's ratio, ν (Rickman et al., 2008). There is also a BI that is related to a specific combination of Lamé parameters λ and μ (Goodway et al., 2010). Another category of definitions of BI is based on mineral content of the rocks (Jarvie et al., 2007) while definition of BI can also be based on strength parameters to derive a BI from a combination of uniaxial compressive strength σ_c and tensile strength σ_t (Altindag, 2003). Whatever system is used, the BI is essentially a lithology (or mineralogy) indicator and the user should make sure to provide the method employed to determine the index thereby mitigating any possible confusion.

Finally, specialized methods to characterize pore structure, storage capacity, and flow characteristics of tight reservoirs need further development. The techniques currently applied to assess the properties of tight formations are methods that were originally developed for conventional reservoir rocks and coal seams using the assumption that the same transport and storage mechanisms occur in tight reservoirs containing natural gas and crude oil. However, tight formations have a rock composition and structure that is different to coal and conventional reservoirs and many of the techniques employed to characterize porosity are not always suitable for tight formations. In addition, most methods for characterization require predrying of the samples which can cause the clay minerals to shrink and, hence, alter the rock structure. As a consequence, the sample and the properties determined therefrom may not to be representative of the in situ formation. Whether or not the errors are small enough to be discounted, in reality small errors in porosity (which is related to gas storage capacity) can eventually lead to errors in the capacity of the reservoir to hold natural gas and crude oil because of the large areal extent of many tight reservoirs. Understanding the limitations of current characterization techniques and the adjustments that are required to calibrate these measurements to the realities of the tight formation are important for accurate assessment of any natural gas or crude oil resource base.

REFERENCES

Aguilera, R., 2008. Role of natural fractures and slot porosity on tight gas sands. Paper No. SPE 114174, In: Proceedings of SPE Unconventional Reservoirs Conference, Keystone, CO. Society of Petroleum Engineers, Richardson, TX.

Aguilera, R., Harding, T.G., 2008. State-of-the-art tight gas sands characterization and production technology. J. Can. Pet. Technol. 47 (12), 37–41.

Ahmed, U., Crary, S.F., Coates, G.R., 1991. Permeability estimation: the various sources and their interrelationships. J. Pet. Technol. 43 (5), 578–587. Paper No. SPE-19604-PA. Society of Petroleum Engineers, Richardson, TX.

Akanji, L.T., Matthai, S.K., 2010. Finite element-based characterization of pore-scale geometry and its impact on fluid flow. Transp. Porous Media 81, 241–259.

Akanji, L.T., Nasr, G.G., Bageri, M., 2013. Core-scale characterization of flow in tight Arabian formations. J. Pet. Explor. Prod. Technol. 3, 233–241.

Al-Kharusi, A.S., Blunt, M.J., 2007. Network extraction from sandstone and carbonate pore space images. J. Pet. Sci. Eng. 56 (4), 219–231.

Alreshedan, F., Kantzas, A., 2015. Investigation of permeability, formation factor, and porosity relationships for Mesaverde tight gas sandstones using random network models. J. Pet. Explor. Prod. Technol. http://dx.doi.org/10.1007/s13202-015-0202-x. Published online September 22, 2015, http://download.springer.com/static/pdf/29/art%253A10.1007%252Fs13202-015-0202-x.pdf?originUrl=http%3A%2F%2Flink.springer.com%2Farticle%2F10.1007%2Fs13202-015-0202-x&token2=exp=1460646961~acl=%2Fstatic%2Fpdf%2F29%2Fart%25253A10.1007%25252Fs13202-015-0202-x.pdf%3ForiginUrl%3Dhttp%253A%252F%252Flink.springer.com%252Farticle%252F10.1007%252Fs13202-015-0202-x*~hmac=342abf10be838dd3f869a372bd753a11debc3698f1c525ec0a689af6b4ff2677 (accessed 10.03.16.).

Altindag, R., 2003. Correlation of specific energy with rock brittleness concepts on rock cutting. J. South Afr. Inst. Min. Metall. 103 (3), 163–172.

Anderson, W.G., 1986. Wettability literature survey: part 1. Rock-oil-brine interactions and the effects of core handling on wettability. J. Pet. Technol. (October), 1125–1144.

ASTM D2638, 2015. Standard test method for real density of calcined petroleum coke by helium pycnometer. Annual Book of Standards. American Society for Testing and Materials, West Conshohocken, PA.

Bakke, S., Øren, P.E., 1997. 3-D pore-scale modelling of sandstones and flow simulations in the pore networks. SPE J. 2, 136–149.

Beaton, A.P., Pawlowicz, J.G., Anderson, S.D.A., Rokosh, C.D., 2009. Rock Eval™ total organic carbon, adsorption isotherms and organic petrography of the Colorado Group: shale gas data release. Open file report no. ERCB/AGS 2008–11, Energy Resources Conservation Board, Calgary, AB.

Berg, R.R., 1970. Method for determining permeability from reservoir rock properties. In: Transactions of the GCAGS, vol. 20. Gulf Coast Association of Geological Societies, Houston, TX, p. 303.

Berg, R.R., 1986. Sandstone Reservoirs. Prentice-Hall, Pearson Education Group, Upper Saddle River, NJ.

Bergaya, F., Theng, B.K.G., Lagaly, G., 2011. Handbook of Clay Science. Elsevier, Amsterdam.

Blatt, H., Tracy, R.J., 1996. Petrology: Igneous, Sedimentary, and Metamorphic, second ed. W.H. Freeman and Company, Macmillan Publishers, New York, NY.

Bloch, S., Lander, R.H., Bonnell, L., 2002. Anomalously high porosity and permeability in deeply buried sandstone reservoirs: origin and predictability. AAPG Bull. 86 (2), 301–328.

Bloomfield, J.P., Williams, A.T., 1995. An empirical liquid permeability—gas permeability correlation for use in aquifer properties studies. Q. J. Eng. Geol. Hydrogeol. 28 (2), S143–S150.

Blunt, M.J., 2001. Flow in porous media—pore-network models and multiphase flow. Curr. Opin. Colloid Interface Sci. 6 (3), 197–207.

Blunt, M.J., Jackson, M.D., Piri, M., Valvatne, P.H., 2002. Detailed physics, predictive capabilities and macroscopic consequences for pore network models of multiphase flow. Adv. Water Resour. 25 (8), 1069–1089.

Bonnell, L.M., Lander, R.H., Sundhaug, C., 1998. Grain coatings and reservoir quality preservation: role of coating completeness, grain size and thermal history. In: Proceedings of AAPG Annual Convention, vol. 7, p. A81.

Bustin, R.M., Bustin, A.M.M., Cui, X., Ross, D.J.K., Murthy Pathi, V.S., 2008. Impact of shale properties on pore structure and storage characteristics. Paper no. SPE 119892, In: Proceedings of SPE Conference on Shale Gas Production, Fort Worth, TX, November 16–18. Society of Petroleum Engineers, Richardson, TX.

Byrnes, A.P., 1997. Reservoir characteristics of low-permeability sandstones in the rocky mountains. Mt. Geol. 34 (1), 39–51.

Byrnes, A.P., 2003. Aspects of permeability, capillary pressure, and relative permeability properties and distribution in low-permeability rocks important to evaluation, damage, and stimulation. In: Proceedings of the Rocky Mountain Association of Geologists—Petroleum Systems and Reservoirs of Southwest Wyoming Symposium, Denver, CO, p. 12.

Byrnes, P.A., Cluff, R.M., Webb, J.C., 2009. Analysis of critical permeability, capillary and electrical properties for Mesaverde tight gas sandstones from Western US basins. Technical report, US Department of Energy and the National Energy Technology Laboratory, Washington, DC.

Caruana, A., Dawe, R.A., 1996a. Effect of heterogeneities on miscible and immiscible flow processes in porous media. Trends Chem. Eng. 3, 185–203.

Caruana, A., Dawe, R.A., 1996b. Flow behavior in the presence of wettability heterogeneities. Transp. Porous Media 25, 217–233.

Chatzis, I., Dullien, F.A.L., 1977. Modelling pore structure by 2-D and 3-D networks with application to sandstones. J. Can. Pet. Technol. 16 (1), 97–108.

Cluff, R.M., Byrnes, A.P., 2010. Relative permeability in tight gas sandstone reservoirs—the permeability jail model. In: SPWLA 51st Annual Logging Symposium. Society of Petrophysicists and Well-Log Analysts, Houston, TX.

Cramer, D.D., 2008. Stimulating unconventional reservoirs: lessons learned, successful practices, areas for improvement. SPE paper no. 114172, In: Proceedings of Unconventional Gas Conference, Keystone, CO, February 10–12, 2008.

Cui, X., Bustin, A.M.M., Bustin, R.M., 2009. Measurements of gas permeability and diffusivity of tight reservoir rocks: different approaches and their applications. Geofluids 9, 208–223.

Dandekar, A.Y., 2013. Petroleum Reservoir Rock and Fluid Properties, second ed. CRC Press, Taylor & Francis Group, Boca Raton, FL.

Davis Jr., R., 1992. Depositional Systems: An Introduction to Sedimentology and Stratigraphy, second ed. Prentice Hall, New York, NY.

Dawe, R.A., 2004. Miscible displacement in heterogeneous porous media. In: Proceedings of Sixth Caribbean Congress of Fluid Dynamics, University of the West Indies, Augustine, Trinidad, January 22–23.

Dong, H., Blunt, M.J., 2009. Pore-network extraction from microcomputerized-tomography images. Phys. Rev. E 80 (3), 036307.

Dutton, S.P., 1993. Major low-permeability sandstone gas reservoirs in the continental United States. Report no. 211, Bureau of Economic Geology, University of Texas, Austin, TX.

Fatt, I., 1956. The network model of porous media. Soc. Pet. Eng. AIME 207, 144–181.

GAO, 2012. Information on shale resources, development, and environmental and public health risks. Report no. GAO-12-732. Report to Congressional Requesters, United States Government Accountability Office, Washington, DC. September.

Ghanizadeh, A., Gasparik, M., Amann-Hildenbrand, A., Gensterblum, Y., Krooss, B.M., 2014. Experimental study of fluid transport processes in the matrix system of the European organic-rich shales: I. Scandinavian alum shale. Mar. Pet. Geol. 51, 79–99.

Goodway, B., Perez, M., Varsek, J., Abaco, C., 2010. Seismic petrophysics and isotropic-anisotropic AVO methods for unconventional gas exploration. Lead. Edge 29 (12), 1500–1508.

Grattoni, C.A., Dawe, R.A., 1994. Pore structure influence on the electrical resistivity of saturated porous media. Paper no. SPE 27044, In: Proceedings of SPE Latin America/Caribbean Petroleum Engineering conference. Society of Petroleum Engineers, Richardson, TX, pp. 1247–1255.

Grattoni, C.A., Dawe, R.A., 2003. Consideration of wetting and spreading in three-phase flow in porous media. In: Lakatos, I. (Ed.), Progress in Mining and Oilfield Chemistry. In: Recent Advances in Enhanced Oil and Gas Recovery, vol. 5. Akad. Kiado, Budapest.

Gruse, W.A., Stevens, D.R., 1960. The Chemical Technology of Petroleum. McGraw-Hill, New York, NY.

Guha, R., Chowdhury, M., Singh, S., Herold, B., 2013. Application of geomechanics and rock property analysis for a tight oil reservoir development: a case study from Barmer Basin, India. In: Proceedings of 10th Biennial International Conference & Exposition—KOCHI 2013, November 23–25. Society of Petroleum Geophysicists (SPG) India, Kochi, Kerala.

Herwanger, J.V., Bottrill, A.D., Mildren, S.D., 2015. Uses and abuses of the brittleness index with applications to hydraulic stimulation. Paper no. URTeC 2172545, In: Proceedings of Unconventional Resources Technology Conference, San Antonio, TX. July 20–22. Society of Petroleum Engineers, Richardson, TX.

Hillier, S., 2003. Clay mineralogy. In: Middleton, G.V., Church, M.J., Coniglio, M., Hardie, L.A., Longstaffe, F.J. (Eds.), Encyclopedia of Sediments and Sedimentary Rocks. Kluwer Academic Publishers, Dordrecht, pp. 139–142.

Huang, H., Larter, S.R., Love, G.D., 2003. Analysis of wax hydrocarbons in petroleum source rocks from The Damintun Depression, Eastern China, using high temperature gas chromatography. Org. Geochem. 34, 1673–1687.

Hunt, J.M., 1996. Petroleum Geochemistry and Geology, second ed. W.H. Freeman and Co., New York, NY.

Hunter, C.D., Young, D.M., 1953. Relationship of natural gas occurrence and production in Eastern Kentucky (Big Sandy Gas Field) to joints and fractures in Devonian Bituminous Shales. AAPG Bull. 37 (2), 282–299.

Jarvie, D.M., Hill, R.J., Ruble, T.E., Pollastro, R.M., 2007. Unconventional shale-gas systems: the Mississippian Barnett Shale of North-Central Texas as one model for thermogenic shale-gas assessment. AAPG Bull. 91, 475–499.

Jarvie, D.M., Jarvie, B., Courson, D., Garza, A., Jarvie, J., Rocher, D., 2011. Geochemical tools for assessment of tight oil reservoirs. Article no. 90122/2011, In: AAPG Hedberg Conference, Austin, TX. December 5–10, 2010. American Association of Petroleum Geologists, Tulsa, OK.

Javadpour, F., Fisher, D., Unsworth, M., 2007. Nanoscale gas flow in shale gas sediments. J. Can. Pet. Technol. 46 (10), 55–61.

Jerauld, G.R., Salter, S.J., 1990. The effect of pore-structure on hysteresis in relative permeability and capillary pressure: pore-level modeling. Transp. Porous Media 5 (2), 103–151.

Jones, F.O., Owens, W.W., 1980. A laboratory study of low-permeability gas sands. J. Pet. Technol. 32 (9), 1631–1640.

Knudsen, M., 1909. Die Gesetze der Molukularstrommung und der inneren. Reibungsstrornung der Gase durch Rohren. Ann. der Phys. 28, 75–130.
Kovscek, A.R., 2002. Heavy and thermal oil recovery production mechanisms. Quarterly technical progress report. Reporting period: April 1 through June 30, 2002. DOE contract number: DE-FC26-00BC15311, July.
Ma, Y.Z., Moore, W.R., Gomez, E., Clark, W.J., Zhang, Y., 2016. Tight gas sandstone reservoirs, part 1: overview and lithofacies. In: Ma, Y.Z., Holditch, S., Royer, J.J. (Eds.), Unconventional Oil and Gas Resources Handbook. Elsevier, Amsterdam (Chapter 14).
Magoon, L.B., Dow, W.G., 1994. The petroleum system—from source to trap. Memoir no. 60, American Association of Petroleum Geologists, Tulsa, OK.
Marshak, S., 2012. Essentials of Geology, fourth ed. W.W. Norton & Company, New York, NY.
Maxwell, S., Norton, M., 2012. The impact of reservoir heterogeneity on hydraulic fracture geometry: integration of microseismic and seismic reservoir characterization. In: Proceedings of AAPG Annual Convention and Exhibition, Long Beach, California. April 22–25. (accessed 15.04.15.). http://www.searchanddiscovery.com/documents/2012/40993maxwell/ndx_maxwell.pdf.
Mehmani, A., Prodanović, M., 2014. The effect of microporosity on transport properties in porous media. Adv. Water Resour. 63, 104–119.
Moore, W.R., Ma, Y.Z., Pirie, I., Zhang, Y., 2016. Tight gas sandstone reservoirs, part 2: petrophysical analysis and reservoir modeling. In: Ma, Y.Z., Holditch, S., Royer, J.J. (Eds.), Unconventional Oil and Gas Resources Handbook. Elsevier, Amsterdam (Chapter 15).
Moslow, T.F., 1993. Evaluating tight gas reservoirs. Development Geology Reference Manual. American Association of Petroleum Geologists, Tulsa, OK.
Musser, B.J., Kilpatrick, P.K., 1998. Molecular characterization of wax isolated from a variety of crude oils. Energy Fuel 12 (4), 715–725.
Neasham, J.W., 1977a. Applications of scanning electron microscopy to the characterization of hydrocarbon-bearing rocks. Scan. Electron Microsc. 7, 101–108.
Neasham, J.W., 1977b. The morphology of dispersed clay in sandstone reservoirs and its effect on sandstone shaliness, pore space and fluid flow properties. Paper no. SPE 6858, In: Proceedings of 52nd Annual Fall Technical Conference and Exhibition of the Society of Petroleum Engineers. October 9–12. Society of Petroleum Engineers, Richardson, TX.
Newsham, K.E., Rushing, J.A., 2001. An integrated work-flow process to characterize unconventional gas resources. Part 1: geological assessment and petrophysical evaluation. Paper no. SPE 71351, In: Proceedings of SPE Annual Technical Conference and Exhibition. New Orleans, Louisiana. September 30–October 3. Society of Petroleum Engineers, Richardson, TX.
Okabe, H., Blunt, M.J., 2005. Pore space reconstruction using multiple point statistics. J. Pet. Sci. Eng. 46 (1), 121–137.
Øren, P.E., Bakke, S., 2002. Process based reconstruction of sandstones and prediction of transport properties. Transp. Porous Media 46 (2–3), 311–343.
Øren, P.E., Bakke, S., 2003. Reconstruction of Berea sandstone and pore-scale modelling of wettability effects. J. Pet. Sci. Eng. 39 (3), 177–199.
Øren, P.E., Bakke, S., Arntzen, O.J., 1998. Extending predictive capabilities to network models. SPE J. 3, 324–336.
Patzek, T.W., 2001. Verification of a complete pore network simulator of drainage and imbibition. SPE J. 6 (02), 144–156.
Pedersen, K.S., Christiansen, P.L., 2006. Phase Behavior of Petroleum Reservoir Fluids. CRC Press, Taylor & Francis Group, Boca Raton, FL.

Piri, M., Blunt, M.J., 2005a. Three-dimensional mixed-wet random pore-scale network modeling of two- and three-phase flow in porous media I. Model description. Phys. Rev. E 71 (2), 026301.

Piri, M., Blunt, M.J., 2005b. Three-dimensional mixed-wet random pore-scale network modeling of two- and three-phase flow in porous media II. Results. Phys. Rev. E 71 (2), 026302.

Pittman, E.D., Larese, R.E., Heald, M.T., 1992. Clay coats: occurrence and relevance to preservation of porosity in sandstones. In: Houseknecht, D.W., Pittman, E.D. (Eds.), In: Origin, Diagenesis, and Petrophysics of Clay Minerals, vol. 47. SEPM Society for Sedimentary Geology, Tulsa, OK. Special Publication.

Radlinski, A.P., Ioannidis, M.A., Hinde, A.L., Hainbuchner, M., Baron, M., Rauch, H., Kline, S.R., 2004. Angstrom-to-millimeter characterization of sedimentary rock microstructure. J. Colloidal Interface Sci. 274, 607–612.

Ramm, M., Bjørlykke, K., 1994. Porosity/depth trends in reservoir sandstones: assessing the quantitative effects of varying pore-pressure, temperature history and mineralogy, Norwegian Shelf Data. Clay Miner. 29, 475–490.

Rickman, R., Mullen, M.J., Petre, J.E., Grieser, W.V., 2008. A practical use of shale petrophysics for stimulation design optimization: all shale plays are not clones of the Barnett Shale. Paper no. SPE 115258, In: SPE Annual Technical Conference and Exhibition, Denver, Colorado. September 21–24. Society of Petroleum Engineers, Richardson, TX.

Rose, P.R., 1992. Chance of success and its use in petroleum exploration. In: Steinmetz, R. (Ed.), The Business of Petroleum Exploration. Treatise of Petroleum Geology, AAPG, Tulsa, OK, pp. 71–86.

Rushing, J.A., Newsham, K.E., 2001. Integrated work-flow process to characterize unconventional gas resources. Part 2: formation evaluation and reservoir modeling. Paper no. SPE 71352, In: Proceedings of SPE Annual Technical Conference and Exhibition. New Orleans, LA. September 30–October 3. Society of Petroleum Engineers, Richardson, TX.

Rushing, J.A., Newsham, K.E., Blasingame, T.A., 2008. Rock typing—keys to understanding productivity in tight gas sands. Paper no. SPE 114164, In: Proceedings of 2008 SPE Unconventional Reservoirs Conference, Keystone, CO. February 10–12. Society of Petroleum Engineers, Richardson, TX.

Shanley, K.W., Cluff, R.M., Robinson, J.W., 2004. Factors controlling prolific gas production from low-permeability sandstone reservoirs: implications for resource assessment, prospect development, and risk analysis. AAPG Bull. 88 (8), 1083–1121.

Soeder, D.J., Randolph, P.L., 1987. Porosity, permeability, and pore structure of the Tight Mesaverde Sandstone, Piceance Basin, Colorado. SPE Form. Eval. 2 (2), 129–136.

Solano, N.A., Clarkson, C.R., Krause, F.F., Aquino, S.D., Wiseman, A., 2013. On the characterization of unconventional oil reservoirs. CSEG Rec. 38 (4), 42–47. http://csegrecorder.com/articles/view/on-the-characterization-of-unconventional-oil-reservoirs (accessed 20.07.15.).

Sone, H., 2012. Mechanical properties of shale gas reservoir rocks and its relation to the in-situ stress variation observed in shale gas reservoirs. A dissertation submitted to the Department of Geophysics and the Committee on Graduate Studies of Stanford University in Partial Fulfillment of the Requirements for the Degree of Doctor of Philosophy. SRB volume 128, Stanford University, Stanford, CA.

Sone, H., Zoback, M.D., 2013a. Mechanical properties of shale-gas reservoir rocks—part 1: static and dynamic elastic properties and anisotropy. Geophysics 78 (5), D381–D392.

Sone, H., Zoback, M.D., 2013b. Mechanical properties of shale-gas reservoir rocks—part 2: ductile creep, brittle strength, and their relation to the elastic modulus. Geophysics 78 (5), D393–D402.

Speight, J.G., 2007. Natural Gas: A Basic Handbook. GPC Books, Gulf Publishing Company, Houston, TX.
Speight, J.G., 2013. The Chemistry and Technology of Coal, third ed. CRC Press, Taylor & Francis Group, Boca Raton, FL.
Speight, J.G., 2014. The Chemistry and Technology of Petroleum, fifth ed. CRC Press, Taylor & Francis Group, Boca Raton, FL.
Speight, J.G., 2015. Fouling in Refineries. Gulf Professional Publishing, Elsevier, Oxford.
Stonecipher, S.A., May, J.A., 1990. Facies controls on early diagenesis: Wilcox Group, Texas Gulf Coast. In: Meshri, I.D., Ortoleva, P.J. (Eds.), Prediction of Reservoir Quality Through Chemical Modeling. American Association of Petroleum Geologists, Tulsa, OK, pp. 25–44. AAPG Memoir No. 49.
Thomas, R.D., Ward, D.C., 1972. Effect of overburden pressure and water saturation on gas permeability of tight sandstone cores. J. Pet. Technol. 24 (2), 120–124.
Tucker, M.E., Wright, V.P., 1990. Carbonate Sedimentology. Blackwell Scientific Publications, Wiley-Blackwell, John Wiley & Sons Inc., Hoboken, NJ.
Valvatne, P.H., Blunt, M.J., 2003. Predictive pore-scale network modeling. Paper no. SPE 84550, In: Proceedings of SPE Annual Technical Conference and Exhibition. Society of Petroleum Engineers, Richardson, TX.
Walderhaug, O., 1996. Kinetic modelling of quartz cementation and porosity loss in deeply buried sandstone reservoirs. AAPG Bull. 80, 731–745.
Weimer, R.J., Sonnenberg, S.A., 1994. Low resistivity pays in J Sandstone, Deep Basin Center Accumulations, Denver Basin. In: Proceedings of AAPG Annual Convention, vol. 3, p. 280.
Wescott, W.A., 1983. Diagenesis of cotton valley sandstone (Upper Jurassic), East Texas: implications for tight gas formation pay recognition. AAPG Bull. 67, 1002–1013.
Wheaton, R., Zhang, H., 2000. Condensate banking dynamics in gas condensate fields: compositional changes and condensate accumulation around production wells. Paper no. 62930, In: Proceedings of SPE Annual Technical Conference and Exhibition, Dallas, TX. October 1–4. Society of Petroleum Engineers, Richardson, TX.
White, D.A., 1993. Geologic risking guide for prospects and plays. AAPG Bull. 77, 2048–2061.
Wilhelms, A., Larter, S.R., 1994a. Origin of tar mats in petroleum reservoirs. Part I: introduction and case studies. Mar. Pet. Geol. 11 (4), 418–441.
Wilhelms, A., Larter, S.R., 1994b. Origin of tar mats in petroleum reservoirs. Part II: formation mechanisms for tar mats. Mar. Pet. Geol. 11 (4), 442–456.
Wilson, M.D., 1982. Origins of clays controlling permeability and porosity in tight gas sands. J. Pet. Technol., 2871–2876.
Wilson, M.D., Pittman, E.D., 1977. Authigenic clays in sandstones: recognition and influence on reservoir properties and paleoenvironmental analysis. J. Sediment. Petrol. 47, 3–31.
Wollrab, V., Streibl, M., 1969. Earth waxes, peat, Montan wax, and other organic brown coal constituents. In: Eglinton, G., Murphy, M.T.J. (Eds.), Organic Geochemistry. Springer-Verlag, New York, NY, p. 576.
Wylie, G., 2012. Shale gas. World Petroleum Council Guide: Unconventional Gas. World Petroleum Council, London. http://www.world-petroleum.org/docs/docs/gasbook/unconventionalgaswpc2012.pdf. pp. 46–51 (accessed 15.03.15.).
Yao, C.Y., Holditch, S.A., 1996. Reservoir permeability estimation from time-lapse log data. Paper no. SPE-25513-PA, In: Proceedings of SPE Symposium on Formation Evaluation, vol. 11(1). Society of Petroleum Engineers, Richardson, TX, pp. 69–74.
Zhang, M., Zhang, J., 1999. Geochemical characteristics and origin of tar mats from the Yaha field in the Tarim Basin, China. Chin. J. Geochem. 18 (3), 250–257.
Ziarani, A.S., Aguilera, R., 2012. Knudsen's permeability correction for tight porous media. Transp. Porous Media 91 (1), 239–260.

CHAPTER THREE

Gas and Oil Resources in Tight Formations

1 INTRODUCTION

Unconventional resources such as natural gas and crude oil from tight formations have evolved into an important resource play in North America and can take on an important energy-source role in Europe and other countries. The shale formations and other tight formations (sandstone formations and carbonate formation) are composed of fine-grained particles with pores that are typically on the nanometer scale. These tight formations, such as the organic-rich shale formations (Table 3.1), have become an attractive target for development and recovery of the resource because within these formations a substantial resource of natural gas, natural gas liquids (NGL), gas condensate exists, and. These (typically) black organic shale formations that were formed millions of years ago (Table 3.2) obtain their black color from the organic matter that was deposited with the primordial form of the shale that developed into shale geologic time. The nonshale sedimentary *tight formation* consists of extraordinarily impermeable, hard rock (usually sandstone or carbonate rock) which exhibits relatively low permeability that can contain oil and gas (Chapter 2). In both cases, some of the organic material was transformed into natural gas and/or crude oil during geologic time (millions of years) and remained in (or was unable to migrate out of) the formation. In other instances, the formed crude oil and natural gas migrated into reservoir formations while the organic material that did not migrate remained in the original rock as kerogen. In the former case (the shale formation or the tight formation that presented migration of the fluids), the source rock and reservoir rock become one and the same.

Generally, a black color in shale formations and sedimentary rock formations almost always indicates the presence of organic material, typically as little as 1–2% (w/w) can impart a dark gray or black color to the shale or sedimentary rock. However, a note of caution: gray shale formations

Deep Shale Oil and Gas
http://dx.doi.org/10.1016/B978-0-12-803097-4.00003-6

Table 3.1 Shale Gas Formations in the United States and Canada

Formation	Geological Period	Location
Antrim shale	Late Devonian	Michigan Basin, Michigan
Baxter shale	Late Cretaceous	Vermillion Basin, Colorado, Wyoming
Barnett shale	Mississippian	Fort Worth and Permian basins, Texas
Bend shale	Pennsylvanian	Palo Duro Basin, Texas
Cane Creek shale	Pennsylvanian	Paradox Basin, Utah
Caney shale	Mississippian	Arkoma Basin, Oklahoma
Chattanooga shale	Late Devonian	Alabama, Arkansas, Kentucky, Tennessee
Chimney Rock shale	Pennsylvanian	Paradox Basin, Colorado, Utah
Cleveland shale	Devonian	Eastern Kentucky
Clinton shale	Early Silurian	Eastern Kentucky
Cody shale	Cretaceous	Oklahoma, Texas
Colorado shale	Cretaceous	Central Alberta, Saskatchewan
Conasauga shale	Middle Cambrian	Black Warrior Basin, Alabama
Dunkirk shale	Upper Devonian	Western New York
Duvernay shale	Late Devonian	West central Alberta
Eagle Ford shale	Late Cretaceous	Maverick Basin, Texas
Ellsworth shale	Late Devonian	Michigan Basin, Michigan
Excello shale	Pennsylvanian	Kansas, Oklahoma
Exshaw shale	Devonian-Mississippian	Alberta, northeast British Columbia
Fayetteville shale	Mississippian	Arkoma Basin, Arkansas
Fernie shale	Jurassic	West central Alberta, northeast British Columbia
Floyd/Neal shale	Late Mississippian	Black Warrior Basin, Alabama, Mississippi
Frederick Brook shale	Mississippian	New Brunswick, Nova Scotia
Gammon shale	Late Cretaceous	Williston Basin, Montana
Gordondale shale	Early Jurassic	Northeast British Columbia
Gothic shale	Pennsylvanian	Paradox Basin, Colorado, Utah
Green River shale	Eocene	Colorado, Utah
Haynesville/Bossier shale	Late Jurassic	Louisiana, east Texas
Horn River shale	Middle Devonian	Northeast British Columbia
Horton Bluff shale	Early Mississippian	Nova Scotia
Hovenweep shale	Pennsylvanian	Paradox Basin, Colorado, Utah
Huron shale	Devonian	East Kentucky, Ohio, Virginia, West Virginia
Klua/Evie shale	Middle Devonian	Northeast British Columbia
Lewis shale	Late Cretaceous	Colorado, New Mexico
Mancos shale	Cretaceous	San Juan Basin, New Mexico, Uinta Basin, Utah

Table 3.1 Shale Gas Formations in the United States and Canada—cont'd

Formation	Geological Period	Location
Manning Canyon shale	Mississippian	Central Utah
Marcellus shale	Devonian	New York, Ohio, Pennsylvania, West Virginia
McClure shale	Miocene	San Joaquin Basin, California
Monterey shale	Miocene	Santa Maria Basin, California
Montney-Doig shale	Triassic	Alberta, northeast British Columbia
Moorefield shale	Mississippian	Arkoma Basin, Arkansas
Mowry shale	Cretaceous	Bighorn and Powder River basins, Wyoming
Muskwa shale	Late Devonian	Northeast British Columbia
New Albany shale	Devonian-Mississippian	Illinois Basin, Illinois, Indiana
Niobrara shale	Late Cretaceous	Denver Basin, Colorado
Nordegg/Gordondale shale	Late Jurassic	Alberta, northeast British Columbia
Ohio shale	Devonian	East Kentucky, Ohio, West Virginia
Pearsall shale	Cretaceous	Maverick Basin, Texas
Percha shale	Devonian-Mississippian	West Texas
Pierre shale	Cretaceous	Raton Basin, Colorado
Poker Chip shale	Jurassic	West central Alberta, northeast British Columbia
Queenston shale	Ordovician	New York
Rhinestreet shale	Devonian	Appalachian Basin
Second White Speckled shale	Late Cretaceous	Southern Alberta
Sunbury shale	Mississippian	Appalachian Basin
Utica shale	Ordovician	New York, Ohio, Pennsylvania, West Virginia, Quebec
Wilrich/Buckinghorse/Garbutt/Moosebar shale	Early Cretaceous	West central Alberta, northeast British Columbia
Woodford shale	Devonian-Mississippian	Oklahoma, Texas

sometimes contain a small amount of organic matter but may also contain calcareous materials or simply clay minerals that result in a gray coloration of the formation. Generally, it can be assumed that the black color of the shale implies that the shale was formed from mud-like sediment which was deposited in an oxygen-deficient environment. Any oxygen that

Table 3.2 The Geologic Timescale[a]

Era	Period	Epoch	Millions of Years Ago
Cenozoic	Quaternary	Holocene	
		Pleistocene	.01
	Tertiary	Pliocene	2
		Miocene	13
		Oligocene	25
		Eocene	36
		Paleocene	58
Mesozoic	Cretaceous		65
	Jurassic		136
	Triassic		190
Paleozoic	Permian		225
	Carboniferous		280
	Devonian		345
	Silurian		405
	Ordovician		425
	Cambrian		500
Precambrian			600

[a]The numbers are approximate (±5%) due to variability of the data in literature sources; nevertheless, the numbers do give an indication of the extent of geologic time.

entered the environment quickly reacted with the decaying organic debris and, if the amount of oxygen was plentiful, the organic debris decayed to produce carbon dioxide and water.

$$[4CH] + 3O_2 \rightarrow 2CO_2 + 2H_2O$$

An oxygen-poor environment also provides the proper conditions for the formation of sulfide minerals such as pyrite (FeS_2), which is also found in many black shale formations. The Barnett shale formation, the Marcellus shale formation, the Haynesville shale formation, the Fayetteville shale formation, and other gas-producing rocks are all dark gray or black shale formations that yield natural gas. The Bakken shale formation of North Dakota and the Eagle Ford shale formation of Texas are examples of shale formations that yield crude oil.

Because of the presence of these shale formations, natural gas production and crude oil production in the United States have grown significantly in recent years as improvements in horizontal drilling and hydraulic fracturing technologies have made it commercially viable to recover gas trapped in tight formations, such as shale and coal. The United States is now the number one natural gas producer in the world and, together with Canada, accounts for more than 25% of global natural gas production (BP, 2015), which will play an ever-increasing role in this resource base and economic outlook of the United States. Furthermore, production of shale gas is projected to increase to 49% (v/v) of total gas production in the United States by 2035, up from 23% (v/v) in 2010, highlighting the significance of shale gas in the future energy mix in the United States (Bonakdarpour et al., 2011).

Determining the amount of gas in place in unconventional reservoirs is complex because of the heterogeneous structure of the reservoirs as well as the potential conflicts in the definitions of resources and reserves (Chapter 1). Moreover, assessing productivity is dependent on a detailed investigation of the characteristics of the reservoir, which can vary not only between reservoirs but also horizontally and vertically in any given reservoir. As an approximate estimate, ultimately recoverable unconventional gas resources, excluding methane hydrates, are estimated close to 12.0 quadrillion cubic feet (12.0×10^{15} ft^3; 340 trillion cubic meters, 340×10^{12} m^3). Of this amount, 24% (v/v) can be found in countries of the Organization for Economic Cooperation and Development (OECD) Americas, 28% (v/v) in Asia Pacific countries, 14% (v/v) in Latin American countries, and 13% (v/v) in Eastern Europe countries and Eurasian countries with smaller shares in African countries, the OECD European countries, and the Middle Eastern countries (IEA, 2012, 2013). With remaining recoverable resources of conventional natural gas at 16.3 quadrillion cubic feet (16.3×10^{15} ft^3) and unconventional gas at 12.0 quadrillion cubic feet, together they can sustain more than 200 years of production at current rates. The future potential for unconventional gas production remains contentious, with questions over the size and recoverability of the physical resource being central to the debate. While interest has focused upon shale gas in recent years, there is also considerable potential for coal bed methane (coalbed methane) and tight gas to contribute to global gas supply. However, despite recent advances, there remains considerable uncertainty over the size of recoverable resources for each type of gas, at both the regional and global level. This even applies to the United States, where the development of shale natural gas resources and shale crude oil resources is relatively advanced.

However, estimating the technically recoverable oil and natural gas resource base in the United States is an evolving process (Chapter 1). For natural gas and crude oil, the evolution of resource estimates is likely to continue for some time. The *true size* (i.e, *estimated true size*) of the technically recoverable natural gas and crude oil resource base in the United States becomes evident only as the gas and oil producers drill into geologic deposits in which natural gas and crude reside and the potential for production of these resources on a commercial basis becomes real. As producers find plays that contain more crude oil or natural gas than expected, resource estimates have to be adjusted to reflect the latest information. Thus, estimates of the technically recoverable resource base will be continuously adjusted as knowledge of the resource base and future technologies and management practices improve. Consequently, it is advisable to recognize that the resource estimates in any current (recently published) report may not be the actual estimates by the time the report is published since, in the interim time period between data acquisitions and publication, the estimates change as more wells are drilled and completed, technologies evolve, and the long-term performance of the wells to produce natural gas and crude oil becomes better established.

Furthermore, not all tight shale formations and tight nonshale formations are newly discovered. In many parts of the natural gas and crude oil producing world, a reexamination of old drilling records is opening up opportunities for the *rediscovery* of natural gas and crude oil resources that were rejected as worthy of development at an earlier time because of lower resources prices and/or more limited recovery technology (i.e., the technology is inadequate to the task of efficient recovery). This is especially true with natural gas and crude oil, which in many instances was a *stranded* resource having little or no market value. Also until quite recently with improvements in recovery technology, natural gas in tight sand or shale reservoirs could not be produced at commercial rates. There are more than 50 shale gas resource for formations in the United States and Canada, some of which are older (known) shale formation and others are more recent and new (Table 3.1).

Shale gas reserves in the United States are considerable and not concentrated in any particular area (USGS, 2014). The estimates place 482 trillion cubic feet (482×10^{12} ft^3) of technically recoverable shale gas resources in the rower 48 states with the largest portions in the Northeastern states (63%, v/v), Gulf Coast states (13%, v/v), and Southwestern states (10%, v/v), respectively. The largest shale gas resources (*plays*) are the Marcellus shale (141 trillion cubic feet, 141×10^{12} ft^3), Haynesville shale (74.7 trillion cubic feet, 74.7×10^{12} ft^3), and Barnett shale (43.4 trillion cubic feet, 43.4×10^{12} ft^3). Activity in new shale resources has increased shale gas production in the United States from 388 billion cubic feet (388×10^9 ft^3) in 2000 to 4944 billion cubic feet (4944×10^9 ft^3) in 2010 (US EIA, 2011a). This production potential has the ability to change the nature of the energy mix in the United States, and the natural gas resource base could support supply for five or more decades at current or greatly expanded levels of use (NPC, 2011). However, in addition to these data and possibly because of the uncertainty of calculating reserve estimates (Chapter 1), there are indications from numbers recently released that the estimated shale gas resource for the continental United States doubled from 2010 to 2011 to approximately 862 trillion cubic feet (862×10^{12} ft^3) and from 2006 to 2010 annual shale gas production in the United States almost quintupled to 4.8 trillion cubic feet (from 1.0×10^{12} ft^3 to 4.8×10^{12} ft^3) (EIUT, 2012).

Finally, it must always be remembered that each of the shale basins is different and each has a unique set of exploration criteria and operational challenges that need consideration in any development program. Because of these differences, the development of shale formation resources and tight formation resources in each of these areas poses potential challenges not only to the resource developers but also to the surrounding communities and ecosystems. For example, the Antrim and New Albany shale formations are shallower shale formations which produce significant volumes of formation water unlike most of the other gas shale formations. This is water that cannot be ignored for recycle use as well as purification so that contamination of aquifers does not occur (Chapter 9).

At this stage, a comment on each of the tight formations in the various countries is warranted, especially the United States and Canada where the majority of the pioneering work has been done.

2 UNITED STATES

Conventional resources of natural gas (or for that matter, any fossil fuel) exist in discrete, well-defined subsurface accumulations (reservoirs), with permeability values greater than a specified lower limit. Such conventional gas resources can usually be developed using vertical wells and generally yield the high recovery factors.

Briefly, permeability is a measure of the ability of a porous medium, such as that found in a hydrocarbon reservoir, to transmit fluids, such as gas, oil, or water, in response to a pressure differential across the medium. In petroleum engineering, permeability is usually measured in units of milliDarcys (mD).

By contrast, unconventional resources are found in accumulations where permeability is low (<0.1 mD). Such accumulations include *tight* sandstone formations, coal beds (coalbed methane), and shale formations (Chapter 2). Tight sandstone formations can be sandstones originally formed as beaches or estuaries with relatively little organic matter but later sealed with cap rock, and thus becoming low permeability traps, which captured hydrocarbons seeping upward from lower source rocks. Also, unconventional resource accumulations tend to be distributed over a larger area than conventional accumulations and usually require advanced technology such as horizontal wells or artificial stimulation in order to be economically productive; recovery factors are much lower—typically of the order of 15–30% of the gas initially in place. While the horizontal wells and multistage fracture completions are more expensive than nonfractured vertical wells, they generate a higher initial rate of oil production than even the best vertical wells. In addition, eight or more horizontal wells can be drilled from the same surface location which simplifies tie-ins and minimizes the cost and environmental impact of having to build more roads, pipelines, and well pads.

The mature, organic-rich shale formations that serve as sources for natural gas and crude oil predominantly in the United States and Canada (Figs. 3.1 and 3.2) and which have received considerable interest, have become an attractive target because they represent a substantial resource of natural gas and crude oil, and are distributed throughout the 48 contiguous

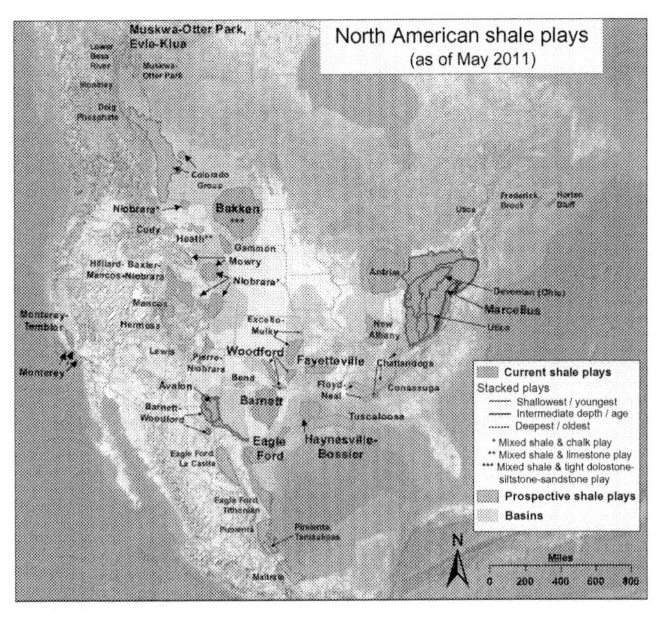

Fig. 3.1 Shale gas resources (shale gas plays) in the contiguous United States, Canada, and Mexico. *(Adapted from US EIA, 2012. U.S. Crude Oil, Natural Gas, and Natural Gas Liquids Proved Reserves, 2010. Energy Information Administration, United States Department of Energy, Washington, DC, Aug).*

Fig. 3.2 Major basins with the potential for tight oil development. *(Adapted from US EIA, 2012. U.S. Crude Oil, Natural Gas, and Natural Gas Liquids Proved Reserves, 2010. Energy Information Administration, United States Department of Energy, Washington, DC, Aug).*

United States as well as Western Canada (Hill and Nelson, 2000; NEB, 2009; IEA, 2012).

Due to the unique nature of shale, every basin, play, well, and pay zone may require a unique treatment. Briefly comparing the characteristics of some of the current hottest plays can help illustrate the impact of these differences throughout development. It is necessary to study and understand key reservoir parameters for gas shale deposits, and these parameters include: (1) thermal maturity, (2) reservoir thickness, (3) total organic carbon (TOC) content, (4) adsorbed gas fraction, (5) free gas within the pores and fractures, and (6) permeability (Chapter 4). The first two parameters are routinely measured. Thermal maturity is commonly measured in core analysis and reservoir thickness is routinely measured with logs. The calculation of the final four parameters requires a novel approach.

Almost all (96%, v/v) of the shale natural gas in proved reserves in 2010 came from the six largest shale plays in the United States (US EIA, 2012). The Barnett again ranked as the largest shale gas play in the United States, significantly higher increases over 2009 proved reserves were registered by the Haynesville/Bossier (which more than doubled 2009 volumes) and the Marcellus (which nearly tripled). Among these six shale plays, the only decline from 2009 volumes was in the Antrim of northern Michigan—a mature, shallow biogenic shale gas play discovered in 1986 that is no longer being developed at the same pace as the other leading shale resources.

The first Barnett shale gas production, by Mitchell Energy and Development Corp, took place in the Fort Worth Basin in 1981. Until Barnett shale successes, it was believed that natural fractures had to be present in gas shale. A low-permeability gas-shale play is presently viewed as a technological play. Advances in microseismic fracture mapping, 3-D seismic, horizontal drilling, fracture stimulation, and multiple fracturing stages have all contributed to successful gas shale wells. By the early part of the 21st century the main gas resources to that point had been: Antrim shale in the northern Michigan Basin; Barnett shale in the Fort Worth Basin, Texas; Lewis shale in the San Juan Basin; New Albany shale in the Illinois Basin; and the Ohio shale in the Appalachian Basin (GAO, 2012). More recent development of gas shale resources includes (alphabetically): (1) the Fayetteville shale in Arkansas, (2) the Eagle Ford shale in Texas, (3) the Haynesville shale in Louisiana, (4) the Marcellus shale in the Appalachian Basin, (5) Utica shale in New York and Quebec, Canada, and (6) the Woodford shale in Oklahoma.

However, there are other gas shale resources that are of increasing importance to the United States energy balance and economics. These must not be ignored and the predominant natural shale gas and crude oil resources in the United States and the formations are formations presented below are not listed in any particular order of preference or importance, other than *alphabetical order* for ease of location.

2.1 Antrim Shale

The Antrim shale, one of the largest unconventional natural gas reserves in North America, is a finely laminated, pyritic, organic matter-rich, thermally immature Devonian black shale located in the Michigan Basin in the northern part of Michigan State and is part of an extensive, organic-rich shale depositional system that covered large areas of the ancestral North American continent in the Middle-to-Late Devonian. The intracratonic Michigan basin was one of several deposition centers situated along the Eastern Interior Seaway. The basin has been filled with more than 17,000 ft of sediment, 900 ft of which comprises the Antrim shale and associated Devonian-Mississippian rocks. The base of the Antrim, near the center of the modern structural basin, is approximately 2400 ft below sea level (Braithwaite, 2009; US EIA, 2011a). The Antrim shale is a black, organic rich bituminous shale which is divided into four members, from base to top: Norwood, Paxton, Lachine, and upper members. The play and ranges from 600 to 2200 ft deep and is 70 to 120 ft think The total area of the Antrim shale play is approximately 12,000 mile^2, which includes the developed and undeveloped area of the play. The shale gas play has an average expected ultimate recovery of approximately 19.9 Tcf of technically recoverable gas. The upper members are overlaid by the Greenish-grey Ellsworth shale.

The stratigraphy of the Antrim shale stratigraphy is relatively straightforward and wells are typically completed in the Lachine and Norwood members of the lower Antrim, whose aggregate thickness approaches 160 ft. The TOC content of the Lachine and Norwood ranges from 0.5% to 24% (w/w). These black shale formations are silica rich (20–41% microcrystalline quartz and wind-blown silt) and contain abundant dolomite and limestone concretions and carbonate, sulfide, and sulfate cements. The remaining lower Antrim unit, the Paxton, is a mixture of lime mudstone and gray shale lithology (Martini et al., 1998) containing 0.3–8% (w/w) TOC and 7–30% (w/w) silica. Correlation of the fossil alga *Foerstia* has established time equivalence among the upper part of the Antrim shale, the Huron Member of the Ohio shale of the Appalachian basin, and the Clegg Creek Member of the New Albany shale of the Illinois basin (Roen, 1993).

Typical depths for the entire Antrim shale unit range from 500 to 2300 ft, and the areal extent is roughly approximately 30,000 mile^2 (Gutschick and Sandberg, 1991; Braithwaite, 2009; US EIA, 2011a). The entire area is overlain by Devonian and Mississippian sediments and hundreds of feet of glacial till. The Antrim mineralogy shows the shale to be laminated with very fine grains. The composition consists mainly of illite and quartz with small quantities of organic material and pyrite.

The Antrim shale has an organic matter up to 20% (w/w) and is mainly made up of algal material. The vitrinite reflectance is in the range of 0.4–0.6, indicating that the shale is thermally immature. The shale is also shallow and there is a high concentration of methane in the composition, which would lead one to assume the gas is of a microbial origin, but $\delta^{13}C$ values indicate a more thermogenic origin (Martini et al., 1996).

For shallow wells in the Antrim shale, the gas is of microbial origin. Deeper wells have a mix of thermogenic gas and microbial gas. For gas compositions with $C1/(C2+C3) < 100$ the gas origin is thermogenic, and this occurs for the gas present in the Niagaran formation which under lays the Antrim shale. Since the Antrim has so many natural fractures, it is reasonable to assume there is migration of gas from the Niagaran formation in to the Antrim shale.

The Antrim shale has two main ways of storing gas: absorption and free gas in the pore volume. The lower Norwood member has a higher adsorption capacity (approximately 115 ft^3 per ton) than the Lachine member (approximately 85 ft^3 per ton) (Kuuskraa et al., 1992). This is an important factor to consider when designing a fracture treatment because it would be more beneficial to have more of the proppant in the zone with the highest gas content. The free gas in the pore space can account for up to 10% of the total gas in place, but it is still not clear on how dependant the free gas is on the water in place. The very low permeability of the matrix could make it very difficult if not impossible to remove a significant portion of the free gas.

Two dominant sets of natural fractures have been identified in the northern producing trend, one oriented toward the northwest and the other to the northeast and both exhibiting subvertical to vertical inclinations. These fractures, generally uncemented or lined by thin coatings of calcite (Holst and Foote, 1981; Martini et al., 1998), have been mapped for several meters in the vertical direction and tens of meters horizontally in surface exposures. Attempts to establish production in the Antrim outside this trend have commonly encountered organic, gas-rich shale but minimal natural fracturing and, hence, permeability (Hill and Nelson, 2000).

Thus, the Antrim shale is highly fractured for a shale reservoir. Fracture spacing can be as close as 1–2 ft, compared to 10–20 ft for the Barnett shale. These fractures can create permeability-thicknesses in the range of 50–5000 md/ft, which increases gas production. But, it also helps water flow, and thus most wells produce large amounts of water which must be disposed (Kuuskraa et al., 1992).

2.2 Avalon and Bone Springs Shale

The Avalon and Bone Springs shale oil play is located in the Permian Basin in Southeast New Mexico and West Texas and has a reported depth from

6000 to 13,000 ft and a thickness ranging from 900 to 1700 ft. The area of the play has been estimated to be approximately 1313 mile^2 with approximately 1.58 billion barrels (1.58×10^9 bbls) of technically recoverable oil.

2.3 Bakken Shale

The Bakken shale oil play is located within the Williston Basin in Montana and North Dakota and has an area on the order of approximately 6522 mile^2 in the United States. This discovery, first made in 1951, was not considered significant until 2004 when the employment of horizontal drilling and hydraulic fracturing (fracking) in combination demonstrated that this "shale oil" could be efficiently and economically produced.

The play has seen a similar growth rate to the Barnett and is another technical play in which the development of this unconventional resource has benefited from the technological advances in horizontal wells and hydraulic fracturing (Cohen, 2008; Cox et al., 2008; Braithwaite, 2009). In Apr. 2008, the United States Geological Survey (USGS) released an updated assessment of the undiscovered technically recoverable reserves for this shale play estimating there are 3.65 billion barrels (3.65×10^9 bbls) of oil, 1.85 trillion cubic feet (1.85×10^{12} ft^3) of associated natural gas, and 148 million barrels (148×10^6 bbls) of NGLs in the play (USGS, 2008; US EIA, 2011a).

The Bakken shale formation differs from other shale plays in that it is an oil reservoir, a dolomite layered between two shale formations, with depths ranging from around 8000–10,000 ft from which oil, gas, and NGLs are produced. The formation contains approximately 3.59 billion barrels (3.59×10^9 bbls) of technically recoverable oil. The shale formation ranges from 4500 to 7500 ft deep with a mean of 6000 ft and an average thickness of 22 ft.

Each succeeding member of the Bakken formation—lower shale, middle sandstone, and upper shale member—is geographically larger than the one below. Both the upper and lower shale formations, which are the petroleum source rocks, present fairly consistent lithology, while the middle sandstone member varies in thickness, lithology, and petrophysical properties.

The Bakken shale formation is not as naturally fractured as the Barnett shale formation and, therefore, requires more traditional fracture geometry with both longitudinal and transverse fractures. Diversion methods are used throughout hydraulic fracture treatments, which primarily use gelled water fracture fluids, although there is a growing trend toward the use of an intermediate strength proppant. Recently, the Bakken gas shale has seen an increase in activity, and the trend is toward longer laterals—up to 10,000 ft for single laterals in some cases. In addition, there is also a trend to drill below the lower Bakken shale and fracture upward.

Interestingly, this shale formation was estimated by industry experts to contain up to 24 billion barrels (24×10^9 bbls) of total oil. Currently, more than 200 drilling rigs are using these techniques developing new wells at the rate of one each eight weeks or less. Production in the Bakken is on the order of 450,000 + barrels of crude oil per day and estimate are for a steady increase in production for the next several years. In terms of transportation, rail is supplementing the pipeline capacity to transport the crude to refining centers. A key driver for the refinery development in the area is the rapidly growing availability of the crude oil, which has less than 0.005% (w/w) sulfur and more on the order of 0.0015% (w/w) sulfur. This crude oil is also light (low in density) at API specific gravity of approximately 42, compared to an API gravity of 39.6 for West Texas Intermediate. The laboratory distillation of Bakken crude shows high yields of naphtha and mid-distillates (kerosene and diesel) and low residual fuel oil and asphaltene constituents.

2.4 Barnett Shale

The Barnett shale gas play is located within the Fort Worth and Permian Basins in Texas. The Barnett shale is divided into two sections: the "Core/Tier I" and the Undeveloped. The Core/Tier I section corresponds to the areas of the Barnett shale that are currently under development. It is primarily located in the Parker, Wise, Johnson, and other neighboring counties.

The wedge-shaped Fort Worth basin covers approximately 15,000 mile^2 in North-Central Texas and is centered along the north-south direction, deepening to the north and outcropping at the Liano uplift in Liano County (Bowker, 2007a,b; Jarvine et al., 2007). The Cambrian Riley and Hickory formations are overlaid by the Viola-Simpson and Ellenburger groups. The Viola-Simpson limestone group is found in Tarrant and Parker counties and acts as a barrier between the Barnett and the Ellenburger formation. The Ellenburger formation is a very porous aquifer (Zuber et al., 2002) that if fractured will produce copious amounts of highly saline water, effectively shutting down a well with water disposal cost.

Geochemical and reservoir parameters for the Barnett shale in the Fort Worth basin differ markedly from those of other gas-productive shale formations, particularly with respect to gas in place. For example, Barnett shale gas is thermogenic in origin and hydrocarbon generation began in the Late Paleozoic, continued through the Mesozoic, and ceased with uplift and cooling during the Cretaceous (Jarvie et al., 2001, 2007). In addition, organic matter in the Barnett shale formation has generated liquid hydrocarbons and Barnett-sourced oils in other formations, ranging from Ordovician to Pennsylvanian in age, in the western Fort Worth basin (Jarvie et al., 2001, 2007)—cracking of this oil may have contributed to the gas-in-place resource.

The Mississippian-age Barnett shale overlies the Viola-Simpson group. The Barnett shale varies in thickness from 150 to 800 ft and is the most productive gas shale in Texas. The permeability ranges from 7 to 50 nanoDarcys and the porosity from 4% to 6% (Montgomery et al., 2005; Cipolla et al., 2010). In addition, well performance of the Barnett shale changes significantly with changing produced fluid type, depth, and formation thickness (Hale and William, 2010) and on the type of completion method implemented and the large hydraulic fracture treatments (Ezisi et al., 2012).

The three most important production-related structures in the basin include both major and minor faulting, fracturing, and karst-related collapse features (Frantz et al., 2005). Fracturing is important to gas production because it provides a conduit for gas to flow from the pores to the wellbore, and it also increases exposure of the well to the formation. The Barnett shale formation exhibits complex fracture geometry which often creates difficulty in estimating fracture length and exposure to the formation due to the complex geometry. The fracturing is believed to be caused by the cracking of oil into gas. This cracking can cause a 10-fold increase in the hydrocarbon volume, increasing the pressure until the formation breaks. The precipitation of calcium carbonate in the fractures can cut down on the conductivity of the fractures. This precipitation is hard to detect on logs and can cause a well location that appears to be good on seismic into an unproductive well. This precipitation is also hard to treat with acidization due to the long distances the acid is required to travel before making a noticeable impact on production.

Change in gas content with pressure occurs in the Barnett shale with a typical reservoir pressure in the range of 3000–4000 psi (Frantz et al., 2005). In low permeability formations, pseudo radial flow can take over 100 years to be established. Thus, most gas flow in the reservoir is a linear flow from the near fracture area toward the nearest fracture face. Faulting and karst-related collapse features are important mainly in the Ellenburger formation.

In addition to drilling longer laterals, current trends in the Barnett are toward bigger hydraulic fracturing projects and more stages. Infills are being drilled and testing of spacing is down to 10 acres, while refracturing of the first horizontal wells from 2003 and 2004 has commenced; both infills and refracturing methods are expected to improve the *estimated ultimate recovery* from 11–18% (v/v). In addition, pad drilling (Chapter 5), especially in urban areas, and recycling of water (Chapter 5) are growing trends in the Barnett shale, as elsewhere.

The total area for the Barnett formation, as estimated by USGS, is 6458 mile^2. This area is subdivided into two sections—(1) the Greater Newark East Frac-Barrier Continuous Barnett shale Gas (1555 mile^2) and (2) the Extended Continuous Barnett shale Gas (4903 mile^2). As the development of the Barnett extended beyond the Newark East field, the active section of the Barnett was also extended. The remaining area is considered to be undeveloped section of the Barnett. The Barnett shale gas play, including the active and undeveloped areas, is approximately 43.37 Tcf of technically recoverable gas.

The Barnett-Woodford shale gas play is located in the Permian Basin in West Texas and has an area of approximately 2691 mile^2 and ranges from 5100 to 15,300 ft deep and 4 to 800 ft thick with an estimated recovery of approximately 32.2 Tcf of technically recoverable gas.

2.5 Baxter Shale

The Baxter shale is stratigraphically equivalent to the Mancos, Cody, Steele, Hilliard, and Niobrara/Pierre formations (Braithwaite, 2009; Mauro et al., 2010; US EIA, 2011a) and was deposited in hundreds of feet of water in the Western Interior Seaway from about 90–80 million years ago (Coniacean to lower Campanian) and consists of about 2500 ft of dominantly siliceous, illitic, and calcareous shale that contains regionally correlative coarsening-upward sequences of quartz-and carbonate-rich siltstones several tens of feet thick. The TOC content ranges from 0.5% to 2.5% in the shale and from 0.25% to 0.75% in the siltstones. Measured porosities in both the shale and siltstones typically range from 3% to 6% with matrix permeability of 100–1500 nanoDarcys.

Gas production has been established from the Baxter shale in 22 vertical wells and 3 horizontal wells in the Vermillion Basin of northwestern Colorado and adjacent Wyoming. Production comes mainly from the silt-rich intervals as determined by production logs. The productive area in the Baxter shale has vitrinite reflectance values approaching 2% and is in the dry gas generation window.

The resource area is defined by numerous wells with gas shows and overpressuring in the Baxter shale with gradients ranging from 0.6 to 0.8 psi/ft at depths greater than 10,000 ft.

A challenge within this reservoir is the ability to economically access this large unconventional gas accumulation. This is not a classic 100- to 300-foot-thick organic-rich shale gas reservoir. Instead it is a very large hydrocarbon resource stored in 2500 ft of shale with interbedded siltstone intervals. 3-D seismic data have proved useful in helping define potential fracture networks in the Baxter shale that can be targeted with horizontal wells.

The Hilliard-Baxter-Mancos shale gas play is located in the Greater Green River Basin in Wyoming and Colorado and the depth for the shale ranges from 10,000 to 19,500 mile^2 and is 2850 to 3300 ft thick. The total active area for the play is 16,416 mile^2 with approximately 3.77 Tcf of technically recoverable gas.

2.6 Big Sandy

The Devonian Big Sandy shale gas play includes the Huron, Cleveland, and Rhinestreet formations located within the Appalachian Basin in Kentucky, Virginia, and West Virginia.

The USGS has estimated a total area for the Big Sandy shale play as 10,669 mile^2 (6,828,000 acres). The shale play has an average expected ultimate recovery of approximately 7.4 Tcf of technically recoverable gas. Big Sandy has a total active area of approximately 8675 mile^2 and an undeveloped area of 1994 mile^2 with a well spacing of 80 acres per well. The formation ranges from 1600 to 6000 ft deep and has a thickness of 50 to 300 ft.

2.7 Caney Shale

The Caney shale (Arkoma Basin, Oklahoma) is the stratigraphic equivalent of the Barnett shale in the Fort Worth Basin (Boardman and Puckette, 2006; Andrews, 2007; Jacobi et al., 2009; Andrews, 2012). The formation has become a gas producer since the large success of the Barnett shale formation.

The Caney shale, Chesterian age, was deposited in the Oklahoma part of the Arkoma Basin one of a series of foreland basins that formed progressively westward along the Ouachita Fold Belt from the Black Warrior Basin in Mississippi to basins in southwest Texas. The Arkoma Basin in Oklahoma is in the Southeast corner of the state north and northwest of the Ouachita Mountains. The formation dips southward from a depth of 3000 ft in northern McIntosh County, Oklahoma to 12,000 ft north of the Choctaw thrust. The Caney formation thickens toward the southeast from 90′ at its northwest edge to 220′ along the Choctaw fault in the south. It can be subdivided into six intervals based on characteristics of the density and resistivity logs.

Reported average TOC values the Caney formation range from 5% to 8% (w/w), which show a linear correlation with density. Mud log gas shows have a strong correlation with desorbed gas values that range from 120–150 cubic feet per ton of shale. Estimates of gas in place for the Caney range from 30–40 billion cubic feet ($30–40 \times 10^9$ ft^3).

2.8 Chattanooga Shale

The Chattanooga shale (Black Warrior Basin) and has been considered as a rich shale formation (Rheams and Neathery, 1988). The shale formation sits within the thermogenic gas window in much of the Black Warrior Basin (Carroll et al., 1995) and may thus contain significant prospects for natural gas. The Chattanooga disconformity overlies Ordovician through Devonian strata, and the time value of the disconformity increases northward (Thomas, 1988). The Chattanooga is overlain sharply by the Lower Mississippian Maury shale, which is commonly thinner than 2 ft, and the Maury is in turn overlain by the micritic Fort Payne Chert. The Chattanooga shale in Alabama was apparently deposited in dysoxic to anoxic subtidal environments and can be considered as a cratonic extension of the Acadian foreland basin (Ettensohn, 1985).

The thickness of the Chattanooga varies significantly within the Black Warrior Basin. The shale is thinner than 10 ft and is locally absent in much of Lamar, Fayette, and Pickens Counties, which is the principal area of conventional oil and gas production in the Black Warrior Basin. For this reason, the Chattanooga has not been considered to be the principal source rock for the conventional oil and gas reservoirs in this area. The shale is thicker than 30 ft in a belt that extends northwestward from Blount County into Franklin and Colbert Counties. A prominent deposition center is developed along the southwestern basin margin in Tuscaloosa and Greene Counties. Here, the shale is consistently thicker than 30 ft and is locally thicker than 90 ft.

The Chattanooga shale is in some respects analogous to the Barnett shale of the Fort Worth Basin in that it is an organic-rich black shale bounded by thick, mechanically stiff limestone units that may help confine induced hydrofractures within the shale (Hill and Jarvie, 2007; Gale et al., 2007).

2.9 Conasauga Shale

The Conasauga shale gas formation (Conesauga shale gas formation, Alabama) continues to be developed primarily in northeast Alabama (US EIA, 2011a). With the exception of one well in Etowah County and one well in Cullman County, all of the development has been in St. Clair County. Etowah and St. Clair Counties are located northeast of Birmingham in the Valley and Ridge Province of Alabama. Cullman County is north of Birmingham in the Cumberland Plateau Province.

This shale formation it represents the first commercial gas production from shale in Alabama, but because it is geologically the oldest and most structurally complex shale formation from which gas production has been established. The Conasauga differs from other gas shale formations in several respects. The productive lithology is thinly interbedded shale and micritic limestone that can contain more than 3% TOC.

The Conasauga shale is of Middle Cambrian age and can be characterized as a shoaling-upward succession in which shale passes vertically into a broad array of inner ramp carbonate facies. The shale was deposited on the outer ramp, and the shale is thickest in basement grabens that formed during late Precambrian to Cambrian Iapetan rifting (Thomas et al., 2000).

The shale facies of the Conasauga is part of the weak litho-tectonic unit that hosts the basal detachment of the Appalachian Thrust Belt in Alabama (Thomas, 2001; Thomas and Bayona, 2005). The shale has been thickened tectonically into antiformal stacks that have been interpreted as giant shale duplexes, or mushwads (Thomas, 2001). In places, the shale is thicker than 8000 ft, and the shale is complexly folded and faulted at outcrop scale.

Surface mapping and seismic exploration reveal that at least three Conasauga antiforms are preserved in the Alabama Appalachians. Exploration has focused primarily on the southeastern portion of the Gadsden antiform, which is in St. Clair and Etowah Counties. The Palmerdale and Bessemer antiforms constitute the core of the Birmingham anticlinorium. The Palmerdale and Bessemer structures are overlain by a thin roof of brittle Cambrian-Ordovician carbonate rocks, and Conasauga shale facies are exposed locally. The Palmerdale structure is in the heart of the Birmingham metropolitan area and thus may be difficult to develop, whereas the southwestern part of the Bessemer structure is in rural areas and may be a more attractive exploration target. Additional thick shale bodies may be concealed below the shallow Rome thrust sheet in Cherokee and northeastern Etowah Counties and perhaps in adjacent parts of Georgia (Mittenthal and Harry, 2004).

2.10 Eagle Ford Shale

The Eagle Ford shale (Eagleford shale) (discovered in 2008) is a sedimentary rock formation from the Late Cretaceous age underlying much of South Texas which covers 3000 square miles and consists of an organic-rich marine shale that also has been found to appear in outcrops (Braithwaite, 2009; US EIA, 2011a). The Eagle Ford is a geological formation directly beneath the Austin Chalk and is considered to be the source rock of hydrocarbons that are contained in the Austin Chalk above the play.

The play is located within the Texas Maverick Basin and contains a high liquid component and takes its name from the town of Eagle Ford Texas where the shale outcrops at the surface in clay form. The Eagle Ford is an extremely active shale play with over 100 active rigs in operation. This hydrocarbon-producing formation rich in oil and natural gas extends from the Texas-Mexico border in Webb and Maverick counties and extends 400 miles toward East Texas. The formation is 50 miles wide and an average of 250 ft thick at a depth between 4000 and 12,000 ft. The shale contains a high amount of carbonate which makes it brittle and easier to apply hydraulic fracturing to produce the oil or gas.

The formation is best known for producing variable amounts of dry gas, wet gas, NGLs, gas condensate, and crude oil. The most active area lies above the Edwards Reef Trend where the formation yields a gas-condensate production stream. The Eagle Ford shale formation is estimated to have 20.81 trillion cubic feet (20.81 × 1012 ft3) of natural gas and 3.351 billion barrels (3351 × 109 bbls) of oil.

2.11 Fayetteville Shale

The Fayetteville shale gas play is a geologic formation of Mississippian age (354–323 million years ago) composed of tight shale within the Arkoma basin of Arkansas. The play is divided into two main units, Central and Western based on the location of the shale. The shale ranges from 1000 to 7000 ft deep and 20 to 200 ft thick. With productive wells penetrating the Fayetteville shale (Arkoma Basin) at depths between a few hundred and 7000 ft, this formation is somewhat shallower than the Barnett shale formation (Braithwaite, 2009; US EIA, 2011a). Mediocre production from early vertical wells stalled development in the vertically fractured Fayetteville, and only with recent introduction of horizontal drilling and hydraulic fracturing has drilling activity increased.

In the most active Central Fayetteville shale, horizontal wells are drilled using oil-based mud in most cases, and water-based mud in others. In addition, 3-D seismic is gaining importance as longer laterals of 3000-plus feet are drilled and more stages are required for hydraulic fracturing. With growing numbers of wells and a need for more infrastructure—pad drilling is another trend emerging in the Fayetteville.

The total area for the Fayetteville shale play, including Central and Western Fayetteville, is 9000 mile^2. Fayetteville Central is 4000 mile^2 and the remaining shale, Fayetteville Western, is approximately 5000 mile^2. The shale gas play is estimate to have approximately 31.96 Tcf of technically recoverable gas.

2.12 Floyd Shale

The Upper Mississippian Floyd shale is an equivalent of the prolific Barnett shale of the Fort Worth Basin. The shale is an organic-rich interval in the lower part of the Floyd shale that is informally called the Neal shale, which is an organic-rich, starved-basin deposit that is considered to be the principal source rock for conventional hydrocarbons in the Black Warrior basin.

The Floyd shale is a black marine shale located stratigraphically below the Mississippian Carter sandstone and above the Mississippian Lewis sandstone (US EIA, 2011a). Although the Carter and Lewis sandstones have historically been the most prolific gas-producing zones in the Black Warrior Basin Region of Alabama, there has been no prior production history reported for the Floyd shale. The Chattanooga shale is located below the Floyd and is separated from it in most areas by the Tuscumbia Limestone and the Fort Payne Chert.

The Mississippian Floyd shale is an equivalent of the prolific Barnett shale of the Fort Worth Basin and the Fayetteville shale of the Arkoma Basin and has thus been the subject of intense interest. The Floyd is a broadly defined formation that is dominated by shale and limestone and extends from the Appalachian Thrust Belt of Georgia to the Black Warrior Basin of Mississippi.

Usage of the term, Floyd, can be confusing. In Georgia, the type Floyd shale includes strata equivalent to the Tuscumbia Limestone, and in Alabama and Mississippi, complex facies relationships place the Floyd above the Tuscumbia Limestone, Pride Mountain Formation, or Hartselle Sandstone and below the first sandstone in the Parkwood Formation.

Importantly, not all Floyd facies are prospective as gas reservoirs. Drillers have long recognized a resistive, organic-rich shale interval in the lower part of the Floyd shale that is called informally the Neal shale (Cleaves and Broussard, 1980; Pashin, 1994). In addition to being the probable source rock for conventional oil and gas in the Black Warrior Basin, the Neal has the greatest potential as a shale-gas reservoir in the Mississippian section of Alabama and Mississippi. Accordingly, usage of the term, Neal, helps specify the facies of the Floyd that contains prospective hydrocarbon source rocks and shale-gas reservoirs.

The expected ultimate recovery is on the order of 4.37 Tcf of technically recoverable gas. The shale ranges from 6000 to 10,000 ft deep and 80 to 180 ft thick with a well spacing of 2 well per square mile (320 acres per well).

2.13 Gammon Shale

The Gammon shale is a Cretaceous marine mudstone that is typical of gas-bearing rocks and contains large quantities of biogenic methane (Gautier, 1981). Organic matter in the low-permeability reservoirs served as the source of biogenic methane, and capillary forces acted as the trapping mechanism for gas accumulation. The Gammon member of the Pierre shale of the northern Great Plains, USA, contains abundant siderite concretions (Gautier, 1982).

At Little Missouri field, southwestern North Dakota, Gammon reservoirs consist of discontinuous lenses and laminae of siltstone, less than 10 mm thick, enclosed by silty clay shale. Large amounts of allogenic clay, including highly expansible mixed-layer illite-smectite cause great water sensitivity and high measured and calculated water-saturation values. The shale layers are practically impermeable, whereas siltstone microlenses are porous (30–40%) and have a permeability on the order of 3–30 mDs. Reservoir continuity between siltstone layers is poor and, overall, reservoir permeability is probably less than 0.4 mDs. Connecting passageways between siltstone lenses are 0.1 μm or less in diameter. The reservoirs and non-reservoirs cannot be distinguished on the basis of lithology, and much of the Gammon interval is potentially economic.

2.14 Haynesville Shale

The Haynesville shale (also known as the Haynesville/Bossier) is situated in the North Louisiana Salt Basin in northern Louisiana and eastern Texas with depths ranging from 10,500 to 13,500 ft (Braithwaite, 2009; Parker et al., 2009;

US EIA, 2011a). The Haynesville is an Upper Jurassic-age shale bounded by sandstone (Cotton Valley Group) above and limestone (Smackover Formation) below.

The Haynesville shale covers an area of approximately 9000 mile^2 with an average thickness of 200–300 ft. The thickness and areal extent of the Haynesville has allowed operators to evaluate a wider variety of spacing intervals ranging from 40 to 560 acres per well. Gas content estimates for the play are 100 to 330 scf/ton. The Haynesville formation has the potential to become a significant shale gas resource for the United States with original gas-in-place estimates of 717 trillion cubic feet (717×10^{12} ft^3) and technically recoverable resources estimated at 251 trillion cubic feet (251×10^{12} ft^3).

Compared to the Barnett shale, the Haynesville shale is extremely laminated, and the reservoir changes over intervals as small as 4 in. to 1 foot. In addition, at depths of 10,500–13,500 ft, this play is deeper than typical shale gas formations creating hostile conditions. Average well depths are 11,800 ft with bottomhole temperatures averaging 155°C (300°F) and wellhead treating pressures that exceed 10,000 psi. As a result, wells in the Haynesville require almost twice the amount of hydraulic horsepower, higher treating pressures and more advanced fluid chemistry than the Barnett and Woodford shale formations.

The high-temperature range, from 125°C (260°F) to 195°C (380°F), creates additional problems in horizontal wells, requiring rugged, high-temperature/high-pressure logging evaluation equipment. The formation depth and high-fracture gradient demand long pump times at pressures above 12,000 psi. In deep wells, there is also concern about the ability to sustain production with adequate fracture conductivity. In large volumes of water for fracturing, making water conservation and disposal a primary issue.

The Bossier shale often linked with the Haynesville shale is a geological formation that produces hydrocarbon and delivers large amounts of natural gas when properly treated. While there is some confusion when distinguishing Haynesville shale from the Bossier shale, it is a relatively simple comparison—the Bossier shale lies directly above the Haynesville shale but lies under the Cotton Valley sandstones. However, some geologists still consider the Haynesville shale and the Bossier shale one in the same. The thickness of the Bossier shale is approximately 1800 ft in the area of interest. The productive zone is located in the upper 500–600 ft of the shale. The Bossier shale is located in eastern Texas and northern Louisiana.

The Upper Jurassic (Kimmeridgian to Lower Tithonian) Haynesville and Bossier shale formations of East Texas and northwest Louisiana are currently one of the most important shale-gas plays in North America, exhibiting overpressure and high temperature, steep decline rates, and resources estimated together in the hundreds of trillions of cubic feet. These shale-gas resources have been studied extensively by companies and academic institutions within the last year, but to date the depositional setting, facies, diagenesis, pore evolution, petrophysics, best completion techniques, and geochemical characteristics of the Haynesville shale formation and the Bossier shale formations are still not fully characterized and understood. Our work represents new insights into Haynesville and Bossier shale facies, deposition, geochemistry, petrophysics, reservoir quality, and stratigraphy in light of paleographic setting and regional tectonics.

Haynesville and Bossier shale deposition was influenced by basement structures, local carbonate platforms, and salt movement associated with the opening of the Gulf of Mexico basin. The deep basin was surrounded by carbonate shelves of the Smackover/Haynesville Lime Louark sequence in the north and east and local platforms within the basin. The basin periodically exhibited restricted environment and reducing anoxic conditions, as indicated by variably increased molybdenum content, presence of framboidal pyrite, and TOC-S-Fe relationships. These organic-rich intervals are concentrated along and between platforms and islands that provided restrictive and anoxic conditions during Haynesville and part of Bossier times.

The mudrock facies range from calcareous-dominated facies near the carbonate platforms and islands to siliceous-dominated lithology in areas where deltas prograded into the basin and diluted organic matter (e.g., northern Louisiana and northeast Texas). These facies are a direct response to a second-order transgression that lasted from the early Kimmeridgian to the Berriasian. The Haynesville shale formation and the Bossier shale formation each compose three upward-coarsening cycles that probably represent third-order sequences within the larger second-order transgressive systems and early highstand systems tracts, respectively. Each Haynesville shale formation is characterized by unlaminated mudstone grading into laminated and bioturbated mudstone. Most of the three Bossier third-order cycles are dominated by varying amounts of siliciclastic mudstones and siltstones. However, the third Bossier formation exhibits higher carbonate and an increase in organic productivity in a southern restricted area (beyond the basinward limits of Cotton Valley progradation), creating another productive gas-shale opportunity. This organic-rich Bossier formation extends across the Sabine Island complex and the Mt. Enterprise Fault Zone in a narrow trough from Nacogdoches County, Texas, to Red River Parish, Louisiana. Similar to the organic-rich Haynesville cycles, each third-order cycle grades from unlaminated into laminated mudstone and is capped by bioturbated, carbonate-rich mudstone facies. Best reservoir properties are commonly found in facies with the highest TOC, lowest siliciclastics, highest level of maturity, and highest porosity. Most porosity in the Haynesville and Bossier is related to interparticle nano- and micropores and, to a minor degree, by porosity in organic matter.

The Haynesville shale formation and the Bossier shale formation are distinctive on wireline logs: high gamma ray, low density, low neutron porosity, high sonic travel time, moderately high resistivity. A multiline log model seems to predict the total organic content of the formations from the logs. Persistence of distinctive log signatures is similar for the organic-rich Bossier shale formation and the Haynesville shale formation across the study area, suggesting that favorable conditions for shale-gas production extend beyond established producing areas.

2.15 Hermosa Shale

The black shale of the Hermosa Group (Utah) consists of nearly equal portions of clay-sized quartz, dolomite and other carbonate minerals, and various clay minerals. The clay is mainly illite with minor amounts of chlorite and mixed-layer chlorite-smectite (Hite et al., 1984).

The area of interest for the Hermosa Group black shale is the northeast half of the Paradox basin, the portion referred to as the fold and fault belt. This is the area of thick halite deposits in the Paradox Formation, and consequently narrow salt walls and broad interdome depressions. To the southwest of this stratigraphically controlled structural zone the black shale intervals are fewer and thinner, and they lack the excellent seals provided by the halite. The area encompasses eastern Wayne and Emery Counties, southern Grand County, and the northeast third of San Juan County (Schamel, 2005, 2006). The kerogen in the shale is predominantly gas-prone humic type III and mixed type II-III (Nuccio and Condon, 1996).

Numerous factors favor the possible development of shale gas in the black shale intervals of the Hermosa Group. First, the shale is very organic-rich, on the whole the most carbonaceous shale in Utah, and they are inherently gas-prone. Second, they have reached relatively high degrees of thermal maturity across much of the basin. Third and perhaps most significant, the shale is encased in halite and anhydrite which retard gas leakage, even by diffusion. Yet it is curious that the Paradox basin is largely an oil province (Morgan, 1992; Montgomery, 1992) in which gas production is historically secondary and associated gas, which relates to the concentration of petroleum development in the shallower targets on the southwest basin margin and in the salt-core anticlines.

2.16 Lewis Shale

The Lewis shale gas play is located in the San Juan Basin in Colorado and New Mexico and has an area of approximately 7506 mile^2. The depth of the Lewis shale ranges from 1640 to 8202 ft deep and is 200–300 ft thick with approximately 11.6 Tcf of technically recoverable gas.

The Lewis shale (San Juan Basin) is a quartz-rich mudstone that was deposited in a shallow, off-shore marine setting during an early Campanian transgression southwestward across shoreline deposits of the underlying progradational Cliffhouse Sandstone Member of the Mancos Formation (Nummendal and Molenaar, 1995; US EIA, 2011a). The gas resources of the Lewis shale are currently being developed, principally through recompletions of existing wells targeting deeper, conventional sandstone gas reservoirs (Dube et al., 2000; Braithwaite, 2009). Current estimates place the expected ultimate recovery of gas on the order of 21.02 Tcf of technically recoverable gas.

The 1000–1500 ft thick Lewis shale is lowermost shore-face and prodelta deposits composed of thinly laminated (locally bioturbated) siltstones, mudstones, and shale. The average clay fraction is just 25%, but quartz is 56%. The rocks are very tight. Average matrix gas porosity is 1.7% and the average gas permeability is 0.0001 mD. The rocks also are organically lean, with an average TOC content is only 1.0%; the range is 0.5–1.6%. The reservoir temperature is 46°C (140°F). Yet the adsorptive capacity of the rock is 13–38 scf/ton, or about 22 billion cubic feet per quarter section (i.e., per 160 acres) (Jennings et al., 1997).

Four intervals and a conspicuous, basin-wide bentonite marker are recognizable in the shale. The greatest permeability is found in the lowermost two-thirds of the section, which may be the result of an increase in grain size and micro-fracturing associated with the regional north-south/east-west fracture system (Hill and Nelson, 2000).

2.17 Mancos Shale

The Mancos shale formation (Uintah Basin) is an emerging shale-gas resource (US EIA, 2011a). The thickness of the Mancos (averaging

4000 ft in the Uinta Basin) and the variable lithology present drillers with a wide range of potential stimulation targets. The area of interest for the Mancos shale is the southern two-thirds of the greater Uinta Basin, including the northern parts of the Wasatch Plateau. In the northern one-third of the basin there have been two few well penetrations of the Mancos shale, and it is too deep to warrant commercial exploitation of a "low-density" resource such as shale gas. The area is within Duchesne, Uintah, Grand, Carbon, and the northern part of Emery Counties (Schamel, 2005, 2006; Braithwaite, 2009).

The Mancos shale is dominated by mudrock that accumulated in offshore and open-marine environments of the Cretaceous Interior seaway. It is 3450–4150 ft thick where exposed in the southern part of the Piceance and Uinta Basins, and geophysical logs indicate the Mancos to be about 5400 ft thick in the central part of the Uinta Basin. The upper part of the formation is intertongues with the Mesaverde Group—these tongues typically have sharp basal contacts and gradational upper contacts. Named tongues include the Buck and the Anchor Mine Tongues. An important hydrocarbon-producing unit in the middle part of the Mancos was referred to as the Mancos B Formation, which consists of thinly interbedded and interlaminated, very fine grained to fine-grained sandstone, siltstone, and clay that was interpreted to have accumulated as north-prograding fore slope sets within an open-marine environment. The Mancos B has been incorporated into a thicker stratigraphic unit identified as the Prairie Canyon Member of the Mancos, which is approximately 1200 ft thick (Hettinger and Kirschbaum, 2003).

At least four members of the Mancos have shale-gas potential: (1) the Prairie Canyon (Mancos B), (2) the Lower Blue Gate shale, (3) the Juana Lopez, and (4) the Tropic-Tununk shale. Organic matter in the shale has a large fraction of Terrigenous material derived from the shorelines of the Sevier belt. The thickness of the organic-rich zones within individual system tracts exceeds 12 ft. Vitrinite reflectance values from a limited number of samples at the top of the Mancos range from 0.65% at the Uinta Basin margins to >1.5% in the central basin.

Across most of Utah the Mancos shale has not been sufficiently buried to have attained the levels of organic maturity required for substantial generation of natural gas, even in the humic kerogen-dominant (type II-III) shale that characterize this group (Schamel, 2005, 2006). However, vitrinite reflectance values beneath the central and southern Uinta Basin are well within the gas generation window at the level of the Tununk shale, and even the higher members of the Mancos shale. In addition to the in situ gas within the shale, it is likely that some of the gas found a reservoir in the silty shale intervals has migrated from deeper source units, such as the Tununk shale or coals in the Dakota.

The Mancos shale warrants consideration as the significant gas reservoir and improved methods for fracture stimulation tailored to the specific rock characteristics of the Mancos lithology are required. The well completion technologies used in the sandstones cannot be applied to the shale rocks without some reservoir damage.

2.18 Marcellus Shale

The Marcellus shale (Appalachian Basin), also referred to as the Marcellus Formation, is located in the Appalachian Basin across the Eastern Part of the United States. The Marcellus shale formations are 400 million years old and extend from western Maryland to New York, Pennsylvania, and West Virginia and encompassing the Appalachian region of Ohio along the Ohio River. The formation is a Middle Devonian black, low density, carbonaceous (organic-rich) shale that occurs in the subsurface beneath much of Ohio, West Virginia, Pennsylvania, and New York. Small areas of Maryland, Kentucky, Tennessee, and Virginia are also underlain by the Marcellus shale (Braithwaite, 2009; Bruner and Smosna, 2011; US EIA, 2011a). It has been estimated that the Marcellus shale formation could contain as much as 489 trillion cubic feet of natural gas, a level that would establish the Marcellus as the largest natural gas resource in North America and the second largest in the world.

Throughout most of its extent, the Marcellus is nearly a mile or more below the surface. These great depths make the Marcellus Formation a very expensive target. Successful wells must yield large volumes of gas to pay for the drilling costs that can easily exceed a million dollars for a traditional vertical well and much more for a horizontal well with hydraulic fracturing. There are areas where the thick Marcellus shale can be drilled at minimum depths and tends to correlate with the heavy leasing activity that has occurred in parts of northern Pennsylvania and western New York.

Natural gas occurs within the Marcellus shale in three ways: (1) within the pore spaces of the shale, (2) within vertical fractures (joints) that break through the shale, and (3) adsorbed on mineral grains and organic material. Most of the recoverable gas is contained in the pore spaces. However, the gas has difficulty escaping through the pore spaces because they are very tiny and poorly connected. The gas in the Marcellus shale is a result of its contained organic content. Logic therefore suggests that the more organic material there is contained in the rock the greater its ability to yield gas. The areas with the greatest production potential might be where the net thickness of organic-rich shale within the Marcellus Formation is greatest. Northeastern Pennsylvania is where the thick organic-rich shale intervals are located. The Marcellus shale ranges in depth from 4000 to 8500 ft, with gas currently produced from hydraulically fractured horizontal wellbores. Horizontal lateral lengths exceed 2000 ft, and, typically, completions involve multistage fracturing with more than three stages per well.

Before the year 2000, many successful natural gas wells had been completed in the Marcellus shale. The yields of these wells were often unimpressive upon completion. However, many of these older wells in the Marcellus have a sustained production that decreases slowly over time and many of them continued to produce gas for decades. To exhibit the interest in this shale formation, The Pennsylvania Department of Environmental Protection reports that the number of drilled wells in the Marcellus shale has been increasing rapidly. In 2007, 27 Marcellus shale wells were drilled in the state; however, in 2011 the number of wells drilled had risen to more than 2000.

For new wells drilled with the new horizontal drilling and hydraulic fracturing technologies the initial production can be much higher than what was seen in the old wells. Early production rates from some of the new wells have been over 1 million cubic feet of natural gas per day. The technology is so new that long-term production data are not available. As with most gas wells, production rates will decline over time, however, a second hydraulic fracturing treatment could stimulate further production.

2.19 Monterey/Santos Shale

The Monterey/Santos shale oil play includes the Lower Monterey shale formation and the Santos shale formation and is located in the San Joaquin and Los Angeles Basins in California. The active area for the Monterey/Santos shale play is approximately 1752 mile^2. The depth of the shale ranges from 8000 to 14,000 ft deep and is between 1000 and 3000 ft thick with approximately 15.42 billion barrels (15.42×10^9 bbls) of technically recoverable oil.

2.20 Neal Shale

The Neal shale is an organic-rich facies of the Upper Mississippian age Floyd shale formation. The Neal shale formation has long been recognized as the principal source rock that charged conventional sandstone reservoirs in the Black Warrior Basin (Telle et al., 1987; Carroll et al., 1995; US EIA, 2011a) and has been the subject of intensive shale-gas exploration in recent years.

The Neal shale is developed mainly in the southwestern part of the Black Warrior Basin and is in facies relationship with strata of the Pride Mountain Formation, Hartselle Sandstone, the Bangor Limestone, and the lower Parkwood Formation. The Pride Mountain-Bangor interval in the northeastern part of the basin constitutes a progradational parasequence set in which numerous stratigraphic markers can be traced southwestward into the Neal shale. Individual parasequences tend to thin southwestward and define a clinoform stratal geometry in which near-shore facies of the Pride Mountain-Bangor interval pass into condensed, starved-basin facies of the Neal shale.

The Neal formation maintains the resistivity pattern of the Pride Mountain-Bangor interval, which facilitates regional correlation and assessment of reservoir quality at the parasequence level. The Neal shale and equivalent strata were subdivided into three major intervals, and isopach maps were made to define the depositional framework and to illustrate the stratigraphic evolution of the Black Warrior Basin in Alabama. The first interval includes strata equivalent to the Pride Mountain Formation and the Hartselle Sandstone and thus shows the early configuration of the Neal basin. The Pride Mountain-Hartselle interval contains barrier-strand plain deposits (Cleaves and Broussard, 1980; Thomas and Mack, 1982). Isopach contours define the area of the barrier-strand plain system in the northeastern part of the basin, and closely spaced contours where the interval is between 25 and 225 ft thick define a southwestward slope that turns sharply and faces southeastward in western Marion County. The Neal starved basin is in the southwestern part of the map area, where this interval is thinner than 25 ft.

The second interval includes strata equivalent to the bulk of the Bangor Limestone. A generalized area of inner ramp carbonate sedimentation is defined in the northeastern part of the formation where the interval is thicker than 300 ft. Muddy, outer-ramp facies are concentrated where this interval thins from 300 to 100 ft, and the northeastern margin of the Neal starved basin is marked by the 100-foot contour. Importantly, this interval contains the vast majority of the prospective Neal reservoir facies, and the isopach pattern indicates that the slope had prograded more than 25 miles southwestward during Bangor deposition.

The final interval includes strata equivalent to the lower Parkwood Formation. The lower Parkwood separates the Neal shale and the main part of the Bangor Limestone from carbonate-dominated strata of the middle Parkwood Formation, which includes a tongue of the Bangor that is called the *Millerella* limestone. The Lower Parkwood is a succession of siliciclastic deltaic sediment that prograded onto the Bangor ramp in the northeastern part of the study area and into the Neal basin in the southern part and contains the most prolific conventional reservoirs in the Black Warrior Basin (Cleaves, 1983; Pashin and Kugler, 1992; Mars and Thomas, 1999). The lower Parkwood is thinner than 25 ft above the inner Bangor ramp and includes a variegated shale interval containing abundant slickensides and calcareous nodules, which are suggestive of exposure and vertical soil formation. The area of deltaic sedimentation is where the lower Parkwood is thicker than 50 ft and includes constructive deltaic facies in the Neal basin and destructive, shoal-water deltaic facies along the margin of the Bangor ramp. In the southern part of the study area, the 25-foot contour defines a remnant of the Neal basin that persisted through lower Parkwood deposition. In this area, condensation of lower Parkwood sediment brings middle Parkwood carbonate rocks within 25 ft of the resistive Neal shale.

2.21 New Albany Shale

The New Albany shale gas play (Illinois Basin, Illinois, Indiana, and Kentucky) is organic-rich shale located over a large area in southern Indiana and Illinois and in Northern Kentucky (Zuber et al., 2002; Braithwaite, 2009; US EIA, 2011a). The depth of the producing interval varies from 500 ft to 2000 ft depth, with thicknesses of approximately 100 ft. The shale is generally subdivided into four stratigraphic intervals: from top to bottom, these are (1) Clegg Creek, (2) Camp Run/Morgan Trail, (3) Selmier, and (4) Blocher intervals.

The total area for the New Albany shale play is approximately 43,500 mile^2. The total area includes an active and undeveloped area of the play—the total active area is on the order of 1600 mile^2 and the remaining area (41,900) square miles and is characterized as undeveloped area. The New Albany formation has an expected ultimate recovery of 10.95 Tcf of technically recoverable gas. The depth of the formation ranges from 1000 to 4500 and is 100 to 300 ft thick.

The New Albany shale can be considered to be a *mixed source rock* in which some parts of the basin produced thermogenic gas, and other parts produced biogenic gas. This is indicated by the vitrinite reflectance in the basin, varying from 0.6 to 1.3 (Faraj et al., 2004). It is not known whether circulating ground waters recently generated this biogenic gas or whether it is original biogenic gas generated shortly after the time of deposition.

Most gas production from the New Albany comes from approximately 60 fields in northwestern Kentucky and adjacent southern Indiana. However, past and current production is substantially less than that from either the Antrim shale or Ohio shale. Exploration and development of the New Albany shale was spurred by the spectacular development of the Antrim shale resource in Michigan, but results have not been as favorable (Hill and Nelson, 2000).

Production of New Albany shale gas, which is considered to be biogenic, is accompanied by large volumes of formation water (Walter et al., 2000). The presence of water would seem to indicate some level of formation permeability. The mechanisms that control gas occurrence and productivity are not as well understood as those for the Antrim and Ohio shale formations (Hill and Nelson, 2000).

2.22 Niobrara Shale

The Niobrara shale formation (Denver-Julesburg Basin, Colorado) is a shale rock formation located in Northeast Colorado, Northwest Kansas, Southwest Nebraska, and Southeast Wyoming. Oil and natural gas can be found deep below the earth's surface at depths of 3000–14,000. Companies drill these wells vertically and even horizontally to get at the oil and natural gas in the Niobrara Formation.

The Niobrara shale is located in the Denver-Julesburg basin which is often referred to as the DJ Basin. This resource exciting oil shale play is being compared to the Bakken shale resource, is located in North Dakota.

2.23 Ohio Shale

The Devonian shale in the Appalachian Basin was the first produced in the 1820s. The resource extends from Central Tennessee to Southwestern NewYork and also contains the Marcellus shale formation. The Middle and Upper Devonian shale formations underlie approximately 128,000 mile^2 and crop out around the rim of the basin. Subsurface formation thicknesses exceed 5000 ft and organic-rich black shale exceeds 500 ft (152 m) in net thickness (Dewitt et al., 1993).

The Ohio shale (Appalachian Basin) differs in many respects from the Antrim shale petroleum system. Locally, the stratigraphy is considerably more complex as a result of variations in depositional setting across the basin (Kepferle, 1993; Roen, 1993). The shale formations can be further subdivided into five cycles of alternating carbonaceous shale formations and coarser grained clastic materials (Ettensohn, 1985). These five shale cycles developed in response to the dynamics of the Acadian orogeny and westward progradation of the Catskill delta.

The Ohio shale, within the Devonian shale, consists of two major stratigraphic intervals: (1) the Chagrin shale and (2) the underlying Lower Huron shale.

The Chagrin shale consists of 700 to 900 ft of gray shale (Curtis, 2002; Jochen and Lancaster, 1993), which thins gradually from East to West. Within the lower 100–150 ft, a transition zone consisting of interbedded black and gray shale lithology announces the underlying Lower Huron formation. The Lower Huron shale is 200–275 ft of dominantly black shale, with moderate amounts of gray shale and minor siltstone. Essentially all the organic matter contained in the lower Huron is thermally mature for hydrocarbon generation, based on vitrinite reflectance studies.

The vitrinite reflectance of the Ohio shale varies from 1% to 1.3%, which indicates that the rock is thermally mature for gas generation (Faraj et al., 2004). The gas in the Ohio shale is consequently of thermogenic origin. The productive capacity of the shale is a combination of gas storage and deliverability (Kubik and Lowry, 1993). Gas storage is associated with both classic matrix porosity as well as gas adsorption onto clay and nonvolatile organic material. Deliverability is related to matrix permeability although highly limited (10^{-9}–10^{-7} mDs) and a well-developed fracture system.

2.24 Pearsall Shale

The Pearsall shale is a gas bearing formation that garnered attention near the Texas-Mexico border in the Maverick Basin before development of the Eagle Ford shale truly commenced. The Pearsall shale formation is found below the Eagle Ford formation at depths of 7000–12,000 ft with a thickness of 600–900 ft (Braithwaite, 2009).

The formation does have the potential to produce liquids east of the Maverick Basin. As of 2012, only a few wells had been drilled in the play outside of the Maverick Basin but early results indicate there is potential that has largely been overlooked.

2.25 Pierre Shale

The Pierre shale, located in Colorado, produced 2 million cubic feet of gas in 2008. Drilling operators are still developing this rock formation, which lies at depths that vary between 2500 and 5000 ft, and will not know its full potential until more wells provide greater information about its limits (Braithwaite, 2009).

The Pierre shale formation is a division of Upper Cretaceous rocks laid down from approximately 146 million to 65 million years ago and is named for exposures studied near old Fort Pierre, South Dakota. In addition to Colorado, the formation also occurs in South Dakota, Montana, Colorado, Minnesota, New Mexico, Wyoming, and Nebraska.

The formation consists of approximately 2000 ft of dark gray shale, some sandstone, and many layers of bentonite (altered volcanic-ash falls that look and feel much like soapy clays). In some regions the Pierre shale may be as little as 700 ft) thick.

The lower Pierre shale represents a time of significant changes in the Cretaceous Western Interior Seaway, resulting from complex interactions of tectonism and eustatic sea level changes. The recognition and redefinition of the units of the lower Pierre shale has facilitated understanding of the dynamics of the basin. The Burning Brule Member of the Sharon Springs Formation is restricted to the northern part of the basin and represents tectonically influenced sequences. These sequences are a response to rapid subsidence of the axial basin and the Williston Basin corresponding to tectonic activity along the Absoroka Thrust in Wyoming. Unconformities associated with the Burning Brule Member record a migrating peripheral bulge in the Black Hills region corresponding to a single tectonic pulse on the Absoroka Thrust. Migration of deposition and unconformities supports an elastic model for the formation and migration of the peripheral bulge and its interaction with the Williston Basin (Bertog, 2010).

2.26 Utah Shale

There are five kerogen-rich shale units as having reasonable potential for commercial development as shale gas reservoirs. These are (1) four members of the Mancos shale in northeast Utah—the Prairie Canyon, the Juana Lopez, the Lower Blue Gate, and the Tununk and (2) the black shale facies within the Hermosa Group in southeast Utah (Schamel, 2005).

The Prairie Canyon and Juana Lopez Members are both detached mudstone-siltstone-sandstone successions embedded within the Mancos shale in northeast Utah. The Prairie Canyon Member is up to 1200 ft thick, but the stratigraphically deeper Juana Lopez Member is less than 100 ft. Both are similar in lithology and basin setting to the gas-productive Lewis shale in the San Juan basin. As in the Lewis shale, the lean, dominantly humic, kerogen is contained in the shale interlaminated with the siltstone-sandstone. The high quartz content is likely to result in a higher degree of natural fracturing than the enclosing clay-mudstone rocks. Thus, they may respond well to hydraulic fracturing. Also the porosity of the sandstone interbeds averaging 5.4% can enhance gas storage. Both units extend beneath the southeast Uinta basin reaching depths sufficient for gas generation and retention from the gas-prone kerogen. Although not known to be producing natural gas at present, both units are worthy of testing for add-on gas, especially in wells that are programmed to target Lower Cretaceous or Jurassic objectives.

The Lower Blue Gate shale formation and the Tropic-Tununk shale formation generally lack the abundant siltstone-sandstone interbeds that would promote natural and induced fracturing, but they do have zones of observed organic richness in excess of 2.0% that might prove to be suitable places for shale gas where the rocks are sufficiently buried beneath the southern Uinta basin and perhaps parts of the Wasatch Plateau.

The black shale facies in the Hermosa Group of the Paradox basin is enigmatic. These shale formations contain mixed type II–III kerogen that should favor gas generation, yet oil with associated gas dominate current production. They are relatively thin, just a few tens of feet thick on average, yet they are encased in excellent sealing rocks, salt and anhydrite. In the salt walls (anticlines) the shale formations are complexly deformed making them difficult to develop even with directional drilling methods, but where they are likely less deformed in the interdome areas (synclines) they are very deep. Yet in these deep areas one can expect peak gas generation. The shale formations are over-pressured, which suggests generation currently or in the recent past. Prospects are good that shale gas reservoirs can be developed in the Paradox basin, but it may prove to be technically and economically challenging (Schamel, 2005).

2.27 Utica Shale

The Utica shale is a rock unit located approximately 4000–14,000 ft below the Marcellus shale and has the potential to become an enormous natural gas resource. The boundaries of the deeper Utica shale formation extend under the Marcellus shale region and beyond. The Utica shale encompasses New York, Pennsylvania, West Virginia, Maryland, and even Virginia. The Utica shale is thicker than the Marcellus and has already proven its ability to support commercial gas production.

The geologic boundaries of the Utica shale formation extend beyond those of the Marcellus shale. The Utica formation, which was deposited 40–60 million years ($40–60 \times 10^6$ years) before the Marcellus formation during the Paleozoic Era, is thousands of feet beneath the Marcellus formation. The depth of Utica shale in the core production area of the Marcellus shale formation production area creates a more expensive environment in which to develop the Utica shale formations. However, in Ohio the Utica shale formation is as little as 3000 ft below the Marcellus shale, whereas in sections of Pennsylvania the Utica formation is as deep as 7000 ft below the Marcellus formation creating a better economic environment to achieve production from the Utica shale formation in Ohio. Furthermore, the investments in the infrastructure to extract natural gas from the Marcellus shale formation also increase the economic efficiency of extracting natural gas from the Utica shale. Although the Marcellus shale is the current unconventional shale drilling target in Pennsylvania. Another rock unit with enormous potential is a few thousand feet below the Marcellus formation.

The potential source rock portion of the Utica shale is extensive and underlies portions of Kentucky, Maryland, New York, Ohio, Pennsylvania, Tennessee, West Virginia, and Virginia. It is also present beneath parts of Lake Ontario, Lake Erie, and part of Ontario, Canada. In keeping with this areal extent, the Utica shale has been estimated to contain (at least) 38 trillion cubic feet (38×10^{12} ft$^{3)}$ of technically recoverable natural gas (at the mean estimate) according to the first assessment of this continuous (unconventional) natural gas accumulation by the US Geological Survey.

In addition to natural gas, the Utica shale is also yielding significant amounts of NGLs and oil in the western portion of its extent and has been estimated to contain on the order of 940 million barrels (940×10^6 bbls) of unconventional oil resources and approximately 208 million barrels (208×10^6 bbls) of unconventional NGLs. A wider estimate place gas resources of the Utica shale, from 2 trillion cubic feet to 69 trillion cubic feet ($2–69 \times 10^{12}$ ft^3), which put this shale on the same resource level as the Barnett shale, the Marcellus shale, and the Haynesville shale formations.

2.28 Woodford Shale

The Woodford shale, located in south-central Oklahoma, ranges in depth from 6000 to 11,000 ft (Abousleiman et al., 2007; Braithwaite, 2009; Jacobi et al., 2009; US EIA, 2011a). This formation is a Devonian-age shale bounded by limestone (Osage Lime) above and undifferentiated strata below. Recent natural gas production in the Woodford shale began in 2003 and 2004 with vertical well completions only. However, horizontal drilling has been adopted in the Woodford, as in other shale gas plays, due to its success in the Barnett shale.

The Woodford shale play encompasses an area of nearly 11,000 $mile^2$. The Woodford play is in an early stage of development and is occurring at a spacing interval of 640 acres per well. The average thickness of the Woodford shale varies from 120 to 220 ft across the play. The gas content in the Woodford shale is higher on average than some of the other shale gas plays at 200 to 300 scf/ton. The original gas-in-place estimate for the Woodford shale is similar to the Fayetteville shale at 23 trillion cubic feet (23×10^{12} ft^3) while the technically recoverable resources are estimate at 11.4 trillion cubic feet (11.4×10^{12} ft^3).

Woodford shale stratigraphy and organic content are well understood, but due to their complexity compared to the Barnett shale, the formations are more difficult to drill and fracture. As in the Barnett, horizontal wells are drilled, although oil-based mud is used in the Woodford shale and the formation is harder to drill. In addition to containing chert and pyrite, the Woodford formation is more faulted, making it easy to drill out of the interval; sometimes crossing several faults in a single wellbore is required.

Like the Barnett shale, higher silica rocks are predominant in the best zones for fracturing in the Woodford formation, although the Woodford has deeper and higher fracture gradients. Due to heavy faulting, 3-D seismic is extremely important, as the Woodford formation trends toward longer laterals exceeding 3000 ft with bigger fracture projects and more stages. Pad drilling also will increase as the Woodford shale formation continues expanding to the Ardmore Basin and to West Central Oklahoma in Canadian County.

The Cana Woodford formation is an emerging gas play located within the Oklahoma Anadarko Basin, about 40 miles west of Oklahoma City. It has been estimated that the Cana Woodford play contains a high liquid content of about 65% (v/v) gas, 30% (v/v) NGLs, and 5% (v/v) crude oil. The active area for Cana Woodford is approximately 688 $mile^2$ with depths that range from about 11,500 to 14,500 ft.

3 CANADA

A number of tight oil formations are found in the Western Canada Sedimentary Basin, and most have seen production in the past at relatively low rate and low recovery, while others are just recently attracting attention for potential development with horizontal multistage fracturing techniques. Generally, these crude oil resources were known to exist but large areas of the shale/tight formations were considered to be uneconomic when compared to production through conventional vertical wells.

Recent estimates (NEB, 2009) in indicate that there is the potential for one quadrillion cubic feet (1×10^{15} ft^3) of gas in place in shale formation in Canada located in different areas but predominantly in the Western Canada Sedimentary basin which includes (1) the Cardium Group, (2) the Colorado Group, (3) the Duvernay shale, (4) the Horn River Basin, and (5) the Montney shale and which also show significant coalbed methane resources (WCSB) (Fig. 3.1). As more shale and tight formations are investigated and identified, estimates of the resources are expected to show significant increases (NEB, 2009).

3.1 Cardium Group

The Cardium formation consists of tight interbedded shale layers and sandstone layers and is found in much of west central area of the Province of Alberta (Figs. 3.1 and 3.2). The formation depth varies from 3900 to 7500 ft and the average thickness of the oil-bearing strata is on the order of 3–10 ft (Peachey, 2014). Some higher quality parts of the formation in the Pembina Oil Field (one of the largest oil fields in Canada) have higher porosity beds of sands and gravel conglomerates, which have already seen production.

The original conventional Cardium oil fields contained 10.6 billion barrels (10.6×10^9 bbls) of light oil, or about 16% of all of the conventional oil resources found in Alberta. The areas around the main pools, currently being developed, may contain an additional 1–3 billion barrels (($1–3 \times 10^9$ bbls) of light oil bbls), although estimates vary and so far 130 million barrels [(130×10^6 bbls)], or 5–10% of the original oil in place, have been claimed as proved and probable reserves which might be technically and economically produced.

At the onset of oil and gas formation, the Cardium, may not have contained much organic matter but the formations now contain natural gas and crude oil which may have migrated into the formations from deeper shale formations. Within the Cardium formation there are areas of the reservoir which lack the highly permeable conglomerate zones, or where there are tight zones above or below the conglomerates which are not connected to the flow system leading to a producing vertical well. While these zones contain considerable amounts of light crude oil, they did not allow for commercial rates of production with vertical wells. However, the introduction of long horizontal wells within the formation, and a series of fracture stages to connect the horizontal wells to various layers in the formation results in economic oil production rates, even in relatively poor reservoir rock. Some of the oil found in the Cardium Group may be referred to as *halo oil* which is oil that exists on the fringe regions of existing oil fields that surround the areas of historical production. Traditional technology cannot produce this oil because of the low permeability of the reservoir matrix.

3.2 Colorado Group

The Colorado Group consists of various shale-containing horizons deposited throughout southern Alberta and Saskatchewan during the middle Cretaceous Period—including the Medicine Hat and Milk River shale-containing sandstones—when globally high sea levels brought about the deposition of these formations, which have been producing natural gas for over 100 years (NEB, 2009). There is also the Second White Speckled shale, which has been producing natural gas for decades (Beaton et al., 2009).

In the Wildmere area of Alberta, the Colorado shale is approximately 650 ft thick, from which natural gas has potential to produce from five intervals. Unlike shale formations from the Horn River Basin and the Utica Group of Quebec, shale from the Colorado Group produces through thin sand beds and lamina, making it a hybrid gas shale like the Montney shale. Furthermore, the gas produced in the Colorado has biogenic rather than thermogenic origins. This would suggest very low potential for NGLs and an under-pressured reservoir, which is more difficult to hydraulically fracture. Colorado Group shale formations are sensitive to water, which makes them sensitive to fluids used during hydraulic fracturing.

The total volume of gas in the Colorado Group is very difficult to estimate given the wide lateral extent of the shale and variability of the reservoir and the absence of independent and publicly available analyses. However, there could be at least 100 trillion cubic feet (100×10^{12} ft^3) of gas in place.

3.3 Duvernay Shale

The Devonian Duvernay shale is an oil and natural gas field located in Alberta, Canada (in the Kaybob area) which extends into British Columbia. The Devonian formation is considered the source rock for the Leduc reefs light oil resources, the discovery of which in 1947 was one of the defining moments in the past-present-and-future Western Canadian oil and gas industry.

The Duvernay Formation (Devonian-Frasnian) of Alberta, Canada is a proven source rock of marine origin which has yielded much of the crude oil and natural gas to the adjacent classical Devonian, conventional fields in carbonate reefs and platform carbonates. Production in these conventional fields is in decline and exploration and development has now shifted to their source, the Duvernay shale.

The deeper Duvernay shale formation underlies much of the Cardium and the portions of the Duvernay formation that underlie the Cardium formation appear to be more likely to contain natural gas reserves than crude oil reserves (NEB, 2009; Peachey, 2014). The Duvernay shale, which can be found just north of the Montney shale, is distributed over most of central Alberta and absent in areas of Leduc reef growth, except beneath the Duhamel reef, where it may be represented by a thin development of calcilutite (a dolomite or limestone formed of calcareous rock flour that is typically nonsiliceous). In the East shale Basin, the formation is approximately 175 ft thick and thickens to 246 ft east and southeastward toward the Southern Alberta Shelf. In a northeast direction, the formation reaches approximately 395 ft at its truncation in the subsurface at the precretaceous unconformity. In the West shale Basin, the formation is approximately 195 ft thick and thickens northward, attaining a thickness in excess of 820 ft to the east of Lesser Slave Lake.

The formation consists of interbedded dark brown bituminous shale sediments, dark brown, black, and occasionally grey-green calcareous shale sediments and dense argillaceous limestone sediments. The shale formations are characteristically petroliferous and exhibit plane parallel millimeter lamination. The formation is characterized by (1) a porosity on the order of 6.0–7.5%, (2) a permeability on the order of 236–805 nanoDarcys, and (3) TOC content 2.0–7.5% (w/w). X-ray diffraction results from core and cuttings samples indicate it is likely very brittle with a low clay content (26%, w/w), amorphous biogenic silica (47%, w/w), and a calcite ($CaCO_3$) and dolomite ($CaCO_3$ $MgCO_3$) matrix (Switzer et al., 1994; Fowler et al., 2003).

3.4 Horn River Basin

The Horn River Basin encompasses approximately 2.5 million acres of land in northeastern British Columbia, north of Fort Nelson and south of the Northwest Territories border. This is an unconventional shale play targeting dry gas from mid-Devonian aged over pressured shale formations of the Muskwa, Otter Park, and Evie Formations. The Horn River Basin is confined to the west by the Bovie Lake Fault Zone and to the East and South by the time equivalent Devonian Carbonate Barrier Complex (NEB, 2009; BCOGC, 2014).

Stratigraphically, the organic-rich siliciclastic Muskwa, Otter Park, and Evie shale formations of the Horn River group are overlain by the Fort Simpson shale and underlain by the Keg River platform carbonates. The Evie shale consists of dark grey to black, organic-rich, pyritic, variably calcareous, and siliceous shale. This shale exhibits relatively high gamma ray readings and high resistivity on well logs. The unit is at its thickest immediately west of the barrier reef complex, generally thinning westward toward the Bovie Lake Fault Structure. The Otter Park shale thickens considerably in the southeast corner of the Basin, characterized by increasingly argillaceous and calcareous facies. Limestone marls were deposited at the expense of shale. The unit thins to the north and west, exhibiting higher gamma ray readings. The Muskwa shale consists of grey to black, organic-rich, pyritic, siliceous shale. A gradational contact exists between the overlying silt-rich shale of the Fort Simpson Formation. Generally, the Muskwa Formation thickens westward toward the Bovie Lake Structure. This shale thins and extends over the top of the barrier reef complex and continues eastward into Alberta, stratigraphically equivalent to the Duvernay shale. Muskwa and Otter Park formations were mapped in combination and analyzed as one interval. From a geomechanical perspective, the Muskwa and Otter Park Formations are considered as one flow unit after hydraulic fracturing with few barriers to fracture propagation. The Evie formation was evaluated and mapped as a separate unit. Mapping completed thus far has defined areas of reservoir variability within the Basin, particularly within the Otter Park Formation.

The Devonian Horn River Basin shale formations were deposited in deep waters at the foot of the Slave Point carbonate platform in northeast British Columbia. The shale formations are silica-rich (approximately 55% (v/v) silica) and approximately 450 ft thick. The total organic content is 1–6% (w/w) and the formations have been producing conventional natural gas for many decades. The Horn River shale Formation located in British Columbia, is the largest shale gas field in Canada and contain an estimated 250 trillion cubic feet (250×10^{12} ft^3) of natural gas (Ross and Bustin, 2008). This shale gas play also includes the Cordova Embayment and the whole formation extends into both the Yukon Territory and the Northwest Territories, although its northward extent beyond provincial/territorial borders is poorly defined.

The advent of horizontal drilling combined with multistage hydraulic fracturing increased interest in unlocking the potential of shale gas. Prior to 2005, operators were targeting Devonian pinnacle reefs, with the basin shale formations then considered a seal and source rock for gas. After 2005, operators began applying horizontal drilling and multistage hydraulic fracturing technology from the analogous Barnett shale in Texas to investigate economic recovery in the Horn River Basin.

3.5 Horton Bluff Group

Lacustrine muds of the Horton Bluff Group of the Canadian Maritime Provinces were deposited in the Early Mississippian Period (approximately 360 million years ago) (Table 3.2) during regional subsidence (NEB, 2009). The silica content in the Frederick Brook shale of the Horton Bluff Group in New Brunswick is approximately 38% (v/v) but the clay content is also high, on the order of 42% (v/v). There are indications that organic contents of the Frederick Brook member in Nova Scotia are significantly higher than other Canadian gas shale formations, at 10% (v/v), and the pay zone appears to be over 500 ft thick, sometimes exceeding 2500 ft in New Brunswick.

There are also indications that most of the gas is adsorbed onto clay and organic matter, and it will take very effective reservoir stimulation to achieve significant production from Nova Scotia shale formations. It is unclear at this time at what proportion of gas is adsorbed onto clay and organic matter in the New Brunswick shale formations.

Analysis indicates that 67 trillion cubic feet (67×10^{12} ft^3) of free gas in place is present in the Frederick Brook shale of the Sussex/Elgin subbasins of southern New Brunswick and 69 trillion cubic feet (69×10^{12} ft^3) of gas is present on the Windsor land block in Nova Scotia.

3.6 Montney Shale

The Montney shale is a unique resource play in that it is a hybrid between tight gas and traditional shale. The formation is rich in silt and sand (characteristics similar to tight gas) but the source of the natural gas originated from its own organic matter like shale plays. Due to the presence of siltstone and sand, the Montney formation has extremely low permeability and requires higher levels of fracture stimulation. However, the reservoir characterization is complicated because the upper and lower Montney zones in the same area have different mineralogy (hence, different properties) which affects the formation evaluation data. The lower Montney is especially difficult as conventional open hole logs have historically led investigators to believe that the lower Montney was a very tight formation but the porosity was higher than expected but still lower than the porosity of the upper Montney formation (Williams and Kramer, 2011; NEB, 2009).

The Montney shale formation is a shale rock deposit located deep below British Columbia, Canada and is located in the Dawson Creek area just south of the Horn Rover shale formation as well as the Duvernay shale formation. The Triassic Montney formation spans a wide variety of depositional environments, from shallow-water sands in the east to offshore muds to the west. Natural gas is currently produced from conventional shallow-water shore-face sandstones at the eastern edge of the Montney and from deep-water tight sands at the foot of the ramp. However, hybrid shale gas potential is being realized in two other zones: (1) the Lower Montney, in sandy, silty shales of the offshore transition and offshore-marine parts of the basin and (2) the Upper Montney, below the shore-face, where silt has buried the tight sands at the foot of the ramp. The Montney is so thick (well over 1000 ft in some places) that it lends itself to the stacked horizontal well concept, where horizontal legs are drilled at two elevations in the same well, penetrating and fracturing both the Upper Montney formation and the Lower Montney formation. The TOC in Montney shale is on the order of 7% (w/w) (maximum) the rocks were heated until they were well into the thermogenic gas window.

As a result, natural gas can be found in large quantities trapped in this tight shale formation (Williams and Kramer, 2011). The formation is a hybrid between a tight gas and shale gas resource and the sandy mudstone formation dates back to the Triassic period and is located beneath the Doig formation at depths ranging from 5500–13,500 ft and is up to 1000 ft thick in places. With these parameters, the Montney shale has the potential to become one of the most significant sources of shale gas in Canada.

The play has been estimated to contain as much as 50 trillion cubic feet (50×10^{12} ft^3) of natural gas trapped within low-permeability shale and siltstone formation. Horizontal wells are drilled at depths from 5500–13,500 ft and hydraulic fracturing enables the gas to flow more easily. Microseismic monitoring techniques can be used to assess fracture stimulations by locating events along each stage of the fracture and calculating the dimensions, geometry and effective fracture volume.

3.7 Utica Group

The Upper Ordovician Utica shale is located between Montreal and Quebec City and was deposited in deep waters at the foot of the Trenton carbonate platform (NEB, 2009). Over geologic time, the shale formation evolved and was changed due to early Appalachian Mountain growth which resulted in faulting and folding and formation on its southeastern side. The formation is approximately 500 ft thick and has a total organic content of 1–3% (w/w) and was identified in the early-to-mid 20th century as a source rock for associated conventional crude oil reservoirs.

However, unlike other Canadian gas shale formations, the Utica has higher concentrations of calcite ($CaCO_3$), which occur at the expense of some silica (SiO_2) (Theriault, 2008). While the calcite in the formation is brittle, hydraulic fractures do not transmit as well through it.

4 OTHER COUNTRIES

Significant amounts of shale gas occur outside of the United States and Canada and the potential natural gas and crude oil resources in tight formations rivals the resource estimates from conventional gas and oil accumulations and could help meet the world's burgeoning demand for energy, which is forecast to increase 60% by 2035 (Table 3.3) (Khlaifat et al., 2011; Aguilera et al., 2012; Islam and Speight, 2016). However, the technical and environmental challenges as well as the economic challenges involved in the commercialization of these largely untapped vast resources requires a multidisciplinary approach involving geoscience, engineering and economics. Nevertheless, natural gas and crude oil from tight formations have the potential to contribute a significant volume of the gas that is needed to satisfy global primary energy consumption up to and beyond 2035.

In addition, to the United States and Canada, China, and Argentina are currently (at the time of writing) the only countries in the world that are producing commercial volumes of either natural gas from tight formation (shale formations) or crude oil from tight formations (tight oil). The United States is by far the dominant producer of both shale gas and tight oil. In China, Sinopec and PetroChina have reported commercial production of

Table 3.3 Estimated of Shale Gas and Tight Oil in Various Countries

Country	Shale Gas (10^{12} ft^3)	Tight Oil (10^9 bbls)
Algeria	706.9	5.7
Argentina	801.5	27
Australia	429.3	15.6
Bolivia	36.4	0.6
Brazil	244.9	5.3
Bulgaria	16.6	0.2
Chad	44.4	16.2
Chile	48.5	2.3
China	1115.2	32.2
Colombia	54.7	6.8
Denmark	31.7	0
Egypt	100	4.6
France	136.7	4.7
Germany	17	0.7
India	96.4	3.8
Indonesia	46.4	7.9
Jordan	6.8	0.1
Kazakhstan	27.5	10.6
Libya	121.6	26.1
Lithuania/Kaliningrad	2.4	1.4
Mexico	545.2	13.1
Mongolia	4.4	3.4
Morocco	11.9	0
Netherlands	25.9	2.9
Oman	48.3	6.2
Pakistan	105.2	9.1
Paraguay	75.3	3.7
Poland	145.8	1.8
Romania	50.7	0.3
Russia	284.5	74.6
South Africa	389.7	0
Spain	8.4	0.1
Sweden	9.8	0
Thailand	5.4	0
Tunisia	22.7	1.5
Turkey	23.6	4.7
Ukraine	127.9	1.1
United Arab Emirates	205.3	22.6
United Kingdom	25.8	0.7
Uruguay	4.6	0.6
Venezuela	167.3	13.4
West Sahara	8.6	0.2
Total[a]	6381.2	331.8

[a]Excluding the United States and Canada.

shale gas from fields in the Sichuan Basin. In Argentina, tight oil production comes mainly from the Neuquen Basin where the national oil company YPF (Yacimientos Petrolíferos Fiscales) is producing approximately 20,000 barrels of tight oil per day from the Loma Campana area.

The initial estimate of shale gas resources in the 42 countries (other than the United States and Canada) is on the order of 6381 trillion cubic feet (6381×10^{12} ft^3) (US EIA, 2011b; DECC, 2013; NRF, 2013; US EIA, 2015). Adding the US estimate of the shale resources of results in a total shale gas resource base estimate of 7576 trillion cubic feet (7576×10^{12} ft^3) for the United States and the other 42 countries assessed. To put this shale gas resource estimate in context, the technically recoverable gas resources worldwide are approximately 16,000 trillion cubic feet ($16,000 \times 10^{12}$ ft^3), largely excluding shale gas (US EIA, 2011b). Thus, adding the identified shale gas resources to other gas resources increases total world technically recoverable gas resources by more than 40% to 22,600 trillion cubic feet ($22,600 \times 10^{12}$ ft^3). In terms of light oil from tight formations, estimates (excluding Canada and the United States) are on the order of 332 billion barrels (332×10^9 bbls) spread over 42 countries (Table 3.3).

At a country level, there are two country groupings that emerge where shale gas development appears most attractive. The first group consists of countries that are currently highly dependent upon natural gas imports, have at least some gas production infrastructure, and their estimated shale gas resources are substantial relative to their current gas consumption. For these countries, shale gas development could significantly alter their future gas balance, which may motivate development. The second group consists of those countries where the shale gas resource estimate is large (>200 trillion cubic feet, $>200 \times 10^{12}$ ft^3) and there already exists a significant natural gas production infrastructure for internal use or for export. Existing infrastructure would aid in the timely conversion of the resource into production, but could also lead to competition with other natural gas supply sources. For an individual country the situation could be more complex.

The predominant shale gas resources are found in countries (Table 3.3). The resources of crude oil in these countries is still under investigation and estimates are tentative but show promise for the future and there is also the need to define the methods by which other countries determine the amount of resources of shale gas and tight oil and to assure the reliability comparative data (US EIA, 2013). As an example of different terminology for the definitions of the reservoirs, while many countries use the reservoir properties to define a tight reservoir, some countries may use the flow rate of the natural gas or crude oil to define a tight formation whether or not the flow rate is due to the reservoir properties.

In addition, the nomenclature and methods used in reserve estimation is also subject to question, especially the issue of the volumetric capacity of the reservoir to hold natural gas and crude oil and the means by which the capacity was determined reservoir volumetric (Chapter 2). For example, it is important to distinguish between a technically recoverable resource and an economically recoverable resource (US EIA, 2013, 2015). Technically recoverable resources represent the volumes of natural gas and crude oil gas that can be recovered using current technology, regardless of the prices for natural gas and crude oil and natural gas as well as production costs. On the other hand, economically recoverable resources are resources that can be profitably produced under the current market price. Moreover, economic recoverability can be significantly influenced not only by subterranean geology but also by above-the-ground factors which include (1) private ownership of subsurface rights, (2) the availability of many independent operators and supporting contractors with critical expertise and suitable drilling rigs, (3) the preexisting gathering and pipeline infrastructure, and (4) the availability of water resources for use in hydraulic fracturing (US EIA, 2013, 2015). Not all of these factors are equal in all countries.

5 THE FUTURE OF RESOURCES IN TIGHT FORMATIONS

The storage properties of shale formations and tight formations for natural gas and crude oil are quite different to the storage properties of conventional reservoirs (Chapters 1 and 2). In addition to having gas present in the matrix system of pores similar to that found in conventional reservoir rocks, shale formations and tight formations also have gas or oil bound or adsorbed not only to the inorganic surface of the formation but also bound or adsorbed to surface of organic material in the shale. The relative contributions and combinations of free gas from matrix porosity and from desorption of adsorbed gas and oil is a key determinant of the production profile of the well.

The amount and distribution of gas within the shale is determined by, amongst other things, (1) the initial reservoir pressure, (2) the petrophysical properties of the rock, and (3) the adsorption characteristics of the reservoir rock. Thus, during production there are three main processes at play. First, the initial rate of gas or oil production may be dominated by the depletion of the gas or liquid from the fracture network. This form of production declines rapidly due to limited storage capacity. Second, after the initial decline rate of production stabilizes, the depletion of gas or oil stored in the matrix becomes the primary process involved in the production of the resource. The amount of gas held in the matrix is dependent on the particular properties of the shale reservoir which can be hard to estimate. Third, but secondary to the depletion process, is desorption whereby adsorbed gas or oil is released from the rock as pressure in the reservoir declines.

At that point, the rate of resource production by means of the desorption process is highly dependent on there being a significant drop in reservoir pressure. In addition, pressure changes and any effects on the rate of depletion typically advance through the reservoir rock at a very slow rate due to low permeability. Tight well spacing can therefore be required to lower the reservoir pressure enough to cause significant amounts of adsorbed gas to be desorbed. These overlapping production processes result in the characteristic hyperbolic production profile that can decline very sharply (by a reduction in the production rate on the order of 60–80%) within a year (or two) after commencement of production.

Due to these particular properties, the ultimate recovery of the gas or oil in place surrounding a particular well can be on the order of 28–40% (v/v), whereas the recovery per conventional well may be as high as 60–80% (v/v). Thus, the development of shale plays and tight plays differs significantly from the development of conventional resources. With a conventional reservoir, each well is capable of draining natural gas or crude oil over a relatively large area (dependent on reservoir properties). As such, only a few wells (normally vertical wells) are required to produce commercial volumes from the field. With shale and tight formation projects, a large number of relatively closely spaced wells are required to produce sufficiently large volumes to make the plays economic propositions. As a result, many wells must be drilled in a shale play to drain the reservoir sufficiently; for example, in the Barnett shale play (Texas, United States), the drilling density can exceed one well per 60 acres.

In 2014, the United States increased crude oil production by 1.6 million barrels per day (1.6×10^6 bbls/day) and became the first country ever to increase production by at least 1 million barrels per day (1.0×10^6 bbls/day) for 3 consecutive years (BP, 2015). This was due, in no small part, to the production of crude oil from tight formations. As a result, the United States has replaced Saudi Arabia as the world's largest oil producer and the United States has also overtaken Russia as the world's largest producer of crude oil and natural gas.

In terms of the future, the most significant development on the supply side of natural gas and crude oil is the continuing development of resources held in shale formation and in tight formations in the United States. However, the development of natural gas and crude oil fields will unfortunately, contribute to environmental pollution, including acid rain, the greenhouse effect, and *allegedly* global warming (global climate change).

REFERENCES

Abousleiman, Y., Tran, M., Hoang, S., Bobko, C., Ortega, A., Ulm, F.J., 2007. Geomechanics field and laboratory characterization of woodford shale: the next gas play. Paper Bo. SPE 110120, In: Proceedings of SPE Annual Technical Conference and Exhibition, Anaheim, California, Nov. 11–14.

Aguilera, R.F., Harding, T.G., Aguilera, R., 2012. Tight gas. World Petroleum Council Guide: Unconventional Gas. World Petroleum Council, London, United Kingdom, pp. 58–63. http://www.world-petroleum.org/docs/docs/gasbook/unconventionalgaswpc2012.pdf (accessed 15.03.15.).

Andrews, R.D., 2007. Stratigraphy, production, and reservoir characteristics of the Caney shale in Southern Oklahoma. Shale Shaker 58, 9–25.

Andrews, R.D., 2012. My favorite outcrop—Caney shale along the South flank of the Arbuckle mountains, Oklahoma. Shale Shaker 62, 273–276.

BCOGC, 2014. Horn River Basin Unconventional Shale Gas Play Atlas. British Columbia Oil and Gas Commission, Victoria, British Columbia.

Beaton, A.P., Pawlowicz, J.G., Anderson, S.D.A., Rokosh, C.D., 2009. Rock Eval™ Total Organic Carbon, Adsorption Isotherms and Organic Petrography of the Colorado Group: Shale Gas Data Release. Open File Report No. ERCB/AGS 2008-11, Energy Resources Conservation Board, Calgary, Alberta.

Bertog, J., 2010. Stratigraphy of the lower Pierre shale (campanian): implications for the tectonic and eustatic controls on facies distributions. J. Geol. Res. 2010, 910243. http://dx.doi.org/10.1155/2010/910243. http://www.hindawi.com/journals/jgr/2010/910243/cta/ (accessed 09.05.13.).

Boardman, D., Puckette, J., 2006. Stratigraphy and paleontology of the upper Mississippian Barnett shale of Texas and Caney shale of southern Oklahoma. OGS Open-File Report No. 6-2006, Oklahoma Geological Survey, Norman, Oklahoma.

Bonakdarpour, M., Flanagan, R., Holling, C., Larson, J.W., 2011. The Economic and Employment Contributions of Shale Gas in the United States. Prepared for America's Natural Gas Alliance. IHS Global Insight (USA), Washington, DC. December.

Bowker, K.A., 2007a. Barnett shale gas production, Fort Worth basin, issues and discussion. AAPG Bull. 91, 522–533.

Bowker, K.A., 2007b. Development of the Barnett shale play, Fort Worth basin. W. Tex. Geol. Soc. Bull. 42 (6), 4–11.

BP, 2015. Statistical Review of World Energy 2015. BP PLC, London, United Kingdom June. http://www.bp.com/content/dam/bp/pdf/energy-economics/statistical-review-2015/bp-statistical-review-of-world-energy-2015-full-report.pdf; (accessed 07.04.16.).

Braithwaite, L.D., 2009. Shale-deposited natural gas: a review of potential. Report No. CEC-200-2009-005-SD. Electricity Analysis Office, Electricity Supply Analysis division, California Energy Commission, Sacromento, California, May.

Bruner, K.R., Smosna, R., 2011. A comparative study of the Mississippian Barnett shale, Fort Worth basin, and Devonian Marcellus shale, Appalachian basin. Report No. DOE/

NETL-2011/1478, United States Department of Energy, Morgantown Energy Technology Center, Morgantown, West Virginia.

Carroll, R.E., Pashin, J.C., Kugler, R.L., 1995. Burial history and source-rock characteristics of Upper Devonian through Pennsylvanian strata, Black Warrior Basin, Alabama. Circular No. 187, Alabama Geological Survey, Tuscaloosa, Alabama.

Cipolla, C.L., Lolon, E.P., Erdle, J.C., Rubin, B., 2010. Reservoir modeling in shale-gas reservoirs. SPE Paper No. 125530, SPE Reserv. Eval. Eng. 13 (4), 638–653.

Cleaves, A.W., 1983. Carboniferous terrigenous clastic facies, hydrocarbon producing zones, and sandstone provenance, Northern shelf of the Black Warrior Basin. Trans. Gulf Coast Assoc. Geol. Soc. 33, 41–53.

Cleaves, A.W., Broussard, M.C., 1980. Chester and pottsville depositional systems outcrop and subsurface in the Black Warrior Basin of Mississippi and Alabama. Trans. Gulf Coast Assoc. Geol. Soc. 30, 49–60.

Cohen, D., 2008. Energy Bulletin. An Unconventional Play in the Bakken, Apr.

Cox, S.A., Cook, D.M., Dunek, K., Daniels, R., Jump, C., Barree, B., 2008. Unconventional resource play evaluation: a look at the Bakken shale play of North Dakota. Paper No. SPE 114171, In: Proceedings of SPE Unconventional Resources Conference, Keystone, Colorado, Feb. 10–12.

Curtis, J.B., 2002. Fractured shale-gas systems. AAPG Bull. 86 (11), 1921–1938.

DECC, 2013. The Unconventional Hydrocarbon Resources of Britain's Onshore basins—Shale Gas. Department of Energy and Climate Change, United Kingdom, London.

DeWitt Jr., W., Roen, J.B., Wallace, L.G., 1993. Stratigraphy of Devonian Black Shales and associated rocks in the Appalachian basin. Bulletin No. 1909, In: Petroleum Geology of the Devonian and Mississippian Black Shale of Eastern North America. U.S. Geological Survey, pp. B1–B57.

Dube, H.G., Christiansen, G.E., Frantz Jr., J.H., Fairchild Jr., N.R., 2000. The Lewis Shale, San Juan Basin: What We Know Now. SPE Paper No. 63091, Society of Petroleum Engineers, Richardson, TX.

EIUT, 2012. Fact-Based Regulation for Environmental Protection in Shale Gas Development Summary of Findings. The Energy Institute, University of Texas at Austin, Austin, Texas. http://energy.utexas.edu.

Ettensohn, F.R., 1985. Controls on the development of Catskill Delta complex basin-facies. Special Paper No. 201, Geological Society of America, Boulder, Colorado, pp. 65–77.

Ezisi, L.B., Hale, B.W., William, M., Watson, M.C., Heinze, L., 2012. Assessment of probabilistic parameters for Barnett shale recoverable volumes. SPE Paper No. 162915, In: Proceedings of SPE Hydrocarbon, Economics, and Evaluation Symposium, Calgary, Canada, Sep. 24–25.

Faraj, B., Williams, H., Addison, G., McKinstry, B., 2004. Gas Potential of Selected Shale Formations in the Western Canadian Sedimentary Basin. GasTIPS (Winter). Hart Energy Publishing, Houston, Texas, pp. 21–25.

Fowler, M.G., Obermajer, M., Stasiuk, L.D., 2003. Rock-Eval and TOC Data for Devonian Potential Source Rocks, Western Canadian Sedimentary Basin. Open File No. 1579, Geologic survey of Canada, Calgary, Alberta.

Frantz, J.H., Waters, G.A., Jochen, V.A., 2005. Evaluating Barnett shale production performance using an integrated approach. SPE Paper No. 96917, In: Proceedings of SPE ATCE Meeting, Dallas, Texas, Oct. 9–12.

Gale, J.F.W., Reed, R.M., Holder, J., 2007. Natural fractures in the Barnett shale and their importance for hydraulic fracture treatments. Am. Assoc. Pet. Geol. Bull. 91, 603–622.

GAO, 2012. Information on shale resources, development, and environmental and public health risks. Report No. GAO-12-732, Report to Congressional Requesters. United States Government Accountability Office, Washington, DC. September.

Gautier, D.L., 1981. Lithology, reservoir properties, and burial history of portion of Gammon shale (cretaceous), Southwestern North Dakota. AAPG Bull. 65, 1146–1159.
Gautier, D.L., 1982. Siderite concretions: indicators of early diagenesis in the Gammon shale (cretaceous). J. Sediment. Petrol. 52, 859–871.
Gutschick, R.C., Sandberg, C.A., 1991. Late Devonian history of the Michigan basin. In: Catacosinos, P.A., Daniels, P.A. (Eds.), Early Sedimentary Evolution of the Michigan Basin, pp. 181–202. Geological society of America special paper No. 256.
Hale, B.W., William, M., 2010. Barnett shale: a resource play—locally random and regionally complex. Paper No. SPE 138987, In: Proceedings of SPE Eastern Regional Meeting, Morgantown, West Virginia, Oct. 12–14.
Hettinger and Kirschbaum, 2003. Stratigraphy of the Upper Cretaceous Mancos Shale (Upper Part) and Mesaverde Group in the Southern Part of the Uinta and Piceance Basins, Utah and Colorado. In: Petroleum Systems and Geologic Assessment of Oil and Gas in the Uinta-Piceance Province, Utah and Colorado. USGS Uinta-Piceance Assessment Team. U.S. Geological Survey Digital Data Series DDS–69–B. USGS Information Services, Denver Federal Center Denver, Colorado. Chapter 12.
Hill, R.J., Jarvie, D.M., 2007. Barnett shale. Am. Assoc. Pet. Geol. Bull. 91, 399–622.
Hill, D.G., Nelson, C.R., 2000. Gas Productive Fractured Shales: An Overview and Update. GasTIPS (Summer), Hart Energy Publishing, Houston, TX.
Hite, R.J., Anders, D.E., Ging, T.G., 1984. Organic-rich source rocks of Pennsylvanian age in the paradox basin of Utah and Colorado. In: Woodward, J., Meissner, F.F., Clayton, J.L. (Eds.), Hydrocarbon Source Rocks of the Greater Rocky Mountain Region. Guidebook, Rocky Mountain Association of Geologists Guidebook, Denver, Colorado, pp. 255–274.
Holst, T.B., Foote, G.R., 1981. Joint orientation in Devonian rocks in the Northern portion of the lower Peninsula of Michigan. Geol. Soc. Am. Bull. 92 (2), 85–93.
IEA, 2012. Golden Rules for a Golden Age of Gas. OECD Publishing, International Energy Agency, Paris, France.
IEA, 2013. Resources to Reserves 2013: Oil, Gas and Coal Technologies for the Energy Markets of the Future. OECD Publishing, International Energy Agency, Paris, France.
Islam, M.R., Speight, J.G., 2016. Peak Energy—Myth or Reality? Scrivener Publishing, Beverly, Massachusetts.
Jacobi, D., Breig, J., LeCompte, B., Kopal, M., Mendez, F., Bliven, S., Longo, J., 2009. Effective geochemical and geomechanical characterization of shale gas reservoirs from wellbore environment: Caney and the Woodford shale. Paper No. SPE 124231, In: Proceedings of SPE Annual Technical Meeting, New Orleans, Louisiana, Oct. 4–7.
Jarvie, D.M., Claxton, B.L., Henk, F., Breyer, J.T., 2001. Oil and shale gas from the Barnett shale, Ft. Worth basin, Texas. Proceedings of AAPG Annual Meeting. A100.
Jarvie, D.M., Hill, R.J., Ruble, T.E., Pollastro, R.M., 2007. Unconventional shale-gas systems: the Mississippian Barnett shale of North central Texas, as one model for thermogenic shale-gas assessment. AAPG Bull. 9, 475–499.
Jarvine, D., Hill, R.J., Ruble, T.E., Pollastro, R.M., 2007. Unconventional shale-gas systems: the Mississippian Barnett shale of North-central Texas as one model for thermogenic shale-gas assessment. AAPG Bull. 91, 475–499.
Jennings, G.L., Greaves, K.H., Bereskin, S.R., 1997. Natural gas resource potential of the Lewis Shale, San Juan Basin, New Mexico and Colorado. SPE Paper No. 9766, Society of Petroleum Engineers, Richardson, Texas.
Jochen, J.E., Lancaster, D.E., 1993. Reservoir characterization of an Eastern Kentucky Devonian shale well using a naturally fractured, layered description. SPE Paper No. 26192, In: Proceedings of SPE Gas Technology Symposium, Calgary, Alberta, Canada, Jun. 28–30.

Kepferle, R.C., 1993. A depositional model and Basin analysis for the gas-bearing Black shale (Devonian and Mississippian) in the Appalachian basin. Bulletin No. 1909, In: Roen, J.B., Kepferle, R.C. (Eds.), Petroleum Geology of the Devonian and Mississippian Black Shale of Eastern North America. United States Geological Survey, Reston, Virginia, pp. F1–F23.

Khlaifat, A.L., Qutob, H., Barakat, N., 2011. Tight gas sands development is critical to future world energy resources. Paper No. SPE 142049-MS, In: Proceedings of SPE Middle East Unconventional Gas Conference and Exhibition, Muscat, Oman, Jan. 31–Feb. 2. Society of Petroleum Engineers, Richarson, Texas.

Kubik, W., Lowry, P., 1993. Fracture identification and characterization using Cores, FMS, CAST, and borehole camera: Devonian shale, Pike County, Kentucky. SPE Paper No. 25897, In: Proceedings of SPE Rocky Mountains Regional-Low Permeability Reservoirs Symposium. Denver, Colorado, Apr. 12–14.

Kuuskraa, V.A., Wicks, D.E., Thurber, J.L., 1992. Geologic and reservoir mechanisms controlling gas recovery from the Antrim shale. SPE Paper No. 24883, In: Proceedings of SPE ATCE Meeting, Washington, DC, Oct. 4–7.

Mars, J.C., Thomas, W.A., 1999. Sequential filling of a late paleozoic foreland basin. J. Sediment. Res. 69, 1191–1208.

Martini, A.M., Budal, J.M., Walter, L.M., Schoell, N.M., 1996. Microbial generation of economic accumulations of methane within a shallow organic-rich shale. Nature. Sep. 12.

Martini, A.M., Walter, L.M., Budai, J.M., Ku, T.C.W., Kaiser, C.J., Schoell, M., 1998. Genetic and temporal relations between formation waters and biogenic methane: Upper Devonian Antrim shale, Michigan basin, USA. Geochim. Cosmochim. Acta 62 (10), 1699–1720.

Mauro, L., Alanis, K., Longman, M., Rigatti, V., 2010. Discussion of the Upper Cretaceous Baxter shale gas reservoir, Vermillion Basin, Northwest Colorado and Adjacent Wyoming. AAPG Search and Discovery Article #90122©2011, In: Proceedings of AAPG Hedberg Conference, Austin, Texas, Dec. 5–10.

Mittenthal, M.D., Harry, D.L., 2004. Seismic interpretation and structural validation of the Southern Appalachian fold and thrust belt, Northwest Georgia. Georgia Geological Guidebook, 42, University of West Georgia, Carrollton, Georgian, pp. 1–12.

Montgomery, S., 1992. Paradox basin: cane creek play. Pet. Front. 9, 66.

Montgomery, S.L., Jarvie, D.M., Bowker, K.A., Pollastro, R.M., 2005. Mississippian Barnett shale, Fort Worth basin, north-central Texas: gas-shale play with multi-trillion cubic foot potential. AAPG Bull. 89 (2), 155–175.

Morgan, C.D., 1992. Horizontal drilling potential of the Cane Creek Shale, Paradox Formation, Utah. In: Schmoker, J.W., Coalson, E.B., Brown, C.A. (Eds.), Geological studies relevant to horizontal drilling: examples from Western North America. Rocky Mountain Association of Geologists, pp. 257–265.

NEB, 2009. A Primer for Understanding Canadian Shale Gas. National Energy Board, Calgary, Alberta. Nov.

NPC, 2011. Prudent Development: Realizing the Potential of North America's Abundant Natural Gas and Oil Resources. National Petroleum Council, Washington, DC.www.npc.org.

NRF, 2013. Shale Gas Handbook: A Quick-Reference Guide for Companies Involved in the Exploitation of Unconventional Gas Resources. Norton Rose Fulbright LLP, London, United Kingdom.

Nuccio, V.F., Condon, S.M., 1996. Burial and Thermal History of the Paradox Basin, Utah and Colorado, and Petroleum Potential of the Middle Pennsylvanian Paradox Formation. Bulletin No. 2000-O, United States Geological Survey, Reston, Virginia.

Nummendal, D., Molenaar, C.M., 1995. Sequence stratigraphy of ramp-setting strand plain successions: The Gallup Sandstone, New Mexico. In: Van Wagoner, J.C., Bertram, G.T.

(Eds.), Sequence Stratigraphy of Foreland Basin Deposits, pp. 277–310. AAPG Memoir No. 64.
Parker, M., Buller, D., Petre, E., Dreher, D., 2009. Haynesville shale petrophysical evaluation. Paper No. SPE 122937, In: Proceedings of SPE Rocky Mountain Petroleum Technology Conference, Denver, Colorado, Apr. 14–16.
Pashin, J.C., 1994. Cycles and stacking patterns in Carboniferous rocks of the Black Warrior Foreland basin. Trans. Gulf Coast Assoc. Geol. Soc. 44, 555–563.
Pashin, J.C., Kugler, R.L., 1992. Delta-Destructive Spit Complex in Black Warrior Basin: Facies Heterogeneity in Carter Sandstone (Chesterian), North Blowhorn Creek Oil Unit, Lamar County, Alabama. Trans. Gulf Coast Assoc. Geol. Soc. 42, 305–325.
Peachey, B., 2014. Mapping Unconventional Resource Industry in the Cardium Play Region: Cardium Tight Oil Play Backgrounder Report. Petroleum Technology Alliance Canada (PTAC), Calgary, Alberta. May 1.
Rheams, K.F., Neathery, T.L., 1988. Characterization and Geochemistry of Devonian Oil Shale, North Alabama, Northwest Georgia, and South-Central Tennessee (A Resource Evaluation). Bulletin No. 128, Alabama Geological Survey, Tuscaloosa, Alabama.
Roen, J.B., 1993. Introductory review—Devonian and Mississippian Black shale, eastern North America. Bulletin No. 1909, In: Roen, J.B., Kepferle, R.C. (Eds.), Petroleum Geology of the Devonian and Mississippian Black Shale of Eastern North America. United States Geological Survey, Reston, Virginia, pp. A1–A8.
Ross, D.K., Bustin, R.M., 2008. Characterizing the shale gas resource potential of Devonian-Mississippian strata in the Western Canada sedimentary basin: application of an integrated formation evaluation. AAPG Bull. 92, 87–125.
Schamel, S., 2005. Shale gas reservoirs of Utah: survey of an unexploited potential energy resource. Open-File Report No. 461, Utah Geological Survey. Utah Department of Natural Resources, Salt Lake City, Utah. Sep.
Schamel, S., 2006. Shale Gas Resources of Utah: Assessment of Previously Undeveloped Gas Discoveries. Open-File Report No. 499, Utah Geological Survey, Utah Department of Natural Resources, Salt Lake City, Utah. Sep.
Switzer, S.B., et al., 1994. Devonian Woodbend-Winterburn strata of the Western Canadian sedimentary basin. Geological Atlas of the Western Canadian Sedimentary Basin. CSPG/ARC, Calgary, Alberta, pp. 165–202. Chapter 12.
Telle, W.R., Thompson, D.A., Lottman, L.K., Malone, P.G., 1987. Preliminary burial-thermal history investigations of the Black Warrior Basin: implications for coalbed methane and conventional hydrocarbon development: Tuscaloosa, Alabama, University of Alabama. In: Proceedings of 1987 Coalbed Methane Symposium, pp. 37–50.
Theriault, R., 2008. Characterisation Geochimique et Mineralogique et Evaluation du Potentiel Gazeifere des Shales De l'Utica et du Lorraine, Basses-Terres du Saint-Laurent. Quebec Exploration 2008. Quebec City, Quebec, Nov.
Thomas, W.A., 1988. The Black Warrior basin. In: Sloss, L.L. (Ed.), Sedimentary Cover—North American Craton. In: The Geology of North America, vol. D-2. Geological Society of America, Boulder, Colorado, pp. 471–492.
Thomas, W.A., 2001. Mushwad: ductile duplex in the Appalachian thrust belt in Alabama. Am. Assoc. Pet. Geol. Bull. 85, 1847–1869.
Thomas, W.A., Bayona, G., 2005. The Appalachian Thrust Belt in Alabama and Georgia: Thrust-Belt Structure, Basement Structure, and Palinspastic Reconstruction. Geological Survey Monograph No. 16, Alabama Geological Society Tuscaloosa, Alabama.
Thomas, W.A., Mack, G.H., 1982. Paleogeographic relationship of a Mississippian Barrier-island and Shelf-Bar system (Hartselle Sandstone) in Alabama to the Appalachian-Ouachita orogenic belt. Geol. Soc. Am. Bull. 93, 6–19.
Thomas, W.A., Astini, R.A., Osborne, W.E., Bayona, G., 2000. Tectonic framework of deposition of the Conasauga formation. In: Osborne, W.E., Thomas, W.A.,

Astini, R.A. (Eds.), The Conasauga Formation and Equivalent Units in the Appalachian Thrust Belt in Alabama. AlabamaGeological Society 31st Annual Field Trip Guidebook, Alabama Geological Society, Tuscaloosa, Alabama, pp. 19–40.
US EIA, 2011a. Shale Gas and Shale Oil Plays. Energy Information Administration, United States Department of Energy, Washington, DC. Jul.
US EIA, 2011b. World Shale Gas Resources: An Initial Assessment of 14 Regions Outside the United States. Energy Information Administration, United States Department of Energy, Washington, DC.
US EIA, 2012. U.S. Crude Oil, Natural Gas, and Natural Gas Liquids Proved Reserves, 2010. Energy Information Administration, United States Department of Energy, Washington, DC. Aug.
US EIA, 2013. EIA/ARI World Shale Gas and Shale Oil Resource Assessment: Technically Recoverable Shale Gas and Shale Oil Resources: An Assessment of 137 Shale Formations in 41 Countries Outside the United States. Energy Information Administration, United States Department of Energy, Washington, DC. May 17. http://www.adv-res.com/pdf/A_EIA_ARI_2013%20World%20Shale%20Gas%20and%20Shale%20Oil%20Resource%20Assessment.pdf (accessed 27.02.16.).
US EIA, 2015. Technically Recoverable Shale Oil and Shale Gas Resources. Energy Information Administration, United States Department of Energy, Washington, DC.
USGS, 2008. Assessment of Undiscovered Oil Resources in the Devonian-Mississippian Bakken Formation, Williston Basin Province, Montana and North Dakota. Fact Sheet No. 2008-3021, United States Geological Survey, Reston, Virginia.
USGS, 2014. Map of assessed tight-gas resources in the United States, 2014. Digital Data Series DDS-69-HH, U.S. Geological Survey National Assessment of Oil and Gas Resources Project. United States Geological Survey, Reston, Virginia.
Walter, L.M., McIntosh, J.C., Budai, J.M., Martini, A.M., 2000. Hydrogeochemical controls on gas occurrence and production in the New Albany Shale. GasTIPS 6 (2), 14–20.
Williams, J., Kramer, H., 2011. Montney shale formation evaluation and reservoir characterization case study well comparing 300 m of core and log data in the upper and lower Montney. In: Proceedings of 2011 CSPG CSEG CWLS Convention, Calgary, Alberta, Canada.
Zuber, M.D., Williamson, J.R., Hill, D.G., Sawyer, W.K., Frantz, J.H., 2002. A comprehensive reservoir evaluation of a shale gas reservoir—the New Albany shale. SPE Paper No. 77469, In: Proceedings of Annual Technical Conference and Exhibition, San Antonio, Texas, Sep. 20–Oct. 2.

CHAPTER FOUR

Development and Production

1 INTRODUCTION

The volumes of natural gas and crude oil gas in place in tight reservoirs have been given various estimates, and while production is underway in the United States and Canada, production of these resources by other countries has not yet searched the same level of activity. In fact, outside North America, only the few companies proficient in the complex techniques required to produce this gas have shown an active interest in the resources in tight reservoirs. For this reason, improving the recovery factor of any reservoir and driving down operating costs pose strategic challenges for developers and, as such, constitute the main targets of their efforts. The resources of natural gas and crude oil in tight formations have the potential to expand and facilitate the current wave of growth in the natural gas and crude oil gas industries. However, this is not to be taken from granted and issues arise when producing and refining these two resource's, but nevertheless, the future for the development and production of tight reservoirs holds much promise and may even help discourage the continuing myths of peak energy (i.e., peak oil and gas) (Islam and Speight, 2016).

Although tight reservoirs world, sometimes associated with conventional natural gas and crude resources, they were long considered a secondary target because of the difficulty of producing natural gas and crude oil from these reservoirs. In the case of the United States, not until conventional resources began to decline did operators—spurred on by fiscal incentives—turn to the potential of tight gas and tight oil. Currently, the United States produces approximately 40% (v/v) of its gas from unconventional tight reservoirs. Elsewhere, the volumes of natural gas and crude oil in place, coupled with improvements in production techniques (such as horizontal drilling and hydraulic fracturing) (Chapter 5), have made it economically feasible to produce other natural gas and crude oil from other tight reservoirs (Chapter 3).

Deep Shale Oil and Gas
http://dx.doi.org/10.1016/B978-0-12-803097-4.00004-8

In the present context, a tight gas reservoir and a tight oil reservoir are the terms commonly used to refer to low-permeability formations (i.e., low-permeability reservoirs) that produce dry natural gas or light tight oil (Chapters 1 and 2). Many of the low-permeability reservoirs that have been developed in the past are sandstone, but significant quantities of gas are also produced from low-permeability carbonates, shales, and coal seams. Production of gas from coal seams is not the focus of this book but needs to be acknowledged as another unconventional source of natural gas (coalbed methane) (Chapter 1). Also, natural fractures affect both the overall level of permeability in a reservoir. If a reservoir is naturally fractured, it is possible that a horizontal well or multilateral wellbores will be more effective in producing gas than a vertical well with a hydraulic fracture. If a hydraulic fracturing is performed in a reservoir containing an abundance of natural fractures, problems with multiple hydraulic fractures near wellbore, issues related to tortuosity, and excessive fluid leak-off can occur during the fracturing treatment.

Tight formation reservoirs have a common feature insofar as recovery of the gas or oil requires that a well drilled and completed in the tight reservoir must be successfully stimulated to produce at commercial gas flow rates and produce commercial gas volumes. In addition, a vertical well is not always the most effective type of well to drill into a tight formation to recover the gas or fluid resource, and other actions are required to commercial development of the resource. Thus, in order to sustain a ready supply of energy, unconventional sources of natural gas and crude oil are being continually investigated and developed.

Recent development of new technologies of formation evaluation and simulating by horizontal drilling and hydraulic fracturing (Chapter 5), especially in the United States and Canada, has made low productive unconventional tight shale formations and other tight formations (including tight sandstone formations and tight carbonate formations) as well as coalbed methane formations (Chapter 1) as attractive resources for production of natural gas, gas condensate, and crude oil. However, economic production of tight natural gas and tight crude oil reservoirs is challenging since the permeability of these reservoirs is in the microDarcy range (Fig. 4.1)—although shale reservoirs and tight sandstone (or carbonate) reservoirs could have a permeability on the same order of magnitude. Nevertheless, in spite of the shortcoming of low permeability and the inability of the fluids to be mobile within the reservoir system, these reservoirs and their natural gas and/or crude oil content do offer a huge potential for current and future energy production.

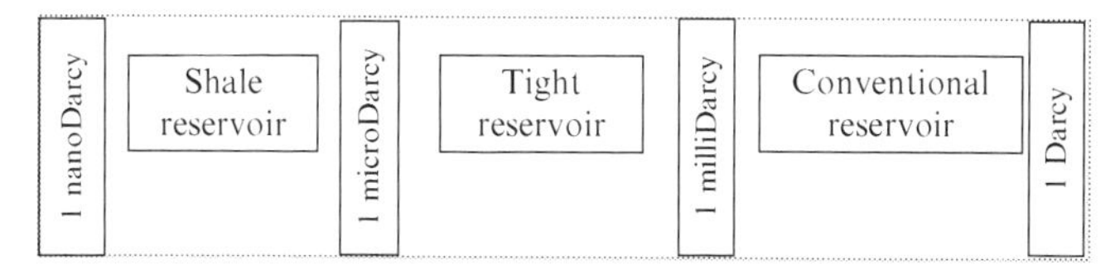

Fig. 4.1 Representation of the differences in permeability of shale reservoirs, tight reservoirs, and conventional reservoirs.

Unlike conventional natural gas and crude oil reservoirs (Chapter 2), shale formations and tight sandstone or carbonate formations are considered to be the source rock in which the organic material evolved and also, because of the low permeability, became the reservoir rock. It is considered likely that the natural gas and crude oil remained in the formation in which they were originally produced. Both the shale reservoir system and the tight sandstone (or carbonate) reservoir systems have free gas stored within the pores of the rock matrix, although the tight reservoir system can differ from the conventional reservoir system in possessing the characteristic of gas or oil adsorption on the surface areas associated with organic content and clay (Chapter 2). The relative importance of adsorbed vs free gas and/or oil varies as a function of (1) the amount of organic matter present, (2) the mineralogy, (3) the pore size distribution, (4) processes that contribute to diagenesis, (5) the rock texture, as well as (6) the reservoir pressure and the reservoir temperature (Bustin et al., 2008).

Geologically shale is a sedimentary rock that is predominantly comprised of very fine-grained clay particles deposited in a thinly laminated texture while the term *tight formation* refers to a formation consisting of extraordinarily impermeable, hard sandstone or carbonate rock. Like the shale formations, tight sandstone and carbonate formations are relatively low-permeability, sedimentary formations that can contain natural gas and crude oil. When a significant amount of organic matter has been deposited with the sediments, the shale rock can contain organic solid material, which is referred to as *kerogen* (Scouten, 1990; Lee, 1991; Speight, 2008, 2012, 2014, 2016).

The shale formations and other tight formations were originally deposited as mud in low-energy depositional environments, such as tidal flats and swamps, where the clay particles and other inorganic particles were deposited from the suspension. During the deposition of these sediments, organic matter was also deposited and deep burial of the sediment resulted in a multimineral layered rock (*shale*, *sandstone*, or *carbonate*) that contributed to the heterogeneous nature and laminar nature of the sediment and which can differ significantly between shale formations.

2 TIGHT RESERVOIRS AND CONVENTIONAL RESERVOIRS

Briefly, *conventional gas reservoirs* and conventional crude oil reservoirs contain *free* gas and crude oil, respectively, in interconnected pore spaces that can flow easily to the wellbore, that is, natural flow is possible (Chapters 1 and 2). These reservoirs are identified by distinguishing between the porous rock that contains the natural gas or crude oil and the cap rock as well as the basement rock that constitute a permeability barrier.

On the other hand, *unconventional gas reservoirs* (i.e., shale gas reservoirs) produce from low-permeability (tight and now ultratight) formations (Chapters 1 and 2). The gas is often sourced from the reservoir rock itself and adsorbed onto the matrix. The low productivity typical of tight gas reservoirs is usually insufficient to ensure the economic viability of their development. The solution is to connect as much of the reservoir volume as possible to the well (provided this can be done cost-effectively), thereby limiting the number of boreholes needed to produce the reserves of natural gas and/or crude oil. The best well design (economic optimum) can be determined based on the identification and appraisal of the reservoirs. Due to the low permeability of these formations, it is necessary to stimulate the reservoir by creating a fracture network to give enough surface area to allow sufficient production from the additional *enhanced* reservoir permeability. Options include vertical wells with single or multiple hydraulic fractures, horizontal or sharply deviated wells, multilateral drains, and multiple fractures in multilateral completions (Chapter 5).

Because of the uncertainties related to the identification of tight reservoirs and the differences that exist between such reservoirs, development of a natural gas field or a crude oil field always poses challenges. For example, meandering formation that are the reservoirs may overlap to varying extents and may also be interconnected to varying degrees. As a result, drilling a well to produce a natural gas or crude oil reservoir with limited or no connections to other zones will yield only a small volume of gas. The key is to site the wells in zones known to have extensive connections to other reservoirs in order for each well to produce enough gas to be economically viable. Meeting this challenge requires comprehensive imaging of the spatial distribution of the reservoir system. The analysis of productivity test data provided valuable information concerning the extent of the sandstone lenses. Based on these data, the geologists and reservoir engineers can replicate the reservoir system in the form of a reservoir model and highlight the differences in well productivity according to the spatial configuration and interconnection of the reservoirs.

Thus, the analysis of any reservoir, including a tight reservoir, should always begin with a thorough understanding of the geologic characteristics of the formation (Chapter 2). The important geologic parameters for a play or basin are: (1) the structural and tectonic regime, (2) the regional thermal gradients, and (3) the regional pressure gradients. Furthermore, knowing the stratigraphy in a basin is an important aspect of resource development and can affect drilling evaluation, completion, and stimulation operations. For example, important (if not, essential) geologic parameters that should be studied for each stratigraphic unit are: (1) the depositional system, (2) the genetic facies, (3) the textural maturity, (4) the mineralogy, (5) any diagenetic processes, (6) reservoir dimensions, and last, but certainly not least, (6) the presence of natural fractures. When most sandstone sediments are deposited, the pores and pore throats are connected, thereby resulting in high permeability of the sediment. The original porosity and permeability of sandstone is determined by characteristics such as mineral composition, pore type, grain size, and texture. After deposition and burial, the grains and matrix are commonly altered by the physical effects of compaction and chemical changes (diagenesis).

To complete, fracture, and produce a natural gas or crude oil from a tight reservoir, each layer of the formation and the formations above and below the gas-containing or oil-containing formation must be thoroughly evaluated in term of: (1) formation thickness, (2) porosity, (3) permeability, (4) water saturation, (5) pressure, (6) in-situ stress, and (7) Young's modulus using data from: logs, cores, well tests, and any drilling records. In as much as tight reservoirs are normally also low-porosity reservoirs, the importance of detailed log analyses becomes critical to understanding the reservoir.

One of the most difficult parameters to evaluate in tight gas reservoirs is the drainage area size and shape of a typical well. In tight reservoirs, months or years of production are normally required before the pressure transients are affected by reservoir boundaries or well-to-well interference. As such, an estimation of the drainage area size and shape for a typical well has to be made in order to estimate reserves. Furthermore, knowledge of the depositional system and the effects of diagenesis on the rock are needed to estimate the drainage area size and shape for a specific well. In tight reservoirs, the typical drainage area of a well largely depends on the number of wells drilled, the size of the fracture treatments pumped on the wells, and the time frame being considered. In lenticular or compartmentalized tight gas or oil reservoirs, the average drainage area is likely a function of the average size of the sand lens or the average size of the compartment and may not be a strong function of the size of the fracture treatment.

Generally, reservoir drainage per well is small in continental deposits and larger in marine deposits. Fluvial systems tend to be more lenticular. Barrier-strand-plain systems tend to be more blanket and continuous. Marine deposits tend to be more blanket and continuous. Most of the more successful tight gas plays are those in which the formation is a thick, continuous, marine deposit. By way of explanation, a strand plain (strandplain) is a broad belt of sand along a shoreline with a surface exhibiting well-defined parallel or semiparallel sand ridges separated by shallow swales. A strand plain differs from a barrier island in that it lacks either the lagoons or tidal marshes that separate a barrier island from the shoreline to which the strand plain is directly attached. Also, the tidal channels and inlets which cut through barrier islands are absent. Strand plains typically are created by the redistribution by waves and longshore currents of coarse sediment on either side of a river mouth.

The best way to determine the depositional system is to cut and analyze cores (Chapter 2). Taking cores from the tight formations and from the non-reservoir rock above and below the main pay interval is recommended. From the test data, important information about the depositional system becomes available, and the core analyses can be correlated with open-hole logging data to determine the various depositional environments. Once these correlations are made, logs from additional wells can be analyzed to generate maps of the depositional patterns in a specific area from which field optimization plans can be developed.

Natural gas and crude oil from shale formations and other tight formations are becoming an increasingly important energy source for meeting rising energy demands in the next several decades. Development of horizontal drilling and hydraulic fracturing is crucial for economic production of tight reservoirs, but it must be performed with caution and as a multidisciplinary approach (King, 2010; Speight, 2016). The success of the Barnett shale in the United States has illustrated that gas can be produced economically from the rock that was previously thought to be source rock and/or cap rock, not reservoir rock, leading to the development of many other tight reservoirs, including (alphabetically but not by preferences) the Fayetteville formation, Haynesville formation, Marcellus formation, and the Woodford formation.

These commercial successes in the Barnett shale, which is currently the largest producing natural gas field, and other shale plays in the United States have made exploration of tight reservoirs not only possible but also economic and development has begun to spread all around the world. In most cases, economic production is possible only if a very complex, highly nonlinear fracture network can be created that effectively connects a large reservoir surface area to the wellbore (Upolla et al., 2009).

However, the economic viability of many unconventional gas developments requires effective stimulation of the extremely low-permeability rock by means of horizontal drilling followed by hydraulic fracturing (Chapter 5). In the process, the drill path is, at first, vertical and then the drill stem is caused to deviate (hence the alternate name of deviated drilling) and into the reservoir. Once reservoir penetration is achieved satisfactorily, hydraulic fractures are created in the reservoir to effectively connect a large gas-bearing or oil-bearing area to the wellbore when the wellbore is drilled in the direction of minimum horizontal stress. Maximizing the total stimulated reservoir volume (SRV) plays a major role in successful economic gas production (Yu and Sepehrnoori, 2013). Despite the success of recent tight reservoir development recently, it is difficult to predict well performance and evaluate economic viability for other shale resources with certainty because it involves high risk and uncertainties.

Thus, the uncertainties of reservoir properties and fracture parameters have significant effect on production of natural gas and crude oil from tight shales and tight formations, making the process of the design of the hydraulic fracturing process and optimization of the process, for economic gas production of gas or oil, much more complex. Thus, it is extremely important to identify the necessary and important process parameters and also to evaluate the effects of these parameters on well performance. As part of the optimization procedure, the detailed reservoir properties that affect each wellbore should be assessed by a multidisciplinary team of scientists and engineers. For example, the optimization of critical hydraulic fracture parameters such as (1) fracture spacing, (2) fracture half-length, and (3) fracture conductivity, which control well performance, is important to obtain the most economical scenario. The cost of hydraulic fracturing of horizontal wells is expensive. Therefore, the development of a method quantifying uncertainties and optimization of natural gas and crude oil production with economic analysis in an efficient and practical way is clearly desirable (Zhang et al., 2007).

In general, the ultralow permeability of tight formations ranges from as low as 1 nanoDarcy (10^{-6} Darcy) to 1 milliDarcy, illustrating that tight reservoirs require to be artificially fractured in order to make low-permeability formations produce economically. Typically, the Barnett shale reservoir exhibits a net thickness of 50–600 ft, porosity of 2–8% (v/v), total organic carbon (TOC) of 1–14% found at depths ranging from 1000 to 13,000 ft (Cipolla et al., 2010). Furthermore, reported that fracture spacing varied in the range from 100 to 700 ft in actual hydraulic fracturing operations in three Barnett shale wells (Grieser et al., 2009) and well performance of Barnett shale changes significantly with (1) the changing properties of the produced fluid, (2) the type of fluid, (3) the depth of the formation, and (4) the thickness of the formation (Hale and William, 2010). Also, to add further complications, the productivity of wells in the Barnett shale is highly dependent on (1) the type of completion method implemented and (2) the large hydraulic fracture treatments (Ezisi et al., 2012). To optimize the development of a tight gas reservoir, the scientists and engineers must optimize the number of wells drilled, as well as the drilling and completion procedures for each well. To further complicate matters, it does not follow that each well will be equal to or equivalent to an adjacent well in terms of behavior and productivity. Often, more data and more scientific and engineering data required to understand and develop tight reservoirs than are required for the higher permeability, conventional reservoirs. For example, on an individual basis, a well in a tight gas reservoir will produce less gas over a longer period of time than that can be expected from a well completed in a higher permeability, conventional reservoir. As such, many more wells (with smaller well spacing between wells) must be drilled into a tight shale reservoir or into another form of tight reservoir (sandstone or carbonate) to recover an effective amount of the original gas in place (OGIP) or the original oil in place (OOIP) when compared to a conventional reservoir. In summary, the optimum drilling, completion, and stimulation methods for each well are a function of the reservoir characteristics as well as economic parameters. The costs to drill, complete, and stimulate the wells, plus the gas price or oil price and the gas market or oil market serve to affect how tight reservoirs are developed.

3 WELL DRILLING AND COMPLETION

As stated earlier (Chapter 1), shale formations and tight formations have very low permeability (measured in nanoDarcys or, at best, milliDarcys) (Fig. 4.1). In addition, natural gas and crude oil in tight formations will not flow readily to vertical wells because of the low permeability of the formation. This can be overcome by drilling horizontal wells where the drill bit is steered from its downward trajectory to follow a horizontal trajectory for the necessary distance—often 1–2 miles–thereby exposing the wellbore to as much of the reservoir as possible. As a result, many wells are required to create acceptable and efficient well productivity and special well design and well stimulation techniques are required to deliver production rates of sufficient levels to make a development economic (Schweitzer and Bilgesu, 2009). Thus, it is not surprising that horizontal drilling and stimulation by hydraulic fracturing have both been crucial in the development of the resources in tight reservoirs (Houston et al., 2009).

3.1 Drilling

The most important part of drilling a well in a tight gas reservoir or a tight oil reservoir is to drill a gauge hole, which is required to obtain an adequate suite of open-hole logs and to obtain an adequate primary cement job. In low porosity, shale reservoirs, the analyses of gamma ray, spontaneous potential, porosity, and resistivity logs to determine accurate estimates of shale content, porosity, and water saturation can be difficult. If the borehole is washed out (*out of gauge*), the log readings will be affected, and it will be even more difficult to differentiate the pay from the nonpay portions of the formation. If the borehole is washed out, obtaining a primary cement seal is difficult, which could affect zonal isolation and cause the well to have to be cement squeezed prior to running tests or pumping stimulation treatments.

Formation damage and drilling speed should be a secondary concern. Some wells are drilled underbalanced to increase the bit penetration rate or to minimize mud filtrate invasion. However, if the wellbore is severely washed out because the well was drilled underbalanced, it is probable that a lot of money will be wasted because the logs are not accurate and the primary cement job might not be adequate. It is best to drill a tight gas well near balanced to minimize borehole washouts and mud filtrate invasion.

Natural gas and crude oil in tight formations will not flow readily to a vertical well drilled through it because of the low permeability of the shale. This can be overcome to some extent by drilling horizontal wells, where the drill bit is steered from its downward trajectory to follow a horizontal trajectory for 1 mile or more to thereby exposing the wellbore to as much reservoir as possible. By drilling horizontally, the wellbore may intersect a greater number of naturally existing fractures in the reservoir—the direction of the drill path is chosen based on the known fracture trends in each area.

However, some shale formations can only be drilled with vertical wells because of the risk of the borehole collapsing.

The use of horizontal drilling in conjunction with hydraulic fracturing has greatly expanded the ability of producers to recover natural gas and crude oil efficiently from low-permeability geologic plays, particularly, shale resources (EIA, 2011). This history of the application of hydraulic fracturing techniques to stimulate crude oil and natural gas production goes back well into the 20th century and even into the 19th century but began to grow rapidly in the 1950s. Since this time, the oil and gas industry has been completing and fracture treating low-permeability wells in the United States. However, it was the natural gas price increase in the 1970s that spurred significant activity in low-permeability gas reservoirs and since then sustained increases in natural gas prices and crude oil prices along with advances in evaluation, completion, and stimulation technology have led to substantial development of tight formations containing natural gas and crude oil.

Furthermore, commencing in the mid-1970s, a partnership of private operators with the United States Department of Energy and predecessor agencies and with the Gas Research Institute endeavored to develop technologies for the commercial production of natural gas from the relatively shallow Upper Devonian (Huron) shale (which lies just above the Marcellus Shale—the Marcellus is part of the Middle Devonian formation), in the eastern United States. This partnership greatly assisted in the development of technologies that eventually became crucial to the production of natural gas from shale formations, including drilling horizontal wells as well as multistage hydraulic fracturing and slick-water fracturing (Chapter 5). Following from this development, the practical application of horizontal drilling to production of crude oil from tight formations began in the early 1980s by which time the advent of improved downhole drilling motors and the development of other necessary supporting equipment, materials, and new technologies (particularly, downhole telemetry equipment) had opened the path to bring some applications into the realm of commercial operations (EIA, 2011). By this time, it had been recognized that tight reservoirs require *fracture stimulation* to connect any natural fracture network to the well bore (Gale et al., 2007). The fissures created by the hydraulic fracturing process are held open by the sand particles so that the reservoir fluids from within the tight (but fractured) formation can flow up through the well. Once released through the well, the natural gas, crude oil, and water are captured, stored, and transported to the relevant on-site processing operations.

The recovery technology was primarily developed in the Texas Barnett shale and applied to other shale lay resources, often with a one-method-fits-all approach. However, the tight formations encountered in any operation to produce natural gas or crude oil are different and each has a unique set of exploration criteria and operational challenges, and there is now a realization that the Barnett shale technology needs to be adapted to other tight resources in a scientifically technologically structured manner. Thus, while it might be thought that there is a rule-of-thumb that unconventional resources (resources in tight formations) need the application of unconventional recovery techniques, it is clear that a poorer quality of the reservoir needs improved technology as well as highly accurate data to be able to fully characterize and develop each reservoir (resource) in an efficient and effective manner (Grieser and Bray, 2007).

In fact, over the past decade, resources from tight formations have emerged as a viable energy source and the accurate characterization of the whole resource (the reservoir plus the reservoir fluids) using geophysical methods has gained high significance (Chopra et al., 2012). The natural gas and crude oil reserves contained in these formations, which are estimated from the TOC content of the formation, influence the compressional and shear velocity as well as the density and anisotropy in the formation. Consequently, detecting changes in the TOC content from the surface seismic response is a necessary step in reservoir (and resource) characterization (Chapter 2). And, in addition to the TOC content, different shale formations have different properties in terms of maturation, gas-in-place, permeability, and brittleness. Thus, there has also been the realization that typical tight reservoirs (if there are such reservoirs) are more expensive and labor intensive than conventional reservoirs, and the expertise needed to characterize reservoir and stimulation treatments is much more specialized.

In the process of reservoir characterization and prior to the onset of recovery operations, a number of vertical wells (perhaps two or three exploratory wells) are drilled and the hydraulic fracturing process applied to the formation to determine if natural gas or crude oil is present and can be extracted. This exploration stage may include an appraisal phase where more wells (perhaps 10–15) are drilled and fractured to characterize the formation. At this time, there is (or should be) an examination of the means in which the fracture pattern develops and the means by which the fractures will propagate and whether or not the natural gas or crude oil can be produced economically. As a final step, more wells may be drilled (perhaps reaching a total of 30 wells) to ascertain the long-term producibility and economic viability of the shale. Once the reservoir properties and contents have been defined, the drilling program and recovery operations can commence with a relatively high degree of success.

The leader and prime mover in the development of natural gas and crude oil recovery from tight formations has been the Barnett shale formation with the results that, as of the current time, the focus of tight resource development has shifted to other tight reservoirs in the United States and Canada such as (alphabetically and not by timing or preference) (1) Fayetteville shale (in Arkansas), (2) the Haynesville (on the Texas-Louisiana border), (3) the Horn River Basin shale (in British Columbia, Canada), and (4) Marcellus shale (in the north-eastern United States). The physical and geophysical properties vary significantly both among and within tight shale gas plays as well as in other tight formations.

Furthermore, insofar as tight formations (especially shale plays) cover large areas, they also require that more wells are drilled more closely together than during the development of those of conventional reservoirs which leads to a much larger surface area being affected by the drilling and production operations. For example, some areas may require that wells be drilled have been drilled with wells every 15–20 acres. In many cases, 20–30 wells are often drilled from a single surface location, and long-reach horizontal wellbores of up to 1–2 miles are drilled to reduce the environmental impact on the surface. Future technological breakthroughs that further reduce the surface environmental impact would facilitate the development of more tight formations, especially in the more densely populated or environmentally sensitive areas, and this will be a key factor in resource development in many countries.

3.1.1 Horizontal Drilling and Pad Drilling

During the 100 years or so of the existence of the petroleum and natural gas industries (Mokhatab et al., 2006; Speight, 2007, 2014), drilling technology has progressed from the point of downward (vertical drilling) only to the modern system where the drill stem can deviate from the vertical and (literally) turn corners and progress on a horizontal track while accurately staying within a narrow directional window. Because the horizontal portion is easily controlled, the well is able to drain natural gas and crude oil resources from a wide geographical area that is much larger than the area accessed by a single vertical well in the same tight formation.

Thus, horizontal drilling is a technique that allows the wellbore to come into contact with significantly larger areas of hydrocarbon bearing rock than in a vertical well. As a result of this increased contact, production rates and recovery factors can be increased. As the technology for horizontal drilling and hydraulic fracturing has improved, the use of horizontal drilling has increased significantly. For example, in the Barnett shale (United States), the number of horizontal wells drilled in the 2001–2003 period totaled 76. In the 2007–2008 period, the number of wells drilled had risen to 1810, while over the same period the number of new vertical wells in the Barnett declined from 2001 to 131 wells.

Many tight formations containing natural gas and crude oil resources are located at depths of 6000 ft or more below ground level and can be relatively thin (e.g., the Marcellus shale formation is between 50 and 200-ft thick, depending upon the location of any particular part of the formation). The efficient extraction of natural gas or crude oil from such a thin layer of the formation rock is not achievable by use of a vertically drilled well and requires drilling horizontally through the shale. This is accomplished by drilling vertically downward to one side of the formation until the drill bit reaches a distance of approximately 900 ft from the shale formation. At this point, a directional drill is used to create a gradual 90-degree curve (a 90-degree elbow) so that the *wellbore* becomes horizontal as optimal distance from the formation is reached. The drilling is then programed so that wellbore follows the formation horizontally for several thousand feet (typically 5000 ft) or more. As a result, multiple horizontal wells accessing different parts of the tight formation can be drilled from a single drilling pad, and thus, horizontal drilling reduces the environmental footprint of the drilling operations by enabling a large area of the tight formation shale to be accessed from a single pad. Thus, in the process, a large number of fractures are created mechanically in the formation, thereby allowing the natural gas and/or crude oil trapped in subsurface formations to move through those fractures to the wellbore from where it can then flow to the surface.

When multiple wells are drilled from the same pad, it is often referred to as *pad drilling* in which as many as 6–8 horizontal wells can originate from the same pad. Briefly, a drilling pad is a location which houses the wellheads for a number of (vertically drilled or horizontally drilled) wells. The benefit of a drilling pad is that drilling operators can drill multiple wells in a shorter time than they might be the case when only one well is drilled per site. For example, moving a drilling rig between two well sites (*drilling rig mobility*) previously involved disassembling the rig (*rigging down*) and reassembling it at the new location (*rigging up*) even if the new location was only a short distance (even a few yards) away. A drilling pad may have 5–10 wells, which are horizontally drilled in different directions, spaced fairly close together at the surface. At the completion of the first phase of drilling, the fully constructed rig can be lifted and moved a few yards over to the next well location using hydraulic walking or skidding systems.

In terms of efficient drilling into tight formations, it has become increasingly common to use a single drill pad to develop as large an area of the reservoir as possible, which also serves to reduce the environmental footprint of the surface operation. In addition, when one surface location is used to drill multiple wells, the operational efficiency of natural gas production and crude oil production is increased. Moreover, the infrastructure costs are reduced and, perhaps more important to some observers, land use is also decreased, thereby reducing and mitigating any negative environmental impact on the land surface as well as reducing the potential for leaching of chemicals into the groundwater.

Typically, the well pad drains an area that is rectangular (known as the *spacing unit*, *unit*, or *pool*), which is typically a size on the order of approximately 1.5 miles wide by 2 miles long with the drilling pad actually positioned at the center of the surface rectangle. The space required for the well pad is typically on the order of 4–5 acres that has to be cleared, leveled, and surfaced over to accommodate the drilling rig, trucks, and various other pieces of ancillary equipment that are required for the drilling and well-completion activities. However, the majority of the surface area in the rectangle is not required for the well pad and will be left completely undisturbed. This approach offers an added benefit by allowing the drilling company to simultaneously develop two separate formations on two separate spacing units, thereby increasing production efficiency of the gas and/or oil and reducing surface environmental effects while, at the same time, allowing the production company to recover more of the available natural gas or crude oil resources from the reservoir.

Pad drilling may be accomplished through the use of a movable flex or suitable-for-the-purpose drilling rigs with the intent to drill as many wells on a pad as are economically feasible. Drilling more wells on a pad is considered to help minimize the environmental impact (*environmental footprint*) of the drilling operation. In summary, for tight reservoirs it is becoming increasingly common to use a single drill pad to develop as large an area of the subsurface as possible. One surface location may be used to drill multiple wells. Pad drilling increases the operational efficiency of gas production and reduces infrastructure costs and land use. Any negative impact upon the surface environment is therefore mitigated. Such technologies and practices developed by industry serve to reduce environmental impacts from operations involving the recovery of resources from tight shale formations and other tight formations.

As stated earlier, shales have very low permeability (measured in nanoDarcys). As a result of this, many wells are required to deplete the reservoir, and special well design and well stimulation techniques are required to deliver production rates of sufficient levels to make a development economic. Stimulation through the application of horizontal drilling and hydraulic fracturing has both been crucial in the development of the natural gas and crude oil from tight formations. Using the Marcellus shale resource in Pennsylvania as an example, a vertical well may only drain a cylinder of shale 1320 ft in diameter and as little as 50 ft high. By comparison, a horizontal well may extend from 2000 to 6000 ft in length and drain a volume up to 6000 ft by 1320 ft by 50 feet in thickness, an area about 4000 times greater than that drained by a vertical well. The increase in drainage creates a number of important advantages for horizontal drilling over vertical wells, particularly with respect to associated environmental issues.

Hydraulic fracturing can both (1) increase production rates from the formation and (2) increase the total amount of gas or oil that can be recovered from a given volume of the formation. Pump pressure causes the rock to fracture, and water carries sand (*proppant*) into the hydraulic fracture to prop it open allowing the flow of gas. While water and sand are the main components of hydraulic fracture fluid, chemical additives are often added in small concentrations to improve fracturing performance. And it is often these additives that give cause for environmental concern arising from hydraulic fracturing (Chapters 5 and 9).

The primary differences between the modern development of tight formations and the development of conventional natural gas and crude oil reservoirs are the extensive uses of horizontal drilling and high-volume hydraulic fracturing. The use of horizontal drilling has not introduced any new environmental concerns. In fact, the reduced number of horizontal wells needed coupled with the ability to drill multiple wells from a single pad has significantly reduced surface disturbances and associated impacts to wildlife, dust, noise, and traffic. Where tight formation development of natural gas and crude oil resources has intersected with urban and industrial settings, regulators and industry have developed special practices to alleviate nuisance impacts, impacts to sensitive environmental resources, and interference with existing businesses.

The Barnett shale of Texas, the Fayetteville Shale of Arkansas, the Haynesville Shale of Louisiana and Texas, and the Marcellus Shale of the Appalachian Basin are examples of gas resources in tight shale formations (Chapter 2). In these rock units, the challenge is to recover gas from very tiny pore spaces in a low-permeability rock unit (Gubelin, 2004). To stimulate the productivity of wells in organic-rich shale, companies drill horizontally through the rock unit and then use hydraulic fracturing to produce artificial permeability. Done together, horizontal drilling and hydraulic fracturing can make a productive well where a vertical well would have produced only a small amount of gas. In fact, the productive potential of the Haynesville Shale was not fully realized until horizontal drilling and hydraulic fracturing technologies were demonstrated in other unconventional shale reservoirs. The hydraulic fracturing process—which is accomplished by sealing off a portion of the well and injecting water or gel under very high pressure into the isolated portion of the hole creating high pressure to fracture the rock open the fractures—helps liberate natural gas and crude oil from the tight formation and horizontal drilling allows a single well to drain a much larger volume of rock than a traditional vertical well.

In some geological settings, it is more appropriate to directionally drill S-shaped wells from a single pad to minimize surface disturbance. S-shaped wells are drilled vertically several thousand feet and then extend in arcs beneath the earth's surface. During drilling, mobile drilling units are moved between wells on a single pad. This avoids dismantling and reassembling drilling equipment for each well, making the process quicker and saving resources.

3.1.2 Stacked Drilling and Multilateral Drilling

Most horizontal wells begin at the surface as a vertical well and drilling progresses until the drill bit is a few hundred feet above the target rock unit. At that point the pipe is pulled from the well and a hydraulic motor is attached between the drill bit and the drill pipe. The hydraulic motor is powered by a flow of drilling mud down the drill pipe, and the motor is used to rotate the drill bit without rotating the entire length of drill pipe between the bit and the surface. This allows the bit to drill a path that deviates from the orientation of the drill pipe. After the motor is installed, the bit and pipe are lowered back down the well, and the bit drills a path that steers the well bore from the vertical direction to a horizontal direction over a distance of several hundred feet—the actual distance is geologically dependent on the charter of the intervening formations and the actual location of the tight formation.

Once the well has been steered to the desired angle, straight-ahead drilling resumes and the well follows the target rock unit. Keeping the well in a thin rock unit requires careful navigation. Downhole instruments are used to determine the azimuth and orientation of the drilling. However, issues can occur during the drilling program, and it is at this stage the decision may be made to use the concept of drilling and positioning stacked wells.

Drilling stacked horizontal wells may be possible where the formation is sufficiently thick or multiple shale rock strata are found layered on top of each other. One vertical wellbore can be used to produce natural gas or crude oil from horizontal wells at different depths. One area where this technology is being employed is in the Pearsall and Eagle Ford plays in southern Texas. Cost savings and efficiencies can be achieved as surface facilities are shared. As in pad drilling, the environmental impact on the surface is mitigated as a result of reduced land use. This technology can be particularly beneficial in the thicker shale formations.

Further operational definitions related to the term *stacked drilling* and which included here to avoid any confusion are (1) warm stacked drilling and (2) cold stacked drilling. The former (warm stacked drilling, also called *hot stacked drilling* or *ready stacked drilling*) implied that a drilling rig is deployable (warm) but idle (stacked). Warm stacked rigs are typically standing by and ready for work if a contract can be obtained. Routine rig maintenance is continued, and daily costs may be modestly reduced but are typically similar to levels incurred in drilling mode. Therefore, rigs are generally held in a ready stacked state if a contract is expected to be obtained relatively quickly. On the other hand, a cold stacked drilling ring (sometimes referred to as a mothballed drilling rig) refers to a drilling rig that has been shut down and stored in a designated nonoperational drilling area. Also referred to as mothballing, cold stacking is a cost reduction step taken when the contracting prospects for a drilling rig are minimal or the available contract terms do not justify an adequate return on the investment needed to make the unit work ready (e.g., repairs or refurbishment). Cost savings primarily come from crew reductions to skeletal levels. However, steps must be taken to protect the drilling rig and include (1) application of protective coatings, (2) filling the motors with protective fluids, and (3) installing dehumidifiers. With the costs of crewing up, inspection, deferred maintenance, and potentially refurbishment acting as deterrents to reactivation, cold stacked rigs may be out of service for extended periods of time and may not be actively marketed. A return to service can be a costly proposition, often requiring tens of millions of dollars for refitting costs.

Multilateral drilling involves drilling of two or more horizontal wells from the same vertical wellbore. With multilateral drilling, the horizontal wells access different areas of the formation at the same depth, but in different directions. Drilling multilateral wells makes it possible for production rates to be increased significantly for a reduced incremental cost.

More specifically, a multilateral well is a single well with one or more wellbore branches radiating from the main borehole. It may be an exploration well, an infill development well, or a reentry into an existing well. It may be as simple as a vertical wellbore with one sidetrack or as complex as a horizontal, extended-reach well with multiple lateral and sublateral branches. General multilateral configurations include multibranched wells, forked wells, wells with several laterals branching from one horizontal main wellbore, wells with several laterals branching from one vertical main well bore, wells with stacked laterals, and wells with dual-opposing laterals. These wells generally represent two basic types: vertically staggered laterals and horizontally spread laterals in fan, spine-and-rib, or dual-opposing T shapes.

Vertically staggered wells usually target several different producing horizons to increase production rates and improve recovery from multiple zones by commingling production. Wells in the Austin Chalk play in Texas (USA) are typically of this type (below right). Their production is a function of the number of natural fractures that the wellbore encounters. A horizontal well has a better chance of intersecting more fractures than a vertical well, but there is a limit to how far horizontal wells can be drilled. By drilling other laterals from the same wellbore, twice the number of fractures can often be exposed at a much lower cost than drilling long horizontal sections or another well.

Horizontal fan wells and their related branches usually target the same reservoir interval. The goal of this type of well is to increase production rates, improve hydrocarbon recovery, and maximize production from that zone. Multiple thin formation layers can be drained by varying the inclination and vertical depth of each drain hole. In a naturally fractured rock with an unknown or variable fracture orientation, a fan configuration can improve the odds of encountering fractures and completing an economic well. If the fracture orientation is known, however, a dual-opposing T well can double the length of lateral wellbore exposure within the zone. In nonfractured, matrix-permeability reservoirs, the spine-and-rib design reduces the tendency to cone water. Lateral branches are sometimes curved around existing wells to keep horizontal wellbores from interfering with the production from a vertical well. A successful multilateral well that replaces several vertical wellbores can reduce overall drilling and completion costs, increase production, and provide more efficient drainage of a reservoir. Furthermore, multilaterals can make reservoir management more efficient and help increase recoverable reserves.

Thus, multilateral completion systems allow the drilling and completion of multiple wells within a single wellbore. In addition to the main wellbore, there are one or more lateral wells extending from the main wellbore. This allows for alternative well-construction strategies for vertical, inclined, horizontal, and extended-reach wells. Multilaterals can be constructed in both new and existing oil and gas wells. A typical multilateral installation includes two laterals; the number of laterals would be determined by: (1) the number of targets, (2) well depth, (3) well pressure, and (4) well-construction parameters.

The most fundamental multilateral system consists of an open-hole main bore with multiple drainage legs (or laterals) exiting from it. The junction in this design is left with no mechanical support or hydraulic isolation, and the integrity of the junction is dependent on natural borehole stability, but it is possible to land a slotted liner in the lateral or the main bore to help keep the hole open during production. The fluids production from this type of simple multilateral system must be commingled, and zonal isolation or selective control of production is not possible. Reentry into either the main bore or the lateral may be difficult or impossible should well intervention be required in the future.

3.2 Well Completion

The stimulation strategy and well-completion strategy required for a tight shale reservoir or a tight sandstone (or carbonate) reservoir very much depends on the number of layers of net gas pay and the overall economic assessment of the reservoir. In almost every case, a well in a tight gas reservoir is not economic to produce unless the optimum fracture treatment is both designed and pumped into the formation. The well can be perfectly drilled, cased, and perforated but will be uneconomic until the optimum fracture treatment is pumped. As such, the entire well prognosis should be focused on how to drill and complete the well so that it can be successfully fracture treated. The hole sizes, casing sizes, tubing sizes, wellhead, flowlines, and perforation scheme should be designed to accommodate the fracture treatment.

By drilling horizontally into the formation, the wellbore may intersect a greater number of naturally existing fractures in the reservoir—the direction of the drill path is chosen based on the known fracture trends in each area. However, some shale formations can only be drilled with vertical wells because of the risk of the borehole collapsing. Thus, as drilling is completed, multiple layers of metal casing and cement are placed around the wellbore. After the well is completed, a fluid composed of water, sand, and chemicals is injected under high pressure to crack the shale, increasing the permeability of the rock, and easing the flow of natural gas. A portion of the fracturing fluid will return through the well to the surface (*flow back*) due to the subsurface pressures. The volume of fluid will steadily reduce and be replaced by natural gas production.

Once the well has been drilled, the final casing cemented in place across the gas-bearing rock has to be perforated in order to establish communication between the rock and the well (Leonard et al., 2007; Britt and Smith, 2009; LeCompte et al., 2009). The pressure in the well is then lowered so that hydrocarbons can flow from the rock to the well, driven by the pressure differential. With tight reservoirs, the rate of the flow of fluids to the well is very low because of the low permeability of the rock. And, since the rate of fluid flow through the reservoir is a direct determination of the economic viability of the well, low flow rates can mean there is insufficient revenue to pay for operating expenses and provide a return on the capital invested. Without additional measures to accelerate the flow of hydrocarbons to the well, the operation is then not economic.

Several technologies have been developed over the years to enhance the flow from low-permeability reservoirs. Acid treatment, involving the injection of small amounts of strong acids into the reservoir to dissolve some of the rock minerals and enhance the permeability of the rock near the wellbore, is probably the oldest and is still widely practiced, particularly in carbonate reservoirs. Wells with long horizontal or lateral sections (horizontal wells) can increase dramatically the contact area between the reservoir rock and the wellbore and are likewise effective in improving project economics. Hydraulic fracturing, developed initially in the late 1940s, is another effective and commonly practiced technology for low-permeability reservoirs. When rock permeability is extremely low, as in the case of natural gas (shale gas) or crude oil from tight formations (light tight oil), it often takes the combination of horizontal wells and hydraulic fracturing to achieve commercial rates of production.

Even though the well casing is perforated, little natural gas will flow freely into the well from the shale. Fracture networks must be created in the shale to allow gas to escape from the pores and natural fractures where it is trapped in the rock. This is accomplished through the process of hydraulic fracturing. In this process, typically several million gallons of a fluid composed of 98–99% (w/w) water and *proppant* (usually sand) is pumped at high pressure into the well. The rest of the fracturing fluid (0.5–2% by volume) is composed of a blend of chemicals, often proprietary, that enhance the fluid's properties. These chemicals typically include acids to clean the shale to improve gas flow, *biocides* to prevent organisms from growing and clogging the formation fractures, corrosion and scale inhibitors to protect the integrity of the well, gels or gums that add viscosity to the fluid and suspend the proppant, and friction reducers that enhance flow and improve the ability of the fluid to infiltrate and carry the proppant into small fractures in the shale.

This fluid pushes through the *perforations* in the well casing and forces the fractures open in the formation—connecting pores and existing fractures and creating a pathway for natural gas to flow back to the well. The proppant lodges in the fractures and keeps them open once the pressure is reduced and the fluid flows back out of the well. Approximately 1000 ft of wellbore is hydraulically fractured at a time, so each well must be hydraulically fractured in multiple stages, beginning at the furthest end of the wellbore. Cement plugs isolate each hydraulic fracture stage and must be drilled out to enable the flow of natural gas up the well after all hydraulic fracturing is complete. Thus, production of natural gas and crude oil from shale formations is a multiscale and multimechanism process and, especially, a multidisciplinary process that requires cross-cooperation of several scientist and engineering discipline. Without the cooperation, the project may be doomed to technical failure with the consequences of serious environmental issues (Chapter 9).

Once the pressure is released, fluid (commonly referred to as *flowback water*) flows back out the top of the well. The fluid that is recovered not only contains the proprietary blend of chemicals present in the hydraulic fracturing fluid but may also contain chemicals naturally present in the reservoir, including hydrocarbons, salts, minerals, and naturally occurring radioactive materials (NORM) that leach into the fluid from the shale or result from mixing of the hydraulic fracturing fluid with brine (e.g., salty water) already present in the formation. The chemical composition of the water produced from the well varies significantly according to the formation and the time after well completion, with early flowback water resembling the hydraulic fracturing fluid but later converging on properties more closely resembling the brine naturally present in the formation.

Briefly and by way of explanation, flowback is a water-based solution that flows to the surface during and after the completion of the hydraulic fracturing process. The water consists of the fluid used to fracture the formation and contains clay minerals, chemical additives, dissolved metal ions, as well as dissolved solids (measured as total dissolved solids, TDS). Typically, flowback water recovery is on the order of 0% to 40% (v/v) of the volume that was initially injected into the well. In contrast, produced water is naturally occurring water found in shale formations that flows to the surface throughout the entire lifespan of the gas well. This water has high levels of TDS and leaches minerals from the formation such as minerals containing barium, calcium, iron, and magnesium. The produced water may also contain dissolved hydrocarbons such as methane, ethane, and propane along with NORM such as radium isotopes.

In many cases, flowback water can be reused in subsequent hydraulic fracturing operations; this depends upon the quality of the flowback water and the economics of other management alternatives. Flowback water that is not reused is managed through disposal. While past disposal options sometimes involved direct dumping into surface waters or deposit at ill-equipped wastewater treatment plants, most disposal now occurs at *Class II injection wells* as regulated by the US Environmental Protection Agency. These injection wells place the flowback water in underground formations isolated from drinking water sources.

The hydraulic fracturing operation (Chapter 5) provides the permeable channels in the formation for natural gas or crude oil to flow to the wellbore but contributes little to the overall storage capacity for natural gas or crude oil. The porosity of the matrix provides most of the storage capacity, but the matrix has very low permeability and gas or oil flow in the fractures occurs in a different flow regime than gas or oil flow in the matrix. Because of these differing flow regimes, the modeling of production performance in fractured tight formations is much more complex than modeling of production performance in conventional reservoirs, and scaling any modeling results up to the field operations is not only challenging but can also be extremely difficult in spite of many claims of success. This, in turn, makes it difficult (sometimes impossible) to confidently predict production performance and devise optimal depletion strategies for tight resources. The proof is in the outcome of the field operations and the balance of the field data with the data predicted from the modeling ventures.

Thus, in order to ensure the optimal development of tight natural gas and crude oil resources, it is necessary to build a comprehensive understanding of geochemistry, geological history, multiphase flow characteristics, fracture properties (including an understanding of the fracture network), and production behavior across a variety of shale plays. It is also important to develop knowledge that can enable the scaling up of pore-level physics to reservoir-scale performance prediction and make efforts to improve core analysis techniques to allow accurate determination of the recoverable resource.

For example, unconventional resources require a high well density for full development. Technology that can reduce well costs and increase wellbore contact with the reservoir can make a significant impact on costs, production rates, and ultimate recovery. Multilateral drilling, whereby a number of horizontal sections can be created from a single vertical wellbore, and coiled tubing drilling to decrease costs represent potential options for future unconventional gas development.

As already noted, a combination of steel casing and cement in the well provides an essential barrier to ensure that high-pressure natural gas or crude oil liquids recovered from the deeper tight formations cannot escape into shallower rock formations or water aquifers. The casing and the support barrier must be designed to withstand the various cycles of stress (without suffering any cracks or induced faults) that will undoubtedly arise endure during the subsequent hydraulic fracturing (Chapter 5). In this respect, the design aspects that are most important to ensure a leak-free well include (1) the drilling of the wellbore to specification (without any additional twists, turns, or cavities), (2) the positioning of the casing in the center of the wellbore before it is cemented in place, which is achieved by use of centralizers placed at regular intervals along the casing as it is run in the drill hole, to keep it away from the rock face, and, equally important, (3) the correct choice of cement. The cement design needs to be suitable for both its liquid properties during pumping—which will ensure that the cement gets to the right place—and then for its mechanical strength and flexibility. It is essential that the cement remains intact. The setting time of the cement is also a critical factor—cement that takes too long to set may have reduced strength—and equally important, cement that sets before it has been fully pumped into place requires difficult remedial action.

A recent innovation in completion technology has been the addition of 3% (v/v) hydrochloric acid to induced fracturing in the Barnett shale, which appears to increase the daily flow rate by enhancing matrix permeability and may add to the estimated ultimate recovery as long as environmental constraints are satisfied (Grieser et al., 2007). In addition, refracturing the reservoir is an option that is becoming more and more commonplace (Cramer, 2008) and can yield additional recoverable reserves.

3.3 Well Integrity

Any well drilled into the Earth creates a potential pathway for liquids and gases trapped underground to reach the surface. The same technologies that contribute to the recovery on unconventional natural gas and crude oil resources—horizontal drilling and hydraulic fracturing—also create challenges for maintaining well integrity. In terms of the recovery of natural gas and crude oil from tight formations, the wells are typically longer and must deviate from the vertical plane to the horizontal plane (1) to travel laterally, (2) access substantially (at least, potentially) overpressured reservoirs, (3) withstand more intense hydraulic fracturing pressures, and (4) withstand larger water volumes than traditional conventional natural gas and crude oil wells. Thus, poor well integrity can have a serious impact human health and the environment (Chapter 9).

For example, as a result of well leakage, formation fluids (liquids or gases) and injected fluids can migrate through holes or defects in the steel casing, through joints between casing, and through defective mechanical seals or cement inside or outside the well. In fact, a buildup of pressure inside the well annulus (*sustained casing pressure*) can force fluids out of the wellbore and into the surrounding formations. As a result of such leaks, fluids escape between the tubing and the rock wall where cement is absent or incompletely applied and the leaking fluids can then reach shallow groundwater or the atmosphere. Furthermore, well operations and the passage of time can seriously affect well integrity. For example, perforations, hydraulic fracturing, and pressure-integrity testing can cause thermal and pressure changes that damage the bond between cement and the adjacent steel casing or rock or that can fracture the cement or surrounding cap rock. In addition, chemical wear and tear can also degrade steel and cement through reactions with brines or other fluids that form corrosive acids in water (such as carbonic or sulfuric acids derived from carbon dioxide or hydrogen sulfide):

$$CO_2 + H_2O \rightarrow H_2CO_3$$

$$2H_2S + 3O_2 \rightarrow 2SO_2 + 2H_2O$$

$$SO_2 + O_2 \rightarrow SO_3$$

$$SO_2 + H_2O \rightarrow H_2SO_3$$

$$SO_3 + H_2O \rightarrow H_2SO_4$$

In the atmosphere, these acids (along with nitrous and nitric acids formed by the reaction of water with nitrogen oxides, NO and NO_2) contribute to the formation of the environmentally harmful acid rain.

3.4 Production, Abandonment, and Reclamation

Once wells are connected to processing facilities, the main production phase can begin. During production, wells will produce hydrocarbons and waste streams, which have to be managed. But the well site itself is now less visible insofar as a *Christmas tree* of valves, typically 3–4 ft high, is left on top of the well, with production being piped to processing facilities that usually serve several wells; the rest of the well site can be reclaimed (Speight, 2014).

In some cases, the operator may decide to repeat the hydraulic fracturing procedure at later times in the life of the producing well, a procedure called refracturing. This was more frequent in vertical wells but is currently relatively rare in horizontal wells, occurring in less than 10% of the horizontal shale gas wells drilled in the United States. The production phase is the longest phase of the lifecycle. For a conventional well, production might last 30 years or more. For an unconventional development, the productive life of a well is expected to be similar, but wells in tight shale typically exhibit a burst of initial production and then a steep decline, followed by a long period of relatively low production. Output typically declines by between 50% and 75% in the first year of production, and most recoverable gas is usually extracted after just a few years.

During production, gas that is recovered from the well is sent to small-diameter gathering pipelines that connect to larger pipelines that collect gas from a network of production wells. Because large-scale natural gas production and light tight oil production from tight formations have only been occurring within the past decade, the production lifetime of wells in the tight formations is not yet fully established.

Although there is substantial debate on the issue, it is generally observed that wells in tight formations experience quicker production declines than conventional natural gas production. In the Fayetteville play in north-central Arkansas, it has been estimated that half-life of a production well, or half of the estimated ultimate recovery, occurs within its first 5 years of the well. Once a well no longer produces at an economic rate, the well-head is removed, the wellbore is filled with cement to prevent leakage of gas into the air, the surface is reclaimed (either to its prewell state or to another condition agreed upon with the landowner), and the site is abandoned to the holder of the land's surface rights.

Like any other well, a well drilled into a tight formation is abandoned once it reaches the end of the producing life when extraction is no longer economic or possible. As with any gas-producing wells, at the end of their economic life, wells need to be safely abandoned, facilities dismantled, and land returned to its natural state or put to new appropriate productive use. Long-term prevention of leaks to aquifers or to the surface is particularly important—sections of the well are filled with cement to prevent residual gas or residual oil flowing into water-bearing zones or up to the surface.

Since much of the abandonment will not take place until production has ceased, the regulatory framework needs to ensure that the companies concerned make the necessary financial provisions and maintain technical capacity beyond the economic life of the reservoir to ensure that abandonment is completed satisfactorily, and well integrity maintained over the long term.

4 PRODUCTION TRENDS

Many of the low-permeability reservoirs that have been developed in the past are sandstone reservoirs but, currently, significant quantities of natural gas and crude are also produced from low-permeability shale reservoirs, carbonate reservoirs, and coal seams. The tight gas reservoirs have one thing in common: a vertical well drilled and completed in the tight gas reservoir must be successfully stimulated to produce at commercial volumes and flow rates. Typically, a large hydraulic fracture treatment is required to produce the natural gas or crude oil and horizontal wells and/or multilateral wells are used to provide the stimulation required for natural gas or crude oil recovery.

To optimize the development of a tight reservoir, the recovery team must optimize the number of wells drilled, as well as the drilling and completion procedures for each well. On an individual well basis, a well in a tight reservoir will produce less gas over a longer period of time than one expects from a well completed in a higher permeability, conventional reservoir. As such, many more wells (or smaller well spacing) must be drilled in a tight gas reservoir to recover a large percentage of the OGIP or the OOIP when compared to a conventional reservoir.

Thus, economic natural gas production from unconventional tight formation reservoirs is achieved by the combination of horizontal drilling and reservoir stimulation by multistage slick-water fracturing. Ideally, every fracture treatment at every stage of the well is successful, but experiences from the Barnett shale and other shale gas reservoirs have shown that not all stages are stimulated equally. The regions of SRV may be different between stages in size and shape, sometimes confined along a plane or sometimes dispersed widely in the reservoir (Waters et al., 2006).

Operators also have observed that the fracturing pressure (fracture gradient) can vary between stages, sometimes to a point where pump pressures cannot reach the fracturing pressure required to propagate a fracture (Daniels et al., 2007). These variations in the outcome of hydraulic fracturing are caused by the heterogeneity in the rock mechanical/deformation properties, presence of natural fractures, and/or the variations in in-situ stress within the reservoir.

However, the increasing participation of major oil companies in the exploitation of North American shale gas and tight oil resources has had positive implications for the use of best practices and technologies in drilling and processing. Continued development of shale gas and tight oil resources in North America and other countries with significant resources will have an impact on the global gas markets; however this impact is expected to remain moderate in the short to medium term, nothing comparable to what happened in the United States (LEGS, 2011).

The increasing use of shale gas and light tight oil will primarily impact power generation, transport fuels, and the petrochemical industry. In fact, estimates of proven reserves of shale gas are increasing globally and will continue to do so as exploration continues. Furthermore, exploitation of the shale gas and light tight oil can affect natural gas and crude oil availability and prices, particularly in North America.

4.1 Technology

Optimizing the production from tight reservoirs calls for complex development geometries, sometimes requiring several dozen multidrain or horizontal wells. Preserving the permeability of the reservoir in the vicinity of the wells—low by definition in tight reservoirs—is a necessity for drilling projects. One option is the use of *underbalanced drilling* which is a procedure used to drill oil and gas wells where the pressure in the wellbore is kept lower than the fluid pressure in the formation being drilled. As the well is being drilled, formation fluid flows into the wellbore and up to the surface which is the opposite of the usual situation, where the wellbore is kept at a pressure above the formation to prevent formation fluid entering the well. In conventional *overbalanced drilling*, if the well is not shut-in, it can lead to a blowout (gusher) which can be a hazardous situation. In underbalanced drilling, however, there is a rotating head at the surface, which is a seal that diverts produced fluids to a separator while allowing the drill string to continue rotating. If the formation pressure is relatively high, using a lower density mud will reduce the wellbore pressure below the pore pressure of the formation. Sometimes an inert gas is injected into the drilling mud to reduce its equivalent density and hence the hydrostatic pressure throughout the well depth. This gas is commonly nitrogen, as it is noncombustible and readily available, but air, reduced oxygen air, processed flue gas, and natural gas have all been used. On the other hand, coiled tubing drilling uses the established concept of coiled tubing and combines it with directional drilling using a mud motor to create a system for drilling reservoirs which allows for continuous drilling and pumping, and therefore underbalanced drilling can be utilized which can increase the rate of penetration.

Thus, in the underbalanced drilling technique, the mud pressure is kept lower than the formation pressure, thereby preventing invasive formation damage and the associated risk of clogging. This complex technique, implemented on the Hassi Yakour tight gas field in Algeria during the second half of 2006, demands expert skill: the mud properties must be constantly adjusted to accommodate pressure variations caused by differences in subsoil characteristics. Bringing the wells into production is another critical juncture. Tight matrices need to be stimulated by fracturing to ensure sufficient flow rates. Gas gathering is improved by high-pressure injection and pumping of an aqueous fluid to break up the rock. However, the capillary pressure of the fluid can modify reservoir permeability; the fracturing fluid can become trapped, preventing efficient flow of the gas. To get around this drawback, total adds volatile agents such as methanol to the fracturing fluid, to help the liquid recede faster. The final step entails the injection of granular supporting agents (proppants, usually a material such as sand) to maintain the properties of the induced fractures that allow the natural gas or crude oil to pass to the production well(s).

The development of natural gas and crude oil resources in tight formations is a modern technologically driven process for the production of natural gas resources. Currently, the drilling and completion of wells in tight shale reservoirs and in other tight reservoirs includes both vertical and horizontal wells. In both kinds of wells, casing and cement are installed to protect fresh and treatable water aquifers. The emerging shale gas and tight oil basins are expected to follow a trend similar to the Barnett shale play with increasing numbers of horizontal wells as the plays mature. Operators are increasingly relying on horizontal well completions to optimize recovery and well economics. Horizontal drilling provides more exposure to a formation than does a vertical well.

This increase in reservoir exposure creates a number of advantages over vertical wells drilling. About 6–8 horizontal wells drilled from only 1 well pad can access the same reservoir volume as 16 vertical wells. Using multi-well pads can also significantly reduce the overall number of well pads, access roads, pipeline routes, and production facilities required, thus minimizing habitat disturbance, impacts to the public, and the overall environmental footprint (Chapter 5).

The other technological key to the economic recovery of natural gas and crude oil from tight reservoirs is hydraulic fracturing, which involves the pumping of a fracturing fluid under high pressure into a shale formation to generate fractures or cracks in the target rock formation. This allows the natural gas to flow out of the shale to the well in economic quantities. Ground water is protected during the hydraulic fracturing process by a combination of the casing and cement that is installed when the well is drilled and the thousands of feet of rock between the fracture zone and any fresh or treatable aquifers. For development of natural gas and crude oil from tight reservoirs, fracture fluids are primarily water-based fluids mixed with additives that help the water to carry sand proppant into the fractures. Water and sand make up over 98% of the fracture fluid, with the rest consisting of various chemical additives that improve the effectiveness of the fracture job. Each hydraulic fracture treatment is a highly controlled process that must be designed to the specific conditions of the target formation.

A combination of improved technology and shale-specific experience has also led to improvements in recovery factors and reductions in decline rates. Each shale resource requires its own specific completion techniques, which can be determined through careful analysis of rock properties. The correct selection of well orientation, stimulation equipment, fracture size, and fracturing fluids can all affect the performance of a well. The initial production rate from a particular well is highly dependent on the quality of the fracture and the well completion. In the United States it has been seen that initial production rates have been augmented over time as the resource matures. Initial production rates can be increased by several techniques, in particular by increasing the number of fracture stages and increasing the number of perforations per fracture stage. The quality of the fracture is also improved as fluid properties are developed. Microseismic data can also be used to improve the efficiency of the hydraulic fracturing process.

The primary differences between modern natural gas and crude oil development from tight reservoirs and the development of natural gas and crude oil from conventional reservoirs are the extensive uses of horizontal drilling and high-volume hydraulic fracturing. The use of horizontal drilling has not introduced any new environmental concerns. In fact, the reduced number of horizontal wells needed coupled with the ability to drill multiple wells from a single pad has significantly reduced surface disturbances and associated impacts to wildlife, dust, noise, and traffic. Where shale gas and tight oil development has intersected with urban and industrial settings, regulators and industry have developed special practices to alleviate nuisance impacts, impacts to sensitive environmental resources, and interference with existing businesses.

Hydraulic fracturing has been a key technology in making shale gas and tight oil an affordable addition to the US energy supply, and the technology has proved to be an effective stimulation technique. While some challenges exist with water availability and water management, innovative regional solutions are emerging that allow shale gas and tight oil development to continue while ensuring that the water needs of other users are not affected and that surface and ground water quality is protected. Taken together, state and federal requirements along with the technologies and practices developed by industry serve to reduce environmental impacts from shale gas and tight oil operations.

4.2 The Future

Development of shale gas and tight oil resources in Western Europe, Scandinavia, and Poland has the potential to cut the heavy dependence by western European countries on Russian resources, unless, of course, Russian companies gain control of these resources through company actions or invasive actions. Furthermore, discoveries of tight resources in South America have the potential of realigning the energy relationships on the continent. Argentina, Brazil, and Chile are likely beneficiaries decreasing their dependence on Bolivian gas.

Notwithstanding the environmental moratoriums in some countries, shale gas and tight oil will be used in many parts of the world. This will place downward pressure on prices of natural gas and lower natural gas prices may lead to significant shifting in power generation and transport fuels.

Finally, shale gas liquids are having a significant impact on the petrochemical industry in North America, which has spillover effects in Europe, the Middle East, and Asia. Further, the liquids are making shale gas more profitable than many traditional dry gas reservoirs.

Despite the availability of proven production technologies, environmental impacts are still being queried; in particular, the impact on ground water resources and the possible methane releases associated with current production techniques. These issues are the subject of intense scrutiny at the moment. The supply and use of shale gas and tight crude oil is already showing an impact on the fossil fuels energy sector which is not restricted to the global natural gas pricing outlook but its development has become entwined with the global energy mix and emerging nexus in energy-climate-water and its impact on the global energy supplies environment.

Finally, property and mineral rights differ across the world. In the United States, individuals can own the mineral rights for land they own. In many parts of Asia, Europe, and South America this is not the case. Therefore, unresolved legal issues remain obstacles for shale gas and tight oil exploitation in many countries across the globe. Given the investment requirements needed to develop shale basins that presently have little to no infrastructure, the legal issues are important to attract the investment needed for exploration and exploitation.

In summary, the recovery of natural gas and crude oil from tight reservoirs has heterogeneous geological and geomechanical characteristics that pose challenges to accurate prediction of the response to hydraulic fracturing. Experience in tight formations that contain natural gas and light tight oil formations shows that stimulation often results in formation of a complex fracture structure, rather than the planar fracture aligned with the maximum principal stress. The fracture complexity arises from intact rock and rock mass textural characteristic and the in-situ stress and their interaction with applied loads. Open and mineralized joints and interfaces and contact between rock units play an important role in fracture network complexity which affects the rock mass permeability and its evolution with time.

Currently, the mechanisms that generate these fracture systems are not completely understood and can generally be attributed to lack of in-situ stresses within the reservoir rock, rock brittleness, shear reactivation of mineralized fractures, and textural heterogeneity. This clearly indicates the importance of linking the mineralogy, rock mechanics, and geomechanics to determine the prospectively of an unconventional shale resource.

REFERENCES

Britt, L.K., Smith, M.B., 2009. Horizontal well completion, stimulation optimization, and risk mitigation. Paper No. SPE 125526, In: Proceedings of SPE Eastern Regional Meeting, Charleston, WV, September 23–25.

Bustin, A.M.M., Bustin, R.M., Cui, X., 2008. Importance of fabric on the production of gas shales. SPE Paper No. 114167, In: Proceedings of Unconventional Gas Conference, Keystone, Colorado, February 10–12.

Chopra, S., Sharma, R.K., Keay, J., Marfurt, K.J., 2012. Shale gas reservoir characterization workflows. In: Proceedings of SEG Annual Meeting, Las Vegas, Nevada. Society of Exploration Geophysicists, Tulsa, Oklahoma.

Cipolla, C.L., Lolon, E.P., Erdle, J.C., Rubin, B., 2010. Reservoir modeling in shale-gas reservoirs. Paper No. SPE 125530, SPE Res. Eval. Eng. 13 (4), 638–653.

Cramer, D.D., 2008. Stimulating unconventional reservoirs: lessons learned, successful practices, areas for improvement. SPE Paper No. 114172, In: Unconventional Gas Conference, Keystone, Colorado, February 10–12.

Daniels, J., Waters, G., LeCalvez, J., Lassek, J., Bentley, D., 2007. Contacting more of the Barnett shale through an integration of real-time microseismic monitoring, petrophysics and hydraulic fracture design. Paper No. 110562, In: Proceedings of SPE Annual Technical Conference and Exhibition, Anaheim, California.

EIA, 2011. Review of Emerging Resources: US Shale Gas and Shale Oil Plays. Energy Information Administration, United States Department of Energy, Washington, DC.

Ezisi, L.B., Hale, B.W., William, M., Watson, M.C., Heinze, L., 2012. Assessment of probabilistic parameters for Barnett shale recoverable volumes. Paper No. SPE 162915, In: Proceedings of SPE Hydrocarbon, Economics, and Evaluation Symposium, Calgary, Canada, September 24–25.

Gale, J.F.W., Reed, R.M., Holder, J., 2007. Natural fractures in the Barnett shale and their importance for hydraulic fracture treatments. AAPG Bull. 91, 603–622.

Grieser, B., Bray, J., 2007. Identification of production potential in unconventional reservoirs. Paper No. SPE 106623, In: Proceedings of SPE Production and Operations Symposium, Oklahoma City, Oklahoma, March 31–April 3.

Grieser, B., Wheaton, B., Magness, B., Blauch, M., Loghry, R., 2007. Surface reactive fluid's effect on shale. SPE Paper No. 106815, In: Proceedings of SPE Production and Operations Symposium, Society of Petroleum Engineers, Oklahoma City, Oklahoma, March 31–April 3.

Grieser, B., Shelley, B., Soliman, M., 2009. Predicting production outcome from multi-stage, horizontal Barnett completions. Paper No. SPE 120271, In: Proceedings of SPE Production and Operation Symposium, Oklahoma City, OK, April 4–8.

Gubelin, G., 2004. Improving gas recovery factor in the Barnett shale through the application of reservoir characterization and simulation answers. In: Proceedings of Gas Shales: Production & Potential, Denver, Colorado, July 29–30.

Hale, B.W., William, M., 2010. Barnett shale: a resource play—Locally random and regionally complex. Paper No. SPE 138987, In: Proceedings of SPE Eastern Regional Meeting, Morgantown, WV, October 12–14.

Houston, N., Blauch, M., Weaver III, D., Miller, D.S., O'Hara, D., 2009. Fracture-stimulation in the Marcellus shale-lessons learned in fluid selection and execution. Paper No. SPE 125987, In: Proceedings of SPE Regional Meeting, Charleston, West Virginia, September 23–25.

Islam, M.R., Speight, J.G., 2016. Peak Energy—Myth or Reality? Scrivener Publishing, Beverly, Massachusetts.

King, G.E., 2010. Thirty years of gas shale fracturing: what have we learned?. Paper No. SPE 133456, In: Proceedings of SPE Annual Technical Conference and Exhibition Florence, Italy, September.

LeCompte, B., Franquet, J.A., Jacobi, D., 2009. Evaluation of Haynesville shale vertical well completions with a mineralogy based approach to reservoir geomechanics. Paper No. SPE 124227, In: Proceedings of SPE Annual Technical Meeting, New Orleans, Louisiana, October 4–7.

Lee, S., 1991. Oil Shale Technology. CRC Press, Taylor & Francis Group, Boca Raton, Florida.

LEGS, 2011. An Introduction to Shale Gas. LEGS Resources, Isle of Man, United Kingdom.

Leonard, R., Woodroof, R.A., Bullard, K., Middlebrook, M., Wilson, R., 2007. Barnett shale completions: a method for assessing new completion strategies. Paper No. SPE 110809, In: Proceedings of SPE Annual Technical Conference and Exhibition, Anaheim, California, November 11–14.

Mokhatab, S., Poe, W.A., Speight, J.G., 2006. Handbook of Natural Gas Transmission and Processing Elsevier. Netherlands, Amsterdam.

Schweitzer, R., Bilgesu, H.I., 2009. The role of economics on well and fracture design completions of Marcellus shale wells. Paper No. SPE 125975, In: Proceedings of SPE Eastern Regional Meeting, Charleston, WV, September 23–25.

Scouten, C.S., 1990. Oil shale. In: Speight, J.G. (Ed.), Fuel Science and Technology Handbook. Marcel Dekker Inc., New York, pp. 795–1053. Chapters 25 to 31.

Speight, J.G., 2007. Natural Gas: A Basic Handbook. GPC Books, Gulf Publishing Company, Houston, Texas.

Speight, J.G., 2008. Synthetic Fuels Handbook: Properties, Processes, and Performance. McGraw-Hill, New York.

Speight, J.G., 2012. Shale Oil Production Processes. Gulf Professional Publishing, Elsevier, Oxford, United Kingdom.

Speight, J.G., 2014. The Chemistry and Technology of Petroleum, fifth ed. CRC Press, Taylor & Francis Group, Boca Raton, Florida.

Speight, J.G., 2016. Handbook of Hydraulic Fracturing. John Wiley & Sons Inc., Hoboken, New Jersey.

Upolla, C.L., Lolon, E., Erdle, J., Tathed, V.S., 2009. Modeling Well Performance in Shale Gas Reservoirs. Paper No. SPE 125532, In: Proceedings of SPE/EAGE Reservoir Characterization and Simulation Conference, Abu Dhabi, United Arab Emirates, October 19–21. Society of Petroleum Engineers, Richardson, Texas.

Waters, G., Heinze, J., Jackson, R., Ketter, A., Daniels, J., Bentley, D., 2006. Use of horizontal well image tools to optimize Barnett shale reservoir exploitation. SPE Paper No. 103202, In: Proceedings of SPE Annual Technical Conference and Exhibition, San Antonio, Texas.

Yu, W., Sepehrnoori, K., 2013. Optimization of multiple hydraulically fractured horizontal wells in unconventional gas reservoirs. Paper No. SPE 164509, In: Proceedings of SPE Production and Operations Symposium, Oklahoma, OK, March 23–26.

Zhang, J., Delshad, M., Sepehrnoori, K., 2007. Development of a framework for optimization of reservoir simulation studies. J. Petro. Sci. Eng. 59, 135–146.

CHAPTER FIVE

Hydraulic Fracturing

1 INTRODUCTION

Hydraulic fracturing (hydraulic fracture stimulation, fraccing, fracking) has been referenced in an earlier chapter (Chapter 4) and, because it is one of the key techniques that allow access to natural gas and crude oil from tight formations, it is presented here in more detail.

Extraction of natural gas and crude oil from conventional reservoirs involves drilling through impervious rock that traps concentrated underground reservoirs of petroleum and natural gas and may even involve a mining operation for heavy oil and tar sand bitumen (Chapter 3) (Speight, 2014, 2016a). With conventional petroleum, extraction occurs simply due to the change in pressure caused by the drilling. Conventional petroleum reservoirs depend on the pressure of their gas cap and oil-dissolved gas to lift the oil to the surface (i.e., *gas drive*) (Chapter 3). Water trapping the petroleum from below also exerts an upward hydraulic pressure (i.e., *water drive*). The combined pressure in petroleum reservoirs produced by the natural gas and water drives is known as the *conventional drive.* As a reservoir's production declines, lifting further petroleum to the surface, like the lifting of water, requires pumping or *artificial lift.*

However, not all of the petroleum and natural gas is conveniently located in conventional and accessible reservoirs. Many oil and gas resources are trapped in the pore spaces and cracks within impermeable sedimentary rock formations—shale formations, tight sandstone formations, and tight carbonate formations are examples of such reservoirs (Chapter 2). These reservoirs can vary in thickness—the shale formations are relatively thin layers (albeit deep under the ground) but cover extensive horizontal areas and a vertically drilled well will only access a small area of the reservoir and, by inference due to the impermeable nature of the formation, a minimal part of the resource. However, when the drilling operation can deviate from the conventional

Deep Shale Oil and Gas
http://dx.doi.org/10.1016/B978-0-12-803097-4.00005-X

vertical plane and move in the horizontal plane much more of the reservoir resource becomes accessible (Ely, 1985; Gidley et al., 1990).

In the past three-to-four decades, hydraulic fracturing (Table 5.1) has been increasingly used in formations that were known to be rich in natural gas that was locked so tightly in the rock that it was technologically and economically difficult to produce. Application of fracturing techniques to stimulate oil and gas production began to grow rapidly in the 1950s, although experimentation dates back to the 19th century. Starting in the mid-1970s, partnerships of various private operators, the U.S. Department of Energy and predecessor agencies, and the Gas Research Institute endeavored

Table 5.1 Highlights in the Development of Hydraulic Fracturing

Date	Comment
Early 1900s	Natural gas extracted from shale wells
	Vertical wells fractured with foam
1947	Klepper gas unit No. 1; first well to be fractured to increase productivity
1949	Stephens County, Oklahoma; first commercial fracturing treatment
1950	Fracturing with cement pumpers
1950s	Evolution of fracture geometry
	Increasing well productivity
1960s	Fracturing pumpers and blenders
1970s	Massive hydraulic fracturing
	Increase recoverable reserves
	Hydraulic fracturing in Europe
1983	First gas well drilled in Barnett shale in Texas
1980s	Evolution of proppant transport
	Fracture conductivity testing
	Crosslinked gel fracturing fluids developed; used in vertical wells
1990s	First horizontal well drilled in Barnett shale
	Orientation of induced fractures identified
	Foam fracturing
1996	Slickwater fracturing fluids introduced
1996	Microseismic postfracturing mapping developed
1997	Hydraulic fracturing in Barnett shale
	Slickwater fracturing developed
1998	Slickwater refracturing of originally gel-fractured wells
2002	Multistage slickwater fracturing of horizontal wells
2003	First hydraulic fracturing of Marcellus shale
2004	Horizontal wells become dominant
2005	Increased emphasis on improving the recovery factor
2007	Use of multiwell pads and cluster drilling

to develop technologies for the commercial production of natural gas from the relatively shallow Devonian (Huron) shale in the eastern United States. This partnership helped foster technologies that eventually became crucial to the production of natural gas from shale rock, including horizontal wells, multistage fracturing, and slickwater fracturing. Practical application of horizontal drilling to oil production began in the early 1980s, by which time the advent of improved downhole drilling motors and the invention of other necessary supporting equipment, materials, and technologies (particularly, downhole telemetry equipment) had brought some applications within the realm of commercial viability.

Also key to shale gas development is the presence of natural fractures and planes of weakness that can result in complex fracture geometries during stimulation (Reddy and Nair, 2012). Furthermore, the presence and ability to open and maintain flow in both primary and secondary natural fracture systems are critical to shale gas production (King, 2010). The technology involves pumping water, a proppant such as sand (Table 5.2) to keep the fractures open, and a small amount of one or more chemical additives into the well to assist the fracture process, by which the natural gas or crude oil is enabled to flow to the wellbore. In fact, the use of horizontal drilling in conjunction with hydraulic fracturing has greatly expanded the ability of producers to profitably recover natural gas and oil from low-permeability geologic plays—particularly, shale plays and other tight reservoirs (Speight, 2016b). In the process, the pressure exceeds the rock strength and the fluid opens or enlarges fractures in the rock. As the formation is fractured, a propping agent, such as sand or ceramic beads, is pumped into the fractures to keep them from closing as the pumping pressure is released. The fracturing fluids (water and chemical additives) are then returned back to the surface. Natural gas and crude oil will then flow from pores and fractures in the rock into the well for subsequent extraction to the surface.

Table 5.2 Proppant Type Definition

Sand	Includes all raw sand types
Resin-coated sand	Includes only resin-coated proppants for which the substrate is sand; does not include any double-counting with the "Sand category" described above
Ceramic	Any proppant for which the substrate is a ceramic or otherwise manufactured proppant, resin-coated ceramic proppant is included in this category

Thus, hydraulic fracturing is a process through which a large number of fractures are created mechanically in the rock, thus allowing the natural gas and/or crude oil trapped in subsurface formations to move through those fractures to the wellbore from where it can then flow to the surface. Hydraulic fracturing can both increase production rates and increase the total amount of gas that can be recovered from a given volume of shale. Pump pressure causes the rock to fracture, and water carries sand ("proppant") into the hydraulic fracture to prop it open allowing the flow of gas. While water and sand are the main components of hydraulic fracture fluid, chemical additives are often added in small concentrations to improve fracturing performance.

The development of large-scale natural gas and crude oil production from tight formations is changing the US energy market—as well as the energy markets in various other countries (Chapter 3)—by generating expanded interest in the usage of natural gas in sectors such as electricity generation and transportation. At the same time, there is much uncertainty of the environmental implications of hydraulic fracturing and the rapid expansion of natural gas production from shale formations. For example, water for the hydraulic fracturing process can come from surface water sources (such as rivers, lakes, or the sea), or from local boreholes (which may draw from shallow or deep aquifers and which may already have been drilled to support production operations), or from further afield (which generally requires trucking) (Chapter 6). Transportation of water from its source and to disposal locations can be a large-scale activity. Nevertheless, hydraulic fracturing has been a key technology in making shale gas an affordable addition to the national energy supply, and the technology has proved to be an effective stimulation technique (Arthur et al., 2009; Spellman, 2013, 2016).

While some challenges exist with water availability and water management (Chapter 6), innovative regional solutions are emerging and continually being sought that allow natural gas and crude oil development to continue while ensuring that the water needs of other users are not affected and that surface and ground water quality is protected. In the late 1940s, drilling companies began inducing hydraulic pressure in wells to fracture the producing formation. This stimulated further production by effectively increasing the contact of a well with a formation. Moreover, advances in directional drilling technology have allowed wells to deviate from nearly vertical to extend horizontally into the reservoir formation, which further increases contact of a well with the reservoir. Directional drilling technology (sometimes referred to as *deviated drilling technology*) also enables drilling a number of wells from a single well pad, thus cutting costs while reducing environmental disturbance. Combining hydraulic fracturing with directional drilling has opened up the production of tight (less permeable) petroleum and natural gas reservoirs, particularly unconventional gas shales such as the Marcellus shale formation.

The application of hydraulic fracturing to tight sands revitalized old fields and allowed establishment of new fields. Subsequently, the application of hydraulic fracturing to shale opened up new areas to development, including the Marcellus shale in the eastern United States, the Barnett shale in Texas, and the Fayetteville shale in Arkansas, amongst others. In fact, the rise in production of natural gas and crude oil from these and other shale plays has effects on the movement of natural gas and crude oil prices to lower (currently) more stable levels (Fisher, 2012; Scanlon et al., 2014; US EIA, 2014). In practice, the well is fractured in stages and a plug set between each stage. When all of the stages have been completed and plugged, the plugs are removed (drilled out), which allows the natural gas or crude oil to flow up through the well to begin production.

Briefly, the process involves use of a *perforating gun* which is lowered into a newly drilled well and lined up precisely within the target formation (a tight shale, sandstone, or carbonate formation) using seismic images, well logs, global positioning systems, and other indicators to target the spots from which natural gas and crude oil are most likely to occur. When fired, the gun punches small holes into the well casing, cement, and rock after which the fracturing fluid is forced out of the perforations under high pressure. This creates fractures (small cracks) in the formation that allow the natural gas and crude oil to flow from the reservoir into the wellbore. The fracturing fluid contains proppants such as sand or other similarly sized materials in order to maintain the fractures created by the pressure treatment in the open position, thus preventing closure when the pressure treatment is terminated. Although the fracturing fluid (*slickwater*) is predominantly water, it does contain chemicals (in addition to the proppant) which can pose an environmental risk (Green, 2014, 2015).

Thus, hydraulic fracturing has become an essential part of natural gas production and crude oil production, especially, production of natural gas and crude oil that is trapped in low-permeability (shale, sandstone, and carbonate) formations (Agarwal et al., 1979). The procedure significantly improves the recovery from the reservoir by stimulating the movement of petroleum and natural gas. When used in conjunction with horizontal drilling, an advanced drilling technology, hydraulic fracturing has made it possible to develop vast unconventional resources.

Without hydraulic fracturing and horizontal drilling, resources contained in tight formations would remain largely undeveloped. In fact, many modern oil-field production operations would not exist without hydraulic fracturing and as the global balance of supply and demand forces the natural gas and crude oil industries toward more unconventional resources including shale formations such as the Barnett, Haynesville, Bossier, and Marcellus plays, hydraulic fracturing will continue to play a major role in unlocking these otherwise unobtainable reserves.

To stimulate gas and/or oil flow from tight sand formation or from shale formations, where gas is trapped in tiny pores in the rock (rather than accumulated in large pools or more porous rock), hydraulic fracturing is applied. In spite of the various negative attitudes to hydraulic fracturing, the technique is a proven technique and has been used for decades in many kinds of natural gas and crude oil recovery operations. However, hydraulic fracturing must be applied with diligence and caution and use of multidisciplinary team, like all reservoir management operations (Speight, 2016b). Starting the process on the basis that *one person knows all* is guaranteed to cause a multitude of problems and results in failure.

Finally, naturally fractured reservoirs contain secondary or induced porosity in addition to their original primary porosity. Induced porosity is formed by tension or shear stresses causing fractures in a competent or brittle formation. Fracture porosity is usually very small. Values between 0.0001 and 0.001 of rock volume are typical (0.01–0.1%) Fracture-related porosity, such as solution porosity in granite or carbonate reservoirs, may attain much larger values, but the porosity in the actual fracture is still very small. There are, of course, exceptions to all rules of thumb. In rare cases, such as the cooling of intrusive minerals or surface lava flows, natural fracture porosity may exceed 10%. When buried and later filled with hydrocarbons, they form very interesting reservoirs.

2 RESERVOIR EVALUATION

Reservoir characteristics, as well as the characteristics of the surrounding formations, must always be considered when designing any hydraulic fracturing project (Veatch, 1983; Smith and Hannah, 1996; Reinicke et al., 2010). More specifically, in tight formations, fracture length is the overriding factor for increased productivity and recovery. Typical conventional crude oil and natural gas reservoirs have a permeability deposits boast a permeability level on the order of 0.01–1 Darcy, but the formations termed *tight formations* (*tight reservoirs*) typically have permeability levels on the order of milliDarcys (Darcy × 10^{-3}) or microDarcy (Darcy × 10^{-6}) even down to nanoDarcys (Darcy × 10^{-9}) (Fig. 5.1), although the permeability data given for shale reservoirs and tight sandstone or carbonate reservoirs are not always as sharp as this illustration would indicate. Furthermore, there is no true or regular relationship between porosity and permeability, specifically for shale rock although some general trends may appear when the two are compared (Fig. 5.2).

In addition, and from a reservoir-development standpoint, understanding the fracture geometry and orientation is crucial for determining well spacing (Holditch et al., 1978). Thus, field-development strategies designed to extract more hydrocarbons petroleum or natural gas and, furthermore, natural fractures, which are often the primary means for fluid flow in low-permeability reservoirs, can (and often do) compromise the predictability of the geometry of hydraulic fractures and the effect on production and drainage.

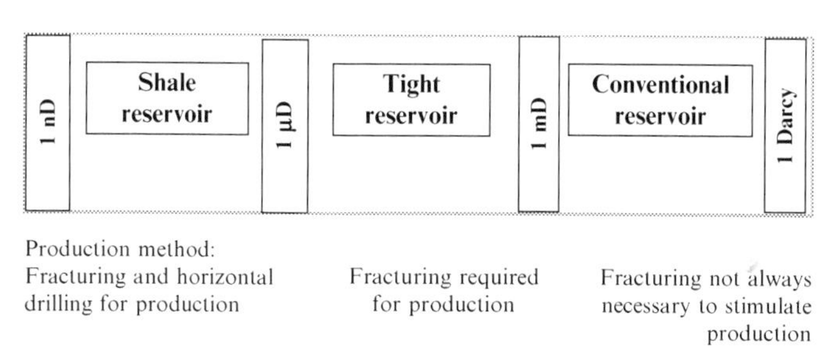

Fig. 5.1 Illustration of reservoir types based on permeability and production methods.

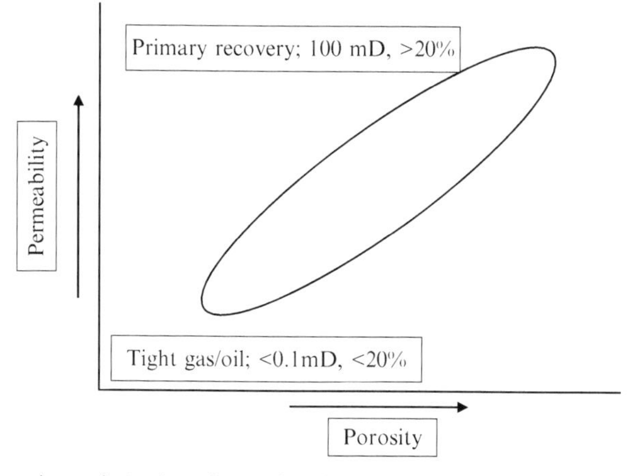

Fig. 5.2 General trends in the relationship between porosity and permeability.

Finally, success or failure of a hydraulic fracture treatment often depends on the quality of the candidate well selected for the treatment. Choosing an excellent candidate for stimulation often ensures success, while choosing a poor candidate normally results in economic failure. To select the best candidate for stimulation, the design engineer must consider many variables. The most critical reservoir parameters for hydraulic fracturing that need consideration are (1) formation permeability, (2) the in situ stress distribution, (3) reservoir fluid viscosity, (4) reservoir pressure, (5) reservoir depth, (6) the condition of the wellbore, and (7) the skin factor, which refers to whether the reservoir is already stimulated or, perhaps, damaged. Typical values for the skin factor range from −6 for an infinite-conductivity massive hydraulic fracture to more than 100 for a poorly formed gravel pack.

Tight formations vary considerably and for this reason no single technique for hydraulic fracturing has universally worked. Each play has unique properties that need to be addressed through fracture treatment and fluid design. For example, numerous fracture technologies have been applied in the Appalachian basin alone, including the use of carbon dioxide, nitrogen, carbon dioxide foam, and slickwater fracturing. The composition of fracturing fluids must be altered to meet specific reservoir and operational conditions. Slickwater hydraulic fracturing, which is used extensively in shale basis of the United States and Canada, is suited for complex reservoirs that are brittle and naturally fractured and are tolerant of large volumes of water. Ductile reservoirs require more effective proppant placement to achieve the desired permeability. Other fracture techniques, including carbon dioxide polymer foam and nitrogen foam, are occasionally used in ductile rock (for instance, in the Montney shale in Canada).

One of the most important aspects of drilling for any petroleum is predetermining the success rate of the operation. Operators do not just drill on a whim or a hunch and extensive seismic data is gathered and analyzed to determine where to drill and just what might be located below the earth's surface. These seismic surveys can help to pinpoint the best areas to tap tight gas reserves. A survey might be able to locate an area that portrays an improved porosity or permeability in the rock in which the gas is located. If drilling is sufficiently accurate to directly penetrate the best area to develop the reservoir, the cost of development and recovery can be minimized.

The changes in porosity and permeability of shale matrix occur when production starts. This variation in porosity and consequently permeability is because of two reasons: gas desorption from shale surface (unlike conventional reservoirs) and increasing effective stresses by pressure depletion.

When gas molecules leave the surface of the shale rock and move toward the pore spaces, the pore volume is increased as the matrix volume is decreased. This will result in an increase in porosity of the shale matrix. Unlike desorption effect, the porosity of shale tends to decrease with production as a result of increased net stress effect. The change in porosity and subsequent permeability of the shale has not been studied in the reservoir simulation while focusing on its effect on a long-term production outlook.

Generally, in early stages of production from shale formations as the reservoir pressure is considerably high, there is no significant desorption from shale surface to contribute to production. This means that up to some stages of production, reservoir is encountered with porosity reduction due to increased net stress and compaction. Once a critical lifetime of a shale reservoir is reached, the porosity changes due to desorption must be considered in which porosity will enhance due to increase of pore volume of shale and consequent reduction in rock volume. This critical lifetime of a shale reservoir depends on its isothermal desorption behavior that should be measured experimentally. After the critical period of shale production is passed, three different scenarios are possible. First, the compaction effect on porosity dominates the porosity change against desorption and the total effect tends to reduce shale porosity. Second, the two effects may not overcome each other that means the porosity reduction due to compaction is balanced by its enhancement due to desorption. Last, porosity increases several orders of magnitude more than its reduction due to compaction.

Most tight gas and oil formations are found onshore, and land seismic techniques are undergoing transformations to better map out where drilling and development of these unconventional plays take place. Typical land seismic techniques include exploding dynamite and *vibroseis*, or measuring vibrations produced by purpose-built trucks. While these techniques can produce informational surveys, advancements in marine seismic technologies are now being applied to land seismic surveys, enhancing the information available about the world below.

Fracturing a well involves breaking the rocks in the formation apart. Performed after the well has been drilled and completed, hydraulic fracturing is achieved by pumping the well full of fracturing fluids under high pressure to break the rocks in the reservoir apart and improve permeability, or the ability of the gas to flow through the formation. Additionally, acidizing the well is employed to improve permeability and production rates of tight gas formations, which involves pumping the well with acids that dissolve the limestone, dolomite, and calcite cement between the sediment grains of the reservoir rocks. This form of production stimulation helps to reinvigorate permeability by reestablishing the natural fissures that were present in the formation before compaction and cementation.

While vertical wells may be easier and less expensive to drill, they are not the most conducive to developing tight gas. In a tight gas formation, it is important to expose as much of the reservoir as possible, making horizontal and directional drilling a necessity. Here, the well can run along the formation, opening up more opportunities for the natural gas to enter the wellbore. A more common technique for developing tight gas reserves includes drilling more wells. The more the formation is tapped, the more the gas will be able to escape the formation. This can be achieved through drilling myriad directional wells from one location, which lessens the drilling footprint and lowers the costs. After seismic data has illuminated the best well locations, and the wells have been drilled, production stimulation is employed on tight gas reservoirs to promote a greater rate of flow. Production stimulation can be achieved on tight gas reservoirs through both fracturing and acidizing the wells.

Formation evaluation is the process of interpreting a combination of measurements taken inside a wellbore to detect and quantify oil and gas reserves in the rock adjacent to the well. A formation consists of rock layers (strata) that have similar properties (Chapter 2) and, thus, formation evaluation is an important aspect of the fracturing process, especially in low-permeability reservoirs because of the presence of alternating layers with various properties. Hence, it is necessary to define these properties, which include (but are not limited to) thickness, fluid saturation, porosity, Young's modulus, in situ stress, permeability, formation conductivity (Holditch et al., 1987).

Thus, formation evaluation is used to determine the ability of a well (borehole) to produce crude oil and/or natural gas. Typically, the well is drilled by a rotary drill which uses a heavy mud (drilling mud) as a lubricant and as a means of producing a confining pressure against the formation face in the borehole, preventing blowouts. A blowout is an uncontrolled release of fluids during drilling, completion, or production of crude oil and natural gas. In former times, the blow out may have been referred to as a gusher which was spectacular but (Hollywood movies not withstanding) was a waste of crude oil and an environmental nightmare.

When a blowout occurs, it is typically when unexpectedly high pressures are encountered in the subsurface or due to valve failure or other type of mechanical failure. Blowouts may take place at the surface (wellhead or elsewhere) or subsurface (naturally high pressure, or may be artificially induced in the wellbore during hydraulic fracturing during completion operations, but not during pumping). A high percentage of blowouts occur due to casing or cement failure, allowing high-pressure fluids to escape up the wellbore and flow into subsurface formations. The potential environmental consequences of a blowout depend mostly on (1) the timing of the blowout relative to well activities, which determines the nature of the released fluid such as natural gas or pressurized fracturing fluid, (2) the occurrence of the escape of containments through the surface casing or deep in a well, and (3) the risk receptors, such as freshwater aquifers or water wells that are impacted.

Controlling a blowout can have disadvantages such as (1) mud filtrate soaking into the formation in the near vicinity of the borehole and (2) a mud cake plasters the sides of the hole. These factors obscure the possible presence of oil or gas in even very porous formations and further complications arise with the occurrence of small amounts of petroleum in the rocks of many sedimentary formations (sedimentary provinces). In fact, if a sedimentary formation does not exhibit any evidence for the presence of natural gas or petroleum, drilling operations will be terminated.

2.1 Geological and Geotechnical Evaluation

A primary step in the evaluation of feasibility of fracturing is characterization of the *geological parameters* by examination of at least one continuous core boring sample in order to characterize major and minor changes in lithology as well as an analysis of the geological cross sections which show sediment layering in the target zone, and the contaminant characteristics present in the target zone. Cores (cylinders of rock, approximately 3–4 in. in diameter and up to 60 ft long) collected during continuous and depth-specific sampling should also be examined for factors contributing to secondary permeability such as coarse-grained sediment inclusions and naturally occurring fractures. More specifically, the geologic parameters to be evaluated include (1) the type of soil/rock, (2) the type of deposition, (3) the groundwater locale, including depth and dimensions of the aquifer, (4) the possibility of prefracturing contamination, (5) the type of contamination, and (6) the depth and extent of the contamination. The last three categories are necessary to establish the base-case condition of the area before fracturing commences and is the control by which the hydraulic fracturing process and any ensuing effects is measured.

Geotechnical characterization of the formations involves determining two general factors: (1) the lower limits for porosity, permeability, and upper limits for water saturation that permit profitable production from a particular formation or pay zone and (2) whether or not the formations under consideration exceed these lower limits. Thus, target zone samples should be submitted for evaluation of (1) the grain size analysis, (2) the liquid and plastic limits of the formations, (3) the moisture content, and (4) the unconfined compressive strength.

Grain size analysis recognizes that although fractures can be created in sediments and rock of variable grain size, the highest degree of permeability improvement can be expected from the finer grained materials. Grain size analysis can be performed by using the sieve analysis method and/or the hydrometer analysis method (ASTM D421, ASTM D422). The *liquid and plastic limits* of the formations (the Atterberg limits—a measure of the critical water contents of a fine-grained soil, such as its shrinkage limit, plastic limit, and liquid limit) characterize the plasticity of a formation (ASTM D4318). The *moisture content* (ASTM D2216) can influence the process insofar as permeability improvements are achievable with fracturing but vapor flow in particular is also controlled by the presence of moisture. Improvements in vapor flow through highly saturated formations (at or near capacity) will not always be achieved by fracturing alone and additional means of moisture removal may be required to obtain the desired effect through fracturing.

Data from measurement of the *unconfined compressive strength* (ASTM D2166) can be used for predicting the orientation and direction of propagation of fractures. Since hydraulic fracturing is generally applied at sites with characteristically low permeability, a baseline estimate of *permeability* (vapor and/or liquid) must be available, usually from testing concluded at the site during site investigations. This baseline estimate of permeability provides a basis for evaluating the necessity, benefit, and effectiveness of the fracturing process. In general, greater improvement of vapor or fluid flow and radial influence is observed in formations with lower initial permeability. In terms of cohesion, the more cohesive the soil is, more amenable it will be to fracturing. Longevity of the fractures, upon relaxation of fracture stress, is high in cohesive formations and fracturing in cohesive formations (such as silty clays) has been particularly successful.

2.2 Formation Integrity

The formation integrity test (FIT) is carried out to confirm the strength of formation and well casing shoe by increasing the bottomhole pressure to a design pressure. There is considerable confusion in industry nomenclature, such as the use of leak-off tests (LOTs) (API RP 13M-4). The LOTs, also known as pressure integrity tests, are used to determine the fracture gradient of a formation (from stress estimates). Low leak-off (fluid loss) rate is the property that permits the fluid to physically open the fracture and one that controls its areal extent. The rate of leak-off to the formation is dependent upon the viscosity and the wall-building properties of the fluid (API RP 13M). Postfracture breakdown is necessary such that the injected fluids do not hinder the passage of oil and gas in the formation. However, FITs are conducted to show that the formation below the casing shoe will not fail while drilling subsequent sections with a higher bottom hole pressure (Lee and Holditch, 1981). The test is generally conducted soon after drilling resumes after an intermediate casing string has been set and the purpose of the test is to determine the maximum pressures that may be safely applied without the risk of formation breakdown and the maximum wellbore pressure does not exceed the least principal stress or was not sufficient to initiate a fracture of the wellbore wall in an open hole test (Zoback, 2010).

2.3 Permeability and Porosity

Permeability is of critical importance in determining wells applicable for hydraulic fracturing. The main reason fracturing is done is to extract natural gas or crude oil that would not flow naturally to the wellbore. The permeability of the formation also affects the formation breakdown pressure in hydraulically fractured wells and permeable rock typically has a lower breakdown pressure than impermeable rock under similar conditions (Postler, 1997).

The standard method of obtaining permeability in routine core analysis is by allowing dry gas, usually nitrogen, helium, or air to flow through the samples. It has the following advantages over using liquid permeability: reduced fluid-rock interaction, easier to execute, faster and less expensive although the validity of the gas permeability method has been questioned (Unalmiser and Funk, 2008). Another shortcoming of using dry gas to obtain permeability is that it has to be corrected for gas slippage (the Klinkenberg effect), which arises with the type of gas used and the mean existing pressures in the core sample at the time of the measurement.

Porosity (the ratio of void volume to total) is obtained by measurement of either two of the three variables: pore volume (PV), bulk volume (BV), and grain volume (GV). It is important that calibration to standard temperature and barometric pressure is performed when measuring grain density for determination of the grain volume. The type of porosity test to be used out depends on the formation being sampled, for instance in vug formations (formations with small cavern or cavity within a reservoir rock, such as in carbonate formations) special procedures are required and, in addition, porosity measurement, like permeability measurement, is also sensitive to drying time.

As stated previously, there is no true or regular relationship between porosity and permeability, specifically for shale rock although some general trends may appear when the two are comparted graphically (Fig. 5.2). However, the porosity does tend to increase with the increase of clay minerals possibly because pore volumes in these formations mostly reside within the clay mineral aggregates and solid organics in the formation (Loucks et al., 2009; Sondergeld et al., 2010).

2.4 Residual Fluid Saturation and Capillary Pressure

Measurement of the *residual fluid saturation* was originally obtained by (1) the use of high-powered vacuum distillation to recover oil and water or (2) distillation extraction, which divides the extraction process into two parts in which the water was distilled and then the oil was extracted using suitable solvents for oil solubility (Keelan, 1982; Speight, 2015). Currently, fluid tracer studies, displaced-miscible fluid analyses (reducing damage to clays), and improved geochemical techniques are used to obtain saturation (Unalmiser and Funk, 2008). Fluid saturations are normally reported as a percent of the pore volume, and the accuracy of measurements is largely determined by conditions during sample recovery (Holditch et al., 1987).

A parameter of interest that influences most of the properties in the first phase is the wettability of the sample as it relates to fracture properties (Fernø et al., 2008). It is a measure of the preferred inclination of a fluid, that is, water or oil to spread on the rock surface (Unalmiser and Funk, 2008). It combines the interaction of the rock surface, fluid interfaces, and pore shape. Another category of the analysis involves measurement of formation geomechanical properties such as Poisson's ratio, Young's modulus, and fracture toughness (Holditch et al., 1987).

The *capillary pressure* is the difference in pressure across the interface between two phases. Relative permeability and capillary pressure relationships are used for estimating the amount of oil and gas in a reservoir and for predicting the capacity for flow of oil, water, and gas throughout the life of the reservoir. The relative permeability and capillary pressure are complex functions of the structure and chemistry of the fluids and solids in a producing reservoir and, as a result, will vary within a reservoir.

Thus, capillary pressure is used to characterize the reservoir by indicating water saturation, size of pore channels, and differentiating productive from nonproductive intervals (Keelan, 1982; Slattery, 2001; Unalmiser and Funk, 2008).

2.5 Mechanical Properties

Because of the high interest in producing natural gas and crude oil from unconventional tight reservoirs, there is an increasing need to understand the petrophysical and mechanical properties of these rocks (Sone and Zoback, 2013a,b). In fact, there is a wide variation in the properties of the formations (between reservoirs and within a reservoir) and, as might be expected, the properties of these rocks are a strong function of their material composition. Furthermore, the general characterization of these organic-rich rocks (shale, sandstone, and carbonates) can be challenging because these rocks vary significantly.

For example, formations in the Barnett formation are silica-rich whereas the Ford shale formation is carbonate-rich and contains relatively smaller amounts of silica minerals and clay minerals. In fact, many (tight) shale reservoirs typically have significant differences in lithology and petrophysical properties (Passey et al., 2010). Another source of complexity in these reservoirs is the mechanical anisotropy which is due to the organized distribution of the presence of the clay minerals (Hornby et al., 1994; Johnston and Christensen, 1995; Sondergeld and Rai, 2011) and compliant organic materials (Vernik and Nur, 1992; Vernik and Liu, 1997; Vernik and Milovac, 2011). There are also indications that it is not only the amount of clay minerals or organic constituents, but also the maturity of the shale formations that may influence the various properties of these organic-rich shales (Vanorio et al., 2008; Ahmadov et al., 2009). Understanding the anisotropy and its causes is crucial because they strongly influence analyses/interpretations of seismic surveys, sonic logs, and microseismic monitoring (Sone and Zoback, 2013a,b).

Many tight gas reservoirs are thick, layered systems that must be hydraulically fracture treated to produce at commercial gas flow rates (Chapter 4) and to optimize production it is necessary to understand the mechanical properties of all the layers above, within, and below the natural gas or crude oil pay zones. Basic rock properties such as in situ stress, Young's modulus, and Poisson's ratio are needed to design a fracture treatment. The in situ stress of each rock layer affects how much pressure is required to create and propagate a fracture within the layer. The values of Young's modulus relate to the stiffness of the rock and help determine the width of the hydraulic fracture. The values of Poisson's ratio relate to the lateral deformation of the rock when stressed. Poisson's ratio is a parameter required in several fracture design formulas.

The most important mechanical property is in situ stress, often called the minimum compressive stress or the fracture closure pressure and to optimize gas or oil production it is very important to know the values of in situ stress in the various strata. Generally, when the pressure inside the fracture is greater than the in situ stress, the fracture is open but when the pressure inside the fracture is less than the in situ stress, the fracture is closed. The necessary data can be obtained by using core sample or injection tests.

3 THE FRACTURING PROCESS

Hydraulic fracturing is not a method for drilling or constructing a well but it is the process for creating a fracture or fracture system in a porous medium by injecting a fluid under pressure through a wellbore in order to overcome the natural and inherent stresses in a formation. To fracture a formation, energy must be generated by injecting a fluid down a well and into the formation. The effectiveness of hydraulically created fractures is measured both by the orientation and areal extent of the fracture system and by the postfracture enhancement of vapor or liquid recovery.

The process is applied after well completion to facilitate movement of the reservoir fluids to the well and thence to the surface. This process creates access to more natural gas and crude oil but requires the use of large quantities of water and fracturing fluids, which are injected underground at high volumes and pressure. Thus, the sequence of fracturing a particular formation typically consists of (1) an acid stage, (2) a pad stage, (3) a prop sequence stage, and (4) a flushing stage.

The *acid stage* consists of several thousand gallons of water mixed with a dilute acid, such as hydrochloric or muriatic acid, which serves to clear cement debris in the wellbore by dissolving carbonate minerals and opening fractures near the wellbore. The *pad stage* consists of the use of approximately one hundred thousand gallons of water (or more) and 100,000 gallons of slickwater without proppant material. In this slickwater pad stage, the slickwater solution fills the wellbore and opens the formation which helps to facilitate the flow and placement of proppant material. The *prop sequence stage* may consist of several substages of the use of water combined with proppant material, which consists of a fine mesh sand or ceramic material, and is intended to keep open (prop) the fractures created and/or enhanced during the fracturing operation after the pressure is reduced; this stage may collectively use several hundred thousand gallons of water. The proppant material may vary from a finer particle size to a coarser particle size throughout this sequence. The *flushing stage* consists of a volume of fresh water sufficient to flush the excess proppant from the wellbore; the amount of water used is dependent upon the site characteristics (including the character of the subterranean formations).

Most of the fluids used in hydraulic fracturing are water and chemicals (typically 1% *v/v* of the water). The formulas for fracturing fluids vary, partly depending on the composition of the gas-bearing or oil-bearing formations, remembering that all gas-bearing and oil-bearing formations are not the same even when the formations are composed of the same minerals (shale, sandstone, or carbonate). In addition, some of the chemical additives can be hazardous if not handled carefully (Table 5.3) and caution is advised since the amount of the chemical(s) must not exceed the amount specified in regulatory requirements related to handling hazardous materials. Even if the chemical is one that is indigenous to the subsurface (and supposedly benign because it is found naturally), the amount used must not exceed the indigenous amount—in some case, exceeding the indigenous amount of a chemical can cause environmental problems.

Safe handling of all water and other fluids on the site, including any added chemicals, must be a high priority and compliance with all regulations regarding containment, transport, and spill handling is essential. When it comes to disposal of the fracturing fluid, there are options. For example, the fluid, when possible without causing adverse effects to the environment, can be reused for additional wells in a single field—this reduces the overall use of fresh water and reduces the amount of recovered water and chemicals that must be sent for disposal. However, in such cases, recognition of the geological or mineralogical similarities or difference within a site must have been determined to assure minimal environmental damage. In addition, tanks (or *lined* storage pits) for the storage of recovered water are also a necessity until the water can be sent for disposal to a permitted saltwater injection disposal well or taken to a treatment plant for processing. The linings of such pits must be in accordance with local environmental regulations.

All injection wells must be designed to meet the regulations set by the national agency (for example, the United States Environmental Protection Agency) or any local agency to protect the groundwater. In addition,

Table 5.3 Additives Used in the Hydraulic Fracturing Process

Water and sand: approximately 98% v/v			
Water	Expand the fracture and delivers sand	Some stays in the formation while the remainder returns with natural formation water as produced water (actual amounts returned vary from well to well)	Landscaping and manufacturing
Sand (proppant)	Allows the fractures to remain open so that the oil and natural gas can escape	Stays in the formation, embedded in the fractures (used to "prop" fractures open)	Drinking water filtration, play sand, concrete, and brick mortar
Other additives: ~2%			
Acid	Helps dissolve minerals and initiate cracks in the rock	Reacts with the minerals present in the formation to create salts, water, and carbon dioxide (neutralized)	Swimming pool chemicals and cleaners
Antibacterial agent	Eliminates bacteria in the water that produces corrosive byproducts	Reacts with microorganisms that may be present in the treatment fluid and formation; these microorganisms break down the product with a small amount returning to the surface in the produced water	Disinfectant; sterilizer for medical and dental equipment
Breaker	Allows a delayed breakdown of the gel	Reacts with the crosslinker and gel in the formation making it easier for the fluid to flow to the borehole; this reaction produces ammonia and sulfate salts, which are returned to the surface in the produced water	Hair colorings, as a disinfectant and in the manufacture of common household plastics

Table 5.3 Additives Used in the Hydraulic Fracturing Process—cont'd

Clay stabilizer	Prevents formation clays from swelling	Reacts with clays in the formation through a sodium-potassium ion exchange; this reaction results in sodium chlorine (table salt), which is returned to the surface in the produced water	Low-sodium table salt substitutes, medicines, and IV fluids
Corrosion inhibitor	Prevents corrosion of the pipe	Bonds to the metal surfaces, such as pipe, downhole; any remaining product that is not bonded is broken down by microorganisms and consumed or returned to the surface in the produced water	Pharmaceuticals, acrylic fibers, and plastics
Crosslinker	Maintains fluid viscosity as temperature increases	Combines with the "breaker" in the formation to create salts that are returned to the surface in produced water	Laundry detergents, hand soaps, and cosmetics
Friction reducer	Minimizes friction	Remains in the formation where temperature and exposure to the breaker allows it to be broken down and consumed by naturally occurring microorganisms; a small amount returns to the surface with the produced water	Cosmetics including hair, make-up, nail and skin products

Continued

Table 5.3 Additives Used in the Hydraulic Fracturing Process—cont'd

Gelling agent	Thickens the water to suspend the sand	Combines with the breaker in the formation making it easier for the fluid to flow to the borehole and return to the surface in the produced water	Cosmetics, baked goods, ice cream, toothpastes, sauces, and salad dressings
Iron control	Prevents precipitation of metal in pipe	Reacts with minerals in the formation to create simple salts, carbon dioxide, and water, all of which are returned to the surface in the produced water	Food additives; food and beverages; lemon juice
Nonemulsifier	Breaks or separates oil/water mixtures (emulsions)	Generally, returns to the surface with produced water, but in some formations it may enter the gas stream and return to the surface in the produced oil and natural gas	Food and beverage processing, pharmaceuticals, and wastewater treatment
pH adjusting agent	Maintains the effectiveness of other components, such as crosslinkers	Reacts with acidic agents in the treatment fluid to maintain a neutral (nonacidic, nonalkaline) pH; this reaction results in mineral salts, water, and carbon dioxide—a portion of each is returned to the surface in the produced water	Laundry detergents, soap, water softeners, and dish washer detergents

production zones should have multiple confining layers above the zone to keep the injected fluids within the target gas-bearing or oil-bearing formation. In addition, multiple layers of well casing and cement (similar to production wells) should be used with periodic mechanical integrity tests to verify that the casing and cement are holding the liquids. The amount and pressure of the injected fluid (specified in each well permit) should be monitored to maintain the fluids in the target zone and the pressure in the injection well and the spaces between the casing layers (also called the annuluses) should also be monitored to check and verify the integrity of the injection well.

Finally, to be an aid in production, fractures must be connected to a reasonable hydrocarbon bearing reservoir with sufficient volume to warrant exploitation. If there is no reservoir volume, a lot of fractures won't help much unless there is sufficient fracture-related solution porosity to hold an economic reserve. This can be determined by normal log analysis techniques. In reasonable nonfractured reservoirs, it is usually possible to estimate permeability, and hence productivity (Speight, 2014, 2016b), but this is not always possible in fractured reservoirs. Although both the presence of fractures and the presence of a reservoir can be determined from logs, a production test will be needed to determine whether economic production is possible. The test must be analyzed carefully to avoid over optimistic predictions based on the flush production rates associated with the fracture system. Local correlations between fracture intensity observed on logs and production rate are also used to predict well quality.

Sometimes the primary reservoir and the fracture system may be so poorly connected that they are saturated with different fluids. Production from fractures full of hydrocarbons in a water-bearing formation may initially be very good but very short lived. A more desirable scenario is a primary reservoir with appreciable hydrocarbon saturation and a fracture system that is full of water close to the borehole, showing invasion and hence good permeability, but full of hydrocarbon in the uninvaded formation. There are several issues that must be taken into consideration during the fracturing process, these are (1) equipment, (2) well development, (3) fracture patterns, (4) fracture optimization, and (5) fracture monitoring.

3.1 Equipment

Conventional cement-pumping and acid-pumping equipment was used initially to execute fracturing treatments. One to three units equipped with one pressure pump delivering 75–125 hhp were adequate for the small volumes injected at the low rates. However, as treating volumes increased accompanied by a demand for greater injection rates, special pumping and blending equipment was developed and the development continues.

Initially, sand was added to the fracturing fluid by pouring it into a tank of fracturing fluid over the suction. More recently, with less-viscous fluid(s), a ribbon or paddle type of batch blender was used after which a continuous proportioning blender utilizing a screw to lift the sand into the blending tub was developed. As the procedure evolved, blending equipment has also had to evolve to meet the need for proportioning a large number of dry and liquid additives, then uniformly blending them into the base fluid and adding the various concentrations of sand or alternate propping agents. In fact, the hydraulic fracturing treatment follows the actual drilling and completion of the well (Hubbert and Willis, 1957; Hibbeler and Rae, 2005; Arthur et al., 2009).

In the initial stages of a fracturing project, the drilling may be the same as drilling a conventional reservoir. Thus, in the process, a borehole is drilled vertically, then a casing is put in place after which cement and mud is pumped into the annulus to form a barrier between the borehole and adjacent formations. Drilling is then continued, to an adequate depth adjacent (sometimes within) within the producing reservoir. From this point (the *kick-off point*), the wellbore is deviated gradually until it curves into a horizontal plane and drilled a distance on the order of 1000 ft to more than 5000 ft (Arthur et al., 2009).

The hydraulic fracturing procedure is then initiated. The process involves fracturing at isolated intervals along the horizontal well since it is difficult (if not impossible) to apply pressure along the entire length of the wellbore because of loss of pressure efficiency over the distance involved (1000–5000 ft). The fracturing areas are isolated using packers and perforations are created in the wellbore within the interval bounded by packers (Arthur et al., 2009). In some fracture treatments, acid is pushed through the perforated interval to help breakdown any barrier that might be due to the characteristics of the formation(s) surrounding the wellbore.

The design of fracture treatment is a complex task, which involves analysis, planning, experience, and rigorous observation of different stages in the entire process. In order to develop a formation containing natural gas and/or crude oil, the wellbore is drilled in successive sections through the rock layers. Once the desired length of each wellbore section has been drilled, the drilling assembly is removed, and steel casing is inserted and cemented in place. As the well is constructed, concentric layers of steel casing and cement form the barrier to protect groundwater resources from the fluids that will later flow inside the well. In the next step, a section of casing within the formation is perforated at the desired location for gas or oil production.

At this stage, the well is ready for hydraulic fracturing process which involves pumping fluid through the perforations. The fracturing fluid exerts pressure against the rock, creating tiny cracks, or fractures, in the reservoir deep underground. The fluid is predominantly water, proppant (grains of sand or ceramic particles), and a small amount (on the order of 1% *v/v*) of chemical additives. Once fluid injection stops, pressure begins to dissipate, unless the necessary steps are taken, the fractures previously held open by the fluid pressure begin to close. The necessary steps include the injection of proppants which act as wedges to hold open the narrow fractures, thereby creating pathways for the natural gas or crude oil and the fracturing fluids to flow more easily to the well. A plug is set inside the casing to isolate the stimulated section of the well and the perforate-inject-plug cycle is repeated at regular intervals along the targeted section of the reservoir. Finally, the plugs are drilled out, allowing the petroleum, natural gas, and fluids to flow into the well casing and up to the surface. The petroleum/gas/fracturing fluid mixture is separated at the surface, and the fracturing fluid (also known as flowback water) is captured in tanks or lined pits. The fracturing fluids are then disposed of according to the regulatory-approved methods.

Briefly, several definitions exist for the terms *produced water* and *flowback water* and indicate the confusion in the use of the terminology. The common definitions are (1) *produced water* is any of the many types of water produced from oil and gas well and (2) *flowback water* is the hydraulic fracturing fluid that returns to the surface after a hydraulic fracture is completed.

Fractures from both horizontal and vertical wells can propagate vertically out of the intended zone thereby (1) reducing stimulation effectiveness, (2) wasting proppant and fluids, and potentially connecting up with other hydraulic fracturing stages or unwanted water or gas intervals which can also lead to a variety of environmental issues (Chapter 9). The direction of lateral propagation is largely dictated by the horizontal stress regime, but in areas where there is low horizontal stress anisotropy or in reservoirs that are naturally fractured, fracture growth is not always easy to predict (Hammack et al., 2014). In shallow zones, horizontal hydraulic fractures can develop because the weight of the overburden—the vertical stress component—is smallest. A horizontal hydraulic fracture reduces the effectiveness of the stimulation treatment because it most likely forms along horizontal areas of weakness—such as the areas between the formation strata—and is aligned preferentially to formation vertical permeability, which is typically much lower than horizontal permeability.

More specifically, after a hydraulic fracture is initiated, the degree to which the fracture grows laterally or vertically depends on numerous factors, such as confining stress, fluid leak from the fracture, fluid viscosity, fracture toughness, and the number of natural fractures in the reservoir. Thus, prediction of the precise behavior of the fracture is difficult and, in many cases, may even be impossible because of incorrect information and assumptions used in planning the fracturing project.

The extent of a hydraulic fracture is a complex relationship between the strength of the rock and the pressure difference between the rock and the fracturing pressure. The extent is defined by the fracture dimensions—height, depth of penetration (wing length or fracture length), and aperture (width or opening). One measure of the strength of the rock is the Poisson ratio. Thus, when a material is compressed in one direction, it usually tends to expand in the other two directions perpendicular to the direction of compression (the Poisson effect) and the Poisson's ratio (ν, the fraction or percent) of expansion divided by the fraction (or percent) of compression is a measure of this effect. The Poisson ratio is low (0.10–0.30) for most sandstone formations and carbonates—rocks that fracture relatively easily. On the other hand, the Poisson ratio is high (0.35–0.45) for shale, sandstone, and coal—rocks that are more elastic and are harder to fracture (Sone and Zoback, 2013a,b). Shale is often the upper and lower barrier to the height of a fracture in conventional sandstone.

Another aspect of equipment consideration relates to the inducement of fracturing by pneumatic methods. Pneumatic fractures can be generated in geologic formations if air or any other gas is injected at a pressure that exceeds the natural strength as well as the in situ stresses present in the formation. As noted earlier, pneumatic fracture propagation will be predominantly horizontal at over-consolidated formations. However, in shallow recent fills, some upward inclination of the fractures has been observed, the reason for which is attributed to the lack of stratification and consolidation in these formations. The amount of pressure required to initiate pneumatic fractures is dependent on the cohesive or tensile strength of the formation, as well as the overburden pressure (dependent upon the depth and density of the formation). The most important system parameter for efficient pneumatic fracturing is injection flow rate, as it largely determines the dimensions of a pneumatic fracture. Once a fracture has been initiated, it is the high volume airflow which propagates the fracture and supports the formation. The design goal of a pneumatic fracturing system therefore becomes one of providing the highest possible flow rate. Field observations indicate that pneumatic fractures reach their maximum dimension in less than 20 s, after which continued injection simply maintains the fracture network in a dilated state (in essence, the formation is "floating" on a cushion of injected air). Pneumatically induced fractures continue to propagate until they intersect a sufficient number of pores and existing discontinuities, so that leak-off (fluid loss) rate into the formation exactly equals the injection flow rate.

An individual pneumatic fracture is accomplished by (1) advancing a borehole to the desired depth of exploration and withdrawing the auger, (2) positioning the injector at the desired fracture elevation, (3) sealing off a discrete 1 or 2 ft interval by inflating the flexible packers on the injector with nitrogen gas, (4) applying pressurized air for approximately 30 s, and (5) repositioning the injector to the next elevation and repeating the procedure. A typical fracture cycle takes approximately 15 min, and a production rate of 15–20 fractures per day is attainable with one rig.

The pneumatic fracturing procedure typically does not include the intentional deposition of foreign propping agents to maintain fracture stability. The created fractures are thought to be "self propping," which is attributed to both the asperities present along the fracture plane as well as the block shifting which takes place during injection. The aperture or thickness of a typical pneumatically induced fracture is approximately 0.5–1 mm. Testing to date has confirmed fracture viability in excess of 2 years, although the longevity is expected to be highly site-specific.

Without the carrier fluids used in hydraulic fracturing, there are no concerns with fluid breakdown characteristics for pneumatic fracturing. There is also the potential for higher permeability within the fractures formed pneumatically, in comparison to hydraulic fractures, as these are essentially air space and are devoid of propping agents. The open, self-propped fractures resulting from pneumatic fracturing are capable of transmitting significant amounts of fluid flow.

3.2 Well Development and Completion

Well development is an integral part of the hydraulic fracturing process and is typically divided into two stages: (1) the drilling stage and (2) the completion stage. For a successful fracturing operation, it is important that drilling equipments are properly maintained and that their rated capacity is not exceeded. A drilling rig is the most visible part of the drilling operation, however what is important is the underground activity and the main considerations in the selection of a rig are (King, 2012): (1) noise, which can be minimized by using electric rigs, (2) dust—if air drilling is used, control of air and cuttings is required, (3) appearance—most rigs for unconventional well drillings are from 50 ft to over 100 ft tall, which is visually undesirable and take more time to set up; lower profile rigs are preferred on shallower wells but the trade-off is that larger rigs are faster in operation, (4) water and mud storage: requiring determination of size of pits or steel tanks—also storage considerations for chemicals that would be mixed with the mud, and (5) pressure control equipment—the equipment must undergo regular servicing and inspection. Completions involve the final stages of the well development process, which include casing and cementing design.

The rotary drilling process for a vertical well or for a directional well involves (1) application of a force downward on a drill bit, (2) rotation of the drill bit, and (3) circulation of the drilling fluid from the surface through the tubular (drill string), and back to the surface through the annular space, which is the area between drill string and borehole wall or casing (Azar and Samuel, 2007). On the other hand, horizontal drilling involves directing the drill bit to follow a horizontal path, oriented at approximately 90 degree from the vertical, through the reservoir rock (Azar and Samuel, 2007). Over the years, hydraulic fracturing has been performed on vertical, deviated and horizontal wells. However, the coupling of horizontal wells and hydraulic fracturing have been proven to improve well performance in oil and gas reservoirs (Britt et al., 2010; Devereux, 2012) and enhances the recovery of natural gas or crude oil by reducing the number of vertical wells to develop fields of interest. In the current context, horizontal wells have found ready application in the Barnett shale, the Marcellus shale as well as in other shale plays.

Horizontal wellbores allow a far greater exposure to a formation than a conventional vertical wellbore which is particularly useful in tight formations that do not have a sufficiently high permeability to produce natural gas or crude oil economically from a vertically drilled well. Furthermore, the type of wellbore completion used will influence the number of times that a formation is fractured and the locations along the horizontal section of the wellbore that fracturing is necessary. In North America, shale reservoirs such as the Bakken, Barnett, Montney, Haynesville, Marcellus, and (more recently) the Eagle Ford, Niobrara, and Utica shale formations have been drilled, completed, and fractured using this method. The method by which the fractures are placed along the wellbore is most commonly achieved by one of two methods: (1) as the *plug and perf* method and (2) the *sliding sleeve* method.

The wellbore for the *plug and perf* method is generally composed of standard joints of steel casing, either cemented or uncemented, which is set in place at the conclusion of the drilling process. Once the drilling rig has been removed, a perforation is created near the end of the well, following which a fracturing stage is initiated. Once the fracturing stage is completed, a plug is set in the well to temporarily seal off that section of the wellbore. Another stage is then pumped, and the process is repeated as necessary along the entire length of the horizontal part of the wellbore.

On the other hand, the wellbore for the *sliding sleeve* method is different insofar as the sliding sleeves are included at set distances (spacing) in the steel casing at the time the casing is set in place. The sliding sleeves are usually all closed at this time and when the well is ready for application of the fracturing process, the bottom sliding sleeve is opened and the first stage is pumped. Once finished, the next sleeve is opened which concurrently isolates the first stage, and the process is repeated. These completion techniques may allow more than 30 stages to be pumped into the horizontal section of a single well if required, which is many more stages than would typically be pumped into a vertical well (Mooney, 2011).

Finally, to optimize the completion, it is necessary to understand the mechanical properties of all the layers above, within, and below the gas pay intervals. Basic rock properties such as in situ stress, Young's modulus, and Poisson's ratio are needed to design a fracture treatment. The in situ stress of each rock layer affects how much pressure is required to create and propagate a fracture within the layer. The values of Young's modulus relate to the stiffness of the rock and help determine the width of the hydraulic fracture. The values of Poisson's ratio relate to the lateral deformation of the rock when stressed. Poisson's ratio is a parameter required in several fracture design formulas.

The most important mechanical property is in situ stress, often called the minimum compressive stress or the fracture closure pressure. When the pressure inside the fracture is greater than the in situ stress, the fracture is open. When the pressure inside the fracture is less than the in situ stress, the fracture is closed. To optimize the completion, it is very important to know the values of in situ stress in every rock layer.

3.3 Fracturing Fluids

Initially, the fluid is injected that does not contain any propping agent and is injected to create a fracture that is multidirectional and spreads in several directions. This creates a fracture that is sufficiently open for insertion of the proppant, which is injected as a slurry—a mix of the proppant and the carrier fluid and proppant material. In shallow reservoirs, sand is often used and remains the most common proppant but in deep reservoirs, ceramic beads (in place of the usual sand proppant) may be used to prop open the fractures. Once the fracture has initiated, fluid is continually pumped into the wellbore to extend the created fracture and develop a fracture network. However, each formation has different properties and, therefore, different in situ stress forces are operational es so that each hydraulic fracture project is unique and must be designed accordingly by identification of the properties of the target formation including estimating fracture treating pressure, amount of material, and the desired length for optimal economics. Furthermore, the fracturing fluid should have a number of properties that are suited to the properties of the formation, such as (1) compatibility with the formation rock, (2) compatibility with the formation fluid, (3) suitability to generate sufficient pressure drop down the fracture to create a wide enough fracture, and (4) a sufficiently low viscosity to allow cleanup after the treatment. Water-based fluids are commonly used—*slickwater* is the most common fluid used for shale gas fracturing, where the major chemical added is a surfactant polymer to reduce the surface tension or friction, so that water can be pumped at lower treating pressures. Other fluids that have been considered are oil-based fluids, foams, and emulsions but caution is advised when using nonaqueous fluids since these fluids must be allowable for injection (by regulation) and must not have any detrimental effect on the environment (Chapter 9).

Additives for fracturing fluids are chosen according to the properties of the reservoir and include (Table 5.3) (1) polymers, which allow for an increase in the viscosity of the fluid, together with crosslinkers, (2) crosslinkers, which increase the viscosity of the linear polymer base gel, (3) breakers, which are used to break the polymers and crosslink sites at formation temperature, for better cleanup, (4) biocides, which are used to kill bacteria in the mix water, (5) buffers, which are used to control the pH, (6) fluid-loss additives, which are used to control excessive fluid leak-off into the formation, and (7) stabilizers, which are used to keep the fluid viscous at higher temperature. However, it must be emphasized that additives are used for every site and in general as few additives as possible are added to avoid potential environmental contamination (use of the additives must be controlled) and production problems with the reservoir.

Environmental concerns have focused on the fluid used for hydraulic fracturing and the risk of water contamination through leaks of this fluid into groundwater (Chapters 6 and 9). Water, together with sand or ceramic beads (*proppant*), is the most common and acceptable (environmentally acceptable and economically acceptable) typical fracturing fluid, but a mixture of chemical additives is also used to give the fluid the properties that vary according to the type of formation. The additives (not all of which would be used in all fracturing fluids) (Table 5.3) typically assist in the accomplishment of four tasks:

(1) To maintain the proppant in suspension in the fluid by gelling the fluid while it is being pumped into the well and to ensure that the proppant ends up in the fractures being created. In the absence of this effect, the denser proppant particles would succumb to the influence of gravity and remain unevenly distributed in the fluid, thereby losing some of the effectiveness. To ensure suspension of the proppant in the fluid, gelling polymers (such as guar or cellulose) are used at a concentration of about 1% while crosslinking agents, such as borates or metallic salts, may also be used (also at a very low concentration) to form a stronger gel.

(2) To allow a change in the properties of the hydraulic fracturing over time. Generally, the characteristics that are needed to deliver the proppant into subsurface cracks are not desirable at other stages in the injection process and the time-dependent properties that are imparted to the fluid can reduce the viscosity after fracturing, so that hydrocarbons (such natural gas liquids and as crude oil) will flow more easily along the fractures to the production well.

(3) To reduce friction and therefore reduce the energy required to inject the fluid into the well; atypical drag-reducing polymer is polyacrylamide.

(4) To reduce the risk that naturally occurring bacteria in the water affect the performance of the fracturing fluid or proliferate in the reservoir, producing hydrogen sulfide; this is often achieved by using a disinfectant-type (biocide) additive.

3.4 Fracture Patterns

Hydraulic fracturing may be performed on a single reservoir interval in a vertical well. Horizontal wells, however, by virtue of their significant wellbore length in the target formation, are generally isolated into several discrete intervals along the horizontal wellbore—there may be 4–20 intervals for each horizontal well, with each interval requiring its own fracturing stage. This is due to the difficulty in maintaining pressures sufficient to induce fractures over the complete length of the lateral leg.

The most important data for designing a fracture treatment and the resulting fracture patterns are (1) the in situ stress profile, (2) formation permeability, (3) fluid-loss characteristics, (4) total fluid volume pumped, (5) propping agent—type and amount, (6) viscosity of the fracture fluid, (7) injection rate, and (8) formation modulus. The in situ stress profile and the permeability profile of the zone to be stimulated must be quantified and identification of the layers of rock above and below the target zone since these formations will influence fracture height growth. In order to design the optimum treatment, the effect of fracture length and fracture conductivity on the productivity and the ultimate recovery from the well must be determined.

The selection of the fracture fluid for the treatment is a critical decision and is on the basis of factors such as (Economides and Nolte, 2000): (1) reservoir temperature, (2) reservoir pressure, (3) the expected value of fracture half-length, and (4) any water sensitivity. The definition of what comprises a water-sensitive reservoir and what causes the damage is not always evident. Most reservoirs contain water, and most natural gas or crude oil reservoirs can be successfully water-flooded. Thus, most fracture treatments should be pumped with suitable water-base fracture fluids. Acid-base fluids can be used in carbonates but many deep carbonate reservoirs have been stimulated successfully with water-base fluids containing propping agents.

When selecting a propping agent, it is necessary to determine the maximum effective stress on the agent. The maximum effective stress depends on the minimum value of flowing bottomhole pressure expected during the life of the well. To confirm exactly which type of propping agent should be used during a specific fracture treatment, the designer should factor in the estimated values of formation permeability and optimum fracture half-length (Cinco-Ley et al., 1978). The treatment must be designed to create a fracture wide enough, and pump proppants at concentrations high enough, to achieve the conductivity required to optimize the treatment. There is a tendency to compromise fracture length and conductivity in an often unsuccessful attempt to prevent damage to the formation around the fracture and substantial damage to the formation around the fracture can be tolerated as long as the optimum fracture length and conductivity are achieved

(Holditch, 1979). However, damage to the fracture or the propping agents can be very detrimental to the productivity of the fractured well. Ideally, the optimum fracture length and conductivity can be created while minimizing the damage to the formation.

Finally, in horizontal wells, transverse fractures are relatively more difficult to achieve than longitudinal fractures. However, for natural gas and crude oil formations typically characterized by low permeability, transverse fractures in horizontal wells have greater production benefits (Valko et al., 1998). Transverse vertical fractures move along the path of least resistance, which is normal to the minimum horizontal stress. In horizontal wells or deviated wells, there are effects in the immediate vicinity around the wellbore that lead to the transverse fractures taking unpredicted paths before eventually aligning normal to the horizontal stress. These effects are increased by the presence of natural fractures in the formation and the deviation of the horizontal well at an angle from the minimum horizontal stress.

In the design of a hydraulic fracturing procedure, most procedures to optimize well productivity begin with the fracture size. Limitations in the different hydraulic fracture design methods are inherent in their assumptions of (1) fracture geometry, (2) dependence on fracture fluid properties, (3) dependence on reservoir properties, (4) dependence on whether or not the formations are layered, and (5) a variety of other factors, such as stress intensity. Challenges in fracture geometry when fracturing unconventional reservoirs include fracture azimuth and dip, not creating expected length, brittle, and ductile rocks—complex and simple networks, wellbore axis (vertical or horizontal drilling) (Kennedy et al., 2012). In all cases, however, knowledge of existing in situ stress is essential to developing a fracture-propagation model which describes the methods of obtaining a desired hydraulic fracture geometry definitely including the fracture (half) length, width, height, and fracture complexity.

The ideal formation evaluation would be one where the value of the in situ stress obtained from injection tests and those calculated from logs and core analysis all result in a consistent stress profile (Holditch et al., 1987). The hydraulic methods are also the most reliable for determining in situ stress in deep (>160 ft) formations (Amadei and Stephansson, 1997).

Hydraulic fractures are formed in the direction perpendicular to the least stress. Typically, horizontal fractures will occur at depths less than approximately 2000 ft because the overburden at these depths provides the least principal stress. If pressure is applied to the center of a formation under these relatively shallow conditions, the fracture is most likely to occur in the horizontal plane, because it will be easier to part the rock in this direction than in any other. In general, therefore, these fractures are typically parallel to the bedding plane of the formation.

As depth increases beyond approximately 2000 ft, overburden stress increases by approximately 1 psi/ft, making the overburden stress the dominant stress. This means the horizontal confining stress is now the least principal stress and, since hydraulically induced fractures are formed in the direction perpendicular to the least stress, the resulting fractures at depths greater than approximately 2000 ft will be oriented in the vertical direction.

In the case where a fracture might cross over a boundary where the principal stress direction changes, the fracture would attempt to reorient itself perpendicular to the direction of least stress. Therefore, if a fracture propagated from deeper to shallower formations it would reorient itself from a vertical to a horizontal pathway and spread sideways along the bedding planes of the rock strata.

The extent that a created fracture will propagate is controlled by the upper confining zone or formation, and the volume, rate, and pressure of the fluid that is pumped. The confining zone will limit the vertical growth of a fracture because it either possesses sufficient strength or elasticity to contain the pressure of the injected fluids or an insufficient volume of fluid has been pumped. This is important because the greater the distance between the fractured formation and the underground source of drinking water, the more likely it will be that multiple formations possessing the qualities necessary to impede the fracture will occur. However, while it should be noted that the length of a fracture can also be influenced by natural fractures or faults, natural attenuation of the fracture will occur over relatively short distances due to the limited volume of fluid being pumped and dispersion of the pumping pressure regardless of intersecting migratory pathways.

3.5 Fracture Optimization

Hydraulic communication is a key factor for determining hydrocarbon or thermal energy recovery sweep efficiency in an underground reservoir. Sweep efficiency is a measure of the effectiveness of heat, gas, or oil recovery process that depends on the volume of the reservoir contacted by an injected fluid (Britt, 2012). Artificial (stimulated) hydraulic fractures are usually initiated by injecting fluids into the borehole to increase the pressure to the point where the minimal principal stress in the rock becomes tensile. Continued pumping at an elevated pressure causes tensile failure in the rock, forcing it to split and generate a fracture that grows in the direction normal to the least principal stress in the formation. Hydraulic fracturing activities often involve injection of a fracturing fluid with proppants in order to better propagate fractures and to keep them open (Britt, 2012). The design of fracturing treatment should involve the optimization of operational parameters, such as the viscosity of the fracturing fluid, injection rate and duration, as well as proppant concentration, so that a fracture geometry is created that favors increased sweep efficiency. The net present value is the economic criterion that is usually used as an objective for optimal fracturing treatment design. Some studies have been reported to use a sensitivity-based optimization procedure coupled with a fracture-propagation model and an economic model to optimize design parameters leading to maximum net present value (Hareland et al., 1993; Rueda et al., 1994; Aggour and Economides, 1998; Mohaghegh et al., 1999; Chen et al., 2013).

In summary, fracture geometry optimization involves defining the desired fracture half-length, width, and conductivity for maximized production. While there are several optimization methods, all involve a relative comparison of the flow potential of the fracture to that of the reservoir. Thus, a fracture is often considered (or defined) as a high-permeability path in a low-permeability rock formation but if the fracture is filled with a cementing material, such as calcite (calcium carbonate, $CaCO_3$), the result is a fracture with little or no permeability. Thus, in any evaluation of the reservoir, it is important to distinguish between open fractures and healed (plugged or filled) fractures. The total volume of fractures is often small compared to the total pore volume of the reservoir. Thus, natural fractures in reservoir rocks (especially in tight reservoir formations) contribute significantly to natural gas or crude oil production. Therefore, it is important to glean every scrap of information from open hole logs to locate the presence and intensity of fracturing. Even though some modern logs, such as the formation micro-scanner and televiewer, are the tools of choice for fracture indicators, many wells lack this data.

Most natural fractures are vertical—a horizontal fracture may exist for a short distance, propped open by bridging of the irregular surfaces. Most horizontal fractures, however, are sealed by overburden pressure. Both horizontal and semi-vertical fractures can be detected by various logging tools. The vertical extent of fractures is often controlled by thin layers of plastic material, such as shale beds or laminations, or by weak layers of rock, such as stylolites in carbonate sequences. The thickness of these beds may be too small to be seen on logs, so fractures may seem to start and stop for no apparent reason.

The nucleation and propagation of hydraulic rock fractures are chiefly controlled by the local in situ stress field, the strength of the rock (stress level needed to induce failure), and the pore fluid pressure. Temperature, elastic properties, pore water chemistry, and the loading rate also have an influence (Secor, 1965; Phillips, 1972; Sone and Zoback, 2013a,b). Fractures in rock can be classified as tensile, shear or hybrid (a mixture of tensile and shear). If the dominant displacement of the wall rocks on either side of the fracture is perpendicular to the fracture surface, then the fracture is deemed tensile. New tensile fractures form when the pore fluid pressure in the rock exceeds the sum of the stress acting in a direction perpendicular to the fracture wall and the tensile strength of the rock. Note that any preexisting fractures that are uncemented (i.e., have zero cohesion) can be opened at a lower value of pore fluid pressure, when it exceeds the stress acting in a direction perpendicular to the fracture wall.

The formation or reactivation of shear fractures depends on the shear stress, the normal stress, the pore fluid pressure, and the coefficient of friction for the specific rock type. It is important to recognize that the hydraulic fracturing process of pumping large volumes of water into a borehole at a certain depth cannot control the type of fractures that are created or reactivated. The array of fractures created and/or reactivated or reopened depends on a complex interplay of the in situ stress, the physical properties of the local rock volume and any preexisting fractures, and the pore fluid pressure (Phillips, 1972). This could have implications for the risk of ground water contamination by hydraulic fracturing operations, as the fracture network generated by the hydraulic fracturing fluid could be complex and difficult to predict in detail. The orientations, sizes, and apertures of permeable rock fractures created by a hydraulic fracturing operation ultimately control the fate of the hydraulic fracturing fluid and the released shale gas, at least in the deep subsurface. Geomechanical models used to predict these fracture pattern attributes therefore need thorough testing/benchmarking, together with ongoing and future developments.

A common issue encountered in hydraulic fracturing operations in tight formations (especially in shale formations) is the variability and unpredictability of the outcome of the fracturing process. The injection pressure required to fracture the formation (fracture gradient) often varies significantly along a well, and there can be intervals where the formation cannot be fractured successfully by the injection of the fluid. The use of real-time fracture mapping allows real-time observation of changes in fracture design and also allows changes in re-stimulation design to maximize the *effective stimulation volume* (ESV—the reservoir volume that has been effectively contacted by the stimulation treatment). A correlation of microseismic activity with log data allows estimation of fracture geometry to be made after which the data can be used to design a stimulation that has the greatest chance of maximizing production (Fisher et al., 2004; Baihly et al., 2006; Daniels et al., 2007). Shale gas reservoirs also respond to fluid injection in a variety of modes although distribution of activated seismicity can be confined along a macroscopic fracture plane, but most time they are dispersed throughout a wide region in the reservoir reflecting the development of a complex fracture network (Waters et al., 2006; Cipolla et al., 2009; Das and Zoback, 2011; Maxwell, 2011).

In recent years, various attempts have been made to optimize the design of transverse fractures of horizontal wells for shale gas reservoirs. However, these optimization methods may not provide the optimal design. Hence, the optimization of hydraulic fracturing treatment design for shale gas production remains a challenge.

Finally, earlier literature on fracture analysis suggested that fractures might contribute as much as a few to several percent porosity but more recent work using the fracture aperture (calculated from resistivity micro-scanner logs) indicates a much lower contribution to the porosity of the formation. The term *secondary porosity* also includes rock-volume shrinkage due to dolomitization, porosity increase due to solution or recrystallization, and other geological processes. *Secondary porosity* should not be confused with *fracture porosity* which can be determined by processing the formation micro-scanner curves for fracture aperture and fracture frequency (*fracture intensity*). The effect of fracture porosity on reservoir performance, however, is very large due to the substantial contribution to permeability. As a result, naturally fractured reservoirs behave differently than nonfractured reservoirs with similar porosity due to the relatively high flow rate and capacity of the secondary porosity system. This provides high initial production rates, which can lead to extremely optimistic production forecasts and sometimes, economic failures when the small reservoir volume is not properly taken into account.

3.6 Fracture Monitoring

During the hydraulic fracturing process, fluid leak-off which is loss of fracturing fluid from the fracture channel into the surrounding permeable rock can (and often does) occur. If not controlled properly, the fluid loss can exceed 70% *v/v* of the injected volume which may result in formation matrix damage, adverse formation fluid interactions, or altered fracture geometry and thereby decreased production efficiency. Thus, fracture geometry and fracture monitoring are important aspects of the hydraulic fracturing process.

Monitoring technologies are used to map where fracturing occurs during a stimulation treatment and includes such techniques as microseismic fracture mapping, and tiltmeter measurements (Arthur et al., 2008). These technologies can be used to define the success and orientation of the fractures created during a stimulation process. Measurements of the pressure and rate during the growth of a hydraulic fracture, as well as knowing the properties of the fluid and proppant being injected into the well, provide the most common and simplest method of monitoring a hydraulic fracture treatment. This data, along with knowledge of the underground geology, can be used to model information such as length, width, and conductivity of a propped fracture (Fisher, 2012).

Microseismic monitoring is the process by which the seismic waves generated during the fracturing of a rock formation are monitored and used to map the locations of the fractures generated. Monitoring is done using a similar technology to that used to monitor larger naturally occurring seismic events associated with earthquakes and other natural processes. Microseismic monitoring is an active monitoring process performed during a hydraulic fracture treatment. As an active monitoring process microseismic monitoring can be used to develop real-time changes to a fracture program. Microseismic monitoring provides engineers the ability to manage the resource through intelligent placement of additional wells to take advantage of the natural conditions of the reservoir and expected fracture results in new wells.

Microseismic theory and mapping is based on earthquake seismology. Similar to earthquakes, but at a much higher frequency (200–2000 Hz), microseismic events emit elastic P waves (compressional) and S waves (shear waves) (Jones and Britt, 2009). During hydraulic fracture, there is an increase in formation stress proportional to the net fracturing pressure, and an increase in pore pressure due to fracturing fluid leak-off. The increase in stresses at the fracture tip and pore pressure increments cause shear slippages to occur. Microseismic technology thus uses earthquake seismology methodologies to detect and locate these hydraulic fracturing induced shear slippages, which resemble microearthquakes. Microseismic events or microearthquakes occur with fracture initiation and are observed with receivers placed on an offset wellbore like with the downhole tiltmeters.

Microseismic mapping technology involves installing an array of tri-axial geophone or accelerometer receivers into an offset well at approximately the depth of the fracture (like in downhole tiltmeters), orienting the receivers (geophones), recording seismic data, finding microearthquakes in the data and locating them. Locating the earthquake events requires the determination of compressional (P) and shear (S) wave arrivals and consequent acoustic interpretation of the velocity of the P-S waves (Davis et al., 2008; Jones and Britt, 2009). Standard microseismic mapping uses P-S arrival time separation for distance location. Horizontal and vertical plane holograms are used to determine azimuth and inclination (Warpinski et al., 2005).

Tiltmeters are passive monitoring technologies which record the deformation of rocks that are induced by the hydraulic fracture process. Tiltmeters can be placed at the ground surface away from a well or downhole in a nearby wellbore tightly into the rock. Tiltmeters measure changes in inclination in two orthogonal directions, which can then be translated into the strain rotation that results from hydraulic fracturing. Engineers can then determine based on the strain rotation the location of the hydraulic fracturing event that caused the strain rotation.

Downhole tiltmeter mapping technology was developed to circumvent the limitations of the surface tiltmeter by giving estimates of the fracture dimensions. The downhole tiltmeters have the same operational principle as the surface tiltmeters, but instead of being at the surface, the tiltmeters are positioned by wireline in one or multiple offset wellbores at the depth of the hydraulic fracture. Typically, the array consists of 7–12 tiltmeters coupled to the borehole with standard oil-field centralizer springs (Wright et al., 1999). Downhole tiltmeters provide a map of the deformation of the Earth adjacent to the hydraulic fracture. Thus, what is obtained is an estimate of an ellipsoid that best approximates the fracture dimensions. The tiltmeters are located closer to the fracture than the surface tiltmeter and hence more sensitive to fracture dimensions (Cipolla and Wright, 2002). The closer the downhole tiltmeter to the fracture, the better the quality of data obtained to determine fracture height (Jones and Britt, 2009) which may be limited by the volume of the hydraulic fracturing fluid regardless of whether the fluid interacts with faults (Flewelling et al., 2013). The downhole array tilts in a continuous fashion, similar to surface tiltmeter records but the arrays span the same depth interval as the zone being fractured. The total interval covered by a downhole tilt array ranges from 300 ft to more than 1000 ft, depending on the design conditions. Conventionally, surface and downhole tiltmeter analyses are done separately but techniques have been proposed to combine them for evaluating fracture geometry during drill cuttings disposal.

The greatest advantage of both surface and downhole tiltmeter fracture mapping is that for a given fracture geometry, the induced deformation field is almost completely independent of formation properties. Also, the required degree of formation description is lower in tiltmeter mapping than microseismic mapping (such as velocity profiles and attenuation thresholds) as will be described in a later section. Complex fracture growth would yield independent fractures at different orientations or depths but in tiltmeter mapping a simpler analysis is required (Wright et al., 1999).

At the completion of the stimulation process, approximately 20–30% *v/v* of the water flows back up the wellbore, where it is collected and then recycled in a subsequent well completion operation. Over the productive life of the well, additional "produced" water slowly comes to the surface, where it is collected in on-site storage tanks and transported to permitted treatment facilities.

3.7 Fracture Diagnostics

As part of the fracturing process, it is necessary that the fracture be analyzed through the various procedures known as *fracture diagnostics*, which are the techniques used to analyze the original state of the formation (prefracture analysis) and the created fractures and after hydraulic fracture treatment (postfracture analysis) (Barree et al., 2002; Vulgamore et al., 2007). The data will allow the determination of the dimensions of the created fractures and also whether or not the fractures are effectively maintained in an open mode (*propped*). Thus, the fracture diagnostic techniques can be conveniently subdivided into three groups: (1) direct far-field techniques, (2) direct near-wellbore techniques, and (3) indirect fracture techniques.

3.7.1 Direct Far-Field Techniques

The *direct far-field* techniques require the use of a tiltmeter—an instrument designed to measure very small changes from the horizontal level, either on the ground or in subterranean structures—and microseismic fracture mapping techniques which require that the instrumentation is placed in boreholes surrounding, and near to, the well through which fracturing will occur. The microseismic fracture mapping technique relies on the use of a downhole receiver array of geophones, to locate any *microseismic events* that are triggered by movement in any natural fractures surrounding the hydraulic fracture.

3.7.2 Direct Near-Wellbore Techniques

Direct near-wellbore techniques are used in the well that is being fractured to locate the portion of fracture that is very near wellbore. These techniques consist of tracer logs, temperature logging, production logging, borehole image logging, downhole video logging, and caliper logging. The caliper log provides a continuous measurement of the size and shape of a borehole along its depth and is commonly used when drilling wells to detect natural gas and crude oil formations. The measurements that are recorded can be an important indicator of cave ins or swelling in the borehole, which can affect the results of other well logs.

3.7.3 Indirect Fracture Techniques

Indirect fracture techniques consist of modeling the hydraulic fracturing process followed by matching of the net surface treating pressures, together with subsequent pressure transient test analyses and production data analyses. As fracture treatment data and the postfracture production data are normally available on every well, the indirect fracture diagnostic techniques are widely used methods to determine (or estimate) the shape and dimensions of the created fractures and the propped hydraulic fracture.

4 HYDRAULIC FRACTURING IN TIGHT RESERVOIRS

Many tight reservoirs are thick, layered systems that must be hydraulically fracture treated to produce natural gas or crude oil at commercial flow rates. In tight reservoirs, as in conventional reservoirs, to create the fracture, a fluid is pumped into the wellbore at a high rate to increase the pressure in the wellbore. When the pressure reaches a value greater than the breakdown pressure (the sum of the in situ stress and the tensile strength of the rock) of the formation the formation fractures. Once the fracture is created, the fracture can be extended using pressure (the *fracture-propagation pressure*) which is equal to the sum of (1) the in situ stress, (2) the net pressure drop, and (3) the near-wellbore pressure drop. The net pressure drop is equivalent to the pressure drop along the fracture as the result of viscous fluid flow in the fracture, plus any pressure increase caused by other effects. On the other hand, the near-wellbore pressure drop can be a combination of the pressure drop of the viscous fluid flowing through the perforations and/or the pressure drop resulting from tortuosity (such as twists and turns in the system) between the wellbore and the propagating fracture. This emphasizes the importance of the properties of the fracturing fluid in the creation and propagation of the fracture.

Overall, the hydraulic fracturing process is responsible for creating highly conductive channels and paths for the reservoir fluid to flow from the reservoir pay zone to the wellbore provided the correct procedures are followed (Table 5.4). Moreover, stress-induced natural fractures open as a result of the hydraulic fracturing process thereby creating a secondary fracture network in addition to hydraulic fractures due to stress alterations during hydraulic fracturing (Cho et al., 2013). The main difference between primary fractures, which are hydraulic fractures, and secondary fractures is that the secondary fracture network does not contain proppants and is, therefore, unpropped. Therefore, since the natural fractures lack proppant, their conductivities are much more pressure dependent compared to hydraulic fractures (Cipolla et al., 2010; Eshkalak et al., 2014).

Physically, fractures appear in the rocks as narrow zones of structural discontinuity (*loss of cohesion*) that are the product of mechanical rupture. The term *brittle failure* refers to this mode of deformation while at higher temperatures and higher pressures, *ductile failure* (permanent deformation due to flow, but without loss of cohesion) may occur before the point of brittle failure is reached (Lei et al., 2015). Fractures may be dilational, that is, *joints* (mode I fractures), or may exhibit shearing with components parallel (mode II) or perpendicular (mode III) to the direction of propagation of the fracture front. Shear fractures are also known as *faults*.

In carbonate rocks, natural fractures are more common than they are in sandstone rocks and some of the best fractured reservoirs occur in granite formation—often also referred to as unconventional reservoirs. Fractures occur in preferential directions, determined by the direction of regional stress, which is usually parallel to the direction of nearby faults or folds, but in the case of overthrust faults, they may be perpendicular to the fault or there may be two orthogonal directions. Induced fractures usually have a preferential direction, often perpendicular to the natural fractures.

4.1 Reservoir Selection

The success or failure of a hydraulic fracture treatment often depends on the quality of the candidate reservoir selected for the treatment and the choice of an excellent candidate for stimulation often ensures success while the converse (i.e., choosing a poor candidate for hydraulic fracture treatment) typically results in technical and economic failure. To select the best candidate for

Table 5.4 Recommended Practices and Procedures to Follow During Hydraulic Fracturing Operations[a]

- API Spec 4F, Drilling and Well Servicing Structures
- API RP 4G, Recommended Practice for Use and Procedures for Inspection, Maintenance, and Repair of Drilling Well Service Structures
- API RP 5A3, Recommended Practice on Thread Compounds for Casing, Tubing, and Line Pipe
- API RP 5A5, Field Inspection of New Casing, Tubing, and Plain-end Drill Pipe
- API Spec 5B, Specification for Threading, Gauging, and Thread Inspection of Casing, Tubing, and Line Pipe Threads
- API RP 5B1, Gauging and Inspection of Casing, Tubing, and Line Pipe Threads
- API RP 5C1, Recommended Practice for Case and Use of Casing and Tubing
- API TR 5C3, Technical Report on Equations and Calculations for Casing, Tubing, and Line Pipe Used as Casing or Tubing; and Performance Properties Tables for Casing and Tubing
- API RP 5C5, Recommended Practice on Procedures for Testing Casing and Tubing Connections
- API RP 5C6, Welding Connections to Pipe
- API Spec 5CT, Specification for Casing and Tubing
- API Spec 6A, Specification for Wellhead and Christmas Tree Equipment
- API Spec 7B-11C, Specification for Internal Combustion Reciprocating Engines for Oil-Field Service
- API RP 7C-11F, Recommended Practice for Installation, Maintenance, and Operation of Internal-Combustion Engines
- API Spec 10A, Specification for Cements and Materials for Well Cementing
- API RP 10B-2, Recommended Practice for Testing Well Cements
- API RP 10B-3, Recommended Practice on Testing of Deepwater Well Cement Formulations
- API RP 10B-4, Recommended Practice on Preparation and Testing of Foams and Cement Slurries at Atmospheric Pressure
- API RP 10B-5, Recommended Practice on Determination of Shrinkage and Expansion of Well Cement Formulations at Atmospheric Pressure
- API RP 10B-6, Recommended Practice on Determining the Static Gel Strength of Cement Formulations
- API Spec 10D, Specification for Bow Spring Casing Centralizers
- API RP 10D-2, Recommended Practice for Centralizer Placement and Stop Collar Testing
- API RP 10F, Recommended Practice for Performance Testing of Cementing Float Equipment
- API TR 10TR1, Cement Sheath Evaluation
- API TR 10TR2, Shrinkage and Expansion in Oilwell Cements

Continued

Table 5.4 Recommended Practices and Procedures to Follow During Hydraulic Fracturing Operations—cont'd

- API TR 10TR3, Temperatures for API Cement Operating Thickening Time Tests
- API TR 10TR4, Technical Report on Considerations Regarding Selection of Centralizers for Primary Cementing Operations
- API TR 10TR5, Technical Report on Methods for Testing of Solid and Rigid Centralizers
- API RP 11ER, Recommended Practice for Guarding of Pumping Units
- API Bulletin 11K, Data Sheet for Design of Air Exchange Coolers
- API Spec 11N, Specification for Lease Automatic Custody Transfer (LACT) Equipment
- API Spec 12B, Specification for Bolted Tanks for Storage of Production Liquids
- API Spec 12D, Specification for Field Welded Tanks for Storage of Production Liquids
- API Spec 12F, Specification for Shop Welded Tanks for Storage of Production Liquids
- API Spec 12J, Specification for Oil and Gas Separators
- API Spec 12K, Specification for Indirect Type Oilfield Heaters
- API Spec 12L, Specification for Vertical and Horizontal Emulsion Treaters
- API RP 12N, Recommended Practice for the Operation, Maintenance, and Testing of Flame Arresters
- API Spec 12P, Specification for Fiberglass Reinforced Plastic Tanks
- API RP 12R1, Recommended Practice for Setting, Maintenance, Inspection, Operation, and Repair of Tanks in Production Service
- API Spec 13A, Specification for Drilling Fluid Materials
- API RP 13B-1, Recommended Practice for Field Testing Water-based Drilling Fluids
- API RP 13B-2, Recommended Practice for Field Testing Oil-based Drilling Fluids
- API RP 13C, Recommended Practice on Drilling Fluid Processing Systems Evaluation
- API RP-13D, Recommended Practice on the Rheology and Hydraulics of Oil-well Drilling Fluids
- API RP 13I, Recommended Practice for Laboratory Testing Drilling Fluids
- API RP 13J, Testing of Heavy Brines
- API RP 13M, Recommended Practice for the Measurement of Viscous Properties of Completion Fluids
- API RP 13M-4, Recommended Practice for Measuring Simulation and Gravel-pack Fluid Leak-off Under Static
- API RP 19B, Evaluation of Well Perforators
- API RP 19C, Recommended Practice for Measurement of Properties of Proppants Used in Hydraulic Fracturing and Gravel-packing Operations

Table 5.4 Recommended Practices and Procedures to Follow During Hydraulic Fracturing Operations—cont'd

- API RP 19D, Recommended Practice for Measuring the Long-term Conductivity of Proppants
- API RP 49, Recommended Practice for Drilling and Well Servicing Operations Involving Hydrogen Sulfide
- API RP 53, Recommended Practices for Blowout Prevention Equipment Systems for Drilling Operations
- API RP 54, Occupational Safety for Oil and Gas Well Drilling and Servicing Operations
- API RP 55, Recommended Practices for Oil and Gas Producing and Gas Processing Operations Involving Hydrogen Sulfide
- API RP 65, Cementing Shallow Water Flow Zones in Deep Water Wells
- API RP 67, Recommended Practice for Oilfield Explosives Study
- API RP 74, Occupational Safety for Oil and Gas Well Drilling and Servicing Operations
- API RP 75L, Guidance Document for the Development of a Safety and Environmental Management System for Onshore Oil and Natural Gas Production Operation and Associated Activities
- API RP 76, Contractor Safety Management for Oil and Gas Drilling and Production Operations
- API RP 90, Annular Casing Pressure Management for Offshore Wells
- API RP 2350, Overfill Protection for Storage Tanks in Petroleum Facilities
- API Publication 4663, Remediation of Salt-Affected Soils at Oil and Gas Production Facilities
- API Bulletin E2, Bulletin on Management of Naturally Occurring Radioactive Waste Materials (NORM) in Oil and Gas Production
- API Bulletin E3, Environmental Guidance Document: Well Abandonment and Inactive Well Practices for U.S. Exploration and Production Operations
- API Environmental Guidance Document E5, Waste Management in Exploration and Production Operations
- API Guidelines for Commercial Exploration and Production Waste Management Facilities

[a]American Petroleum Institute, Washington, DC (www.api.org).
These practices were created to meet or exceed federal requirements while remaining sufficiently flexible to accommodate the variations in regulatory frameworks that occur due to differences in regional geology and other factors.

stimulation, the design engineer must consider many variables—the most critical parameters for hydraulic fracturing are (1) formation permeability, (2) the in situ stress distribution, (3) viscosity of the reservoir fluid, (4) reservoir pressure, (5) reservoir depth, (6) the condition of the wellbore, and (7) the skin factor. As noted above, the skin factor refers to whether the reservoir is already stimulated or, perhaps, damaged and typical values for the skin factor range from −6 for an infinite-conductivity massive hydraulic fracture to more than 100 for a poorly formed gravel pack. If the skin factor is positive, the reservoir is damaged and will likely be an excellent candidate for stimulation by hydraulic fracturing.

The best candidate wells for hydraulic fracturing treatment to recover natural gas or crude oil from a tight reservoir have a substantial volume of the original gas in place or the original oil in place and good barriers to vertical fracture growth above and below the net pay intervals. It is also understood that it is beneficial if a crude oil reservoir has gas in place to assist in the flow of oil to the wellbore. Such reservoirs have (1) a thick pay zone, (2) medium to high pressure, (3) in situ stress barriers to minimize vertical height growth, and (4) substantial areal extent. In the case of tight gas reservoirs, the reservoirs that are not good candidates for hydraulic fracturing are those with (1) a small volume of gas in place because of the thinness of the reservoir, (2) low reservoir pressure, and (3) small areal extent.

Also, reservoirs that do not have enough clean shale above or below the pay interval to suppress vertical fracture growth are considered to be poor candidates. Furthermore, reservoirs with extremely low permeability might not produce a sufficient amount of natural gas or crude oil to pay for all the drilling and completion costs (recovery costs), even if the reservoirs are successfully stimulated; thus, such reservoirs are not considered to be good candidates for hydraulic fracture stimulation.

4.2 Fracture Treatment Optimization

The goal of the process design engineer is to design the optimum fracture treatment for each and every well by optimization of both the propped fracture length and the drainage area (well spacing) for low-permeability natural gas and crude oil reservoirs (Holditch et al., 1978, 1987). As the propped length of a fracture increases, the cumulative production increased leading to profitability. However, as the fracture length increases, the incremental financial benefit (on the basis of $ of revenue per foot of additional propped fracture length) decreases. As the treatment volume increases, the propped fracture length increases. As the fracture length increases, the incremental cost of each foot of fracture (again, on the basis of $ of cost per foot of additional propped fracture length) increases. When the incremental cost of the treatment is compared to the incremental benefit of increasing the treatment volume, an optimum propped fracture length can be assessed found for every individual situation.

4.3 Design Considerations and Design Procedures

The most important data for designing hydraulic fracturing treatment are (1) the in situ stress profile, (2) the permeability of the formation, (3) any fluid-loss characteristics, (4) the total fluid volume pumped, (5) the propping agent type, (6) the amount of the propping agents, (6) the pad volume, (7) the viscosity of the fracture fluid, (8) the injection rate, and (9) the formation modulus. If the modulus is large, the material is stiff. In hydraulic fracturing, a stiff rock results in more narrow fractures. If the modulus is low, the fractures are wider. The modulus of a rock is a function of the lithology, porosity, fluid type, and other variables. Thus, it is extremely important to quantify the in situ stress profile and the permeability profile of the zone to be stimulated, plus the layers of rock above and below the target zone that influence fracture height growth.

To design the optimum treatment, it is essential to first determine the effect of fracture length and fracture conductivity upon the productivity and the ultimate recovery from the formation. As in all engineering problems, sensitivity runs must be made to evaluate uncertainties, such as estimates of formation permeability and drainage area. Then, a fracture treatment must be designed by using a fracture-propagation model to achieve the desired length and conductivity at minimum cost. A hydraulic fracture-propagation model should be developed to help in the determination of the material that needs to be mixed and pumped into the well to achieve the optimum values of propped fracture length and fracture conductivity—at the same time remembering that all wells may not be (are not) equal. This will help determine which variables are the most uncertain. For example, in many cases the values of in situ stress, Young's modulus, permeability, and fluid-loss coefficient may not be known with any degree of certainty and have to be estimated, which can introduce elements of doubt. These uncertainties must be acknowledged and there is the necessity to determine the sensitivity with the fracture-propagation model in order to note the effect of these uncertainties on the design process. As databases are developed and increase in terms of the data included, the number and magnitude of the uncertainties should diminish. This can only be achieved if there are numerous simulations of the hydraulic fracture treatment of the well which will assist in design improvement while also presenting indications of the means by which specific variables affect the dimensions of both the created fractures and the propped fractures.

4.4 Fracture Fluid Selection

A critical decision for any hydraulic fracturing project is the selection of the fracture fluid for the treatment by developing a flow chart that can be used to select the category of fracture fluid required to stimulate a natural gas or crude oil well on the basis of factors such as (1) reservoir temperature, (2) reservoir pressure, (3) the expected value of fracture half-length, and (4) a determination of whether the reservoir is water sensitive (Economides and Nolte, 2000).

The selection of the propping agent is based on the maximum effective stress that is applied to the propping agent during the life of the well. The maximum effective stress depends on the minimum value of flowing bottomhole pressure that one expects during the life of the well. For example, if the maximum effective stress is less than 6000 psi, sand is usually recommended as the propping agent but if the maximum effective stress is on the order of 6000–12,000 psi, either resin-coated sand or intermediate strength proppant is preferred for use, depending on the temperature. In cases in which the maximum effective stress is greater than 12,000 psi, high-strength bauxite should be used as the propping agent. But there will always be exceptions to these general *rules of thumb*.

For example, even if the maximum effective stress is less than 6000 psi, it may be preferable to choose to use resin-coated sand or other additives that effectively *lock* the proppant in place, especially when proppant flowback becomes an issue. Proppant flowback can result in significant nonproductive time. If a production operator chooses to pump a project with raw sand only, the proppant can flowback, requiring well intervention and causing significant production loss. Also, in high flow-rate gas wells, intermediate strength proppants may be needed because of inertial flow. For fracture treatments in countries that do not mine sand (and thus the sand is not readily available for hydraulic fracturing projects), the largest cost for the proppant may be the transportation costs. Thus, if it becomes necessary to import the propping agent, use of intermediate strength proppants may be the option, even for relatively shallow wells, but only if the cost differential between the intermediate strength proppants and sand is not always a significant factor.

Thus, the ideal properties of a fracturing fluid relate to its compatibility with the formation rock; its compatibility with the formation fluids; its ability to transfer enough pressure throughout the entire fracture to create a wide fracture, and be able to transport the proppant into the fracture, while breaking back down to a low viscosity fluid for cleanup after the treatment. Finally, and most importantly, the fracture treatment must meet necessary performance specifications.

4.5 Postfracture Reservoir Evaluation and Production Data

Analyzing postfracture production and pressure data requires a thorough understanding of the flow patterns in the reservoir. The technique applied to analyze the data must be compatible with the flow regime that is occurring when the data are collected. For a well containing a finite conductivity hydraulic fracture, the flow regimes that occur consist of (1) bilinear flow, (2) linear flow, (3) transitional flow, and (4) pseudoradial flow. These flow regimes can be defined in terms of dimensionless time.

The times that encompass bilinear flow, linear flow, and transitional flow can be termed *transient flow*. The pseudoradial flow data can be analyzed using semi-steady-state methods. In most tight gas reservoirs containing a finite conductivity hydraulic fracture, the flow-rate and pressure data measured during well tests generally fall into the transient flow category. Seldom can semi-steady-state analyses techniques be used to successfully analyze well-test data in tight reservoirs containing a hydraulic fracture. As such, transient flow analyses methods are usually preferred to analyze such data. If long-term (years) production data are available, semi-steady-state methods can be used successfully to analyze the production and pressure data.

In many cases, after a well is fracture treated (especially in the early days of application of the hydraulic fracturing technology), the well fluids were produced to a pit until the fracturing fluid was sufficiently clean to allow the well fluids to be suitable for sales. After the well has ceased to produce proppant and fracturing fluid, a test separator is usually installed to measure the flow rate of the sales fluid—natural gas or crude oil. The flow rate and the flow pressures are subsequently analyzed to estimate the value of the reservoir and fracture properties.

If a finite difference reservoir simulator is used to analyze field, it is clear that the production and pressure transient data, if measured accurately, can lead to a much better characterization of the reservoir and the hydraulic fracture. The very early flow-rate data are mostly affected by the fracture conductivity, often called bilinear flow. Later in the life of the well, during linear flow, the flow-rate data are most affected by the fracture half-length. If pseudoradial flow is reached, the flow-rate data are most affected by the formation permeability. As such, if the early time flow-rate and pressure data, during the first few days and weeks, are not measured accurately, it is possible to completely misunderstand the properties of the hydraulic fracture.

REFERENCES

Agarwal, R.G., Carter, R.D., Pollock, C.B., 1979. Evaluation and performance prediction of low-permeability gas wells stimulated by massive hydraulic fracturing. J. Pet. Technol. 31 (3), 362–372. SPE-6838-PA.

Aggour, T.M., Economides, M.J., 1998. Optimization of the performance of high-permeability fractured wells. Paper no. 39474, In: Proceedings of the SPE International Symposium on Formation Damage Control, Lafayette, Louisiana. Society of Petroleum Engineers, Richardson, TX.

Ahmadov, R., Vanorio, T., Mavko, G., 2009. Confocal laser scanning and atomic-force microscopy in estimation of elastic properties of the organic-rich Bazhenov formation. Lead. Edge 28, 18–23.

Amadei, B., Stephansson, O., 1997. Rock Stress and Its Measurement. Cambridge University Press, Cambridge.

API RP 13M. Recommended Practice for the Measurement of Viscous Properties of Completion Fluids. Industry Guidance/Best Practices on Hydraulic Fracturing (HF). American Petroleum Institute, Washington, DC.

API RP 13M-4, 2015. Recommended Practice for Measuring Stimulation and Gravel-Pack Fluid Leak-Off Under Static Conditions. American Petroleum Institute, Washington, DC.

Arthur, J.D., Bohm, B., Layne, M., 2008. Hydraulic fracturing considerations for natural gas wells of the Marcellus shale. ALL consulting. In: Presented at the GWPC Annual Forum in Cincinnati, OH, September. Groundwater Protection Council, Oklahoma City, OK.

Arthur, J.D., Bohm, B., Coughlin, B.J., Layne, M., 2009. Evaluating implications of hydraulic fracturing in shale gas reservoirs. Paper no. SPE 121038, In: Proceedings of the 2009 SPE Americas Environmental and Safety Conference, San Antonio, Texas, March 23–25. Society of Petroleum Engineers, Richardson, TX.

ASTM D2166 2015. Standard Test Method for Unconfined Compressive Strength of Cohesive Soil. Annual Book of Standards. ASTM International, West Conshohocken, PA.

ASTM D2216 2015. Standard Test Methods for Laboratory Determination of Water (Moisture) Content of Soil and Rock by Mass. Annual Book of Standards. ASTM International, West Conshohocken, PA.

ASTM D421 2015. Standard Practice for Dry Preparation of Soil Samples for Particle-Size Analysis and Determination of Soil Constants. Annual Book of Standards. ASTM International, West Conshohocken, PA.

ASTM D422 2015. Standard Test Method for Particle-Size Analysis of Soils. Annual Book of Standards. ASTM International, West Conshohocken, PA.

ASTM D4318 2015. Standard Test Methods for Liquid Limit, Plastic Limit, and Plasticity Index of Soils. Annual Book of Standards. ASTM International, West Conshohocken, PA.

Azar, J.J., Samuel, G.R., 2007. Drilling Engineering. PennWell Corporation, Tulsa, OK.

Baihly, J., Laursen, P., Ogrin, J., Le Calvez, J.H., Villarreal, R., Tanner, K., Bennett, L., 2006. Using microseismic monitoring and advanced stimulation technology to understand fracture geometry and eliminate screenout problems in the Bossier Sand of East Texas. Paper no. SPE 102493, In: Proceedings of the SPE Annual Conference and Exhibition, San Antonio, Texas, September 24–27.

Barree, R.D., Fisher, M.K., Woodrood, R.A., 2002. A practical guide to hydraulic fracture diagnostics technologies. Paper no. SPE 77442, In: Proceedings of the SPE Annual Technical Conference and Exhibition, San Antonio, TX, USA, September 29–October 2.

Britt, L.K., 2012. Fracture stimulation fundamentals. J. Nat. Gas Sci. Eng. 8, 34–51.

Britt, L.K., Jones, J.R., Miller, W.K., 2010. Defining horizontal well objectives in tight and unconventional gas reservoirs. Paper no. CSUG/SPE 137839, In: Proceedings of the Canadian Unconventional Resources & International Petroleum Conference, Calgary Alberta, Canada, October 19–21. Society of Petroleum Engineers, Richardson, TX.

Chen, M., Sun, Y., Fu, P., Carrigan, C.R., Lu, Z., Tong, C.H., Buscheck, T.A., 2013. Surrogate-based optimization of hydraulic fracturing in pre-existing fracture networks. Comput. Geosci. 58, 69–79.

Cho, Y., Ozkan, E., Apaydin, O.G., 2013. Pressure-dependent natural-fracture permeability in shale and its effect on shale-gas well production. SPE Reserv. Eval. Eng. 16 (02), 216–228.

Cinco-Ley, H., Samaniego, V.F., Dominguez, A.N., 1978. Transient pressure behavior for a well with a finite-conductivity vertical fracture. SPE J. 18 (4), 253–264.

Cipolla, C., Wright, C.A., 2002. Diagnostic techniques to understand hydraulic fracturing: What? Why and How? In: Proceedings of the 2002 SPE/CERI Gas Technology Symposium, Calgary, Canada, April 3–5. Society of Petroleum Engineers, Richardson, TX.

Cipolla, C.L., Lolon, E.P., Mayerhofer, M.J., Warpinski, N.R., 2009. Fracture design considerations in horizontal wells drilled in unconventional gas reservoirs. Paper no. SPE 119366, In: Proceedings of the SPE Hydraulic Fracturing Technology Conference, The Woodlands, Texas, January 19–21.

Cipolla, C.L., Lolon, E.P., Erdle, J.C., Rubin, B., 2010. Reservoir modeling in shale-gas reservoirs. SPE Res. Eval. Eng. 13 (4), 638–653. also paper no. SPE 125530-PA, Society of Petroleum Engineers, Richardson, Texas.

Daniels, J., Waters, G., LeCalvez, J., Lassek, J., Bentley, D., 2007. Contacting more of the Barnett shale through an integration of real-time microseismic monitoring, petrophysics and hydraulic fracture design. Paper no. 110562, In: Proceedings of the SPE Annual Technical Conference and Exhibition, Anaheim, California.

Das, I., Zoback, M.D., 2011. Long-period, long-duration seismic events during hydraulic fracture stimulation of a shale gas reservoir. Lead. Edge 30, 778–786.

Davis, J., Warpinski, N.R., Davis, E.J., Griffin, L.G., Malone, S., 2008. Joint inversion of downhole tiltmeter and microseismic data and its application to hydraulic fracture mapping in tight gas sand formation. Paper no. ARMA 08-344, In: Proceedings of the 42nd US Rock Mechanics Symposium and 2nd US-Canada Rock Mechanics Symposium. San Francisco, June 29–July 2.

Devereux, S., 2012. Drilling Technology in Non-Technical Language, second ed. PennWell Publishing Corporation, Tulsa, OK.

Economides, M.J., Nolte, K.G., 2000. Reservoir Stimulation, third ed. John Wiley & Sons Inc., Hoboken, NJ.

Ely, J.W., 1985. Handbook of Stimulation Engineering. PennWell Publishing, Tulsa, OK.

Eshkalak, M.O., Aybar, U., Sepehrnoori, K., 2014. An integrated reservoir model for unconventional resources, coupling pressure dependent phenomena. Paper no. SPE 171008, In: Proceedings of the 2014 SPE Eastern Regional Meeting, Charleston, West Virginia, October 21–23. Society of Petroleum Engineers, Richardson, TX.

Fernø, M.A., Haugen, Å., Graue, A., Howard, J.J., 2008. The significance of wettability and fracture properties on oil recovery efficiency in fractured carbonates. In: Proceedings of the International Symposium of the Society of Core Analysts held in Abu Dhabi, United Arab Emirates, October 29–November 2. Paper no. SCA2008–22. http://www.researchgate.net/publication/267678303_The_significance_of_wettability_and_fracture_properties_on_oil_recovery_efficiency_in_fractured_carbonates (accessed 16.01.15.).

Fisher, K., 2012. Trends take fracturing back to the future. American Oil and Gas Reporter 55 (8), 86–97.

Fisher, M.K., Heinze, J.R., Harris, C.D., McDavidson, B.M., Wright, C.A., Dunn, K.P., 2004. Optimizing horizontal completion techniques in the Barnett shale using microseismic fracture mapping. Paper no. SPE 90051, In: Proceedings of the SPE Annual Technical Conference and Exhibition, Houston, Texas, September 26–29.

Flewelling, S.A., Tymchak, M.P., Warpinski, N., 2013. Hydraulic fracture height limits and fault interactions in tight oil and gas formations. Geophys. Res. Lett. 40, 3602–3606.
Gidley, J.L., Holditch, S.A., Nierode, D.E., Veatch, R.W., 1990. In: Hydraulic Fracturing to Improve Production. Monograph Series SPE 12, Society of Petroleum Engineers, Richardson, TX.
Green, K.P., 2014. Managing the Risks of Hydraulic Fracturing. Fraser Institute, Vancouver, BC, Canada http://catskillcitizens.org/learnmore/managing-the-risks-of-hydraulic-fracturing.pdf (accessed 15.01.16.).
Green, K.P., 2015. Managing the Risks of Hydraulic Fracturing: An Update. Fraser Institute, Vancouver, BC, Canada https://www.fraserinstitute.org/studies/managing-the-risks-of-hydraulic-fracturing-an-update (accessed 04.04.16.).
Hammack, R., Harbert, W., Sharma, S., Stewart, B., Capo, R., Wall, A., Wells, A., Diehl, R., Blaushild, D., Sams, J., Veloski, G., 2014. An evaluation of fracture growth and gas/fluid migration as horizontal Marcellus shale gas wells are hydraulically fractured in Greene County, Pennsylvania. Report no. NETL-TRS-3-2014. EPAct technical report series, In: National Energy Technology, Laboratory/US Department of Energy, Pittsburgh, PA/Washington, DC.
Hareland, G.I., Rampersad, P., Dharaphop, J., Sasnanand, S., 1993. Hydraulic fracturing design optimization. Paper no. 26950, In: Proceedings of the SPE Eastern Regional Conference and Exhibition, Pittsburgh, Pennsylvania. Society of Petroleum Engineers, Richardson, TX, pp. 493–500.
Hibbeler, J., Rae, P., 2005. Simplifying hydraulic fracturing: theory and practice. Paper no. SPE 97311, In: Proceedings of the 2005 SPE Technical Conference and Exhibition, Dallas, Texas, October 9–12. Society of Petroleum Engineers, Richardson, TX.
Holditch, S.A., 1979. Factors affecting water blocking and gas flow from hydraulically fractured gas wells. J. Pet. Technol. 31 (12), 1515–1524.
Holditch, S.A., Jennings, J.W., Neuse, S.H., 1978. The optimization of well spacing and fracture length in low permeability gas reservoirs. Paper no. SPE-7496, In: Proceedings of the SPE Annual Fall Technical Conference and Exhibition, Houston, Texas, October 1–3. Society of Petroleum Engineers, Richardson, TX.
Holditch, S.A., Robinson, B.M., Whitehead, W.S., 1987. Prefracture and postfracture formation evaluation necessary to characterize the three dimensional shape of the hydraulic fracture. In: Proceedings of the SPE Formation Evaluation, December 1987. Society of Petroleum Engineers, Richardson, TX.
Hornby, B.E., Schwartz, L.M., Hudson, J.A., 1994. Anisotropic effective-medium modeling of the elastic properties of shales. Geophysics 59, 1570–1583.
Hubbert, M.K., Willis, D.G., 1957. Mechanics of hydraulic fracturing. Pet. Trans. AIME 210, 153.
Johnston, J.E., Christensen, N.I., 1995. Seismic anisotropy of shales. J. Geophys. Res. 100, 5991–6003.
Jones, J.R., Britt, L.K., 2009. Design and Appraisal of Hydraulic Fractures. Society of Petroleum Engineers, Richardson, TX.
Keelan, D.K., 1982. Core analysis for aid in reservoir description. Society of Petroleum Engineers (SPE) of AIME Distinguished Author Series. Society of Petroleum Engineers, Richardson, TX.
Kennedy, R.L., Gupta, R., Kotov, S.V., Burton, W.A., Knecht, W.N., Ahmed, U., 2012. Optimized shale resource development: proper placement of wells and hydraulic fracture stages. Paper no. 162534. In: Proceedings of the Abu Dhabi International Petroleum Conference and Exhibition, Abu Dhabi, United Arab Emirates, November 11–14. Society of Petroleum Engineers, Richardson, TX.
King, G.E., 2010. Thirty years of gas shale fracturing: What have we learned? Paper no. SPE 133456, In: Proceedings of the SPE Annual Technical Conference and Exhibition Florence, Italy, September.

King, G.E., 2012. What every representative, environmentalist, regulator, reporter, investor, university researcher, neighbor and engineer should know about estimating frac risk and improving frac performance in unconventional gas and oil wells. Paper no. SPE 152596, In: Proceedings of the SPE Hydraulic Fracturing Technology Conference, Woodlands, Texas, February 6–8. Society of Petroleum Engineers, Richardson, TX.

Lee, W.J., Holditch, S.A., 1981. Fracture evaluation with pressure transient testing in low-permeability gas reservoirs. J. Pet. Technol. 33 (9), 1776–1792. SPE-9975. Society of Petroleum Engineers, Richardson, Texas.

Lei, X., Zhang, S., Guo, T., Xiao, B., 2015. New evaluation method of the ability of forming fracture network in tight sandstone reservoir. Int. J. Environ. Sci. Develop. 6 (9), 688–692.

Loucks, R.G., Reed, R.M., Ruppel, S.C., Jarvie, D.M., 2009. Morphology, genesis, and distribution of nanometer-scale pores in siliceous mudstones of the Mississippian Barnett shale. J. Sediment. Res. 79, 848–861.

Maxwell, S., 2011. Microseismic hydraulic fracture imaging: the path toward optimizing shale gas production. Lead. Edge 30, 340–346.

Mohaghegh, S., Balanb, B., Platon, V., Ameri, S., 1999. Hydraulic fracture design and optimization of gas storage wells. J. Pet. Sci. Eng. 23, 161–171.

Mooney, C., 2011. The truth about fracking. Sci. Am. 305, 80–85.

Passey, Q.R., Bohacs, K.M., Esch, W.L., Klimentidis, R., Sinha, S., 2010. From oil-prone source rocks to gas-producing shale reservoir—geologic and petrophysical characterization of unconventional shale-gas reservoir. Paper no. SPE 131350, In: Proceedings of the CPS/SPE International Oil & Gas Conference and Exhibition. Society of Petroleum Engineers, Richardson, TX.

Phillips, W.J., 1972. Hydraulic fracturing and mineralization. J. Geol. Soc. Lond. 128, 337–359.

Postler, D.P., 1997. Pressure integrity test interpretation. Paper no. SPE/IADC 37589, In: Proceedings of the 1997 SPE/IADC Conference, Amsterdam, Netherlands, March 4–6. Society of Petroleum Engineers, Richardson, TX.

Reddy, T.R., Nair, R.R., 2012. Fracture characterization of shale gas reservoir using connected—cluster DFN simulation. In: Sharma, R., Sundaravadivelu, R., Bhattacharyya, S.K., Subramanian, S.P. (Eds.), Proceedings of the 2nd International Conference on Drilling Technology 2012 (ICDT-2012) and 1st National Symposium on Petroleum Science and Engineering 2012 (NSPSE-2012), December 6–8, pp. 133–136.

Reinicke, A., Rybacki, E., Stanchits, S., Huenges, E., Dresen, G., 2010. Hydraulic fracturing stimulation techniques and formation damage mechanisms: implications from laboratory testing of tight sandstone-proppant systems. Chem. Erde 70 (S3), 107–117.

Rueda, J.I., Rahim, Z., Holditch, S.A., 1994. Using a mixed integer linear programming technique to optimize a fracture treatment design. Paper no. 29184, In: Proceedings of the SPE Eastern Regional Meeting, Charleston, South Carolina. Society of Petroleum Engineers, Richardson, TX, pp. 233–244.

Scanlon, B.R., Reedy, R.C., Nicot, J.P., 2014. Comparison of water use for hydraulic fracturing for oil and gas versus conventional oil. Environ. Sci. Technol. 48, 12386–12393.

Secor, D.T., 1965. Role of fluid pressure in jointing. Am. J. Sci. 263, 633–646.

Slattery, J.C., 2001. Two-phase flow through porous media. AICHE J. 16 (3), 345–352.

Smith, M.B., Hannah, R.R., 1996. High-permeability fracturing: the evolution of a technology. SPE J. Pet. Technol. 48 (7), 628–633.

Sondergeld, C.H., Rai, C.S., 2011. Elastic anisotropy of shales. Lead. Edge 30, 324–331.

Sondergeld, C.H., Ambrose, R.J., Rai, C.S., Moncrieff, J., 2010. Microstructural studies of gas shales. Paper no. 131771, In: Proceedings of the SPE Unconventional Gas Conference. Society of Petroleum Engineers, Richardson, TX.

Sone, H., Zoback, M.D., 2013a. Mechanical properties of shale-gas reservoir rocks—part 1: static and dynamic elastic properties and anisotropy. Geophysics 78 (5), D381–D392.
Sone, H., Zoback, M.D., 2013b. Mechanical properties of shale-gas reservoir rocks—part 2: ductile creep, brittle strength, and their relation to the elastic modulus. Geophysics 78 (5), D393–D402.
Speight, J.G., 2014. The Chemistry and Technology of Petroleum, fifth ed. CRC Press, Taylor & Francis Group, Boca Raton, FL.
Speight, J.G., 2015. Handbook of Petroleum Product Analysis, second ed. John Wiley & Sons Inc., Hoboken, NJ.
Speight, J.G., 2016a. Introduction to Enhanced Recovery Methods for Heavy Oil and Tar Sands, second ed. Gulf Professional Publishing, Elsevier, Oxford.
Speight, J.G., 2016b. Handbook of Hydraulic Fracturing. John Wiley & Sons Inc., Hoboken, NJ.
Spellman, F.R., 2013. Environmental Impacts of Hydraulic Fracturing. CRC Press, Taylor & Francis Group, Boca Raton, FL.
Spellman, F.R., 2016. Handbook of Environmental Engineering. CRC Press, Taylor & Francis Group, Boca Raton, FL.
Unalmiser, S., Funk, J.J., 2008. Engineering core analysis. In: SPE Distinguished Author Series. Society of Petroleum Engineers, Richardson, TX. Paper no. SPE 36780.
US EIA, 2014. United States Energy Information Administration, May 7, 2014, Annual Energy Outlook 2014. Energy Information Administration, US Department of Energy, Washington, DC. http://www.eia.gov/forecasts/aeo/index.cfm (accessed 21.03.15.).
Valko, P.P., Oligney, R.E., Economides, M.J., 1998. High permeability fracturing of gas wells. Pet. Eng. Int. 71 (1), 75–88.
Vanorio, T., Mukerji, T., Mavko, G., 2008. Emerging methodologies to characterize the rock physics properties of organic-rich shales. Lead. Edge 27, 780–787.
Veatch Jr., R.W., 1983. Overview of current hydraulic fracturing design and treatment technology, Part 1. J. Pet. Technol. 35 (4), 677–687. Paper SPE-10039-PA, Society of Petroleum Engineers, Richardson, TX.
Vernik, L., Liu, X., 1997. Velocity anisotropy in shales: a petrophysical study. Geophysics 62, 521–532.
Vernik, L., Milovac, J., 2011. Rock physics of organic shales. Lead. Edge 30, 318–323.
Vernik, L., Nur, A., 1992. Ultrasonic velocity and anisotropy of hydrocarbon source rocks. Geophysics 57, 727–735.
Vulgamore, T., Clawson, T., Pope, C., Wolhart, S., Mayerhofer, M., Machovoe, S., Waltman, C., 2007. Applying hydraulic fracture diagnostics to optimize stimulations in the Woodford shale. Paper no. SPE 110029, In: Proceedings of the SPE Annual Technical Conference and Exhibition, Anaheim, California. November 11–14.
Warpinski, N.R., Engler, B.P., Young, C.J., Peterson, R., Branagan, P.T., Fix, J.E., 2005. Microseismic mapping of hydraulic fractures using multi-level wireline receivers. Paper no. SPE 30507, In: Proceedings of the 2005 SPE Annual Technical Conference and Exhibition, Dallas, Texas, October 22–25. Society of Petroleum Engineers, Richardson, TX.
Waters, G., Heinze, J., Jackson, R., Ketter, A., Daniels, J., Bentley, D., 2006. Use of horizontal well image tools to optimize Barnett shale reservoir exploitation. SPE paper no. 103202, In: Proceedings of the SPE Annual Technical Conference and Exhibition, San Antonio, Texas.
Wright, C.A., Davis, E.J., Wang, G., Weijers, L., 1999. Downhole tiltmeter fracture mapping: a new tool for direct measurement of hydraulic fracturing growth. In: Amadei, B., Krantz, R.L., Scott, G.A., Smeallie, P.H. (Eds.), Rock Mechanics for Industry. Balkema Publishers, Rotterdam, Netherlands.
Zoback, M.D., 2010. Reservoir Geomechanics. Cambridge University Press, Cambridge.

CHAPTER SIX

Fluids Management

1 INTRODUCTION

For the purposes of this chapter, the term *fluids* refer to any fluid material that is either introduced into the formation to aid in the recovery of natural gas or crude oil as well as any fluid material that is recovered from the fractured formation. Also, as a reminder and to avoid any confusion at this stage, the term *oil shale* refers to any source rock that contains solid insoluble hydrocarbonaceous material (kerogen) that is converted to synthetic fuel liquids when the rock is heated to temperatures in excess of 500°C (930°F) in the absence of oxygen (pyrolysis) (Chapter 1) (Scouten, 1990; Lee et al., 2007; Speight, 2012, 2014a, 2016a). In the various oil shale production processes, oil shale is heated to approximately 500°C (930°F) at which temperature the kerogen decomposes into oil vapor and gas. Cooling this vapor stream produces liquid oil and uncondensed, light hydrocarbon produces liquid oil and uncondensed, light hydrocarbon gases. Subsurface retorting (in situ retorting) or surface retorting (ex situ retorting) is needed to produce valuable valuable products from oil shale kerogen and but, however, most development work on oil shale recovery has been directed to mining the shale, bringing it to surface, and then processing it in surface facilities. Various surface retorting processes can be classified according to the method used to provide heat for the retorting reaction (Scouten, 1990; Lee, 1991; Lee et al., 2007; Speight, 2012, 2014a, 2016a). In the present context, crude oil from tight shale formations (tight oil) is not shale oil since it is produced by a recovery process (hydraulic fracturing) rather than by a thermal conversion process that produces synthetic crude oil from kerogen (Chapter 1).

It must be reemphasized that crude oil from tight formations is not produced in the same manner as oil from the thermal decomposition of kerogen. As a result, the processes are different, especially since the crude oil (and natural gas) in tight formations (shale, sandstone, and carbonate) exists in the formation in the form in which it (the natural gas or the crude oil) is

Deep Shale Oil and Gas
http://dx.doi.org/10.1016/B978-0-12-803097-4.00006-1

recovered (Chapter 1). Therefore, water requirements and water usage to recover natural gas or crude oil cannot be considered to be the same as oil production by the conversion of kerogen (to oil) in situ or ex situ.

Until recently the vast quantities of natural gas and crude oil in tight formations were unrecoverable using the older available technology. However, through the use of recently improved horizontal drilling techniques combined with hydraulic fracturing, have unlocked an abundance of natural gas and crude oil from deep tight formations in the United States. Indeed, given the density of the tight formations, hydraulically fracturing the wells is necessary to create pathways for oil and natural gas to flow to the production well (Chapter 5). Moreover, along with the rapid development of these resources, natural gas and crude oil production from tight formations is undergoing increased scrutiny due to various environmental consequences, not the least of which is the relatively large volume of water that is required to hydraulically fracture each tight formation. Innovation in horizontal drilling techniques hydraulic fracturing technology (Chapters 4 and 5) is driving the rapid development of tight resources (which include shale gas, natural gas liquids, and tight oil) across the United States and Canada. In fact, the development of natural gas and crude oil from tight formations has significantly increased the volume of the natural gas and crude oil resources in the United States. On a worldwide scale, many governments (particularly the government of Argentina, Canada, China, Mexico, Poland, South Africa, and the United Kingdom) have started to explore the commercial viability of their shale reserves (Chapter 3). The potential for expansion is real since and newly discovered resources of natural gas and crude oil in tight formations continue to add to the known resources.

However, water is essential to the development of the resources of natural gas and crude oil in these tight formations but as developers escalate the exploration for suitable tight formations, the limited availability of water will be (if it has not already become) a reality. Extracting these resources requires large amounts of water not only for the drilling operations but particularly for the hydraulic fracturing operations (Chapter 5). In most cases, these demands are met by local freshwater—the financial aspects of shipping water to a remote site are not favorable. Nevertheless, the developers of tight resources are significant users and managers of water at local and regional levels and, as a result, water and energy have become interdependent. For example, improving the efficiency of water use reduces the need to develop, transport, pump, treat, and distribute additional water resources, thereby reducing the amount of power or energy required for recovery of the resources. Alternatively, improving energy efficiency reduces the demand on electricity generation and fuel consumption for water transportation which, in turn, reduces the need for water resources for power generation cooling and fuel processing, along with reducing the water resources needed to extract the natural gas and crude oil.

Thus, the development of natural gas and crude oil resources in deep tight, low-permeability shale, sandstone, or carbonate formations, which are typically found thousands of feet below the surface of the Earth, requires copious amount of water than can floral and faunal communities, including human communities (Stillwell et al., 2010).

Finally, hydrological conditions vary spatially and seasonally across the regions where subterranean tight formations occur, with variation among not only between adjacent formation but also within a single formation. This variability causes different demands on the water sources and makes the ability of the developers to meet the water demands for horizontal drilling and hydraulic fracturing highly unpredictable. Thus, estimates based on previous seasonal demands as well as demands by previously developed formations are not always transferable to new tight formations that are under investigation for the near-future development. Each formation becomes a development that must be considered to be site specific and subject not only to the depth and properties of the formation as well as the formation fluids (Chapter 2) but also to the seasonal availability of water. Furthermore, public concern over increased competition and impacts on freshwater availability for domestic purposes (with the guarantee of untarnished drinking water sources) can influence the awarding of a development license and ability of a developer to operate which, in turn, can lead to changes in government regulations that could impact both the short- and long-term financial aspects of the development. Therefore, responsible water management practice is critical to a developer obtaining a license to operate at the site which is particularly true when a resource is located within a water-poor region, such as may be the case in certain regions of Texas and North Dakota.

The quality of water used in the hydraulic fracturing fluid and the effect and impact of the fluid on well production is a critical factor in developing a water management strategy. Furthermore, regulatory requirements (mandated federal, state, and local regulatory authorities) often dictate water management options. Also within the United States, multistate and regional water permitting agencies may also be responsible for maintaining water quality and water supply. In fact, all authorities may dictate water withdrawal and/or disposal options that are available for consideration and use and injection wells that may be used for disposal of flowback water and other produced waters require state or federal permits which are subject to suitable analysis by standardized text methods (Blauch et al., 2009; Lester et al., 2015). The primary objective of any such program, whether administered at the state or federal level, is protection of underground sources of drinking water but, in many cases, the responsible authority is a function of the acquisition or disposal option chosen. For example, surface water discharge may be regulated by a different agency than subsurface injection. Therefore, regardless of the regulatory agency with program authority over subsurface injection, new injection wells will require a permit that meets the appropriate state or federal regulatory requirements.

Thus, it is the purpose of this chapter to introduce the reader to the various aspects of tight resource development and the necessity for water as part of the development of natural gas and crude oil resources in tight formations as well as the effects of this development on water use, water production, and water quality.

2 WATER REQUIREMENTS, USE, AND SOURCES

A significant part of the development of natural gas and crude oil resources in tight formations is the hydraulic fracturing operation, a necessary part of which involves: (1) securing access to reliable sources of water, (2) the timing associated with this accessibility, and (3) the requirements for obtaining permission to secure these supplies (API, 2009, 2010). However, being forewarned is to be forearmed when investigating potential options for securing water supplies to support hydraulic fracturing operations and an awareness of competing water needs, water management issues, and the full range of permitting and regulatory requirements in a region is critical. Thus, consultation with appropriate water management agencies that usually have top level responsibility for the management (including permitting) and protection of water resources is required.

It is necessary to initiate and pursue any form of transparent and proactive communication with local water planning agencies (and the public must also be informed) to ensure that natural gas and crude oil recovery operations do not disrupt the water needs of local communities. If the developer has an understanding of the water needs of local communities, this will help in the evolution of water acquisition and management plans so that the plans are acceptable to the community that is located close to the proposed natural gas and crude oil development area. The water needs for the project and the juxtaposition of the needs next to the needs of the community must be examined and such needs a major part of the development plan. In fact, although the water needed for horizontal drilling and hydraulic fracturing operations may represent a small volume relative to other water requirements, withdrawals associated with large-scale developments, conducted over multiple years, may have a cumulative impact to the groundwater and/or any watersheds. This potential cumulative impact can be minimized or avoided by working with local water resource managers to develop a plan of when and where withdrawals will occur.

Briefly, a *watershed* (the word is sometimes used interchangeably with *drainage basin* or *catchment*) is an area of land that drains all the streams and rainfall to a common outlet such as the outflow of a reservoir, mouth of a bay, or any point along a stream channel. The watershed consists of surface water (lakes, streams, reservoirs, and wetlands) and all of the underlying ground water, and a large watershed may contain many smaller watersheds, depending on the outflow points. In addition, ridges and hills that separate two watersheds are often referred to as the *drainage divide*. All of the land that drains water to the outflow point is the watershed for that outflow location. Watersheds are important because the streamflow and the water quality of a river are affected by events (human-induced or not) that happen in the land area above the river-outflow point.

Thus, it is not surprising that hydraulic fracturing, the technology central to the success development of unconventional resources—the extraction of natural gas and crude oil from tight formations—has been the subject of intense scrutiny come under fire because of concerns about the chemicals used and any other real or unproven effects. The industry is working to improve transparency and collaboration with landowners, mineral rights holders, regulators, and communities.

Thus, when considering a project involving horizontal drilling and a hydraulic fracturing, the location of rivers and the amount of streamflow in rivers must be given serious consideration—the key concept is to determine the watershed of the river and identify the area of land where all of the water that falls in it and drains off of it goes to a common outlet, remembering that a watershed can be as small as a small pond or sufficiently large enough to encompass all the land that drains water into a major bay.

In order to commence a water requirements project, after preproject communications have commenced, the operator(s) should conduct a detailed, documented review of the identified water sources available in an area that could be used to support hydraulic fracturing operations.

Considerations factoring in this review should include: (1) evaluating source water requirements, (2) fluid handling and storage, and (3) transportation considerations.

2.1 Requirements

Although water is used in several stages of the natural gas and crude oil recovery from tight formations, the majority of water is typically consumed during the production stage. This is primarily due to the large volumes of water (2.3–5.5 million gallons) required to hydraulically fracture a tight formation (Clark et al., 2011). In addition, water in amounts on the order of 190,000–310,000 gallons is also used to drill and cement a natural gas and crude oil well during the drilling phase of the project (Clark et al., 2011). Once the natural gas or crude oil is produced, it is processed, transported, and distributed to customers and ultimately used. Water consumption occurs in each of these stages as well, with the most significant non-production consumption potentially occurring during end use.

A rate-limiting factor in further development of the oil tight formation resources, particularly in certain regions of the western United States, is not only the effect of water quality but also the availability of water, which may not be a problem in other regions and countries. Water from a nearby river may be made available for the drilling process and for the hydraulic fracturing process, but an important factor that must be taken into consideration in any water-use plan is the potential salt loading of the river water. With oil tight formation development near the river, the average annual salinity is anticipated to increase, unless some prevention or treatment is implemented. The economic damages associated with these higher salinity levels could be significant and have been the subject of extensive economic studies.

Water requirements for hydraulically fracturing projects may vary widely, but on average required several million gallons of water for deep unconventional tight reservoirs. While these water volumes may seem large, they generally represent a relatively small percentage of total water use in the areas where hydraulic fracturing is operative (Satterfield et al., 2008). Water used for hydraulic fracturing operations can come from a variety of sources, including: (1) surface water bodies, (2) municipal water supplies, (3) groundwater, (4) wastewater sources, or (5) water that has been recycled from other sources including previous hydraulic fracturing operations.

Nevertheless, in order to acquire the water necessary for use in hydraulic fracturing operations can give rise to a series of challenges, particularly when the water is used in arid regions and has to be acquired from arid regions. Water is difficult to transport, so most commercial activities extract the water they use locally. If large amounts are used for fracking, this will impact on the water table and ecology of the area, and it will also limit the amount of water for other activities taking place in the area. In fact, the volume of water required for a hydraulic fracturing operation is becoming more and more challenging, and project operators have to seek new ways to secure reliable, affordable, supplies of water. In some areas, the operator may need to build large reservoirs to capture water during periods of high into local rivers when withdrawal is permitted and monitored by water resource authorities or for future use in storing the flowback water from a hydraulic fracturing project. Operators have also explored the option of using treated produced water from existing wells as a potential supply source of water for hydraulic fracturing operations. In all case of acquiring and using water, the implementation of these practices for fracturing operations (or, more generally, for water acquisition and use) must conform to local regulatory requirements where operations occur.

The management and disposal of water after it is used for hydraulic fracturing operations may present additional challenges for operators. After a hydraulic fracture stimulation is complete, the fluids returning to the surface within the first 7–14 days (often called flow back) will often require treatment for beneficial reuse and/or recycling or be disposed of by injection. This water may contain dissolved constituents from the formation itself along with some of the fracturing fluid constituents initially pumped into the well.

Typically, water for hydraulic fracturing is withdrawn from one location or watershed over a period of several days (Veil, 2010). Additionally, in some cases, the water may be acquired from remote, often environmentally sensitive headwater areas where even small withdrawals can have a significant impact on the flow regime. As a result, while hydraulic fracturing may account for a small fraction of the water supply, there can be more severe local impacts. In addition, much of the water injected underground is either not recovered or unfit for further use once it is returned to the surface and often usually requires disposal in an underground injection well. This water use represents a consumptive use if it is not available for subsequent use within the basin from which it was extracted. Alternatively, the water may be treated and reused for subsequent hydraulic fracturing projects (US GAO, 2012).

In terms of local impact, especially in areas of water scarcity, the extraction of water for drilling and hydraulic fracturing (or even the production of water, as is the case in the production of coalbed methane) can have broad and serious environmental effects. The extraction of the water can (more likely, will) lower the water table, thereby affecting biodiversity (i.e., the variety of the floral and faunal populations) and harm the local ecosystem as well as reducing the availability of water for use by local communities and in other productive activities, such as agriculture. Thus, the limited availability of water for hydraulic fracturing could become a significant constraint on the development of natural gas and crude oil in tight formations in water-stressed areas. In fact, water stress (water scarcity) is increasing around the world, and all domestic and industrial users are becoming subject to intense scrutiny in terms of water consumption. Thus, water management and groundwater protection are vital issues for the natural gas and crude oil industries.

When the time comes (and this should be sooner rather than later) to evaluate the water requirements for a hydraulic fracturing project, it is necessary to conduct a preproject comprehensive evaluation of the cumulative water demand on a stage-by-stage basis as well as the timing of these needs at an individual well site. This should include consideration of the water requirements for: (1) drilling operations, (2) dust suppression, (3) emergency response, and (4) the water requirements for hydraulic fracturing operations. A decision must be made as to whether or not the sources of water are adequate to support the total operation, with water of the desired quality, and can be accessed when needed for the planned development program.

2.2 Use

Hydraulic fracturing—the mainstay of oil and gas production from shale formations and from tight formations—uses between 1.2 and 3.5 million US gallons of water per well, with large projects using up to 5 million US gallons per well. Typically, a well requires 3–8 million US gallons of water over its lifetime. Water consumption for hydraulic fracturing occurs during: (1) drilling, (2) extraction and processing of proppant sands, (3) testing natural gas transportation pipelines, and (4) gas-processing plants. Typically, for most shale basins, water is acquired from local water supplies, including: (1) surface water bodies, such as rivers, lakes, and ponds, (2) groundwater aquifers, (3) municipal water supplies, (4) treated wastewater from municipal and industrial treatment facilities, and (5) produced and/or flowback water that is recovered, treated, and reused. In regions where hydraulic fracturing occurs, the sources of water should be well documented.

Thus, although water is used in several stages of the tight formation development, the majority of water is typically consumed during the production stage. This is primarily due to the large volumes of water (on the order of several million gallons depending upon the depth of the formation and the formation characteristics) required to hydraulically fracture a well (Clark et al., 2011). Water in amounts of several hundred thousand gallons is also used to drill and cement a shale gas well during construction (Clark et al., 2011). After fracturing a well, anywhere from 5% to 20% of the original volume of the fluid will return to the surface within the first 2 weeks as flowback water. An additional volume of water, equivalent to anywhere from 10% to almost 300% of the injected volume, will return to the surface as produced water over the life of the well. It should be noted that there is no clear distinction between the so-called flowback water and produced water, with the terms typically being defined by operators based upon the timing, flow rate, or sometimes composition of the water produced.

Deep shale oil and gas projects use water primarily during drilling and stimulation but produce a tremendous amount of energy over the approximate 20-years lifespan of the natural gas well (Mantell, 2009). Thus, water is an essential component of deep shale natural gas development. Operators use water for drilling, where a mixture of clay and water is used to carry rock cuttings to the surface as well as to cool and lubricate the drill bit. Drilling a typical deep shale well requires between 65,000 and 1 million gallons of water. Water is also used in hydraulic fracturing, where a mixture of water and sand is injected into the deep shale at high pressure to create small cracks in the rock and allows gas to freely flow to the surface. Hydraulically fracturing a typical deep shale well requires an average of 3.5 million gallons of water.

The water supply requirements of deep shale natural gas development are isolated in that the water needs for each well are limited to drilling and development, and the placement of shale gas wells is spread out over the entire shale gas play. In other words, these shale gas wells are not drawing water from one single source. Subsequent hydrofracturing treatments of wells to restimulate production may be applied, though their use is dependent upon the particular characteristics of the producing formation and the spacing of wells within the field. On the other hand, enhanced oil recovery uses relatively large amounts of water due to the waterfloods used to force the oil out of the reservoir. Oil shale and tar sands (oil sands) also use significantly higher amounts of water due to the in situ steam extraction process and the additional water used to process the liquid fuel (Mantell, 2009).

Geography plays an important role in determining fuel source water efficiency. For example, imported fuels such as foreign oil, Alaskan oil and gas, and even off-shore oil and gas, less water efficient depending on location of origin versus location of end use. As a general rule, unconventional oil and synthetic coal extraction and production processes are more water intensive than their conventional counterparts (Mantell, 2009). The most water inefficient fuel sources are irrigated biofuels including ethanol and biodiesel. Irrigation of the biofuel feedstock requires significant volumes of water input per unit of energy that can be derived from the crop. Biofuels also require a significant amount of water to process the raw fuel into a useable energy source. Water use efficiencies could be improved to a level similar to synthetic coal if only nonirrigated feedstock was used in energy development, although this would significantly decrease the amount of feedstock available due to the limited locations where nonirrigated growth is possible (Mantell, 2009).

The majority of water use occurs during the early exploration and drilling phase of gas and oil production from tight formations. Water is a necessary input and, although only a relatively small amount of water is needed for drilling, more significant quantities of water are used during completions, or hydraulic fracturing operations (Chapter 5). Thus, in terms of water use, the issue is the application of hydraulic fracturing to release the gas from the shale formation—high-volume hydraulic fracturing to create fissures in the rock to release gas or oil trapped inside. Thus, water-related issues in shale drilling are leading to growing and complex policy and regulatory challenges and environmental compliance hurdles that could potentially challenge shale gas production expansion and increase operational costs.

Hydraulic fracturing (Chapter 5) is a technique used in oil and natural gas production to stimulate the production of hydrocarbons. Recall, after a well is drilled into reservoir rock that contains natural gas, crude oil, and water, every effort is made to maximize the production of the gas and oil (Speight, 2014a). In hydraulic fracturing, a fluid (usually water containing special high-viscosity fluid additives) is injected under high pressure. The pressure exceeds the rock strength and the fluid opens or enlarges fractures in the rock. These larger, man-made fractures start at the well and extend deep into the reservoir rock. After the formation is fractured, a *propping agent* (usually sand carried by the high-viscosity additives) is pumped into the fractures to keep them from closing when the pumping pressure is released. This allows the natural gas or the crude oil to move more freely from the rock pores to a production well and thence to the surface.

Even though it is often looked upon as a temporary process, hydraulic fracturing (and the accompanying horizontal drilling process) requires considerable amount of water. For example, the drilling and completion (including hydraulic fracturing) phases of a well (leading to initiation of the production of natural gas or crude oil) is a short time period—typically on the order of 2–8 weeks—compared to an expected production lifetime of a well on the order of 20–40 years. Estimates of water consumption during hydraulic fracturing vary from 0.2 to 4 barrels of water for each barrel of oil recovered (with consumption at Bakken toward the low side of these estimates) (IEA, 2013). In terms of the total amount of water, op to 170,000 barrels of water per well may be used for drilling and fracturing and after the fracturing process has been completed, 30–70% (v/v) of the injected water injected flows out of the well, but this amount is dependent upon the properties (such as size and mineralogy) of the reservoir and the returned water must be treated before disposal or before reuse. Such a level of water consumption can put a heavy strain on local water sources and, if projections of natural gas and light tight oil production are to be met, technology development to reduce consumption will be necessary. Another aspect of water use is that 3,000,000–5,00,000 gallons of water are typically necessary to hydraulically fracture a multistage well. The volume and quality of water used is also characteristic of the reservoir properties, which vary by region and by play and also by well design.

2.3 Sources

The types of water sources that are used by the oil and natural gas industry (for hydraulic fracturing) typically fall into three categories: (1) potable water (freshwater), (2) oil and natural gas generated water (produced and flowback water), and (3) the alternative water sources that are generally not usable by the public (brackish or nonpotable water). The choice among sources depends on a variety of factors such as volume, availability, source-water quality, competing water uses, economics, and regulatory requirements.

Potable water sources typically include freshwater from a groundwater source, surface water (from a lake or river), and municipal water supplies. Water is drawn from these sources over time in accordance with agreements and regulatory requirements and stored for use when a greater amount of water is necessary. A regular and slow withdrawal rate can minimize impacts on water sources that are also used by communities.

Produced water (reused and recycled water) used in oil and natural gas operations is typically flowback and produced water from the field operations. There are few instances of operators using other produced water sources, such as municipal and industrial. In an effort to reduce the use of freshwater sources, the industry is increasing the amount of flowback and produced and brackish water in hydraulic fracturing. There are several key factors that operators must consider concerning the reuse of produced water, including the: quantity of produced water, duration and consistency of produced water, produced water quality, target formation characteristics, scale of the operation, and cost to reuse the water. Operational logistics, such as fluid-handling capabilities, transportation considerations, storage capabilities, and access to treatment locations or on-site treatment options, will influence the ability to reuse produced water. The availability of flowback and produced water is dependent on the amount of water that returns from the fracturing process and formation.

Currently, most of the flowback fluid from fracking operations is either transported from well sites for disposal or processed for reuse in further operations. Suspended solids must be removed from the water before reuse. Recycling this water can be costly and is a major focal point of many environmental groups and environmental regulators. New, more efficient, technologies have been developed which allow fracking fluid to be recycled on-site at reduced cost. However, hydraulic fracturing does not require water that is of potable (drinking water) quality. Recycling wastewater helps conserve water use and provide cost-saving opportunities. In gas recovery from the Marcellus tight formation, there are examples of companies reusing up to 96% of the produced water. Other examples of recycling and reuse include (KPMG, 2012):

(1) The use of portable distilling plants to recycle water in the Barnett tight formation, particularly in regions such as the Granite Wash field in North Texas, where water resources are more critical than in other tight formation basins in the United States.

(2) A water purification treatment center can recycle several thousand barrels of flowback and produced water per day generated from extracting oil and natural gas from a tight formation—this approach is being used in the Eagle Ford tight formation and in the Marcellus tight formation.

(3) The Marcellus tight formation also employs vapor recompression technology to reduce the cost of recycling fracturing water by using waste heat. The unit produces water vapor and solid residue that is disposed of in a waste facility. In addition, to reduce contamination risks during tight formation operations, many gas companies in the Marcellus tight formation are reducing the amount of chemical additives used in fracturing fluid while producing natural gas and crude oil.

(4) A wastewater treatment company specializing in the oil and gas industry has designed a mobile integrated treatment system for hydraulic fracturing that allows the reuse of water for future drilling. Using dissolved air flotation technology, the system can treat up to 900 gallons per minute of fracking flowback water. The accelerated water treatment reduces the equipment burdens and logistics of traditional treatment methods and could significantly reduce operational costs.

(5) Produced water can have high total dissolved solids (TDS) concentrations that can be difficult to treat. Thermal distillation, reverse osmosis (RO), and other membrane-based desalination technologies can be deployed to desalinate produced water to a level fit for purpose.

Fluids other than water may be used in hydraulic fracturing process, including carbon dioxide, nitrogen, or propane, although their use is currently much less widespread than water. The use of flowback and produced water is also dependent on the type of fracturing fluids being used for that play. Hydraulic fracturing fluids (fracking fluids) (Chapter 5) are a complex mixture of many ingredients that are designed to perform a diverse set of functions and accommodate a variety of factors, including local geology, well depth, and length of the horizontal segment of the well. Although the precise recipe is unique to the formation, the fluid is typically composed of proppants (typically sand) to hold open the fractures and allow the natural gas to flow into the well, and chemicals that serve as friction reducers, gelling agents, breakers, biocides, corrosion inhibitors, and scale inhibitors. Industry representatives point out that chemicals represent a small percentage of the fracturing fluid; on average, fracking fluid for shale gas consists of more than 99% water and sand. Given the large volume of fluid that is injected underground, however, a small percentage can represent a large quantity of chemicals.

The sources for the supply of water for hydraulic fracturing project will depend on the cumulative amount of water that will be required for the long-term project that is planned. Water sources will need to be appropriate for the forecasted pace and level of development anticipated. Thus, the selection of a water source (or, more likely, because of the amount of water needed, water sources) ultimately depends upon the: (1) volume required, (2) water quality requirements, (3) regulatory issues, (4) physical availability, (5) competing uses for the water, and (6) characteristics of the formation to be fractured, including water quality and compatibility considerations. If possible, wastewater from an industrial facility (or industrial facilities) should be considered as a water source, followed by groundwater sources and surface water sources (with the preference over nonpotable sources over potable sources), with the least desirable (at least for long-term, large-scale development) being any municipal water supplies.

However, in regard to the choice of the water source(s) will depend on local conditions and the availability of groundwater and surface water resources in proximity to planned operations. Most important, not all options may be available for all situations, and the order of preferences can vary from area to area. Furthermore, for water sources such as industrial wastewater, power plant cooling water, or recycled flowback water and/or produced water, additional treatment may be required prior to use for hydraulic fracturing. Contaminants in the water which are not removable to the desired levels for reinjection may render the water as unsatisfactory because it may deliver subterranean contamination in and around the drilling site and the reservoir and open up the real potential for groundwater contamination as well as contamination of aquifers that provide municipal water supplies and such results cannot assure project success.

Thus, water sources for hydraulic fracturing projects consists of: (1) surface water, (2) groundwater, (3) municipal water suppliers, (4) wastewater and power plant cooling water, and (5) reservoir water and recycled flowback water.

2.3.1 Surface Water

Many municipalities draw their principal water supplies from surface water sources, so the large-scale use of this source for hydraulic fracturing operations will undoubtedly impact not only municipal usage but also other competing uses and will, therefore, be of concern to local water management authorities and other public officials. In some circumstances there will be a need to identify water supply sources capable of meeting the needs of water requirements for vertical and horizontal drilling as well as for hydraulic fracturing and, in addition, it must be made clear to the relevant authorities that the water needs of the drilling/fracturing project will not compete or interfere with community needs as well as with currently existing uses.

Thus, necessary considerations in evaluating water supply requirements from surface water sources include not only the volume of water supplies required but also the sequence and scheduling of acquiring these supplies of water. Withdrawal from surface water bodies, such as rivers, streams, lakes, natural ponds, and private stock ponds, will likely require permits from state agencies or multistate regulatory agencies, as well as permission from the relevant landowners—in some regions, water rights are also a key consideration. In addition, the water quality standards and regulations that have been established by the various regulatory authorities may prohibit any alteration in water flow (such as pertains, e.g., to river flow and stream flow as well as flow into lakes) that would impair a high priority use of the fresh (surface) water, which is often defined by the local water management authorities. Also consideration should be given to ensure that water withdrawal from rivers and streams during periods of low river and/or stream flow do not affect fish and other aquatic life, fishing and other recreational activities, municipal water supplies, and the operational demands for water by other existing industrial facilities, such as power plants.

When the options for application of permits to use water are being considered, the applicant must know that water withdrawal permits can require compliance with specific metering, monitoring, reporting, record keeping, and other consumptive use requirements. In addition, compliance could also include specifications for the minimum measured quantity of water that must pass a specific point downstream of the water intake in order for a withdrawal to occur. Furthermore, in the case where stream flow is less than the prescribed minimum quantity, the withdrawal of water may be required to be reduced or even to cease. Thus it is necessary to consider the various issues that can arise with the timing and location of water withdrawals since impacted watersheds may be sensitive, especially in drought years, when the periods of flow are low or diminished during the years of the project or during periods of the year when activities such as agricultural irrigation place additional demands on the surface supply of water.

To be fully aware of the potential for water use, requests for (or proposals for) the use of surface water withdrawal should a consideration the following potential impacts that could control the timing and volume available: (1) ownership, allocation, or appropriation of existing water resources; (2) water volume available for other needs, including public water supply; (3) degradation of a stream's designated best use; (4) impacts to downstream habitats and users; (5) impacts to fish and wildlife; (6) aquifer volume diminishment; and (7) mitigation measures to prevent transfer of invasive species from one surface water body to another (as a result of water withdrawal and subsequent discharge into another surface water body).

In addition, state, regional, or local water management authorities may request that the applicant identifies the source of water to be used for supplying hydraulic fracturing operations and provides detailed information about any newly proposed surface water source that has not been previously approved for use. Information that must be supplied could include the withdrawal location and the size of the upstream drainage area and available stream gauge data, along with demonstration of compliance relative to stream flow standards. In order to obtain approval for water use as well as maintaining a good relationship with the regulatory bodies, nearby local communities that use water from the area under consideration, and other interested parties, it is obvious that requests for water withdrawals from sensitive watersheds should be carefully considered for their overall impact. Furthermore, in some jurisdictions, a variety of permits may be required for the transport of water via pipelines, canals, or streams, as well as transportation by tanker truck. Moreover, equipment or structures used for surface water withdrawal, such as standpipes, may also require permits.

With the continued development of resources in tight formations, additional regulatory requirements are likely to be associated with water use and requirements. For example, water could be sent to storage impoundments during periods of high flow to allow water withdrawal at a time of peak water availability. However, this approach would typically require the development of water storage capabilities to meet the overall demands of drilling and hydraulic fracturing projects over the course of the project which may even be a multiyear period to accommodate possible periods of drought.

Another alternative for ensuring adequate water supply is to use abandoned surface coal mining pits for the storage of water which provide more permanent facilities for the installation of a comprehensive water distribution system. However, the water quality in such storage areas must meet the operational requirements and with all regulatory requirements which would depend upon the nature of the exposed overburden that could allow subject leaching of undesirable chemical species during periods of heavy rain or winter run-off. In keeping with the concept of using surface coal-mining pits, another option is to excavate low lying areas to allow for collection of rain water. Again, such an option must meet with approval from state, regional, or local water management authorities to ensure compliance with storm water runoff program elements.

2.3.2 Groundwater

In order to use groundwater for drilling and hydraulic fracturing projects, many of the same types of considerations for groundwater as for surface water will need to be addressed. The primary concern regarding groundwater withdrawal is volume diminishment and, in some areas, the availability of fresh groundwater is limited, so withdrawal limitations could be imposed. To overcome such issues and where possible, consideration should be given to the use of nonpotable water for drilling and hydraulic fracturing projects. Another that applies to the protection of groundwater sources is to locate water source wells for oil and gas operations at an appropriate distance from municipal wells, public wells, or private water supply wells. Furthermore, public wells or private water supply wells and fresh water springs within a defined distance of any proposed drilling/fracturing project for a water supply well should be identified and the characteristics of the wells evaluated for production capacity and water quality. As part of the well evaluation, there may also be the need to test the water currently available from these sources. This will require locating the public and private water wells and gathering obtaining information about: (1) well depth; (2) completed interval and use, including whether the well is public or private, community or noncommunity; and (3) the type of facility or establishment if it is not a private residence.

Guidance for groundwater protection related to well drilling and hydraulic fracturing operations is available (API, 2009, 2010) and maintaining well integrity is featured as a key design principle of all natural gas and crude oil wells, which is essential for two primary reasons: (1) to isolate the internal conduit of the well from the surface and subsurface environment and (2) to isolate and contain the well's produced fluid to a production conduit within the well.

2.3.3 Municipal Water Supplies

Obtaining water supplies from municipal water suppliers can be considered, but again, the water needs for fracturing would need to be balanced with other uses and community needs. This option might be limited, since some areas may be suffering from current water supply constraints, especially during periods of drought, so the long-term reliability of supplies from municipal water suppliers needs to be carefully evaluated.

2.3.4 Wastewater and Power Plant Cooling Water

Other possible options for source water to support hydraulic fracturing operations that could be considered are municipal wastewater, industrial wastewater, and/or power plant cooling water. However, the properties or specifications of this water source need to be compatible with the target formation and the plan for fracturing as well as whether treating is technically possible and whether treatment can deliver an overall successful project. In some cases, required water specification could be achieved with the proper mixing of supplies from these sources with supplies from surface water or groundwater sources.

2.3.5 Reservoir Water and Recycled Flowback Water

Produced reservoir water and recycled flowback water can be treated and reused for fracturing, depending on the quality of the water. Natural formation water has been in contact with the reservoir formation for millions of years and thus contains minerals native to the reservoir rock. Some of this formation water is recovered with the flowback water after hydraulic fracturing so that both contribute to the characteristics of the flowback water. However, the salinity, total dissolved solids (TDS, sometime referred to as *total dissolved salts*), and the overall quality of this formation/flowback water mixture can vary by geologic basin and specific rock strata. In addition, other water quality characteristics that may influence water management options for fracturing operations include concentrations of organic compounds, usually hydrocarbons, which are detected by analysis (using standard test methods) for oil and grease, suspended solids, soluble organics, iron, calcium, magnesium, and trace constituents such as benzene, boron, silicates, and possibly other constituents (ASTM, 2015).

Finally, whenever water is recycled and/or reused, or additional sources of industrial wastewater are used to supply water for drilling and hydraulic fracturing operations, additional makeup water may be required. In such cases, water management alternatives to be considered will depend on the volume and quality of both the recycled water and the makeup water, to ensure compatibility with each other and with the formation being fractured.

Much of the focus of recent policy decisions has been on the identification and the disclosure of the chemical constituents used in hydraulic fracturing fluid. In addition, if toxic chemical additives are used in the fracturing fluid, many regulatory authorities require open disclosure of the necessary information and in many cases disclosure of the chemicals employed in a fracturing project is mandated. However, some authorities do require the complete disclosure of the chemical constituents of fracturing fluid along with the concentration and volume of these chemicals. In addition, some regulations require disclosure to the state regulatory authorities rather than to the public, and most authorities allow companies to apply for trade secret exemptions. In regard to the concentration and volume of the chemicals used, it is worth noting at this point that many chemicals that have been classed as nontoxic below a specified dosage. For example, table salt (sodium chloride) is used by many persons at meal time, but it is advisable not to try to consume a pound (454 g) at one sitting or it could be lethal. Thus, the laws of the use of toxic chemicals in hydraulic fracturing fluids need tweaking to specify the amount of chemicals used. This also applies to the use of chemicals that are indigenous to the environments since a concentration of the chemicals about the indigenous amount may become toxic to the environment.

The primary hydraulic fracturing fluid systems are slickwater insofar as chemicals are added to the water to increase the fluid flow and increase the speed at which the pressurized fluid can be pumped into the wellbore. Such chemicals include crosslink or crosslinked gels. However, the procedure is not (and cannot be) standardized since the selection of the fracturing fluids is based on a variety of factors and the fluid systems are designed on the basis of several factors such as: (1) the characteristics of the target formation, (2) the formation fluids, and (3) the source of the makeup water source. In addition, the use of recycled flowback water and produced water will be more likely in a slickwater hydraulic fracture.

2.3.6 Other Sources

Brackish or nonpotable water is considered publicly unusable for hydraulic fracturing projects without significant treatment. Typically, brackish water is high in salinity but has a lower salinity than that high saline brine and has TDS levels are on the order of 1000 ppm or even higher. The type and dose of friction reducer can be adjusted to accommodate for higher TDS water.

In the early exploration phase, water with a relatively low TDS content will be used for hydraulic fracturing. However, for tight formations (especially shale plays) that are beyond the phase of early exploration, the use of brackish water and recycled produced water is becoming an increasingly feasible option due to the advancements in the use of chemical additives and water-treatment technologies. Other wastewater sources are also given consideration for hydraulic fracturing projects. For example, acid mine drainage water is often given consideration for use in hydraulic fracturing projects. However, acid mine drainage is one of most serious threats to water since a mine draining acid can devastate rivers, streams, and aquatic life for prolonged periods measuring in decades and additional caution is advised when considering the use of such water. Pretreated seawater is also being explored as an alternative water source for hydraulic fracturing. However, the viability of any alternative water source is dependent upon factors such as the fracturing fluid type considerations, water treatment, and water transportation costs (e.g., proximity to an ocean).

2.4 Water Contamination

Water contamination, especially groundwater contamination from the retrieval of natural gas and crude oil from tight formations, can (and often does) occur through a variety of pathways. While the formations may be located at varying depths, often (but not always) far below underground, caution must be taken to assure that sources of drinking water are not contaminated and there must be, in the early stages of project planning, an accurate geological assessment of the formations—the project must be recognized as a multidisciplinary project and not merely a recovery engineering project. However, if a well from the surface to the formation level must be drilled through any drinking water sources in order to access the natural gas or crude oil, assurance of well integrity has to be an essential part of the drilling operation. Vibrations and pressure pulses associated with drilling can cause (at least) short-term impacts to groundwater quality, including changes in color, turbidity, and odor. Chemicals (from the drilling fluid), natural gas, and crude oil can escape the wellbore if it is not properly sealed and cased (Chapter 4) and even though there are, in most cases, regulatory requirements for well casing and well integrity, accidents and failures can still occur. In addition, older and abandoned wells that have not been sealed correctly can also potentially serve as migration pathways for contaminants to enter groundwater systems. Natural fractures in subterranean formations as well as the fractures created during the fracturing process could also serve as pathways to contamination of groundwater. Finally, coalbed methane (Chapter 1) is generally found at shallower depths and in closer proximity to underground sources of drinking water and therefore accessing the natural gas from this source might pose a greater risk of contamination.

Bacteria in the fracturing fluid can cause formation biofouling, reducing permeability, and gas production. The presence of sulfate-reducing bacteria can form hydrogen sulfide, making the well sour, creating safety issues, and increasing costs. Permeability and gas production can be reduced by metals in water, specifically iron, which can oxidize and form deposits, and by suspended solids in the frac fluid such as sand, silt, clays, and scale particles. Since some formations are a composite of shale, carbonates, and salt crystals, the use of a fracturing fluid low in TDS will increase dissolution of formation salts, potentially increasing reservoir permeability, and gas production. Understanding the relationship between the fracturing fluid water quality and long-term well production is crucial when considering water treatment technologies for a good water management plan.

Another effect, the scaling tendency of source water, usually caused by poor compatibility of source water with formation water and poor compatibility of reuse water, is another consideration. Scaling can occur within the formation, potentially creating reduced permeability, and ultimately reduced gas production. Scaling can also damage equipment casing, reducing functionality. Multivalent ions and chlorides in the water can limit friction reducer effectiveness and drive up horsepower costs for pumping in a hydraulic fracturing project. As brine, oil, and/or gas proceed from the formation to the surface, pressure and temperature change and certain dissolved salts can precipitate (referred to as *self-scaling*). If brine is injected into the formation to maintain pressure and sweep the oil to the producing wells, there will eventually be a commingling with the formation water. Additional salts may precipitate in the formation or in the wellbore (scale from *incompatible waters*) and other factors (Table 6.1). Many of these scaling processes can and do occur simultaneously. Thus, wells producing water are likely to develop deposits of inorganic scales. Scales can and do coat perforations, casing, production tubulars, valves, pumps, and downhole completion equipment, such as safety equipment and gas lift mandrels. If allowed to proceed, this scaling will limit production, eventually requiring abandonment of the well. However, technology is available for removing scale from tubing, flowline, valves, and surface equipment, restoring at least some of the lost production level. Technology also exists for preventing the occurrence or reoccurrence of the scale, at least on a temporary basis.

Scale remediation techniques must be quick and nondamaging to the wellbore, tubing, and the reservoir. If the scale is in the wellbore, it can be removed mechanically or dissolved chemically. Selecting the best scale-removal technique for a particular well depends on knowing the type and quantity of scale, its physical composition, and its texture. As a simple example of scale reduction, seawater can reduce the likelihood of scaling caused by formation brine that has a high calcium carbonate ($CaCO_3$) scaling potential. Seawater is often injected into reservoirs for pressure maintenance and enhancing oil recovery. When comingled with formation brine, the seawater can markedly decrease the calcium carbonate scaling potential of the mixture. As the percentage of seawater in the mixture

Table 6.1 Factors Involved in Scale Formation

Suspended solid accumulation
Residual suspended solids not removed in the pretreatment system can settle at low velocity points or deposit on surfaces; suspended material can also precipitate (coagulate, flocculate) with increases in temperature and concentration
Organic fouling
Organic material in the feed water can precipitate with an increase in concentration and/or temperature; this material can deposit on surfaces and/or coagulate smaller suspended solid particles that would not otherwise be a problem
Calcium carbonate scale
Sufficient concentrations of calcium and carbonate can cause calcium carbonate ($CaCO_3$) scale; the potential for the formation of $CaCO_3$ scale increases as temperature and concentration increase; typically, carbonate scales are white, chalky, and acid-soluble solids
Sulfate scale
The presences of sulfate and barium, strontium, and/or calcium can cause formation of barium sulfate ($BaSO_4$), strontium sulfate ($SrSO_4$), and/or calcium sulfate ($CaSO_4$) scale; the potential for the formation of sulfate-type scale material increases with concentration; sulfate scales are generally white, hard, or glasslike, and acid-insoluble solids
Silica scale
Silica can form solids in four ways: surface deposition (depositing directly to surfaces), bulk precipitation (particles collide to form larger particles), complexing (metal hydroxides bind with silica to form particles), and silica polymerization (silica molecules combine to form long strings); since silica fouling is impacted by metal hydroxide precipitation, the potential for silicate fouling increases with temperature, concentration, and is highly impacted by pH; silica deposits range from quartz-like hard scale to slimly deposits (generally white, gray, green, or brown solids—the color can vary depending on the types of the metal hydroxides involved)

increases, scaling potential decreases and may even become negative. Once the scaling potential becomes negative, the produced water can dissolve the calcium carbonate scale that may have formed in the flow lines.

Considerations related to water disposal should also be given suitable acknowledgement during the early stages of field development with the recognition that the need for water disposal will increase as activity increases through evolution of the project—moreover, demand for water disposal in terms of the volume of water can be surprising. Thus, as project evolves and water sources for long-term supply are considered, equal consideration should be given to disposal of the water respect to the collective volumes of all water use. In fact, water disposal issues are challenging in most project areas. Facilities accepting flowback water may be able to be designed to accept other water disposal streams, thereby reducing overall treatment costs and creating more makeup water for reuse.

3 WATER DISPOSAL

Water that originates from a hydraulic fracturing project often contains chemical additives to help carry the proppant and may become enriched in salts after being injected into tight formations. Therefore, the water that is recovered during natural gas and crude oil production from tight formations must be either treated or disposed of in a safe manner—a suggestion for disposal of such water is typically to inject the water into deep, highly saline formations through one or more wells drilled specifically for that purpose, but the disposal method must follow clearly defined regulations. Flowback water is infrequently reused in hydraulic fracturing because of the potential for corrosion or scaling, where the dissolved salts may precipitate out of the water and block parts of the well or the formation, thereby interfering with and influencing flow of the fluids.

In addition to chemicals that are added to the fracturing fluid, wastewater from natural gas and crude oil extraction may contain high levels of TDS, which can be a complex collection of inorganic compounds. Furthermore, the amount of saline water produced from tight formations can vary widely—from zero to (at least, depending upon the formation characteristics) several hundred barrels per day. The water may originate from the tight reservoir formation or from any adjacent formations that are connected to the reservoir formation through a natural fracture network or, more likely, through the process-induced fracture network. Typically, this water, like flowback water, is highly saline and must be treated or sent for disposal by injection into deep saline formations, which is also subject to clearly defined regulations. In fact, in some regulatory authorities that oversee the production of natural gas and crude oil have implemented regulations regarding the disclosure of chemicals used in the process of hydraulic fracturing to ensure that the subterranean environment is protected when reinjection of these chemicals is the method of disposal.

In terms of water use management, one alternative that states and natural gas and crude oil production projects are pursuing is to make use of seasonal changes in river flow to capture water when surface water flows are greatest. Utilizing seasonal flow differences allows planning of withdrawals to avoid potential impacts to municipal drinking water supplies or to aquatic or riparian communities. Also included is monitoring of stream water quality as well as game and nongame fish species in the reach of river surrounding the intake. In addition, new treatment technologies have made it possible to recycle the water recovered from hydraulic fracturing. The reuse of treated flowback fluids from hydraulic fracturing is being used (or at least, considered for use) for various projects.

Briefly, and by way of explanation, a riparian community is a community that exits in a riparian zone (riparian area), which is the interface between land and a river or between land and a stream. Plant habitats and communities along the river margins and banks, and stream margins and banks (*riparian vegetation*) are characterized by collections of hydrophilic plants. Riparian zones are important in ecology and in environmental management because of their role in soil conservation, their habitat biodiversity, and the influence they have on fauna and aquatic ecosystems, including grasslands, woodlands, wetlands, or even nonvegetative growing areas. In some regions the terms riparian woodland, riparian forest, riparian buffer zone, and riparian strip are used to characterize a riparian zone.

The rate at which water returns to the surface is highly dependent upon the geology of the formation. In some formations, recycle of flowback water may be as high as 95% (v/v) of the flowback water, whereas in other formations the recycle operations may involve as little as 20% (v/v) of the flowback water. Thus, as might be anticipated in the light of similar statements elsewhere in this text, water management and water reuse are issues that are specific to the locality of the project as well as specific to the depth and characteristics of the gas-bearing or oil-bearing formation as well as the quality and quantity of water and the availability and affordability of management options (Veil, 2010). Over a 30-year life cycle, assuming a typical well is hydraulically fractured three times during that time period, construction and production of shale gas typically can consume between 7,000,000 and 17,000,000 gallons of water per well.

Once the natural gas or crude oil gas is produced, it is processed and transported for further processing (cleaning, refining) (Chapters 7 and 8). Water consumption occurs in each of these stages as well, with the most significant nonproduction consumption potentially occurring during end use. Typically, crude oil from a hydraulic fracturing project is transported to a refinery for use as blending stock with other crude oil for the refinery feedstock, which may constitute a blend of three or more other crude oils (Chapter 8). The refinery requires additional water consumption, but this additional consumption is not directly related to the water use during the hydraulic fracturing project. On the other hand, although natural gas can be combusted directly with no additional water consumption, if the end use of the gas is in the form of a transportation fuel, storage in a vehicle tank is required and the natural gas will most likely have to be compressed by means of an electric compressor (King and Webber, 2008; Wu et al., 2011). It then remains to select an option (or options) for treatment of the water either for reuse or for disposal.

Put simply, water treatment is any process which removes contaminants from the water or reduces the concentration of the contaminants so that the water is more acceptable for a specific end-use. In the present context, the end use may be industrial water supply, irrigation, river flow maintenance, water recreation, or many other uses including being safely returned to the environment. In some cases, exceptional and assiduous treatment of the water can produce water suitable for use as drinking water. Thus, management of flowback water can generally be realized by any one (or more) of the management strategies: (1) disposal, (2) reuse, and (3) recycling (Halldorson and Horner, 2012).

In regard to the *disposal* option for the water, the flowback is transported to an injection well for disposal. This option for disposal is often chosen if there is a ready and relatively inexpensive, abundant supply of fresh water nearby, and most important, nearby injection wells can handle the flowback disposal volumes according to the relevant regulatory guidelines or laws. However, as fresh water availability decreases, costs increase and/or distance to injection wells for disposal increases, making the disposal scenario less appealing.

The *reuse* option (which may be the least expensive strategy) involves careful treatment of the flowback water to remove suspended solids and soluble organic constituents (such as naphthenic acids that are present in some crude oils) (Speight, 2014b) followed by blending the treated water with fresh water to generate a fluid that is suitable for use in a hydraulic fracturing project for a new well. This option reduces the amount of fresh water required for the project and eliminates the need for the disposal option if all flowback water can be treated and reused.

Finally, the *recycling* option treatment of the flowback water to produce a product that is typically of freshwater quality. In this option, the recycled water is blended with makeup water from freshwater sources to generate a hydraulic fracturing fluid that is low in TDS. Recycling is used when: (1) fresh water costs are high, a high quality; (2) a fracturing fluid with a low TDS content is required; or (3) other logistics such as hydraulic fracturing schedules do not permit reuse of the water. If the fracturing fluid is transferred by means of temporary above-ground pipelines (*fastlines*) may recycle the water to minimize potential environmental liability from spills or rupture of the pipeline.

The issue of the TDS in water is a continuous issue, and during an active fracking project involving natural gas and crude oil production is continually being subject to inconsistent flowback water containing TDS—ranging from 5000 to 200,000 ppm—and/or total suspended solids (TSS) ranging from 100 to 3000 ppm. In order to combat such impurities, a clarifier pump can be used to pump flowback water from the source (such as a such as a *frac tank* or pit) into a unit which uses a solids settling system which enables suspended solids settle to the bottom of the unit where they are collected and dewatered. A major benefit of reusing water is that it reduces the financial, social, and environmental costs associated with water transportation.

In addition, evaporation technology can be used to recycle contaminated water—the process involves boiling a solution so that contaminants remain in the liquid phase, while pure water vapor evaporates and can be condensed into distilled water. Also, mechanical vapor recompression evaporation is a leading means of treating waste water—the process differs from conventional evaporation insofar as a compressor is used to generate steam rather than a heat source such as a boiler. High energy efficiency is achieved by utilizing the latent heat of the condensing steam as the primary energy source for boiling the wastewater.

4 WASTE FLUIDS

A variety of waste fluids are generated on-site at natural gas and crude oil wells, and one of the biggest challenges for protecting water resources from tight formation natural gas and crude oil activities is the wastewater generated during production. Wastes, such as drill cuttings, and wastewater generated during exploration, development, and production of crude oil and natural gas are categorized by the US Environmental Protection Agency as *special wastes* that are exempted from federal hazardous waste regulations under Subtitle C of the Resource Conservation and Recovery Act.

Fluids management involves the environmentally friendly disposal or reuse of excess fluids and is an integral and essential phase of project involved in the development of (and recovery of natural gas or crude oil from) tight formations because pollutants in the natural waterways (rivers, stream, and some case, the oceans) result in the reduction of the oxygen content of water leading to serious effects on (even elimination of) the aquatic life. This gives rise to the need, not only for water purification but also for water conservation.

Furthermore, an important challenge is fluids management which involves: (1) treatment, (2) recycling, (3) reuse, and (4) disposal of the flowback water and produced water. This water mix often contains residual fracturing fluid and may also contain substances found in the reservoir formations, such as trace elements of heavy metals and even naturally occurring radioactive materials (NORM). Also, the impact of the recovery operations on land and air must also be minimized. Production wells require roadways to connect drilling pads and there will be the need for either pipelines or trucks or both to transport natural gas, crude oil, or process waste as well as the need for storage units, and water treatment facilities. With the potential for several hundred truck trips per well site, pipeline use has the potential to minimize surface disturbance.

Wastewater from natural gas and crude oil exploration is generally classified into flowback and produced waters. Flowback water is the fluid that returns to the surface after the step of hydraulic fracturing and before natural gas and crude oil production begins, primarily during the days to weeks of well completion. This fluid can consist of 10–40% (v/v) of the injected fracturing fluids and chemicals pumped underground that return to the surface mixed with an increasing proportion of natural brines from the tight formation through time.

Produced water is the fluid that flows to the surface during extended natural gas and crude oil production and primarily reflects the chemistry and composition of deep formation waters and capillary-bound fluids. These naturally occurring brines are often saline to hypersaline and contain potentially toxic levels of elements such as barium, arsenic, and radioactive radium. Wastewater from hydraulic fracturing operations is disposed of in several ways. Deep underground injection of wastewater comprises >95% of disposal in the United States. In contrast, deep injection of wastewater is not permitted in Europe unless the water is used to enhance natural gas and crude oil recovery. Wastewater in the United States is also sent to private treatment facilities or, increasingly, is recycled or reused (see above). More recently, wastewater is increasingly being sent to facilities with advanced treatment technologies such as desalination at rates approaching 90% reuse.

Currently, practices are in place for controlling water flow—ponds and pools are provided where drainage water accumulates and where suspended mineral matter (such as shale and clay particles) can ultimately be removed. Sedimentation, with or without the use of flocculants, can be applied to process the water contained in lagoons. In other cases, water conservation programs are mainly aimed at restricting water flow seeping through the porous structure, boreholes, and fractures in the water bearing strata. In addition, the protection of drinking water sources is also necessary.

One of the first steps in water management involves proper well construction to isolate the wellbore by use of cement, and several different sets of steel casing (Chapter 4) is a critical step taken by the natural gas and crude oil industry to protect groundwater sources. The casings are individually cemented into the wellbore to provide barrier which isolate wellbore fluids from the rock formations. In the process, cement is pumped into the center of casing so that it circulates back to the surface in the space outside of the casing (the *annulus*), following the installation of each length (*string*) of casing in the well. After these steps are complete, the cement must be allowed to set prior to the continuation of drilling—a geophysical log is run to determine the integrity of the cement that surrounds the casing. This is an aid to ensure that the wellbore is adequately cemented and capable of withstanding the pressure associated with hydraulic fracturing. Prior to stimulation, the well is pressure tested to ensure the integrity of the casing system that has been installed underground.

Thus, under the various regulations, comprehensive rules are in place to ensure that wells are constructed in a manner (well integrity) (Chapter 4) that protects freshwater supplies. Specific guidelines vary between the various regulatory authorities but, in all cases, steel casing and cement are used to isolate and protect groundwater zones from deeper oil, natural gas, and saline water zones.

4.1 Fracturing Fluid Requirements

The primary factor influencing water management associated with hydraulic fracturing is related to the fluid requirements for a successful hydraulic fracturing operation. All phases of water management ultimately depend on the requirements of the hydraulic fracturing properties needed for conducting a successful fracturing project. These requirements are the result of the geology of the reservoir formation and the formations above and below the reservoir, the operating environment, the design of the hydraulic fracturing process, the scale of the development process, and the results required for total project success.

The primary issue in understanding water management for a hydraulic fracturing project involves knowledge of the reservoir rock need and what will the rock produce after completion of the fracturing process. The choice of the fracturing fluid dictates the fracturing design as well as the types of fracturing fluids and the additives that are required. Furthermore, the choice of the fracturing fluid dictates the transport and ultimate fate of the fracturing fluid used in the fracturing operations as well as the means by which the recovered fluids will need to be managed and disposed.

The typical hydraulic fracturing practice (if there is such a process that can be described as *typical*) is designed to create single fractures or multiple fractures in specific rock formations. These hydraulic fracture treatments are controlled and monitored processes designed for the site-specific conditions of the reservoir. Moreover, the process conditions are guided by: (1) extraction of the target product, that is, natural gas or crude oil, (2) the respective properties of the target product, (3) the properties—including the mineralogy—of the target formation, (4) the rock fracturing characteristics of the formation, (5) the properties of the formation water, (6) the anticipated water production, that is, formation water vs fracturing flowback water, and (7) the type of well drilled (horizontal or vertical) into the formation.

Thus, understanding the in-situ reservoir conditions is critical to successful stimulations, and in the design of the fracture treatment and fluid used as well as to water management. Hydraulic fracturing designs are continually evolving both during the fracture stimulation itself and as over time as more about fracturing the target formation is understood and this understanding evolves. Thus, while the concepts and general practices are similar, the details of a specific fracture operation can vary substantially from resource to resource, from area to area, and even from well to well.

4.2 Fracturing Fluid Composition

There is a wide variety of additives that could be included in the fracturing fluid mix to achieve successful fracturing (Chapter 5). These include proppants, gel and foaming agents, salts, acids, and other fluid additives, and there is a movement (and a need) to maximize the utilization of environmentally benign additives and minimize the amount of additives required.

The characteristics of the resource target determine the required composition of the hydraulic fracturing fluid composition which, in turn, can affect water management. For example, the tight formation may contain various naturally occurring trace metals and compounds that become available because of the induced fractures and are leached from the reservoir rock by acidic water, by conversion to soluble species as a result of oxidation, and by the action of ionic species that occur in brines. Numerous inorganic and organic compounds have been formed naturally and occur in the tight formation, and a stimulation fluid pumped into a well may require various chemicals to counteract any negative effects these compounds may have in the well or the reservoir. For example, iron compounds require an iron sequestering agent so that the compounds of iron will not precipitate out of the fracturing fluid and be deposited within the pore spaces of the reservoir, reducing the permeability of the reservoir and adding further complications to water management.

One of the major aspects of water management is, when developing plans for a hydraulic fracturing project in addition to considerations associated with successfully fracturing the target formations, the fluid management and disposal implications of the project, and the fracture fluid formulations should receive major consideration. The best water management practice is to use additives that pose minimal risk of possible adverse environmental effects to the extent possible in delivering the needed effectiveness of the fracturing operation. While this is a highly desirable option, product substitution may not be possible in all situations because effective alternatives may not be available for all additives.

4.3 Fracturing Fluid Handling and Storage

Fracturing fluid handling and storage is perhaps that major issue that arises when considering water management options. Fracturing fluid requirements and fracturing fluid composition are all aspects of fluid handling and storage that contribute to water management. Fluids handled at the well site both before and after hydraulic fracturing often must be stored on-site and must be transported from the source of supply to the point of ultimate treatment and/or disposal. Fluids used for hydraulic fracturing will generally be stored on-site in tanks or lined surface impoundments. Returned fluids, or flowback, may also be directed to tanks or lined pits. Furthermore, the volume of initial flowback water recovered during the first month or so following the completion of hydraulic fracturing operations may account for less than 10% (v/v) to more than 70% (v/v) of the original volume of the fracturing fluid. The vast majority of fracturing fluid injected is recovered in a very short period of time, usually to a maximum time period of several months.

Thus all of the components of the fracturing fluids, including and especially water, additives, and proppants, should be managed properly on-site before, during, and after the hydraulic fracturing process. If possible, to assist in water management, the components of the fracturing fluid should all be blended into the composite fracturing fluid on an as-needed basis. In addition, any unused products should be removed from the fracturing location as soon as is appropriate. Furthermore, the project planning process should take into consideration the possibility of unexpected delays in the fracturing operation and ensure that water and the additive materials are managed correctly.

If lined impoundments or pits are used for storage of fracturing fluids or flowback water, the pits must comply with applicable rules, regulations, good industry practice, and liner specifications. Thus, these impoundments must be designed and constructed in such a manner as to provide structural integrity for the life of their operation—correct design is imperative to the objective of preventing a failure or unintended discharge. If the fluids are to be stored in tanks, these tanks must meet appropriate state and federal standards, which may be specific to the use of the tank—for example, if the tank is used as a tank for flowback water or for more permanent production tank batteries.

4.4 Transportation

Before the onset of fracturing, water, sand, and any other additives are generally delivered separately to the well site. Water is generally delivered in tanker trucks that may arrive over a period of days or weeks, or via pipelines from a supply source or treatment/recycling facility. Thus, water supply and management approaches should take into consideration the requirements and constraints associated with fluid transport.

Transportation of water to and from a well site can be a major expense and major activity. While trucking costs can be the biggest part of the water management expense, one option to consider as an alternative to trucking is the use of temporary or permanent surface pipelines, but the transport of fluids associated with hydraulic fracturing by surface pipeline may not be practical, cost effective, or even feasible. Moreover, when fracture fluids are to be transported by truck, the project will need development of a basin-wide trucking plan that includes: (1) the estimated amount of trucking required, (2) the hours of operations, (3) the appropriate off road parking/staging areas, and (4) the trucking routes. Furthermore, considerations for the trucking plan for large volumes of fracture fluid include the following: (1) public input on route selection to maximize efficient driving and public safety; (2) avoidance of peak traffic hours, school bus hours, community events, and overnight quiet periods; (3) coordination with local emergency management agencies and highway departments; (4) upgrades and improvements to roads that will be traveled frequently to and from many different well sites; (5) advance public notice of any necessary detours or road/lane closures; and (6) adequate off-road parking and delivery areas at the site.

The use of multiwell pads (Chapter 4) makes the use of central water storage easier, reduces truck traffic, and allows for easier and centralized management of flowback water. In some cases, it can enhance the option of pipeline transport of water. Furthermore, in order to make truck transportation more efficient and have lees impact on the surrounding environment, it is worth considering the construction of storage ponds and drilling source wells in cooperation with any nonproject (but interested) private property owners. The opportunity to construct or improve an existing pond, drilling a water well, and/or improving the roads on their property can be an extremely helpful (perhaps even a *win-win*) situation for the operator and the landowner by providing close access to a water source for the project as well as adding improvements to the nearby property that could benefit the property owner.

During drilling, used mud and saturated cuttings are produced and must be managed. The volume of mud approximately correlates with the size of the well drilled, so a horizontal well may generate twice as much drilling waste as a single vertical well; however, the horizontal well may replace as many as four vertical wells. Drilling wastes can be managed on-site either in pits or in steel tanks. Each pit is designed to keep liquids from infiltrating vulnerable water resources. On-site pits are a standard in the natural gas and crude oil industry but are not appropriate everywhere; they can be large and they disturb the land for an extended period of time. Steel tanks may be required to store drilling mud in some environments to minimize the size of the well site *footprint* or to provide extra protection for a sensitive environment. Steel tanks are not, of course, appropriate in every setting either but in rural areas or pits or ponds, where space is available at the well site, steel tanks are usually not needed.

The drill cuttings are regarded as controlled or hazardous waste and can be disposed of in the following ways: (1) decontamination treatment, (2) injection of the cuttings into the well, or (3) transfer to a controlled hazardous-waste landfill. The lowest environmental effect for solids treatment, especially for offshore operation, is decontamination treatment followed by discharge. However, oil content in the treated cuttings of >1% still exists in the dried solids by conventional decontamination technology, which does not meet strict environmental regulations in some countries.

Horizontal drilling development has the power to reduce the number of well sites and to group them so that management facilities such as storage ponds can be used for several wells. Make-up water is used throughout the development process to drill the well and to form the basis of the hydraulic fracturing fluid. Large volumes of water may be needed and are often stored at the well site in pits or tanks. For example, surface water can be piped into the pit during high-water runoff periods and used during the year for drilling and fracture treatments in nearby wells. Storage ponds are not suitable everywhere in the area of a natural gas and crude oil resource—just as steel tanks are appropriate in some locations but not in others. Finally, it may be opportune for any hydraulic fracturing project to include consideration of utilizing agricultural techniques to transport the water used near the water sources. Large diameter, aluminum agricultural pipe is sometimes used to move the fresh water from the source to locations within a few miles where drilling and hydraulic fracturing activities are occurring. Water use by the natural gas and crude oil industry when working to recover these resources from tight formations has spurred the formation or expansion business involved in the supply of the temporary pipe, pumps, installation, and after-project removal of these amenities.

5 WATER MANAGEMENT AND DISPOSAL

When water returns to the surface from a tight formation drilling operation, it may be disposed of in a variety of ways, depending on the tight formation basin: (1) reused in a new well, with or without treatment; (2) injected into on- or off-site disposal wells regulated by the United States Environmental Protection Agency; (3) taken to a municipal wastewater treatment plant or a commercial industrial wastewater treatment facility—most wastewater treatment plants are not capable of treating the contaminants in natural gas and crude oil wastewater; or (4) discharged to a nearby surface water body.

In the Marcellus tight formation, one of the largest tight formation basins in the United States located in Pennsylvania and New York state, a large proportion of the hydraulic fracturing fluid, is usually recovered after drilling and stored on-site in evaporation pits. Recovered fluid may be trucked off-site for use in another fracking operation or for treatment and disposal in surface waters, underground reservoirs, or at a wastewater treatment facility. The remainder of the fluid remains underground (Veil, 2010). However, in the water-deprived tight formation basins of Texas (such as Eagle Ford), more of the hydraulic fracturing fluid may remain underground. This water is much harder to track than surface water, which may lead to increased short- and long-term risks for natural gas and crude oil companies.

While treatment of produced fluids (including water) from some hydraulic fracturing projects remains an option in some jurisdictions, requirements associated with the use of this option are likely to fall under stringent federal, states, or regional regulations. The project should be sufficiently well planned to accommodate proper management and disposal of fluids associated with hydraulic fracturing operations (Table 6.2). Furthermore, considerations

Table 6.2 Considerations Associated With Water Acquisition, Use, and Management in Hydraulic Fracturing Operations

Source water acquisition
Involves planning for the locale where the water supplies needed for hydraulic fracturing operations be acquired?
Transport
Involves planning for the transportation of the water from the source to the hydraulic fracturing site and, after recovery of the water as flowback water, from the hydraulic fracturing site to the point of water treatment and/or water disposal
Storage
Involves planning for the water requirements and constraints that may exist for water storage on the hydraulic fracturing site; also involves site planning for source water considerations and fracture fluid requirements affect storage requirements
Use
Involves planning for the methods of water use, the volume of water required, and the steps that must be taken (such as the addition of proppant and additives) to achieve the fracturing objectives
Treatment and reuse/recycle
Involves planning for the treatment of the water produced from the fracturing operation and whether or not, after treatment, the water can be recycled for reuse
Treatment and disposal
Involves planning for the treatment and disposal of that water if water is not to be recycled and or reused; also needs decisions of what steps need to be taken either prior to disposal or with any treatment byproducts

for fluid management should include provision for flowback water disposition, including the planned transport of the water from the well pad (truck or piping), and information about any proposed piping as well as the planned for disposal of the water (such as the treatment facility, the disposal well, water reuse, a centralized surface impoundment, or a centralized tank facility). There should also be clear identification and permit numbers for any proposed treatment facility or disposal well, and the location and construction and operational information for any proposed centralized flowback water surface impoundment.

Typically, a well permit will specify (often with the necessary details) that all fluids, including fracture fluids and flowback water, must be removed from the well site. In addition, any temporary storage pits used for fracturing fluids must be removed as part of reclamation. More specifically, water used in the hydraulic fracturing process is usually managed and disposed of in one of three ways: (1) injection into permitted disposal wells under a regulatory program, (2) delivery to a water treatment facility depending on permitting regulations which may, in some regions, allow treatment of the water to remove pollutants and to achieve all regulated specifications and then followed by regulated discharge of the water to the surface, and (3) reuse if sufficiently pure for the purpose or prior passage through a recycling operation before reuse. However, water disposal options are dependent on a variety of factors, including the availability of suitable injection zones and the possibility of obtaining permits for injection into these zones or the capacity of commercial and/or municipal water treatment facilities, and the ability of either operators or such plants to successfully obtain surface water discharge permits.

Part of the project plan should also ensure surface and groundwater quality is described in detail which may include any necessary sampling/analytical programs that will be sued to acquire samples for passage to an analytical facility. This information will provide unbiased analytical data for a better understanding of water quality before extensive drilling and hydraulic fracturing are initiated, and (more appropriately) will provide the baseline data that will assist in informing the local community about existing groundwater quality. There should also be the move to collect any necessary additional site-specific baseline water samples collected from public and private wells near the hydraulic fracturing operations, as well as from nearby surface water bodies (rivers, stream, lakes ponds) prior to drilling specific wells if existing information is not adequate. The actual parameters to be tested and for which analytical data will be required will depend to a great extent on site-specific geology, hydrology, and water chemistry. Typically, the parameters for testing should include but are not limited to TDS; TSS chloride ion concentrations; carbonate concentration; bicarbonate concentration; as well as the concentrations of sulfate, barium, strontium, arsenic, surfactant derivatives, methane, hydrogen sulfide, benzene, and NORM.

The primary potential destinations for flowback/production fluids generally include the following: (1) injection wells, which are regulated under either a state or federal program; (2) municipal waste water treatment facilities; (3) industrial waste treatment facilities; (3) other industrial uses; and (4) fracture flowback water recycling/reuse.

5.1 Injection Wells

The disposal of flowback fluids through injection, where an injection zone is available, is widely recognized as being a means of fluids disposal, providing the procedure is environmentally benign, well regulated, and proven to have been effective (API, 2009, 2010). However, in order to manage the expected amount of water associated with large-scale developments, additional injection wells in an area may need to be drilled and will have to be authorized through the relevant permitting procedures. Injection wells for the disposal of brine associated with natural gas and crude oil recovery operations will require state or federal permits. Therefore, whether the United States Environmental Protection Agency or the state regulatory agency has underground injection control program (UIC program) authority over the subsurface injection, new injection wells will require an injection well permit that meets the appropriate state and/or federal regulatory requirements.

5.2 Municipal Waste Water Treatment Facilities

Municipal wastewater treatment plants or commercial treatment facilities may be available as treatment centers and disposal options for the treatment of fracturing fluid flowback and/or other produced waters. However, sufficient available capacity must exist for treatment of the flowback water and, thus, the availability of municipal treatment plants or commercial treatment plants may be limited to larger urban areas where large treatment facilities are already in operation with. Moreover, the practicality of the transportation of the fluids from underground injection projects must be given serious consideration in any planning process.

If the size of the treatment plant means that treatment of the fluids is in order, the treating plant (especially if the plant is a publicly owned treatment works, POTW) must have a state-approved pretreatment program for accepting any industrial waste. In addition, the publicly owned treatment works must also notify appropriate regulatory authorities of any new industrial waste they plan to receive at their facility and certify that the facility is capable of treating the pollutants that are expected to be in that industrial waste. Furthermore, publicly owned treatment works are generally required to perform specific analyses to ensure the plant can handle the waste without disturbing (unbalancing) the system or causing a problem in the receiving water. Ultimately, approval is required of such analysis and modifications to the permits that are already in place at the publicly owned treatment works are necessary to ensure that water quality standards in receiving waters are maintained at all times. Thus, in an effort to assist in this review, the publicly owned treatment works may require that operators of a hydraulic fracturing project provide information pertaining to the chemical composition of the hydraulic fracturing additives, specifically to examine the potential environmental hazards in the mix as well as the possible toxicity of the mix.

5.3 Industrial Waste Treatment Facilities

It is considered unlikely that future disposal needs will be met by publicly owned treatment works due to current regulatory restrictions and future regulatory restrictions. Thus, the construction of private or industry-owned treating facilities, perhaps built and operated by an industry cooperative or an environmental services company, is an alternative solution. In several regions, the current and evolving practice is to set up temporary treatment facilities located in active drilling development areas or to treat the waste stream on-site with mobile facilities. The temporary facilities can alleviate/reduce the trucking of waste streams by the use of transitory pipeline systems that serve local wells.

5.4 Other Industrial Uses

In terms of fluids disposal there may be other industrial uses for flowback water that are worthy of consideration, but each proposed use will be highly dependent on site-specific considerations, and some level of treatment to match the fluid properties with the site needs would (more than likely) be required. Two such examples are: (1) the use of the flowback water to support drilling operations or (2) the use of this water as source water for water flooding operations, where water is injected into a partially depleted oil reservoir to displace additional oil and increase recovery (Speight, 2014a, 2016a).

Briefly, waterflood operations are regulated under state regulations and/or underground injection control program of the US Environmental Protection Agency which protects water sources such as drinking water sources (US EPA, 2015a). These authorities would review the proposed use of flowback fluids from hydraulic fracturing operations for suitability in a waterflood injection operation. Often, water injection operations and operations where a change in water or water source occurs are usually required to modify their permits to inject water from a new source. The water provider is required to submit an analysis of the water to be injected any time there is a change in the water or even the water source.

5.5 Flowback Water Recycling and Reuse

Effective management of flowback water requires knowledge of the characteristics of the water. Typically, flowback water contains salts, metals, and organic compounds from the formation as well as many of the compounds that were introduced as additives to the influent stream. Thus, there is the need for an information base on the composition and properties of flowback water and on the influent water streams that are used to perform hydraulic fracturing.

There are instances where it may be more practical to treat the flowback water to a specified quality so that it could be reused for a subsequent hydraulic fracturing project rather than treating the water to meet the necessary requirements that make the water suitable for surface discharge. Consequently, options for the recycling of fracture treatment flowback fluid should be an early consideration since water reuse and water recycling can be a key to enable large-scale future natural gas and crude oil recovery operations that use hydraulic fracturing as the method of accessing and freeing the natural gas and crude oil. While this method is already underway in some areas and been successful, the ability to reuse hydraulic fracturing fluid does depend, to a large degree, on the type of treatment required and the volume of makeup water that is necessary for the operation.

The various options to be considered will depend on: (1) the total water volume of water to be treated, (2) the soluble constituents in the water that need to be treated, (3) the concentrations of the soluble constituents, (4) their treatability of the solubility of the constituents and whether or not the treatment will remove these constituents, (5) the water reuse requirements, and

(6) the water discharge requirements. If all of the above issues can be satisfied to allow reuse of the flowback water, such reuse can provide a practical solution that overcomes many of the constraints imposed by the limited source water supplies as well as by difficulty in disposal of the reused water.

For example, the evolution of water treatment technologies is being adapted to work with the high saline water that results from hydraulic fracturing. Such technologies include innovations in RO technology and innovations in membrane technology. In addition, distillation technology is in the process of refinement to improve the 75–80% treating effectiveness of the current water reuse. However, distillation is also a very energy intensive process, and it is possible that it may only become an option for all operations with technological improvements (as a piggy-back technology, i.e., as a secondary ancillary technology) to increase the treatment effectiveness and the overall efficiency of the process.

Pursuing any such alternate option for water treatment does require careful planning and knowledge of the composition of the flowback water and/or the produced reservoir water. Moreover, finding the correct option also requires careful selection of the chemical additives and design that do not bring about major water treatment issues. Success will be measured by the efficiency of the treatment technology and whether or not the technology makes it more economical to treat these hydraulic fracturing fluids and produce better results in terms of the quality of the treated water. The treatment of these hydraulic fracturing fluids may greatly enhance the quantity of acceptable, reusable fluids, and provide more options for ultimate disposal of the reused water. Such treatment facilities either could be run as integral part of the hydraulic fracturing project or the facility could even function as stand-alone, independent commercial enterprise.

In this context, a number of water treatment innovations do approaches exist, and many other options continue to be developed and modified to address the specific treatment needs of flowback water from different sites and in different operating regions. Processes that can be utilized for water treatment (and this are adaptable to treatment of flowback water) include but are not limited to: (1) filtration, (2) aeration and sedimentation, (3) biological treatment, (4) demineralization, (5) thermal distillation, (6) condensation, (7) RO, (8) ionization, (9) natural evaporation, (10) freeze/thaw procedures, (11) crystallization, and (12) ozonization. This is by no means an exhaustive list, and new alternatives are continuously being considered and evaluated.

Because of the complexity of the hydraulic fracturing process and the varying quality of flowback water, it is likely that multiple processes will be required in many cases, if not in most cases. Some of the processes will serve as piggy-back processes (secondary processes) to the major process option ensure efficient treatment of the water that cannot be achieved by one process alone. Key considerations to the selection of the treatment process will be: (1) the efficiency of the primary, (2) the need for a piggy-back secondary process, (3) the performance of the primary process with and without the secondary process, (4) the volume of water processes in a specific time interval, (5) the environmental considerations associated with the resulting treated water, and (6) the cost-effectiveness of the water treatment process in terms of the primary with and without the secondary piggy-back option.

6 FLUIDS ANALYSIS

The fluids from hydraulic fracturing operations are made up of water, a variety of chemicals, and sand that is pumped into the underground shale rock formations under great pressure to recovery the natural gas and crude oil. There are two types of wastewater from hydraulic fracturing operations: (1) flowback water and (2) produced water (Chapter 5). Flowback water refers to the fluid that returns to the surface upon completion of a hydraulic fracturing event prior to production. Produced water refers to the fluid that returns to the surface once the well starts producing natural gas and oil. Both types of water can contain liquid that was resident in the shale layer (formation brine). However, flowback water is primarily composed of fracking fluid and its associated chemicals.

After a well has been subjected to the fracture treatment, some of the fluid flows back to the surface, usually picking up by leaching additional chemicals that occur naturally in the formation. The composition of fracking fluids varies widely, depending primarily on the geology of the area being subjected to hydraulic fracturing. However, these fluids generally consist of a variety of chemicals and high levels of dissolved solids. The solutions are relatively clean prior to introduction into the drill holes, but the post-use solutions contain many more components, including high levels of dissolved organic constituents. As a result, analysis is challenging but, nevertheless, because of the complexity of the fluids, it is important to determine the metal content of both pre-use and post-use fracturing fluids. These analyses are used to evaluate how often the fluids can be reused and the measures that must be taken for safe disposal (Elliott et al., 2016).

The complexity of the water may be increased by reuse. For example, prior to reuse, the water is desalinated to remove salts, which can inhibit the effectiveness of additives and are known to cause scaling. For example, cations such as calcium, barium, and strontium may cause scale formation in water pumps and pipes, thereby reducing extraction efficiency. To counter this, additional antiscaling agents would need to be added to the fracking fluid prior to reuse. Occasionally organic acid derivatives (such as formate and acetate) are added to control the pH (acidity-basicity). These organic acids are sources of carbon for bacterial growth, and bacterial growth in hydraulic fracturing wastewaters can result in the production of the toxic and corrosive hydrogen sulfide (H_2S). As a result of such possibilities, biocides are added to prevent bacterial growth. A variety of standard test methods are available for water testing (ASTM, 2015; API, 2016) which are instrumental in specifying and evaluating the methods and facilities used in examining the various characteristics of and contaminants in water for health, security, and environmental purposes. These water testing standards allow companies, concerned local government authorities, water distribution facilities, and environmental laboratories to test the quality of water and ensure safe consumption. However, rather than discuss the merits of each standard collection of test methods (Tables 6.3–6.5), only the necessary aspects of testing water will be presented here.

In the simplest sense, water testing includes the application of standards test methods for pH, chlorides, sulfates, nitrates, TDS, and a host of methods that can be used to analyze for the presence of various metals. But, since fluids from hydraulic fracturing operations contain a broad range of chemicals which can vary from low to high concentrations, efforts must be made to determine the chemical composition that are present in the fluid. Following from this, test methods should be applied to establish the lower limits of each constituent which can be measured and the method detection limits (MDLs) determined on the basis of the test method employed for the analysis. Moreover, the accuracy of each analytical method should be assessed by spiking the samples and measuring the recovery of the standard chemicals used for spiking. The accuracy of the method is demonstrated by all recoveries being within narrow limits which may be within 5% of the spiked value. Once the MDLs and the accuracy of the method(s) have been established (which are subject to reproducibility and repeatability) (ASTM, 2015; Speight, 2016b), the fluid samples can be analyzed and the data reported with a high degree of confidence.

Table 6.3 Selected EPA Test Methods for the Detection of Organic Compounds in Water

EPA #	Method
600777113	Environmental Pathways of Selected Chemicals in Freshwater Systems: Part I: Background and Experimental Procedures
600479019	Handbook for Analytical Quality Control in Water and Wastewater Laboratories
625673002	Handbook for Monitoring Industrial Wastewater
815B97001	Manual for the Certification of Laboratories Analyzing Drinking Water Criteria and Procedures Quality Assurance
815D03008	Membrane Filtration Guidance Manual
821R96013	Method 1632: Determination of Inorganic Arsenic in Water by Hydride Generation Flame Atomic Absorption
821R95031	Method 1638: Determination of Trace Elements in Ambient Waters by Inductively Coupled Plasma-Mass Spectrometry
821R96005	Method 1638: Determination of Trace Elements in Ambient Waters by Inductively Coupled Plasma-Mass Spectrometry
821R96006	Method 1639: Determination of Trace Elements in Ambient Waters by Stabilized Temperature Graphite Furnace Atomic Absorption
821R96007	Method 1640: Determination of Trace Elements in Ambient Waters by On-Line Chelation Preconcentration and Inductively Coupled Plasma-Mass Spectrometry
600479020	Methods for Chemical Analysis of Water and Wastes
821B96005	Methods for Organic Chemical Analysis of Municipal and Industrial Wastewater
600488039	Methods for the Determination of Organic Compounds in Drinking Water

Table 6.4 Selected EPA Analytical Methods for the Analysis of Contaminants in Water

Method	Technique	Title
150.1	pH, Electrometric	Methods for Chemical Analysis of Water and Wastes (EPA/600/4-79/020)
200.7 Rev 4.4	Metals and Trace Elements by ICP/Atomic Emission Spectrometry	Methods for the Determination of Metals in Environmental Samples Supplement 1 (EPA/600/R-94/111)
200.8 Rev 5.4	Trace Elements by ICP/Mass Spectrometry	Methods for the Determination of Metals in Environmental Samples Supplement 1 (EPA/600/R-94/111)

Table 6.4 Selected EPA Analytical Methods for the Analysis of Contaminants in Water—cont'd

Method	Technique	Title
200.9 Rev 2.2	Trace Elements by Stabilized Temperature Graphite Furnace AA Spectrometry	Methods for the Determination of Metals in Environmental Samples Supplement 1 (EPA/600/R-94/111)
245.1 Rev 3.0	Mercury by Cold Vapor AA Spectrometry—Manual	Methods for the Determination of Metals in Environmental Samples Supplement 1 (EPA/600/R-94/111)
245.2	Mercury by Cold Vapor AA Spectrometry—Automated	Methods for Chemical Analysis of Water and Wastes (EPA/600/4-79/020)
300.0 Rev 2.1	Inorganic Anions by Ion Chromatography	Methods for the Determination of Inorganic Substances in Environmental Samples (EPA/600/R-93/100)
300.1 Rev 1.0	Determination of Inorganic Anions in Drinking Water by Ion Chromatography	Methods for the Determination of Organic and Inorganic Compounds in Drinking Water, Volume 1 (EPA 815-R-00-014)
375.2 Rev 2.0	Sulfate by Automated Colorimetry	Methods for the Determination of Inorganic Substances in Environmental Samples (EPA/600/R-93/100)
525.2 Rev 2.0	Organic Compounds by Liquid-Solid Extraction and Capillary Column GC/Mass Spectrometry	Methods for the Determination of Organic Compounds in Drinking Water-Supplement III (EPA/600/R-95-131)
550	Polycyclic Aromatic Hydrocarbons (PAHs) by Liquid-Liquid Extraction and HPLC With Coupled UV and Fluorescence Detection	Methods for the Determination of Organic Compounds in Drinking Water Supplement I (EPA/600/4-90/020)
550.1	Polycyclic Aromatic Hydrocarbons (PAHs) by Liquid-Solid Extraction and HPLC With Coupled UV and Fluorescence Detection	Methods for the Determination of Organic Compounds in Drinking Water Supplement I (EPA/600/4-90/020)

Table 6.5 Selected Test Methods for Drinking Water Contaminants

Contaminant	EPA	ASTM[a]	AWWA[b]
Aluminum	200.7, 200.8, 200.9		3120 B, 3113 B, 3111 D
Chloride	300.0	D 4327, D 512	4110 B, 4500-Cl-D, 4500-Cl-B
Color			2120 B
Copper	200.7, 200.8, 200.9	D 1688, D 1688	3120 B, 3111 B, 3113 B
Fluoride	300.0	D 4327, D 1179	4110 B, 4500-F-B, D, 4500-F-C, 4500-F-E
Foaming agents			5540 C
Iron	200.7, 200.9		3120 B, 3111 B, 3113 B
Manganese	200.7, 200.8, 200.9		3120 B, 3111 B, 3113 B
Odor			2150 B
PH	150.1, 150.2	D 1293	4500-H+B
Silver	200.7, 200.8, 200.9		3120 B, 3111 B, 3113 B
Sulfate	300.0, 375.2	D 4327, D 516	4110 B, 4500-SO_4 F, 4500-SO_4 C, D, 4500-SO_4 E
Total dissolved solids			2540 C
Zinc	200.7, 200.8		3120 B, 3111 B

[a]ASTM International, Annual Book of ASTM Standards, 2004, West Conshohocken, Pennsylvania.
[b]AWWA, Standard Methods for the Examination of Water and Wastewater, American Water Works Association, Washington, DC.

Quarter-hour, the issues of *quality assurance* (QA) and *quality control* (QC), must also be addressed during the analytical program. Within the environmental industry there seems to have been a great deal of confusion between the terms QA and QC. QA is an umbrella term that is correctly applied to everything that the laboratory does to assure product reliability. As the product of a laboratory is information, anything that is done to improve the reliability of the generated information falls under QA. QCs are single procedures that are performed in conjunction with the analysis to help assess in a quantitative manner the success of the individual analysis. Examples of QCs are blanks, calibration, calibration verification, surrogate additions, matrix spikes, laboratory control samples, performance evaluation samples, and determination of detection limits. The success of the QC is evaluated against an acceptance limit. The actual generation of the acceptance limit is a function of QA; it would not be termed a QC (US EPA, 2015b).

QA includes all the QCs, the generation of expectations (acceptance limits) from the QCs, plus a great number of other activities. A few examples of these other activities include analyst training and certification; data review and evaluation; preparation of final reports of analysis; information given to clients about what tests are needed to fulfill regulatory requirements; use of the appropriate tests in the laboratory; obtaining and maintaining laboratory certifications/accreditations; conducting internal and external audits; preparing responses to the audit results; the receipt, storage, and tracking of samples; and how—and from where—standards and reagents are purchased. The performance of QC is just one small aspect of the QA program.

The functions of QA are embodied in the terms *analytically valid* and *legally defensible*. Analytically valid means that the target analyte has been: (1) correctly identified, (2) quantified using fully calibrated tests, (3) the sensitivity of the test (MDL) has been established, (4) analysts have demonstrated that they are capable of performing the test, (5) the accuracy and precision of the test on the particular sample has been determined, and the possibility of false positive and false negative results has been evaluated through performance of blanks and other test-specific interference procedures.

REFERENCES

API, 2009. Hydraulic Fracturing Operations—Well Construction and Integrity Guidelines. Guidance Document HF1, first ed. American Petroleum Institute, Washington, DC.

API, 2010. Water Management Associated With Hydraulic Fracturing. API Guidance Document HF2. American Petroleum Institute, Washington, DC.

API, 2016. Water quality. http://www.api.org/oil-and-natural-gas/environment/clean-water (accessed Apr. 2016).

ASTM, 2015. Annual Book of Standards. ASTM International, West Conshohocken, PA.

Blauch, M.E., Myers, R.R., Moore, T.R., Houston, N.A., 2009. Marcellus shale post-frac flowback waters—where is all the salt coming from and what are the implications? In: Proceedings of the SPE Regional Meeting, Charleston, West Virginia, Sep. 23–25. Paper No. SPE 125740.

Clark, C., Han, J., Burnham, A., Dunn, J., Wang, M., 2011. Life-Cycle Analysis of Shale Gas and Natural Gas. Argonne National Laboratory, Argonne, IL. Report No. ANL/ESD/11-11.

Elliott, E.G., Ettinger, A.S., Leaderer, B.P., Bracken, M.B., Deziel, N.C., 2016. A systematic evaluation of chemicals in hydraulic-fracturing wastewater for reproductive and developmental toxicity. J. Expo. Sci. Environ. Epidemiol. 1–10.

Halldorson, B., Horner, P., 2012. Shale gas water management. World Petroleum Council Guide: Unconventional Gas. World Petroleum Council, London, United Kingdom. pp. 58–63, http://www.world-petroleum.org/docs/docs/gasbook/unconventionalgaswpc2012.pdf (accessed Mar. 15, 2016).

IEA, 2013. Resources to Reserves 2013: Oil, Gas and Coal Technologies for the Energy Markets of the Future. OECD Publishing, International Energy Agency, Paris, France.

King, C.W., Webber, M.E., 2008. Water intensity of transportation. Environ. Sci. Technol. 42 (21), 7866–7872.

KPMG, 2012. Watered-Down: Minimizing Water Risks in Natural Gas and Crude Oil and Oil Drilling. KPMG Global Energy Institute, KPMG International, Houston, TX.

Lee, S., 1991. Oil Shale Technology. CRC Press, Taylor & Francis Group, Boca Raton, FL.

Lee, S., Speight, J.G., Loyalka, S.K., 2007. Handbook of Alternative Fuel Technologies. CRC-Taylor & Francis Group, Boca Raton, FL.

Lester, Y., Ferrer, I., Thurman, E.M., Sitterley, K.A., Korak, J.S., Kinden, K.G., 2015. Characterization of hydraulic fracturing flowback water in Colorado: implications for water treatment. Sci. Total Environ. 512–513, 637–644.

Mantell, M.E., 2009. Deep shale natural gas: abundant, affordable, and surprisingly water efficient. In: Proceedings of the 2009 GWPC Water/Energy Sustainability Symposium. Salt Lake City, Utah, Sep. 13–16.

Satterfield, J., Mantell, M., Kathol, D., Hiebert, F., Patterson, K., Lee, R., 2008. Managing water resource's challenges in select natural gas shale plays. In: Proceedings of the GWPC Annual Meeting, September. Groundwater Protection Council, Oklahoma City, OK.

Scouten, C.S., 1990. Oil shale. In: Speight, J.G. (Ed.), Fuel Science and Technology Handbook. Marcel Dekker Inc., New York, pp. 795–1053. Chapters 25–31.

Speight, J.G., 2012. Shale Oil Production Processes. Gulf Professional Publishing, Elsevier, Oxford, United Kingdom.

Speight, J.G., 2014a. The Chemistry and Technology of Petroleum, fifth ed. CRC Press, Taylor & Francis Group, Boca Raton, FL.

Speight, J.G., 2014b. High Acid Crudes. Gulf Professional Publishing, Elsevier, Oxford, United Kingdom.

Speight, J.G., 2016a. Introduction to Enhanced Recovery Methods for Heavy Oil and Tar Sands, second ed. Gulf Publishing Company, Taylor & Francis Group, Waltham, MA.

Speight, J.G., 2016b. Handbook of Petroleum Product Analysis, second ed. John Wiley & Sons Inc., Hoboken, NJ

Stillwell, A.S., King, C.W., Webber, M.E., Duncan, I.J., Herzberger, A., 2010. The energy-water nexus in Texas. Ecol. Soc. 16 (1), 2.

US EPA, 2015a. Analysis of Hydraulic Fracturing Fluid Data from the FracFocus Chemical Disclosure Registry 1.0. In: United States Environmental Protection Agency, Washington, DC. Report No. EPA/601/R-14/003.

US EPA, 2015a. Underground injection control regulations and safe drinking water act provisions. https://www.epa.gov/uic/underground-injection-control-regulations-and-safe-drinking-water-act-provisions (accessed May 1, 2016).

US GAO, 2012. Information on the Quantity, Quality, and Management of Water Produced During Oil and Gas Production. United States Government Accountability Office, Washington, DC. Report No. GAO-12-256.

Veil, J.A., 2010. Water Management Technologies Used by Marcellus Shale Gas Producers. Argonne National Laboratory, United states Department of Energy, Argonne, IL/ Washington, DC. Report No. ANL/EVR/R-10/3.

Wu, M., Mintz, M., Wang, M., Arora, S., Chiu, Y., 2011. Consumptive Water Use in the Production of Ethanol and Petroleum Gasoline—2011 Update. Argonne National Laboratory, Argonne, IL. Report No. ANL/ESD/09-1.

CHAPTER SEVEN

Properties Processing of Gas From Tight Formations

1 INTRODUCTION

Natural gas production from shale reservoirs and other tight reservoirs has been proven to be feasible from the numerous operations in various tight reservoirs in North America, but many challenges still remain for the full exploitation of these unconventional reservoirs. The production process requires stimulation by horizontal drilling coupled with hydraulic fracturing because of the extremely low permeability of the reservoir rocks which prohibits natural movement of the gas to a well. Moreover, maximization or optimization of reservoir producibility can only be achieved by a thorough understanding of the occurrence and properties of the shale gas resources (Chapters 1 and 2) as well as the producibility of the gas from the reservoir (Chapter 3) (Kundert and Mullen, 2009). These needs demonstrate the importance of a thorough characterization of shale gas reservoir (Table 7.1) as well as an understanding of the means by which subterranean formations deform during geologic time as well as the effects of such deformation on the stresses within the formations.

As a refresher on the issue of nomenclature (Chapter 1), there are several general definitions that have been applied to natural gas from conventional formations that can also be applied to gas from tight formations. Thus, lean gas is gas in which methane is the major constituent. Wet gas contains considerable amounts of the higher molecular weight hydrocarbons. Sour gas contains hydrogen sulfide, whereas sweet gas contains very little, if any, hydrogen sulfide. Residue gas is natural gas from which the higher-molecular-weight hydrocarbons have been extracted and casinghead gas is derived from crude oil but is separated at the separation facility at the wellhead. Gas condensate (sometimes referred to as condensate) is a mixture of low-boiling hydrocarbon liquids obtained by condensation of hydrocarbon vapors.

http://dx.doi.org/10.1016/B978-0-12-803097-4.00007-3

Table 7.1 Variation in Shale Properties for Different Shale Gas Reservoirs

Sample Source (g/cc)	Density (% w/w)	Carbonate (% w/w)	Clay (% v/v)	Porosity (% w/w)	Kerogen
Barnett	2.37–2.67	0–60	3–39	1–9	2–11
Haynesville	2.49–2.62	20–53	20–39	4–8	3–6
Eagle Ford	2.43–2.54	46–78	6–21	0–5	4–11

Kerogen refers to undefined organic material, typically insoluble in organic solvents (Lee, 1991; Scouten, 1990; Speight, 2008, 2012, 2013)

The condensate is predominately propane (C_3H_8), butane (C_4H_{10}), and pentane (C_5H_{12}) with minor amounts of higher-boiling hydrocarbons (up to C_8H_{18}) but relatively little methane or ethane. Depending upon the source of the condensate, benzene (C_6H_6), toluene ($C_6H_5CH_3$), xylene isomers ($CH_3C_6H_4CH_3$), and ethyl benzene ($C_6H_5C_2H_5$) may also be present (Table 7.1).

The shale gas resources (Chapter 2) represent a major contribution to the resource base of the United States (Nehring, 2008). However, it is important to note that there is considerable variability in the quality of the resources, both within and between gas shale resources. Elevated levels of ethane, propane, carbon dioxide, or nitrogen in certain gases from shale formations are of concern regarding their interchangeability with traditional natural gas supplies. This high level of variability in individual well productivity clearly has consequences with respect to the variability of individual well economic performance. Not all shale gas plays are the same and, as a result, shale gas from one play can be considerably different (in terms of the necessary gas processing requirements), and the processing requirements for shale gas can vary from area to area. The more popular areas are the Barnett, Haynesville, and Fayetteville shales in the South and the Marcellus, New Albany, and Antrim shales in the East and Midwest (Fig. 7.1). These plays represent a large portion of current and future gas production.

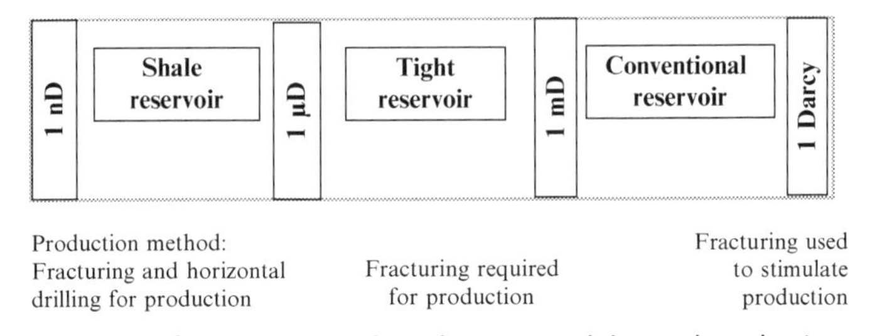

Fig. 7.1 Illustration of reservoir types based on permeability and production methods.

However, because shale is composed primarily of tiny grains of clay minerals and quartz, the mineral components of mud, the composition of the shale reservoir can vary—specifically the rock properties such as porosity, permeability, capillary entry pressure, pore volume compressibility, pore size distribution, and flow path (collectively known as petrophysics) can vary considerably (Sone, 2012). These materials were deposited as sediment in water, which was then buried, compacted by the weight of overlying sediment, and cemented together to form a rock (lithification). Clay minerals are a type of sheet silicate related to mica that usually occurs in the form of thin plates or flakes. As the sediment was deposited, the flakes of clay tended to stack together flat, one on top of another like a deck of cards, and as a result, lithified shale often has the property of splitting into paper-thin sheets. This is a convenient way to identify shale from other fine-grained rocks like limestone or siltstone.

This process, while appearing to be ordered in such a description, is in fact subject to geological disorder and then subject to differing methods of entrapment of the organic material, and different methods or rates of decompositon, different rates of formation of the gas, and hence different composition of the shale gas. The primary point of this subsection is that the geochemical and geological characteristics of each reservoir are relatively unique and must be carefully examined to determine resources. Furthermore, general rules are difficult to apply—innovation in unconventional drilling and completion techniques has added substantial reserves to otherwise uneconomic areas and has also been a key part of safe and efficient development (Cramer, 2008; Grieser et al., 2007). This can affect the economics of shale gas production and gas cleaning (Mokhatab et al., 2006; Speight, 2007; Kidnay et al., 2011).

A major driver of shale economics is the amount of hydrocarbon liquid produced along with gas. Some areas contain wet gas with appreciable amounts of liquid, which can have a considerable effect on the breakeven economics, particularly if the price of oil is high compared to the price of gas. The liquid content of a gas is often measured in terms of the condensate ratio, which is expressed in terms of barrels of liquid per million cubic feet of gas (bbls/MMcf). In some operations, for example, in the case of a condensate ratio in excess of 50 barrels per million cubic feet (50 bbls/MMcf), the liquid production alone can provide an adequate return on the investment, even if the gas cannot realize a fair market value.

It is the purpose of this chapter to present a review of the various gas-processing technologies are appropriate for the variety of gas qualities being produced and planned to be produced. In particular, the focus of this chapter is a brief description of the processes that are an integral part within the concept of production of a pipeline-able product (methane) for sale to the consumer.

2 TIGHT GAS AND GAS COMPOSITION

Briefly, and by way of reintroduction, a gas shale reservoir has an extremely low permeability compared to most conventional reservoirs (typically a sandstone reservoir) (Chapter 2). In fact, the effective bulk permeability in a tight shale reservoir is typically much less than 0.1 milliDarcy—on the order of 1 milliDarcy down to 1 microDarcy (Fig. 7.1), although exceptions exist where the rock is naturally fractured as has been observed in the well-fractured Antrim shale in the Michigan Basin of the United States. However, the role of natural microfractures in producing natural gas or crude oil from a tight reservoir or in assisting artificial fracturing is not fully understood (Chapters 4 and 5). In most cases, it is usual for the reservoir to require artificial stimulation, such as horizontal drilling leading to hydraulic fracturing (Fig. 7.1), to increase permeability to the well in order to produce gas in economical quantities.

However, producing the gas from the reservoir is not the end of the story. Other issues (such as variations in gas composition) arise with the ensuing effects of changing composition on gas-processing operations.

Shale gas is a valuable natural resource that is being eagerly sought out and drilled for using fracturing techniques. The gas is mainly composed of methane, although it does contain constituents that to be separated from the methane to make the gas usable commercially. The other compounds found in shale gas include natural gas liquids (NGLs), which are hydrocarbons of a heavier nature that will be separated in processing plants as liquids. These liquids include ethane (C_2H_6), propane (C_3H_8), butane (C_4H_{10}), pentane (C_5H_{12}), hexane (C_6H_{14}), heptane (C_7H_{16}), and octane (C_8H_{18}) as well as a higher-boiling hydrocarbon mixture (gas condensate, which may include the aforementioned C_{6+} hydrocarbons) and water. The gaseous components of raw shale gas include sulfur dioxide, hydrogen sulfide, helium, nitrogen, and carbon dioxide. Mercury may also be found in smaller concentrations in most reservoirs where natural gas is obtained. The mercury found will be lowered in concentration until it is below the detectable threshold of one part per trillion. Gas in shale deposits varies in composition insofar as it, but in different areas, it may contain different amounts of the same basic compounds. The variability in the composition of natural gas composition depends not only on the region in which the gas originated but also on the properties of the reservoir (Table 7.2).

Tight gas as it is bought to the surface (wellhead gas) is different to the gas that the consumers receive and is much less pure. In fact, differences in natural gas composition occur between different reservoirs, and two wells in the same field may also yield gaseous products that are different in composition (Mokhatab et al., 2006; Speight, 2007; Kidnay et al., 2011). Indeed, there is no single composition of components which might be termed typical natural gas no matter the source of the gas. As with gas from conventional reservoirs (Table 7.3), methane (with varying amounts of and ethane) constitutes the bulk of the combustible components; carbon dioxide (CO_2) and nitrogen (N_2) are the major noncombustible (inert) components (Table 7.4). Other constituents such as hydrogen sulfide (H_2S), mercaptans (thiols; R-SH), as well as trace amounts of other sulfur-containing constituents may also be present.

Table 7.2 Typical Composition (% v/v) of Fluids Produced From Gas Reservoirs

Component	Dry Gas	Wet Gas	Condensate
CO_2	0.1	1.4	2.4
N_2	2.1	0.3	0.3
C_1	86.1	92.5	73.2
C_2	5.9	3.2	7.8
C_3	3.6	1.1	3.6
C_4	1.7	0.3	0.7
C_5	0.5	0.1	0.6
C_6	–	0.1	1.1
C_{7+}	–	0.8	8.2

Table 7.3 Range of Composition of Natural Gas

Methane	CH_4	70–90%
Ethane	C_2H_6	0–20%
Propane	C_3H_8	
Butane	C_4H_{10}	
Pentane and higher hydrocarbons	C_5H_{12}	0–10%
Carbon dioxide	CO_2	0–8%
Oxygen	O_2	0–0.2%
Nitrogen	N_2	0–5%
Hydrogen sulfide, carbonyl sulfide	H_2S, COS	0–5%
Rare gases: Argon, Helium, Neon, Xenon	A, He, Ne, Xe	Trace

Table 7.4 Variation in Composition of Shale Gas From Different Formations (Martini et al., 2003; Hill et al., 2007; Bullin and Krouskop, 2008)*

Source of Gas	Methane (% v/v)	Ethane (% v/v)	Propane (% v/v)	Carbon Dioxide (% v/v)	Nitrogen (% v/v)
Antrim shale	27–86	3–5	0.4–1	0–9	1–65
Barnett shale	80–94	2–12	0.3–3	0.3–3	1–8
Fayetteville shale	97	1	0	1	1
Haynesville shale	95	<1	0	5	0.1
Marcellus shale	79–96	3–16	1–4	0.1–1	0.2–0.4
New Albany shale	97–93	1–2	0.6–3	5–11	0**

*Samples were taken from different wells within each formation and a range of composition indicates variations in composition from different wells within a shale formation.
**The yields of hydrocarbon constituents having a higher molecular weight than propane was not recorded.

2.1 Tight Gas

Tight gas refers to natural gas (mainly methane) found in fine-grained, organic-rich rocks (shale, sandstone, and carbonate rocks) (Chapters 1 and 2). In addition, the descriptor word shale does not refer to a specific type of rock but, in addition to shale (mudstone), has also been used to describes rocks with more fine-grained particles (smaller than sand) than coarse-grained particles, such as (1) siltstone, (2) fine-grained sandstone interlaminated with shale, and (3) carbonate rocks. Thus, shale (including the additional rock types above (tight sand formations and tight carbonate formations) is a source rock that has not released all of the generated hydrocarbons. In fact, source rocks that are tight or inefficient at expelling hydrocarbon gases (and liquids) may be the best prospects for shale gas (or liquid) potential. Thus, in gas shale, shale is a reservoir rock, source rock, and also a trap for natural gas. The natural gas found in these rocks is considered unconventional, similar to coalbed methane.

Shale gas is generated by any combination of (1) primary thermogenic degradation of organic matter, (2) secondary thermogenic decomposition of petroleum, and (3) biogenic degradation of organic matter. Gas generated by thermogenic and biogenic pathways may both exist in the same shale reservoir. After generation, the gas is stored in the shale (tight) formation in three different ways: (1) by adsorption, which refers to adsorbed gas that is physically attached (adsorption) or chemically attached (chemisorption) to organic matter or to clay minerals, (2) nonadsorbed gas, which refers to free gas (also referred to as nonassociated gas) that occurs within the pore spaces in the reservoir rock or in spaces created by the rock cracking (fractures or microfractures), and (3) by solution, which refers to gas (also referred to as associated gas) that exists in solution in liquids such as petroleum, heavy oil, and (in the current context) in gas condensate that occurs in some tight reservoirs with the gas. The amount of adsorbed gas component (typically, methane) usually increases with an increase in organic matter or surface area of organic matter and/or clay. On the beneficial side, a higher free-gas (non-associated) content in unconventional tight reservoirs generally results in higher initial rates of production because the free gas resides in fractures and pores and (when production is commenced) moves easier through the fractures (induced channels) relative to any adsorbed gas. However, the high, initial flow rate will decline rapidly to a low, steady rate as the non-associated gas is produced leaving the adsorbed gas to move to the well as it is slowly released from the shale.

2.2 Gas Composition

Gas composition is a critical issue when choosing the most appropriate methods for gas processing (Mokhatab et al., 2006; Speight, 2007; Kidnay et al., 2011). Indeed, this is an issue that must be resolved if shale gas is to be a major contributor to the US energy resources (or to the energy resource plans of any country). The issue is the amenability of the gas to be included in current gas-processing scenarios (Mokhatab et al., 2006; Speight, 2007; Kidnay et al., 2011). On the understanding that shale gas reservoirs will vary in properties such as origin, permeability, and porosity (Table 7.1), differences in properties of the shale gas, not only from different formations but also even depending upon placement of the well within a formation (Table 7.4) must be anticipated (Chapters 1 and 2) (Bustin et al., 2008).

The composition of natural gas from tight formations is influenced by several variables: (1) the composition of the organic matter in the formation from which the gas is sourced, (2) the thermal maturity at which the gas is generated from the source rock, (3) the gas generation process, specifically whether gas is generated through primary cracking of kerogen, from secondary cracking of oil to gas, or from secondary cracking of wet gas to dry gas, (4) fractionation during gas migration from source to reservoir,

(5) leakage from the reservoir, and (6) bacterial alteration by oxidation of gases. In addition, the influence of the minerals in the tight formation on gas composition by hastening, delaying, or changing any of the maturation reactions cannot be discounted and in fact should be given consideration. The outcome of gas formation is that the multicomponent gases can vary in composition to such an extent that, while some of these processes may be difficult to identify and describe, the end result is the effect that gas composition can have on the options for gas processing.

In addition, gas from tight formation can, by virtue of the immobility of the gas in the low-permeability formation, can exhibit distinct vertical and horizontal variation in composition (Harris et al., 2013). Tight gas has also been claimed to contain less water with depth (i.e., the deeper gas is dryer than the shallower gas), but the composition of the gas may still have a wide range of bulk gas compositions at a given depth. In some cases, there is a distinct shift to a lower gas density at greater depths indicating a preponderance of the lower-molecular-weight hydrocarbon constituents to the point where some tight gases may be almost pure methane (Harris et al., 2013).

Furthermore in addition to processes that influence the formation of tight gas there are also maturation processes (i.e., distinct chemical processes) that can, and have, have influenced the composition of the tight natural gas, which includes: (1) mixing of gases produced by primary cracking with gases from the secondary cracking of oil; however there is no evidence of gases produced by secondary cracking of wet gas to dry gas, (2) gases produced by primary cracking from similar source rocks at varying thermal maturity, (3) gas composition that is altered by bacterial oxidation of hydrocarbon gases, yielding gases that are dry. Furthermore, the obvious stratigraphic variability of the gas composition in some formations and the variation in gas compositions at particular depths suggest complex migration pathways, probably through fracture systems and localized channeling of migrating gases through relatively permeable beds (Cumella and Scheevel, 2008). However, the variation in gas composition cannot be attributed to the ubiquitous vertical diffusion of gases or to the rapid migration of the gas through fractures that have been induced by high pressure effects (Harris et al., 2013).

Thus, while natural gas resources for tight shale formations and from other tight formations does, indeed, represent a significant portion of current and future production, it is essential to recognize that all shale gas is not constant in composition and gas-processing requirements for shale gas will vary depending upon the area from which the gas was produced (Bullin and Krouskop, 2008). In addition, analysis of the composition of the gas from of Devonian shale wells indicates that the composition of produced gas shifts during the production history of the well (Schettler et al., 1989). This may serve to indicate that the composition of the gas changes during production because the different components of natural gas produced have different decline curves and that the total decline curve is the sum of the decline curves of the individual components of the gas. This also suggests that there is fractionation of the gas within the reservoir due to adsorption or absorption phenomena and that the classic mechanisms of viscous flow and ideal-gas void-volume storage do not explain fractionation of the gas within the reservoir and even in the production wells. In fact, in order to explain this apparent fractionation is to assume that total gas from the reservoir to the surface facility arises from several sources within the reservoir, and the gas from each source produced gas has a different compositions and, hence, a decline curve. Moreover, the flow rate of an individual component of a gas mixture can be obtained by multiplying the total flow rate by the mole fraction of the individual component and even if the composition of gas from each source is assumed to remain constant, the composition of the natural gas at the wellbore will change with time if the relative flow rate changes with time (Schettler et al., 1989).

Another alternative that can be used to describe the changing composition of the gas is to assume that the gas composition changes observed at the wellbore reflect changing composition of gas from at least some of the sources themselves. Candidates for such differences in gas composition include: (1) adsorption of the gas components on the reservoir rock, (2) adsorption of the gas components on any nongaseous organic material in the reservoir, (3) solution of the gas components in any organic or inorganic liquids in the reservoir, and (4) absorption and diffusion of the gas components within the pore system of the reservoir rock.

The tendency of the components of the gas for adsorption is associated with the presence of certain types of minerals in the reservoir—clay minerals are especially adept at providing surface (sites) for adsorption to occur. Likewise, the presence of organic material can also provide sites for adsorption to occur. Dissolution of the gas components in any organic or inorganic liquids in the reservoir is associated with the presence of conventional petroleum or heavy oil or even gas condensate, while the absorption/diffusion phenomenon is associated with diffusion of the gas components through small pores such as those present in microporous reservoirs and which give passage to gas components having a specific molecular size. Because these factors are commonly present, any factor or a combination of factors may be involved in explaining fractionation in many reservoirs (Schettler et al., 1989; Mokhatab et al., 2006; Speight, 2007; Kidnay et al., 2011). Thus, as reservoir depletion occurs, the composition of the gas produced may approach the composition of the gas originally in the reservoir. Thus, composition shifts during production will be expected but may not be observed because of the site specificity of the phenomenon. The site specificity arises because of differences such as, but not limited to: (1) gas composition, (2) reservoir mineralogy, (3) reservoir temperature, and (4) reservoir pressure.

3 GAS PROCESSING

Gas processing (gas treating, gas refining) removes one or more components from the recovered (harvested) gas to prepare the gas for use. To include a description of all of the possible process for gas cleaning is beyond the scope of this book.

Gas processing is an instrumental piece of the natural gas value chain and is required to ensure that the natural gas intended for use is as clean and pure as possible, making it the clean burning and environmentally sound energy choice. In the processing sequence, common components removed to meet pipeline, safety, environmental, and quality specifications include hydrogen sulfide, carbon dioxide, nitrogen, hydrocarbons having a higher molecular weight than methane, and water (Mokhatab et al., 2006; Speight, 2007; Kidnay et al., 2011). Once the natural gas has been fully processed and is ready to be consumed, it must be transported from those areas that produce natural gas to those areas that provide a market for natural gas.

Whatever the source (tight shale, tight sandstone, tight carbonate), natural gas needs to be processed before use because of the occurrence of the diluents and contaminants. The processes focus on sulfur removal and carbon dioxide removal. The processes that have been developed to accomplish gas purification vary from a simple once-through wash operation to complex multistep recycling systems. In many cases, process complexities arise because of the need for recovery of the materials used to remove the contaminants or even recovery of the contaminants in the original, or altered, form. However, process complexities in terms of the choice of the process also arise because of the varying composition of gas from tight formations. In fact, there are many variables in treating natural gas from tight formations, and the precise area of application of a given process may be difficult to define. In addition to hydrogen sulfide and carbon dioxide, gas may contain other contaminants, such as mercaptans (R-SH) and carbonyl sulfide (COS). The presence of these impurities may eliminate some of the sweetening processes since some processes remove large amounts of acid gas but not to a sufficiently low concentration. On the other hand, there are those processes that are not designed to remove (or are incapable of removing) large amounts of acid gases. However, these processes also capable of removing the acid gas impurities to very low levels when the acid gases are there in low to medium concentrations in the gas. Process selectivity indicates the preference with which the process removes one acid gas component relative to (or in preference to) another. For example, some processes remove both hydrogen sulfide and carbon dioxide; other processes are designed to remove hydrogen sulfide only. It is important to consider the process selectivity for, say, hydrogen sulfide removal compared to carbon dioxide removal that ensures minimal concentrations of these components in the product, thus the need for consideration of the carbon dioxide to hydrogen sulfide in the gas stream.

In a general processing sense, many chemical processes and physical process are available for processing or refining natural gas (Mokhatab et al., 2006; Speight, 2007; Kidnay et al., 2011). However, there are many variables in the choice of process of the choice of refining sequence that dictate the choice of process or processes to be employed. In this choice, several factors must be considered: (1) the types and concentrations of contaminants in the gas; (2) the degree of contaminant removal desired; (3) the selectivity of acid gas removal required; (4) the temperature, pressure, volume, and composition of the gas to be processed; (5) the carbon dioxide-hydrogen sulfide ratio in the gas; and (6) the desirability of sulfur recovery due to process economics or environmental issues.

The processes that have been developed to accomplish gas purification vary from a simple once-through wash operation to complex multistep recycling systems (Mokhatab et al., 2006; Speight, 2007; Kidnay et al., 2011). In many cases, the process complexities arise because of the need for recovery of the materials used to remove the contaminants or even recovery of the contaminants in the original, or altered, form (Kohl and Riesenfeld, 1985; Newman. 1985; Mokhatab et al., 2006). In addition, it is often necessary to install scrubbers and heaters at or near the wellhead that serve primarily to remove sand and other large-particle impurities. The heaters ensure that the temperature of the natural gas does not drop too low and form a hydrate with the water vapor content of the gas stream (Mokhatab et al., 2006; Speight, 2007; Kidnay et al., 2011).

However, the processes applied to gas processing are subject to several variables that must of necessity be considered: (1) the types of contaminants in the gas, (2) the concentrations of contaminants in the gas, (3) the degree of contaminant removal desired, (4) the selectivity of acid gas removal required, (5) the temperature of the gas to be processed, (6) the pressure of the gas to be processed, (7), the volume of the gas to be processed, (8) the composition of the gas to be processed, (9) the carbon dioxide-hydrogen sulfide ratio in the gas, and (10) the desirability of sulfur recovery due to process economics or environmental issues.

In addition to hydrogen sulfide (H_2S) and carbon dioxide (CO_2), gas may contain other contaminants, such as mercaptans (RSH) and carbonyl sulfide (COS). The presence of these impurities may eliminate some of the sweetening processes since some processes remove large amounts of acid gas but not to a sufficiently low concentration. On the other hand, there are those processes that are not designed to remove (or are incapable of removing) large amounts of acid gases. However, these processes also capable of removing the acid gas impurities to very low levels when the acid gases are there in low to medium concentrations in the gas.

The process units employed to remove the unwanted constituents with the composition of the gas stream. For example, acid-gas removal is commonly by absorption of the hydrogen sulfide, carbon dioxide into aqueous solutions od amine derivatives (such as glycolamine, 2-aminoethanol, 2-hydroxyethylamine, $HOCH_2CH_2NH_2$). This process performs admirably for high-pressure gas streams and those with moderate to high concentrations of the acid-gas components (hydrogen sulfide and carbon dioxide). Physical solvents such as methanol (CH_3OH) or Selexol may also be used in some cases—the Selexol solvent is a mixture of the dimethyl ethers of polyethylene glycol. Removal of acid gases is especially relevant if the tight gas originates from a carbonate ($CaCO_3$) formation.

The Selexol process, unlike the amine-based processes, is a physical solvent that does not rely on a chemical reaction with the acid gases. Since no chemical reactions are involved, Selexol usually requires less energy than the amine-based processes. However, at feed gas pressures below about 300 psi, the Selexol solvent capacity (in amount of acid gas absorbed per volume of solvent) is reduced and the amine-based processes will usually be superior. In the process, the Selexol solvent dissolves (absorbs) the acid gases from the feed gas at relatively high pressure, usually 300–2000 psi. The rich solvent containing the acid gases is then let down in pressure and/or steam stripped to release and recover the acid gases. The Selexol process can operate selectively to recover hydrogen sulfide and carbon dioxide as separate streams so that the hydrogen sulfide can be sent to either a Claus unit for conversion to elemental sulfur wet sulfuric acid process (WSA process) unit for conversion to sulfuric acid while, at the same time, the carbon dioxide can be sequestered or used for enhanced oil recovery. The Rectisol process, which uses refrigerated methanol as the solvent, is similar in principle to the Selexol. However, if the level of carbon dioxide is high, such as in gas from carbon dioxide flooded reservoirs, membrane technology affords bulk carbon dioxide removal in advance of processing with another method. For minimal amounts of hydrogen sulfide in a gas stream, scavengers can be a cost effective approach to hydrogen sulfide removal.

Gas that becomes saturated with water in the reservoir requires dehydration to increase the heating value of the gas and to prevent pipeline corrosion and the formation of solid hydrates. In most cases, dehydration with a glycol is employed. The water-rich glycol can be regenerated by reducing the pressure and applying heat. Another possible dehydration method is use of molecular sieves that contact the gas with a solid adsorbent to remove the water. Molecular sieves can remove the water down to the extremely low levels required for cryogenic separation processes.

Distillation uses the different boiling points of higher-molecular-weight hydrocarbons and nitrogen for separation. Cryogenic temperatures, required for separation of nitrogen and methane, are achieved by refrigeration and expansion of the gas through an expander. Removal of the heavy hydrocarbons is dictated by pipeline quality requirements, while deep removal is based on the economics of the production of NGLs.

In addition to hydrogen sulfide and carbon dioxide, gas may contain other contaminants, such as mercaptans (also called thiols, R-SH) and carbonyl sulfide (COS). The presence of these impurities may eliminate some of the sweetening processes since some processes remove large amounts of acid gas but not to a sufficiently low concentration. On the other hand, there are those processes that are not designed to remove (or are incapable of removing) large amounts of acid gases. However, these processes also capable of removing the acid gas impurities to very low levels when the acid gases are there in low to medium concentrations in the gas.

Process selectivity indicates the preference with which the process removes one acid gas component relative to (or in preference to) another. For example, some processes remove both hydrogen sulfide and carbon dioxide; other processes are designed to remove hydrogen sulfide only. It is very important to consider the process selectivity for, say, hydrogen sulfide removal compared to carbon dioxide removal that ensures minimal concentrations of these components in the product, thus the need for consideration of the carbon dioxide to hydrogen sulfide in the gas stream.

Gas-processing equipment, whether in the field or at processing/treatment plants, assures that these requirements can be met. While in most cases processing facilities extract contaminants and higher-molecular-weight hydrocarbons NGLs from the gas stream. However, in some cases, the higher-molecular-weight hydrocarbons may be blended into the gas stream to bring it within acceptable Btu levels. Whatever the situation, there is the need to prepare the gas for transportation and use in domestic and commercial furnaces. Thus, natural gas processing begins at the wellhead and since the composition of the raw natural gas extracted from producing wells depends on the type, depth, and location of the underground deposit and the geology of the area, processing must offer several options (even though each option may be applied to a different degree) to accommodate the difference in composition of the extracted gas.

In those few cases where pipeline-quality natural gas is actually produced at the wellhead or field facility (Manning and Thompson, 1991), the natural gas is moved directly to the pipeline system. In other instances, especially in the production of nonassociated natural gas, field, or lease facilities referred to as skid-mount plants are installed nearby to dehydrate (remove water) and decontaminate (remove dirt and other extraneous materials) raw natural gas into acceptable pipeline-quality gas for direct delivery to the pipeline system. The skids are often specifically customized to process the type of natural gas produced in the area and are a relatively inexpensive alternative to transporting the natural gas to distant large-scale plants for processing.

Gas processing (Mokhatab et al., 2006; Speight, 2007; Kidnay et al., 2011) consists of separating all of the various hydrocarbons, nonhydrocarbons (such as carbon dioxide and hydrogen sulfide), and fluids from the methane (Table 7.3; Fig. 7.2). Major transportation pipelines usually impose restrictions on the make-up of the natural gas that is allowed into the pipeline. That means that before the natural gas can be transported it must be purified. While the ethane, propane, butanes, and pentanes must be removed from natural gas, this does not mean that they are all waste products. Gas processing is necessary to ensure that the natural gas intended for use is clean-burning and environmentally acceptable. Natural gas used by consumers is composed almost entirely of methane, but natural gas that emerges from the reservoir at the wellhead is by no means as pure methane

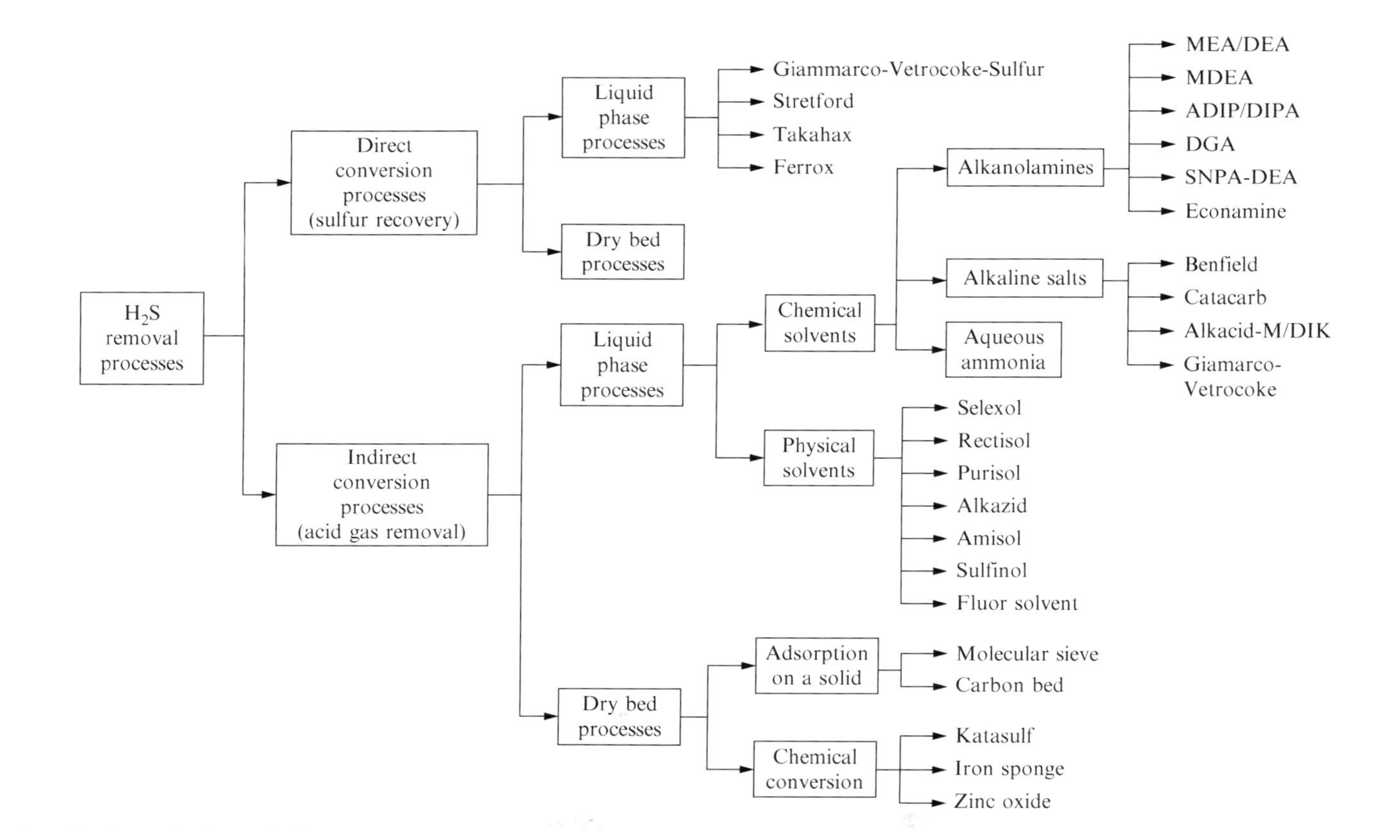

Fig. 7.2 General schematic flow for gas processing.

(Table 7.3; Chapter 3). Although the processing of natural gas is in many respects less complicated than the processing and refining of crude oil, it is equally as necessary before its use by end users.

Raw natural gas comes from three types of wells: oil wells, gas wells, and condensate wells (Chapters 2 and 4). Associated gas (Chapter 1), that is, gas from petroleum wells, can exist separate from oil in the formation (free gas), or dissolved in the crude oil (dissolved gas). Nonassociated gas, that is, gas from gas wells or condensate wells is free natural gas along with a semiliquid hydrocarbon condensate. Whatever the source of the natural gas, once separated from crude oil (if present) it commonly exists in mixtures with other hydrocarbons: principally ethane, propane, butane, and pentanes. In addition, raw natural gas contains water vapor, hydrogen sulfide (H_2S), carbon dioxide, helium, nitrogen, and other compounds. In fact, the associated hydrocarbons (NGLs) can be very valuable by-products of natural gas processing. NGLs include ethane, propane, butane, isobutane, and natural gasoline that are sold separately and have a variety of different uses—including enhancing oil recovery in oil wells, providing raw materials for oil refineries or petrochemical plants, and as sources of energy.

The actual practice of processing natural gas to high quality pipeline gas for the consumer usually involves four main processes to remove the various impurities: (1) water removal, (2) liquids removal, (3) enrichment, (4) fractionation and (5) the process by which hydrogen sulfide is converted to sulfur (the Claus process). Process selectivity indicates the preference with which the process removes one acid gas component relative to (or in preference to) another. For example, some processes remove both hydrogen sulfide and carbon dioxide, while other processes are designed to remove hydrogen sulfide only. It is important to consider the process selectivity for, say, hydrogen sulfide removal compared to carbon dioxide removal that ensures minimal concentrations of these components in the product, thus the need for consideration of the carbon dioxide to hydrogen sulfide ratio in the gas stream (Maddox, 1982; Kohl and Riesenfeld, 1985; Newman, 1985; Soud and Takeshita, 1994).

Initially, natural gas receives a degree of cleaning at the wellhead. The extent of the cleaning depends upon the specification that the gas must meet to enter the pipeline system. For example, natural gas from high-pressure wells is usually passed through field separators at the well to remove hydrocarbon condensate and water (Fig. 7.3). Natural gasoline, butane, and propane are usually present in the gas, and gas-processing plants are required for the recovery of these liquefiable constituents.

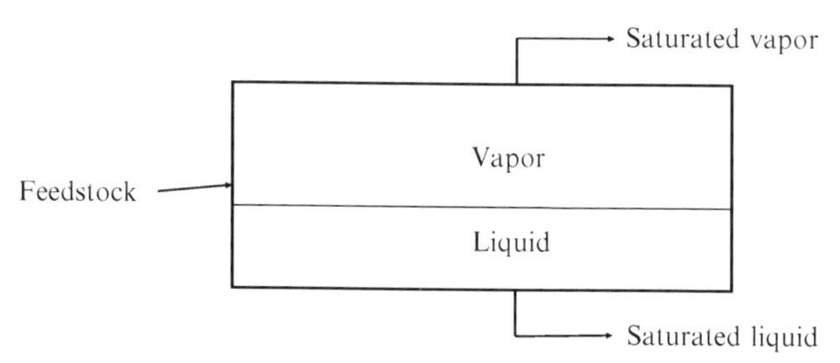

Fig. 7.3 Simplified schematic of s field separator.

An oil/gas separator is a pressure vessel used for separating a wellhead stream into gaseous and liquid components, and the separator is installed close to the wellhead and can be divided into horizontal, vertical, or spherical separators. In teams of fluids to be separated, the oil/gas separators can be grouped into gas/liquid two-phase separator or oil/gas/water three-phase separator. In order to meet the various process requirements, the oil/gas separators are normally designed in stages, in which the first-stage separator is used for preliminary phase separation, while the second- and third-stage separators are applied for further treatment of each individual phase (gas, oil, and water). Depending on a specific application, oil/gas separators are used to remove dispersed droplets from a bulk gas stream or are designed to remove contaminant gas bubbles from the bulk liquid stream.

Throughout this chapter, two terms are used frequently: (1) absorption and (2) adsorption. Absorption is achieved by dissolution (a physical phenomenon) or by reaction (a chemical phenomenon) and is an approach in which the absorbed gas is ultimately distributed throughout the absorbent (liquid). The process depends only on physical solubility and may include chemical reactions in the liquid phase (chemisorption). Common absorbing media used are water, aqueous amine solutions, caustic, sodium carbonate, and nonvolatile hydrocarbon oils, depending on the type of gas to be absorbed. Usually, the gas-liquid contactor designs which are employed are plate columns or packed beds. Absorption is achieved by dissolution (a physical phenomenon) or by reaction (a chemical phenomenon). Chemical adsorption processes adsorb sulfur dioxide onto a carbon surface where it is oxidized (by oxygen in the flue gas) and absorbs moisture to give sulfuric acid impregnated into and on the adsorbent.

Adsorption differs from absorption, in that it is a physical-chemical phenomenon in which the gas is concentrated on the surface of a solid or liquid to remove impurities. Usually, carbon is the adsorbing medium (Mokhatab et al., 2006; Speight, 2007; Kidnay et al., 2011), which can be regenerated upon desorption. The quantity of material adsorbed is proportional to the surface area of the solid, and consequently, adsorbents are usually granular solids with a large surface area per unit mass. Subsequently, the captured gas can be desorbed with hot air or steam either for recovery or for thermal destruction.

The number of steps and the type of process used to produce pipeline-quality natural gas most often depends upon the source and makeup of the wellhead production stream. In some cases, several of the steps (Fig. 7.2) may be integrated into one unit or operation, performed in a different order, or performed at alternative locations, or not required at all (Mokhatab et al., 2006; Speight, 2007; Kidnay et al., 2011).

In many instances pressure relief at the wellhead will cause a natural separation of gas from oil (using a conventional closed tank, where gravity separates the gas hydrocarbons from the higher boiling crude oil). In some cases, however, a multistage gas-oil separation process is needed to separate the gas stream from the crude oil. These gas-oil separators are commonly closed cylindrical shells, horizontally mounted with inlets at one end, an outlet at the top for removal of gas, and an outlet at the bottom for removal of oil. Separation is accomplished by alternately heating and cooling (by compression) the flow stream through multiple steps; some water and condensate, if present, will also be extracted as the process proceeds.

At some stage of the processing, the gas flow is directed to a unit that contains a series of filter tubes. As the velocity of the stream reduces in the unit, primary separation of remaining contaminants occurs due to gravity. Separation of smaller particles occurs as gas flows through the tubes, where they combine into larger particles which flow to the lower section of the unit. Further, as the gas stream continues through the series of tubes, a centrifugal force is generated which further removes any remaining water and small solid particulate matter.

The amount of particulate matter that is present in the gas depends on the properties of the shale reservoir. Over the past several decades since World War II, a range of dry filters containing wood wool, sisal fiber, glass wool, wood chips soaked in oil, and other types of fibrous or granular material were used for removal of the particulate matter (average particle size below 60 μm), but success was very limited. Wet purifiers such as water and oil scrubbers and bubblers are also effective but only within certain limits.

The most effective process for the removal of particulate matter employs cloth filters but normal cloth filters are very sensitive to the gas temperature and at high temperatures these filters are likely to char and decompose in the hot gas stream. Another disadvantage is that the filters are subject to a rapid buildup of particulate matter and need frequent cleaning if not used in conjunction with a pre-filtering step.

The disadvantages of using cloth filters can be partly offset by using woven glass wool filter bags. This material can be used at temperatures up to 300°C (570°F). By heating (insulated) filter housing by means of the hot gas stream coming from the gasifier, temperatures above 100 C (212 F) can be maintained in the filter, thus avoiding condensation and enhanced pressure drop. If a prefiltering step consisting of a cyclone and/or an impingement filter is employed. It is possible to keep the service and maintenance intervals within reasonable limits. Electrostatic filters are also known to have very good particle separating properties, and most probably they could also be used to produce a gas of acceptable quality.

If the gas stream requires cooling, there are three types of gas coolers: (1) natural convection coolers, (2) forced convection coolers, and (3) water coolers. The major factors to be taken into consideration are the sensible heat in the gas, the water vapor content of the gas and its heat of condensation, and the effects of fouling of the cooler. Natural convection coolers consist of a simple length of pipe and are simple to use and clean and require no additional energy input. A forced convection cooler is equipped with a fan which forces the cooling air to flow around the gas pipes and is typically much smaller than the natural convection coolers. The disadvantage of the forced convention cooler is the extra energy input required for the fan and the necessity to use gas cooling pipes of small diameters, which can lead to fouling problems. The former can in some cases be offset by using the cooling air supplied by the engine fan.

Water coolers are available in two types, the scrubber and the heat exchanger; where a water scrubber or bubbler is used, the objective is generally to cool and clean the gas in one and the same operation. Scrubbers of many different types exist, but the principle is always the same: the gas is brought in direct contact with a fluid medium (generally water) which is sprayed into the gas stream by means of a suitable nozzle device. The advantage of this system is its small size, but the disadvantage is the need for fresh water, increased complexity of maintenance, and additional power consumption resulting from the use of a water pump. It is also possible to cool the gas by means of a water-cooled heat exchanger. This is a suitable method in case a source of fresh water is continuously available and the extra investment and power consumption of a suitable water pump can be justified.

4 TIGHT GAS PROPERTIES AND PROCESSING

Shale gas projects have recently contributed significantly to increased production in the United States. There is an expectation that rapid exploitation of shale gas development is also likely to occur in other regions of the world. The shale formations in the United States that presently produce gas commercially exhibit an unexpectedly wide variation in the values of five key parameters: (1) thermal maturity, expressed as vitrinite reflectance, (2) adsorbed-gas fraction, (3) reservoir thickness, (4) total organic carbon (TOC) content, and (5) volume of gas in place (Chapter 2). In addition, the degree of natural fracture development in an otherwise low-matrix-permeability shale reservoir is a controlling factor in gas producibility and, possibly in gas properties.

A wide range of reservoir properties (Chapter 2) controls both the rate and volume of shale-gas production from the gas shale formations, notable: (1) thermal maturity, (2) gas in place, TOC content, (3) reservoir thickness, and (4) proportion of adsorbed gas. Natural fracture networks are required to augment the extremely low shale-matrix permeability. Therefore, geology and geochemistry must be considered together when evaluating a given shale system both before and after drilling as well as gas processing options.

In addition, it is very likely that not only the amount and distribution of gas within the shale but also the composition of the shale gas is determined by, amongst other things: (1) the initial reservoir pressure, (2) the petrophysical properties of the rock, and (3) the adsorption characteristics of the rock—thus during production there are three main processes that can be operative.

Initial gas production is dominated by depletion of gas from the fracture network. This form of production declines rapidly due to limited storage capacity. After the initial decline rate stabilizes, the depletion of gas stored in the matrix becomes the primary process involved in production. The amount of gas held in the matrix is dependent on the particular properties of the shale reservoir which can be hard to estimate. Secondary to this depletion process is desorption, whereby adsorbed gas is released from the rock as pressure in the reservoir declines. The rate of gas production via the desorption process depends on there being a significant drop in reservoir pressure. At the same time, the composition of the gas can, and undoubtedly does, change due to the action of any one of these parameters or due to the interaction of any two of the above parameters or the interaction of all three of the above parameters.

Furthermore, in the Western Canadian Sedimentary Basin (WCSB), Devonian–Mississippian (D–M) and Jurassic shale formations have complex, heterogeneous pore volume distributions as identified by low pressure carbon dioxide and nitrogen sorption (Ross and Bustin, 2009. In fact, high pressure methane isotherms on dried and moisture equilibrated shale samples show a general increase of gas sorption with TOC content. Methane sorption in D–M formations increases with increasing TOC content and micropore volume, indicating that microporosity associated with the organic fraction is a primary control not only on methane sorption but also on shale gas composition.

The sorption capacities for Jurassic shale formations may be unrelated to micropore volume and the large adsorbed gas capacities of organic-rich Jurassic shale formations, independent of surface area, imply a portion of that a portion of the methane is stored by solution in matrix bituminite (Ross and Bustin, 2009). Solute methane does not appear to be an important contributor to gas storage in D–M shale formations (Ross and Bustin, 2009). In fact, it is likely that structural transformation of D–M organic matter has occurred during thermal diagenesis creating and/or opening up microporosity onto which gas can sorb thereby also influencing the composition of the shale gas.

Furthermore, inorganic constituents also influence modal pore size, total porosity, and sorption characteristics of shale formation, thereby adding a further parameter to those parameters (above) that influence changes in the shale gas composition. Clay are known to provide excellent adsorption surfaced for petroleum constituents and are also capable of adsorbing gas into the internal structure, the amount of which is dependent on clay-type.

The uncertainties of reservoir properties and fracture parameters (Chapters 2 and 3) have significant effect on shale gas properties and production, making the process of optimization of hydraulic fracturing treatment design for economic gas production much more complex. It is extremely important to identify reasonable ranges for these uncertainty parameters and evaluate their effects on well performance, because the detailed reservoir properties for each wellbore are difficult to obtain.

Gas production from unconventional shale gas reservoirs has become more common in the past decade, and there are increasing demands to understand the petrophysical and mechanical properties of these rocks. Characterizing these organic-rich shale formations can be challenging as these rocks vary quite significantly (Passey et al., 2010). For instance, formations in the Barnett shale are known to be more silica-rich, whereas Eagle Ford shale rocks are generally carbonate-rich containing relatively smaller amounts of silica and clays. Formations in these shale gas reservoirs also exhibit a wide range of composition within a single reservoir. There are also indications that it is not only the amount of clay or organics, but also the maturity of the shale formations that control the anisotropy of these organic rich shale formations (Ahmadov et al., 2009).

Not surprisingly, gas-bearing shale formations are complex reservoirs (with a porosity on the order of porosity of 4–6 porosity units and a TOC content of $\geq$4%, w/w), which represent significant variety in reservoir characteristics (i.e., mineralogy, porosity, permeability, gas content, and pressure). In addition, the porosity shale porosity changes at very different rates in different regions and formations. Moreover, the gas in these shale formations occurs both as a free phase within pores and fractures and as gas adsorbed onto organic matter. Not surprisingly, and in accordance with the varying shale reservoir properties, it must be anticipated that there will be differences in shale gas composition and properties.

Thus, although shale gas represents a large, new source of natural gas and NGLs, shale gas is certainly not the same everywhere. Produced shale gas observed to date has shown a broad variation in compositional makeup, with some having wider component ranges, a wider span of minimum and maximum heating values, and higher levels of water vapor and other substances than pipeline tariffs or purchase contracts may typically allow. Indeed, because of these variations in gas composition, each shale gas formation can have unique processing requirements for the produced shale gas to be marketable. Ethane can be removed by cryogenic extraction while carbon dioxide can be removed through a scrubbing process. However, it is not always necessary (or practical) to process shale gas to make its composition identical to conventional transmission-quality gases. Instead, the gas should be interchangeable with other sources of natural gas now provided to end-users. The interchangeability of shale gas with conventional gases is crucial to its acceptability and eventual widespread use in the United States.

Although not highly sour in the usual sense of having high hydrogen sulfide content, and with considerable variation from play to resource to resource and even from well to well within the same resource (due to extremely low permeability of the shale even after fracturing) (Chapters 1 and 3), shale gas often contains varying amounts of hydrogen sulfide with wide variability in the carbon dioxide content. The gas is not ready for pipelining immediately after it has exited the shale formation.

The challenge in treating such gases is the low (or differing) hydrogen sulfide/carbon dioxide ratio and the need to meet pipeline specifications. In a traditional gas processing plant, the olamine of choice for content for hydrogen sulfide removal is *N*-methyldiethanolamine (MDEA) (Mokhatab et al., 2006; Speight, 2007, 2013; Kidnay et al., 2011), but whether or not this olamine will suffice to remove the hydrogen sulfide without removal of excessive amounts of carbon dioxide is another issue.

Gas treatment may begin at the wellhead—condensates and free water usually are separated at the wellhead using mechanical separators. Gas, condensate, and water are separated in the field separator. Extracted condensate and free water are directed to separate storage tanks and the gas flows to a gathering system. After the free water has been removed, the gas is still saturated with water vapor, and depending on the temperature and pressure of the gas stream, may need to be dehydrated or treated with methanol to prevent hydrates as the temperature drops. But this may not be always the case in actual practice.

Thus, there is the real need to evaluate gas-processing operations and the ability of a processing plant to treat a variety of shale gases to pipeline specifications. Solvent selection, strength, temperature, and circulation rate, as well as the type and quantity of internals used in the contactor, are some of the process parameters and design variables that must be considered.

4.1 Antrim Shale Formation

The Antrim shale is a shallow shale gas resource in Michigan. The Antrim shale is unique because the gas is predominately biogenic: methane is created as a by-product of bacterial consumption of organic material in the shale. Significant associated water is produced requiring central production facilities for dehydration, compression, and disposal. The carbon dioxide level in these samples varies from 0% to 9% (v/v) (Table 7.4). Carbon dioxide is a naturally occurring by-product of shale gas produced by desorption and, as a result, the carbon dioxide levels in this gas increase during the productive life of a well. Individual well production varies from 50to 60,000 cubic feet per day. Significant associated water is produced, requiring central production facilities for dehydration, compression, and disposal (Bullin and Krouskop, 2008).

The Antrim shale gas, for example, has high nitrogen concentration, as does at least one well tested in the Barnett shale formation. New Albany shale gas shows high carbon dioxide concentrations, while several wells in the Marcellus gas shale have tested up to 16% (v/v). Economically treating and processing these gases requires all the same techniques as conventional gas—plus the ability to handle a great deal of variability in the same field. These differences in quality of shale gas introduce a note of caution into gas processing because of the variability of the composition and properties of the gas (Bullin and Krouskop, 2008).

4.2 Barnett Shale Formation

The Barnett shale formation of north Texas is the more familiar of shale gas resources and grandfather of shale gas plays. Much of the technology used in drilling and production of shale gas has been developed for this resource. The Barnett shale formation lies around the Dallas-Ft. Worth area of Texas and produces gas at depths of 6500–9500 ft. The initial discovery region was in a core area on the eastern side of the play. As drilling has moved westward, the form of the hydrocarbons in the Barnett shale has varied from dry gas prone in the east to oil prone in the west (Table 7.4). As a result of such variations in composition, blending may be the most appropriate methods for equalizing the variations. With the richness of the gas, the Barnett plants remove about substantial amounts of NGLs each day.

The Barnett shale formation is the most well-known of shale gas formations (Chapter 2). Much of the technology used in drilling and production of shale gas has been developed on this resource (Bullin and Krouskop, 2008). The Barnett shale formation produces at depths of 6500–9500 ft with a production rate on the order of 0.5–4 million cubic feet per day (ft^3/d) with estimates of 300–550 cubic feet of gas per ton of shale. The initial discovery region was in a core area on the eastern side of the resource, and as the drilling and gas recovery has moved westward, the composition of the shale gas has changed from dry gas production in the east to wet gas and oil production in the west.

The Barnett shale resource play, for example, contains several hundred parts per million (ppm v/v) of hydrogen sulfide and much higher amounts of carbon dioxide (in the percent v/v range) of carbon dioxide. In other shale resources, such as Haynesville and the Eagleville field (Eagle Ford resource), the hydrogen sulfide content is known to be present. In other gas shale resources, such as the Antrim resource and the New Albany resource, underlying Devonian formations may communicate with and contaminate the shale formations. In addition, some of the shale gas resources have low carbon dioxide content but have a sufficiently high content of hydrogen sulfide to require treating. Thus, even after removing the NGLs, there are reasons for the shale gas needs further treatment to remove hydrogen sulfide and carbon dioxide to meet pipeline specifications.

4.3 Fayetteville Shale Formation

The Fayetteville shale is an unconventional gas reservoir located on the Arkansas side of the Arkoma Basin and ranges in thickness from 50 to 550 ft at a depth of 1500–6500 ft. The gas (Table 7.4) primarily requires only dehydration to meet pipeline specifications. The formation is estimated to hold between 58 and 65 billion cubic feet ($58–65 \times 10^6$ ft^3) per square, and initial production rates were on the order of 0.2–0.6 million cubic feet per day for vertical wells and one to three and one-half million cubic feet per day for horizontal wells. In 2003, production in one area had exceeded 500 million cubic feet per day (Bullin and Krouskop, 2008).

4.4 Haynesville Shale Formation

The Haynesville shale gas resource is the newest shale area to be developed. It lies in northern Louisiana and East Texas. The formation is deep (>10,000 ft), high bottomhole temperature (175 C, 350 F), and high pressure (3000–4000 psi). The wells showed initial production rates up to 20 or more million cubic feet of gas per day with estimates of 100–330 cubic feet of gas per ton of shale (Bullin and Krouskop, 2008).

The gas requires treating for carbon dioxide removal (Table 7.4). Operators in this field are using amine treating to remove the carbon dioxide with a scavenger treatment on the tail gas to remove hydrogen sulfide.

4.5 Marcellus Shale Formation

The Marcellus shale lies in western Pennsylvania, Ohio, and West Virginia. The formation is shallow at depths of 2000–8000 ft and 300–1000 ft thick. The gas composition varies across the field, much as it does in the Barnett—the gas becomes richer from east to west (Table 7.4). Initial production rates have been reported to be on the order of 0.5–4 million cubic feet per day (ft^3/d), with estimates of 60–100 cubic feet of gas per ton of shale (Bullin and Krouskop, 2008).

The Marcellus shale gas has relatively little carbon dioxide and nitrogen. In addition, the gas is dry and does not require removal of NGLs for pipeline transportation. Early indications are that the Marcellus gas has sufficient liquids to require processing.

4.6 New Albany Shale Formation

The New Albany shale is black shale in Southern Illinois extending through Indiana and Kentucky. The formation is 500–4900 ft deep and 100–400 ft thick. The gas composition (Table 7.4) is variable, and low flow rates of wells in the New Albany shale require that production from many wells must be combined to warrant processing the gas. Vertical wells typically produce 25–75,000 cubic feet per day while horizontal wells can have initial production rates up to 2 million cubic feet per day (Bullin and Krouskop, 2008).

5 TIGHT GAS PROCESSING

Natural gas, as it is used by consumers, is much different from the natural gas that is brought from the underground tight formation to the wellhead since the natural gas brought to the wellhead, although still composed primarily of methane, is by no means as pure as need to for sale to the consumer. The natural gas commonly exists in mixtures with other hydrocarbons; principally ethane, propane, butane, and pentanes. In addition, raw natural gas contains water vapor, hydrogen sulfide (H_2S), carbon dioxide, helium, nitrogen, and other compounds.

Whatever the tight reservoir from which the natural gas came, once separated from condensate and crude oil (if present), natural gas commonly exists in mixtures with other hydrocarbons; principally ethane, propane, butane, and pentanes. In addition, raw natural gas contains water vapor, hydrogen sulfide (H_2S), carbon dioxide and may even contain helium, nitrogen, and other compounds. In addition, acid gases corrode refining equipment, harm catalysts, pollute the atmosphere, and prevent the use of hydrocarbon components in petrochemical manufacture. When the amount of hydrogen sulfide is high, it may be removed from a gas stream and converted to sulfur or sulfuric acid.

Gas processing (Mokhatab et al., 2006; Speight, 2007: Kidnay et al., 2011; Speight, 2013) consists of separating all of the various hydrocarbons and fluids from the pure natural gas (Fig. 7.2). Major transportation pipelines usually impose restrictions on the makeup of the natural gas that is allowed into the pipeline and, as a result, the natural gas must be purified before it can be transported. Thus, gas processing is necessary to ensure that the natural gas intended for use is as clean and pure as possible.

The actual practice of processing gas streams to pipeline dry gas quality levels can be quite complex, but usually involves four main processes to remove the various impurities. Gas streams produced during petroleum and natural gas refining, while ostensibly being hydrocarbon in nature, may contain large amounts of acid gases such as hydrogen sulfide and carbon dioxide. Most commercial plants employ hydrogenation to convert organic sulfur compounds into hydrogen sulfide. Hydrogenation is effected by means of recycled hydrogen-containing gases or external hydrogen over a nickel molybdate or cobalt molybdate catalyst (Mokhatab et al., 2006; Speight, 2007; Kidnay et al., 2011).

In summary, refinery process gas, in addition to hydrocarbons, may contain other contaminants, such as carbon oxides (CO_x, where $x=1$ and/or 2), sulfur oxides (So_x, where $x=2$ and/or 3), as well as ammonia (NH_3), mercaptans (R-SH), and carbonyl sulfide (COS). The presence of these impurities may eliminate some of the sweetening processes, since some processes remove large amounts of acid gas but not to a sufficiently low concentration. On the other hand, there are those processes not designed to remove (or incapable of removing) large amounts of acid gases, whereas they are capable of removing the acid gas impurities to very low levels when the acid gases are present only in low-to-medium concentration in the gas (Mokhatab et al., 2006; Speight, 2007, 2014; Kidnay et al., 2011).

There are many variables in treating refinery gas or natural gas. The precise area of application, of a given process is difficult to, define. Several factors must be considered: (1) the types of contaminants in the gas, (2) the concentrations of contaminants in the gas, (3) the degree of contaminant removal desired, (4) the selectivity of acid gas removal required, (5) the temperature of the gas to be processed, (6) the pressure of the gas to be processed, (7) the volume of the gas to be processed, (8) the composition of the gas to be processed, (9) the ratio of carbon dioxide to hydrogen sulfide in the gas stream, and (10) the desirability of sulfur recovery due to process economics or environmental issues.

In addition to hydrogen sulfide (H_2S) and carbon dioxide (CO_2), natural gas from conventional reservoirs and from tight reservoirs may contain other contaminants, such as mercaptans (RSH) and carbonyl sulfide (COS). The presence of these impurities may eliminate some of the sweetening processes since some processes remove large amounts of acid gas but not to a sufficiently low concentration. On the other hand, there are those processes that are not designed to remove (or are incapable of removing) large amounts of acid gases. However, these processes also capable of decreasing the acid gas impurities to very low levels when the acid gases are present in low to medium concentrations in the gas.

Process selectivity indicates the preference with which the process removes one acid gas component relative to (or in preference to) another. For example, some processes remove both hydrogen sulfide and carbon dioxide; other processes are designed to remove hydrogen sulfide only. It is important to consider the process selectivity for, say, hydrogen sulfide removal compared to carbon dioxide removal that ensures minimal concentrations of these components in the product, thus the need for consideration of the carbon dioxide to hydrogen sulfide in the gas stream.

Thus, natural gas processing consists of separating all of the various hydrocarbons and fluids from the pure natural gas, to produce what is known as pipeline quality (sometimes referred to as sales quality) dry natural gas. Major transportation pipelines usually impose restrictions on the makeup of the natural gas that is allowed into the pipeline. Thus, after the natural gas is released from the tight reservoir produced (brought to the surface), water and condensate (higher-hydrocarbon liquids) are typically removed from the raw natural gas at or near the wellhead. Gathering lines then carry the remaining natural gas to a gas-processing facility that removes other constituents so that the processed gas meets pipeline specifications and so maximum value can be obtained for constituents such as NGLs.

5.1 Wellhead Processing

Processing natural gas typically begins at the wellhead—after leaving the gas well, the first step in processing is removing oil, water, and condensates. Heaters and scrubbers are used to prevent the temperature of the gas from dropping too low and remove large-particle impurities, respectively (Manning and Thompson, 1991). Wellhead processing is, in all respects, less complicated than the processing that take place at the natural gas plant where processing, cleaning, and refining is taken to the ultimate step of producing essentially pure methane—the natural gas used by consumers is composed almost entirely of methane.

At the wellhead, the gas stream is prepared for transportation by using processes for: (1) water removal and (2) condensate and crude oil removal.

5.1.1 Water Removal

Water removal from the gas stream at the wellhead is required because of the tendency for the formation of gas hydrates. Process options are available

(Mokhatab et al., 2006, Speight, 2007; Kidnay et al., 2011) and choice is dictated by the amount of the water and process efficiency. An example of water removal process is a mechanical refrigeration plants that chill the gas will promote the formation of gas. Cryogenic liquid recovery plants will freeze up (form hydrates) if the gas water content is above 100 parts per million by volume (ppm v/v). The most prevalent solutions to dry the gas are: (1) contacting the gas with 99% triethylene glycol to dry the gas, (2) injecting 80% (v/v) ethylene glycol into the mechanical refrigeration unit to prevent hydrate formation, and (3) processing the gas in a molecular sieve unit upstream of the cryogenic plant to dry the gas to below 100 ppm v/v.

In addition, heaters and scrubbers are installed, usually at or near the wellhead and serve to remove sand and other particulate matter impurities. The heaters ensure that the temperature of the gas does not drop too low—when natural gas that contains even low quantities of water, natural gas hydrates have a tendency to form at low temperatures (Chapter 1). The gas hydrates are solid or semisolid compounds with the hydrocarbons at the center of an ice-like cage, and an accumulation of the hydrates can impede the flow of the natural gas through the various pipeline valves and through the gathering systems. To reduce the occurrence of hydrates, small natural gas-fired heating units may be installed along the gathering pipe wherever it is possible that that gas hydrates may form.

While the ethane, propane, butane derivatives, pentane derivatives, and natural gasoline (a mixture of higher molecular weight hydrocarbons up to approximately C_{10} and sometimes referred to as gas condensate) must be removed from natural gas. The NGLs are sold separately and have a variety of different uses, including enhancing oil recovery in oil wells, providing raw materials for oil refineries or petrochemical plants, and as sources of energy.

Thus, while some of the needed processing operations can be accomplished at or near the wellhead (field processing), the complete processing of natural gas takes place at a processing plant and the wellhead-treated natural gas is transported to these processing plants through a network of gathering pipelines, which are small-diameter, low pressure pipes. However, in addition to processing done at the wellhead and at centralized processing plants, some final processing is also sometimes accomplished at straddle extraction plants, which are located on major pipeline systems. Although the natural gas that arrives at these straddle extraction plants is already of pipeline quality, in certain instances there may still exist in the gas stream small quantities of NGLs that are extracted at the straddle plants.

Once at the gas processing plant, the natural gas stream is further processed to levels of purity that make the gas suitable for sales to the consumer. To remove the various remaining impurities, the processing sequence has typically four main processes: (1) oil and condensate removal, (2) water removal, (3) separation of NGLs, and (4) sulfur and carbon dioxide removal.

The natural-gas-processing facility is a dedicated separations train that begins with the removal of acid gases (carbon dioxide, hydrogen sulfide. and organosulfur compounds) (Fig. 7.2) (Mokhatab et al., 2006; Speight, 2007; Kidnay et al., 2011); there is still the challenge in treating the gas when the H_2S/CO_2 ratio is low along with the desire to meet pipeline specifications in regard to the carbon dioxide content. Elemental sulfur is often recovered from treatment of the off gas stream from this process. The most appropriate technology for removing acid gas depends on the amount in the feed and the desired contaminant level in the product. The most common processes for removing carbon dioxide are amine treating, membranes, and molecular sieves.

Typically, depending upon the process used for water removal, the density differential between the particulate matter and the gas/condensate/oil would also facilitate removal of the particulate matter with the water. Alternatively, a simple filter system may be installed at the wellhead. Such systems use fabric filters which are typically designed with nondisposable filter bags. As the gas stream flows through the filter media (typically cotton, polypropylene, Teflon, or fiberglass), particulate matter is collected on the bag surface as a dust cake. Fabric filters are generally classified on the basis of the filter bag cleaning mechanism employed. Fabric filters operate with collection efficiencies up to 99.9%, depending upon the size of the particulate matter particles and the size of the filter orifices.

5.1.2 Condensate and Crude Oil Removal

In order to process and transport associated natural gas, it must be separated from the any condensate or crude oil in which it is dissolved and which is typically carried using equipment installed at or near the wellhead. In the next step, after water removal, the condensate or crude oil is separated from the gas with a conventional separator which consists of a closed tank that separates the liquids and solids by the force of gravity. When gravity alone does not separate the two, separators use pressure to cool the gas and then move through a high pressure liquid at a low temperature to knockout any remaining oil and a portion of the water. Typically, the density differential would also facilitate removal of the particulate matter into the condensate/crude oil lay if they have not already been removed during the water-removal process.

However, the actual process used to separate oil from natural gas, as well as the equipment that is used can vary widely, especially since shale gas not only has a varied composition from reservoir to reservoir but also can vary in computation on a well-by-well basis from the same reservoir (Table 7.4). If the natural gas coexists with crude oil in the same tight reservoir, there will be the need to remove the crude oil. Typically, the natural gas can be separated from condensate or crude oil at the wellhead by the use of a separator (Mokhatab et al., 2006; Speight, 2007; Kidnay et al., 2011).

In addition to removing contaminants, shale gas often is produced with NGLs that bring higher value if they are recovered for petrochemical or other uses that exceed their Btu value when left in the natural gas stream. Depending on the inlet gas conditions (such as gas richness, pressure, temperature, and product specifications), the optimum liquid recovery process will be selected. For a production situation that anticipates fairly rich gas and propane-plus recovery, and normally smaller volumes, a suitable choice of process is mechanical refrigeration with injection of ethylene glycol. This process yields high recovery of propane recovery when the gas has five gallons of liquid hydrocarbons per 1000 ft^3 or better. In situations with lean or rich gas and very small proportions of ethane, a Joule–Thomson process with molecular sieve dehydration is a sensible solution. Finally, situations expecting lean or rich gas, high propane or ethane recovery, and medium to large gas volumes should choose a cryogenic process.

In many cases, the separation of liquid crude oil/condensate from the natural gas is relatively straightforward, and the two hydrocarbon streams are then sent (separately) for further processing. The most basic type of separator is a closed tank, where the force of gravity serves to separate the heavier liquids like oil, and the lighter gases, like natural gas. In more difficult situations, a pressure separator or temperature separator may be employed to separate the liquid and gas streams. These separators are typically used for wells producing high pressure or high temperature natural gas along with light crude oil or condensate.

The pressure separators use pressure differentials to cool the wet natural gas and separate the oil and condensate. In the process, wet gas enters the separator, after some cooling has occurred in a heat exchanger. The gas then travels through a high pressure liquid knockout vessel which serves to remove any liquids into a low-temperature separator. The gas then flows into this low-temperature separator through a choke mechanism, which expands the gas as it enters the separator and the rapid expansion of the gas allows for the lowering of the temperature in the separator. After removal of the liquids, the dry gas then travels back through the heat exchanger and is warmed by the incoming wet gas. By varying the pressure of the gas in various sections of the separator, it is possible to vary the temperature, which causes the oil and some water to be condensed out of the wet gas stream. This basic pressure-temperature relationship can work in reverse as well, to extract gas from a liquid oil stream, such as a crude oil stream containing dissolved gas from a tight formation reservoir.

5.2 Gas Plant Operations

Gas processing is an instrumental part of the natural gas value chain insofar as it is instrumental in ensuring that the natural gas is as clean and pure as possible (the gas meets the specifications that render the gas suitable for use) by making the gas the clean burning and an environmentally sound energy choice. Once the natural gas has been fully processed and is ready to be consumed, it must be transported from those areas that produce natural gas to those areas that require it.

It is also at the gas plant that process selectivity plays a major role in gas processing. Process selectivity indicates the preference with which the process removes one acid gas component relative to (or in preference to) another. For example, some processes remove both hydrogen sulfide and carbon dioxide; other processes are designed to remove hydrogen sulfide only. It is important to consider the process selectivity for, say, hydrogen sulfide removal compared to carbon dioxide removal that ensures minimal concentrations of these components in the product, thus the need for consideration of the carbon dioxide to hydrogen sulfide in the gas stream.

Although the processing of natural gas is in many respects less complicated than the processing and refining of crude oil, it is equally as necessary before its use by end users. The natural gas used by consumers is composed almost entirely of methane. However, natural gas found at the wellhead, although still composed primarily of methane, is by no means as pure. Typically, the operations at the gas processing plant involve: (1) water removal, (2) separation of NGLs, and (3) sulfur and carbon dioxide removal.

5.2.1 Water Removal

In addition to separating crude oil and condensate from the gas stream, the water associated with the gas stream must also to be removed. Most of the liquid, free water associated with extracted natural gas can be removed by simple separation methods at or near the wellhead, but the removal of any water that still remains in the natural gas requires a more complex treatment. This treatment consists of a dehydration step at the gas processing plant and can involve one of two processes, either (1) absorption or (2) adsorption. For clarification, absorption occurs when the water vapor is removed for the gas stream by a dehydrating agent, whereas adsorption occurs when the water vapor is condensed and collected on the surface of a dehydrating agent. Natural gas from tight formation does not usually contain as much water as natural gas from conventional reservoirs, but the occasion for water removal at the gas processing plant may still be necessary.

Glycol Dehydration

The glycol dehydration process is an example of a process that provides absorption dehydration, and in the process, a liquid desiccant provides the means to absorb water from the gas stream. Ethylene glycol ($HOCH_2CH_2OH$) was, initially, the principal chemical agent in this process, has a very strong affinity for water and when the glycol is in contact with a stream of water-wet natural gas, the ethylene glycol absorbs the water from the gas stream. Initially, the process used ethylene glycol as the absorbent but, with the advancement of the technology, glycol dehydration now involves the use of an aqueous solution of a glycol derivative in which the glycol is either diethylene glycol (DEG) or triethylene glycol (TEG) (Table 7.5), which is brought into contact with the water-wet gas stream in a contactor. The glycol solution will absorb water from the wet gas and, once absorbed, the glycol sinks to the bottom of the contactor while

Table 7.5 Propoerties of Ethylene Glycol, Diethlene Glycol, and Triethylene Glycol

	Ethylene Glycol	Diethylene Glycol	Triethylene Glycol
Chemical formula	$C_2H_6O_2$	$C_4H_{10}O_3$	$C_6H_{14}O_4$
Acronym molar mass	MEG 62.07 g mol^{-1}	DEG 106.12 g/mol	TEG 150.17 g mol^{-1}
Appearance	Clear, colorless liquid	Colorless liquid	Colorless liquid
Density	1.1132 g/cm^3	1.118 g/cm^3	1.1255 g/cm^3
Melting point	−12.9 C (8.8 F)	−10.45 C (13.19 F)	−7 C (19 F)
Boiling point	197.3 C (387.1 F)	244 C (471 F)	285 C (545 F)
Solubility in water	Miscible	Miscible	Miscible

the natural gas, stripped of most of the water content, is then transported out of the dehydrator. The glycol solution, bearing all of the water stripped from the natural gas, is put through a specialized boiler designed to vaporize only the water out of the solution where the boiling point differential facilitates removal of the water for the makes it relatively easy to remove water from the glycol solution after which the glycol is recycled to the contactor.

In some cases, a flash tank separator-condensers has been added to the unit which, in addition to absorbing water from the gas stream, also regenerates small amounts of methane and other compounds that the glycol solution occasionally carries with it from the contactor stage having also absorbed these constituents from the gas stream. In the past, this methane may have been vented and lost to the product streams as well as making a contribution to atmospheric pollution (Chapter 9). In order to decrease the amount of methane and other compounds that are lost, the flash tank separator-condensers enables removal of the absorbed hydrocarbon constituents before the glycol solution reaches the boiler. In the flash tank separator, the pressure is reduced which allows the lower boiling hydrocarbon constituents (i.e., lower boiling than the glycol solvent), thereby allowing the methane and other hydrocarbons to vaporize (flash) from the solution. The glycol solution is then sent to the boiler, which may also be fitted with air or water-cooled condensers, at which point any remaining hydrocarbons are captured, combined with other hydrocarbon streams, fractionated, and sent to the various product streams. The insertion of a flash-separator-condenser system into the process is valuable adaption for the treatment of natural gas streams from tight formation because of the content of hydrocarbon constituents that are higher molecular weight than methane (Table 7.4).

Solid-Desiccant Dehydration

The solid-desiccant dehydration process is also a well-used process for dehydrating natural gas using adsorption and the unit usually consists of two or more adsorption towers, each of which is filled with a solid desiccant and the typical desiccants are: (1) activated alumina or (2) a granular silica gel material. In the process, water-wet natural gas is passed through the towers, in a top-to-bottom direction and, as the gas stream passes through the desiccant and around the desiccant particles, water is adsorbed on to the surface of the desiccant particles. By this means (i.e., passing the gas stream through the entire desiccant bed), almost all of the water is adsorbed onto the desiccant material, leaving the dry gas stream to exit the bottom of the tower.

The solid-desiccant dehydrator process is typically more effective than the glycol dehydrator process, and solid desiccant units are often installed in a straddle mode along natural gas pipelines and are ideally suited for large volumes of natural gas that is under very high pressure (as is often the case in pipelines) and are, as a result, usually located on a pipeline downstream of a compressor station. The process requires two or more towers due to the fact that after a certain period of use, the desiccant in a particular tower becomes saturated with water and needs to be regenerated. The regeneration stage involves taking the tower off-stream at which time a preheated high-temperature gas is passed through the desiccant to vaporize any residual water in the desiccant after which the dry desiccant is brought back on stream for further use in the process.

5.3 Separation of Natural Gas Liquids

Natural gas coming directly from a well (including wells producing gas from tight formations) contains many hydrocarbon constituents (Tables 7.1, 7.3, and 7.4) that are classed as NGLs (being liquid under conditions where methane remains a gas) that are commonly removed. In most instances, NGLs have a higher value as separate product streams (i.e., as separated ethane, methane, propane, butanes, and pentanes plus) and, as a result, it is economical to remove these constituents from the natural gas stream. The removal of NGLs usually takes place in a relatively centralized processing plant and uses techniques similar to those used to dehydrate natural gas. Wellhead processing of tight gas for removal of the higher molecular weight hydrocarbon constituents is also a possibility but is, as might be excepted dictated by the economics of the separation and the transportation of the individual hydrocarbon streams. In summary, the extraction of NGLs from the natural gas stream produces both cleaner, purer natural gas and the valuable hydrocarbons that are the constituents of the NGLs.

There are two steps to the separation of the NGLs from a natural gas stream: (1) the NGLs, that is, the hydrocarbon constituents, must be extracted from the natural gas and (2) the NGLs must be individually separated to yield the individual constituents.

5.3.1 Absorption Process

The absorption process for the extraction of NGLs from gas streams is similar in principle to the absorption dehydration process with the exception that in order to absorb NGLs an absorption oil is used instead of ethylene glycol or a derivative (diethylene glycol or triethylene glycol). In the process, the natural gas stream is passed through an absorption tower where it is brought into contact with the absorption oil (lean absorption oil) which dissolves a high proportion (if not all) of the NGLs. The absorption oil containing the hydrocarbon constituents (rich absorption oil, fat absorption oil) exits the absorption tower through the base bottom. The absorption oil-hydrocarbon mixture (absorption oil plus ethane, propane, butanes, pentanes, and other higher-molecular-weight hydrocarbons) is fed into lean oil stills where the mixture is heated to a temperature above the boiling point of the highest boiling constituents of the NGLs but below the boiling point of the oil for recovery of the hydrocarbons as a mixture. This process allows for the recovery of approximately 75% (v/v) of the lower boiling hydrocarbons and up to 90% (v/v) of the hydrocarbons having a higher molecular weight (and higher boiling point than pentane originally in the natural has stream).

The basic absorption process as described above can be modified to improve process effectiveness efficiency or to facilitate the extraction of specific hydrocarbons. For example, the refrigerated oil absorption process in which the lean oil is cooled through refrigeration prior to the introduction of the gas stream into the contactor, propane recovery can be upward of 90% (v/v) and approximately 40% (v/v) of the ethane can be extracted from the natural gas stream. Furthermore, the extraction of the other, higher-molecular-weight hydrocarbons can near quantitative (approximately 100%, v/v) using this refrigeration option.

5.3.2 Cryogenic Expansion Process

The cryogenic process is a process designed to extract NGLs from natural gas streams. Absorption processes can extract almost all of the higher-molecular-weight constituents of NGLs but the lower-molecular-weight hydrocarbons, such as ethane, are more difficult to recover from the natural gas stream. However, if it is economically favorable to extract ethane and other lower-molecular-weight hydrocarbons from the gas stream, a cryogenic process typically offers a higher recovery rate.

A cryogenic process involves decreasing the temperature of the gas stream to a temperature on the order of −85°C (−120°F). The reduction in enrapture of the gas stream can be achieved by different methods, but one of the most effective is known as the turbo expander process in which external refrigerants are used to cool the natural gas stream. Then, an expansion turbine is used to rapidly expand the chilled gases which causes the temperature to drop significantly and causes ethane and other hydrocarbons to condense out of the gas stream while maintaining methane in gaseous form.

This process allows for the recovery of about 90–95% (v/v) of the ethane originally in the gas stream and, in addition, the expansion turbine is used to convert of a part of the energy released (when the natural gas stream is expanded) for recompression of the gaseous methane effluent, thus saving energy costs associated with extracting ethane.

5.3.3 Fractionation of Natural Gas Liquids

Once NGLs have been extracted from the natural gas stream, they must be separated into the individual components of the bulk fraction as the next step. The fractionation process that is used to accomplish the separation is based on the different boiling points of the different individual hydrocarbons in the NGLs stream, and the fractionation occurs in stages which consist of boiling off of the different hydrocarbons one by one.

Thus, the entire fractionation process is broken down into stages, starting with the removal of the lower-boiling NGLs from the bulk stream and the particular fractionators are used in the following order: (1) the deethanizer, which separates the ethane from the bulk stream, (2) the depropanizer, which separates the propane from the deethanized stream, (3) the debutanizer, which separates from the deethanized stream and depropanizer but leave the pentane derivatives and the high boiling hydrocarbons in the residue stream. After the debutanization step, the butanes stream is sent to a butane splitter (also referred to as a deisobutanizer) which is used to separate the isobutane from the *n*-butane.

The separation process produced a series of stream that are suitable as petrochemical feedstocks (or alternatively or in addition, depending upon the yield of isobutane); the iso-butane can also be sent to an alkylation unit to produce an alkylated product that can be used as blend stock for gasoline manufacture.

5.4 Sulfur and Carbon Dioxide Removal

In addition to the removal of water, condensate, crude oil, and NGLs, one of the most important aspects of gas processing involves the removal of sulfur and carbon dioxide from the gas stream. Natural gas from some wells contains significant amounts of sulfur and carbon dioxide. In the case of natural gas from tight formation, this is also true but the gas stream may also contain significant amounts of hydrogen sulfide (Tables 7.3 and 7.4). The sulfur derivatives natural gas (the presence of free sulfur in natural is rare) convey on the gas stream properties that can only be termed extremely harmful, even lethal as well as corrosive.

More pertinent to many gas streams and the natural gas stream from tight reservoirs, sulfur exists in natural gas as hydrogen sulfide (H_2S) and the gas is usually considered sour if the hydrogen sulfide content exceeds 5.7 mg of H_2S per cubic meter of natural gas. In addition to the harmful, lethal, and corrosive properties mentioned above, hydrogen sulfide can suffer bouts of explosive oxidation:

$$2H_2S + 3O_2 \rightarrow 2SO_2 + 2H_2O$$

This oxidation process is particular disadvantageous and dangerous during transportation of light crude oil from tight formations when there is a spill (through derailment of a railway tank car carrying the crude oil) when the virulence of the hydrogen sulfide oxidation reaction can contribute to the potential for explosive flammability of the crude oil. Therefore, removal of hydrogen sulfide is an essential part of gas processing.

The process for removal of sulfur compounds from natural gas involves the use of amine (olamine) solutions (the Girbotol process). In the process, the sulfur-containing natural gas stream is run through a tower, which contains the amine solution which has an affinity for hydrogen sulfur and absorbs the hydrogen sulfur in a manner similar to the dehydration process in which the glycol derivative absorbs water. There are several different olamines (Table 7.6), each of which can be used preferentially depending upon the properties of the incoming natural gas stream (Mokhatab et al., 2006; Speight, 2007; Kidnay et al., 2011). However, two of the olamines are more commonly used than the others: (1) monoethanolamine, MEA and (2) diethanolamine, DEA. Either of these compounds, in liquid form, will absorb sulfur compounds from natural gas as it passes through the liquid phase olamine. As a result, the effluent gas is virtually free of sulfur compounds and, similar to the glycol dehydration process and the process for the extraction of NGLs from natural gas streams, the amine solution used can be regenerated (i.e., the absorbed sulfur is removed), thereby allowing the olamine to be reused to treat more sour gas.

Although most sour gas sweetening involves the amine absorption process, it is also possible to use solid desiccants like iron sponges to remove the sulfide and carbon dioxide (Mokhatab et al., 2006; Speight, 2007; Kidnay et al., 2011). In order to recover elemental sulfur (which can be sold and used if reduced to the elemental form.) the sulfur-containing discharge from a gas sweetening process must be further treated. The process used to recover sulfur (the Claus process) involves using thermal and catalytic reactions to

Table 7.6 Olamines Used for Gas Processing

Olamine	Formula	Acronym	Molecular Weight	Specific Gravity	Melting Point (°C)	Boiling Point (°C)	Flash Point (°C)
Ethanolamine (monoethanolamine)	$HOC_2H_4NH_2$	MEA	61.08	1.01	10	170	85
Diethanolamine	$(HOC_2H_4)_2NH$	DEA	105.14	1.097	27	217	169
Triethanolamine	$(HOC_2H_4)_3NH$	TEA	148.19	1.124	18	335[a]	185
Diglycolamine (hydroxyethanolamine)	$H(OC_2H_4)_2NH_2$	DGA	105.14	1.057	−11	223	127
Diisopropanolamne	$(HOC_3H_6)_2NH$	DIPA	133.19	0.99	42	248	127
Methyldiethanolamine	$(HOC_2H_4)_2NCH_3$	MDEA	119.17	1.03	−21	247	127

[a]With decomposition.

extract the elemental sulfur from the hydrogen sulfide solution. In all, the Claus process is usually able to recover 97% (w/w) of the sulfur that has been removed from the natural gas stream. Since it is such a polluting and harmful substance, further filtering, incineration, and tail gas clean-up processes ensure that in excess of 98% (w/w) of the sulfur is recovered.

REFERENCES

Ahmadov, R., Vanorio, T., Mavko, G., 2009. Confocal Laser Scanning and Atomic-Force Microscopy in Estimation of Elastic Properties of the Organic-Rich Bazhenov Formation. The Leading Edge 28, 18–23.

Bullin, K., Krouskop, P., 2008. Composition variety complicates processing plans for US shale gas. In: Proceedings of Annual Forum, Gas Processors Association—Houston Chapter, Houston, Texas.

Bustin, R.M., Bustin, A.M.M., Cui, X., Ross, D.J.K., Pathi, V.S.M., 2008. Impact of shale properties on pore structure and storage characteristics. In: Paper No. SPE 119892. Proceedings of SPE Shale Gas Production Conference, Ft. Worth, Texas.

Cramer, D.D., 2008. Stimulating unconventional reservoirs: lessons learned, successful practices, areas for improvement. In: SPE Paper No. 114172. Proceedings of Unconventional Gas Conference, Keystone, Colorado.

Cumella, S.P., Scheevel, J., 2008. The influence of stratigraphy and rock mechanics on Mesaverde gas distribution, Piceance Basin, Colorado. In: Cumella, S.P., Shanley, K.W., Camp, W.K. (Eds.), Understanding, Exploring, and Developing Tight-Gas Sands.Proceedings of 2005 Vail Hedberg Conference: AAPG Hedberg Series, No. 3, pp. 137–155.

Grieser, B., Wheaton, B., Magness, B., Blauch, M., Loghry, R., 2007. Surface reactive fluid's effect on shale. In: SPE Paper No. 106825. SPE Production and Operations Symposium, Society of Petroleum Engineers, Oklahoma City, Oklahoma.

Harris, N.B., Ko, T., Philp, R.P., Lewan, M.D., Ballentine, C.J., Zhou, Z., Hall, D.L., 2013. Geochemistry of Natural Gases From Tight- Gas-Sand Fields in the Rocky Mountains. RPSEA Report No. 07122-09. Research Partnership to Secure Energy for America. National Energy Technology Laboratory, Unites States Department of Energy, Washington, DC. https://www.netl.doe.gov/file%20library/Research/Oil-Gas/Natural%20Gas/shale%20gas/07122-09-final-report.pdf; accessed May 2, 2016.

Hill, R.J., Jarvie, D.M., Zumberge, J., Henry, M., Pollastro, R.M., 2007. Oil and Gas Geochemistry and Petroleum Systems of the Fort Worth Basin. AAPG Bull. 91 (4), 445–473.

Kidnay, A., McCartney, D., Parrish, W., 2011. Fundamentals of Natural Gas Processing. CRC Press, Taylor & Francis Group, Boca Raton, Florida.

Kohl, A.L., Riesenfeld, F.C., 1985. Gas Purification, fourth ed. Gulf Publishing Company, Houston, Texas.

Kundert, D., Mullen, M., 2009. Proper evaluation of shale gas reservoirs leads to a more effective hydraulic-fracture stimulation. In: Paper No. SPE 123586. Proceedings of SPE Rocky Mountain Petroleum Technology Conference, Denver, Colorado, pp. 14–16.

Lee, S., 1991. Oil Shale Technology. CRC Press, Taylor & Francis Group, Boca Raton, Florida.

Maddox, R.N., 1982. Gas Conditioning and Processing. In: Gas and Liquid Sweetening-Volume 4. Campbell Publishing Co, Norman, Oklahoma.

Manning, F., Thompson, R., 1991. Oilfield Processing of Petroleum Volume One: Natural Gas. PennWell Books, Tulsa, Oklahoma, pp. 339–340. vol. 1.

Martini, A.M., Walter, L.M., Ku, T.C.W., Budai, J.M., McIntosh, J.C., Schoell, M., 2003. Microbial production and modification of gases in sedimentary basins: a geochemical case study from a Devonian shale gas play, Michigan basin. AAPG Bull. 87 (8), 1355–1375.
Mokhatab, S., Poe, W.A., Speight, J.G., 2006. Handbook of Natural Gas Transmission and Processing. Elsevier, Amsterdam, Netherlands.
Nehring, R., 2008. Growing and indispensable: the contribution of production from tight-gas sands to U.S. gas production. In: Cumella, S.P., Shanley, K.W., Camp, W.K. (Eds.), Understanding, Exploring, and Developing Tight-Gas Sands.Proceedings of 2005 Vail Hedberg Conference: AAPG Hedberg Series, No. 3, pp. 5–12.
Newman, S.A., 1985. Acid and Sour Gas Treating Processes. Gulf Publishing, Houston, Texas.
Passey, Q.R., Bohacs, K.M., Esch, W.L., Klimentidis, R., Sinha, S., 2010. From oil-prone source rocks to gas-producing shale reservoir—geologic and petrophysical characterization of unconventional shale-gas reservoir. In: SPE Paper No. 131350. Proceedings of CPS/SPE International Oil & Gas Conference and Exhibition, Beijing, China, Jun. pp. 8–10.
Ross, D.J.K., Bustin, R.M., 2009. The importance of shale composition and pore structure upon gas storage potential of shale gas reservoirs. Marine Petrol. Geol. 26 (6), 916–927.
Schettler Jr., P.D., Parmely, C.R., Juniata, C., 1989. Gas composition shifts in Devonian shales. SPE Reserv. Eng. 4 (3), 283–287.
Scouten, C.S., 1990. Oil shale. In: Speight, J.G. (Ed.), Fuel Science and Technology Handbook. Marcel Dekker Inc., New York, pp. 795–1053. Chapters 25 to 31.
Sone, H., 2012. Mechanical Properties of Shale Gas Reservoir Rocks and Its Relation to the In-Situ Stress Variation Observed in Shale Gas Reservoirs. A Dissertation Submitted to the Department of Geophysics and the Committee on Graduate Studies of Stanford University in Partial Fulfillment of the Requirements for the Degree of Doctor of Philosophy. SRB Volume 128. Stanford University, Stanford, California.
Soud, H., Takeshita, M., 1994. FGD Handbook. No. IEACR/65. International Energy Agency Coal Research, London, England.
Speight, J.G., 2007. Natural Gas: A Basic Handbook. GPC Books, Gulf Publishing Company, Houston, Texas.
Speight, J.G., 2008. Synthetic Fuels Handbook: Properties, Processes, and Performance. McGraw-Hill, New York.
Speight, J.G., 2012. Shale Oil Production Processes. Gulf Professional Publishing, Elsevier, Oxford, United Kingdom.
Speight, J.G., 2013. Shale Gas Production Processes. Gulf Professional Publishing, Elsevier, Oxford, United Kingdom.

CHAPTER EIGHT

Properties and Processing of Crude Oil From Tight Formations

1 INTRODUCTION

Generally, unconventional tight oil resources are found at considerable depths in tight sedimentary rock formations (tight shale formations, tight siltstone formations, tight sandstone formations, and tight carbonate formations) that are characterized by very low permeability (Chapters 1–3). These tight formations scattered thought North America and other parts of the world have the potential to produce considerable reserves of crude oil (*tight oil*) (US EIA, 2011, 2013; Deepak et al., 2014).

Briefly and for reference, a shale play is a defined geographic area containing an organic-rich fine-grained sedimentary rock that underwent physical and chemical compaction during diagenesis to produce the following characteristics: (1) clay to silt sized particles; (2) high % of silica, and sometimes carbonate minerals; (3) thermally mature; (4) hydrocarbon-filled porosity—on the order of 6–14%; (5) low permeability—on the order of <0.1 mD; (6) large areal distribution; and (7) fracture stimulation required for economic production. The crude oils found in such reservoirs are typically light crude oils (high API gravity) with a low sulfur content (Fig. 8.1).

In a conventional crude oil reservoir (e.g., a conventional sandstone reservoir), the pore spaces are interconnected so that natural gas and crude oil can flow relatively easily from the reservoir rock to a wellbore. In reservoirs composed of tight formations, the pores are smaller and are poorly connected by very narrow capillaries which results in low permeability and these formations typically have a permeability on the order of 1 mD or less (<1 mD) which require stimulation by hydraulic fracturing to produce crude oil (Fig. 8.2).

The most notable tight oil plays in North America include the Bakken shale, the Niobrara formation, the Barnett shale, the Eagle Ford shale, and

Deep Shale Oil and Gas
http://dx.doi.org/10.1016/B978-0-12-803097-4.00008-5

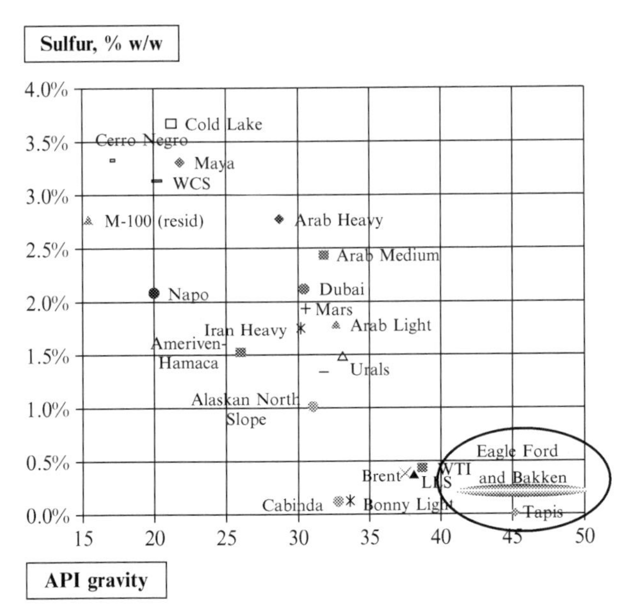

Fig. 8.1 API gravity and sulfur content of various crude oils, including Eagle Ford crude oil and Bakken crude oil.

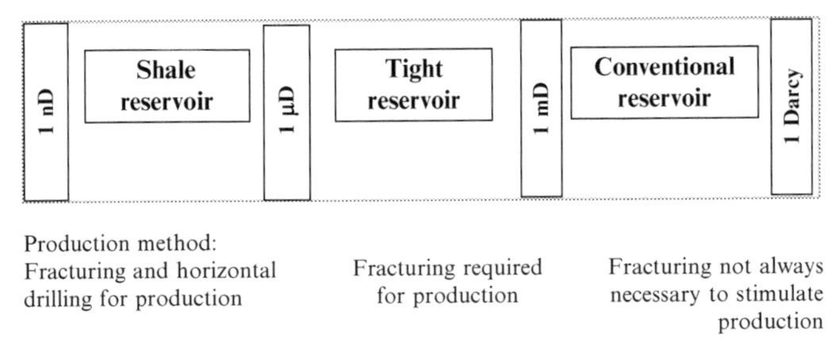

Fig. 8.2 Illustration of reservoir types based on permeability and production methods.

the Miocene Monterey play of the San Joaquin Basin (California) and the Cardium play (Alberta, Canada). In many of these tight formations, the existence of large quantities of crude oil has been known for decades and efforts to commercially produce those resources have occurred sporadically with typically disappointing results. However, starting in the mid-2000s, advancements in well drilling (such as horizontal drilling) and stimulation technologies (such as hydraulic fracturing) (Chapters 4 and 5) combined with favorable economics have turned tight oil resources into one of the most actively explored and produced crude oil resources and crude oil resources in North America (Chapter 3). The production method for Eagle

Ford crude oil and Bakken crude oil is achieved by means of horizontal drilling followed by hydraulic fracturing of the tight formation.

The crude oil comes from the *Bakken formation*, a rock unit from the Late Devonian and early Mississippian age that occupies approximately 200,000 square miles (520,000 km^2) of the subsurface of the Williston Basin that underlies parts of Montana, North Dakota, Saskatchewan, and Manitoba (Fig. 8.3) (Nordquist, 1953; USGS, 2008). The oil is produced from the Bakken formation through horizontal wells in the natural fractures in the rock formation or

Fig. 8.3 Major basins with the potential for tight oil development. *Source: (Adapted from US EIA, 2012. Energy Information Administration, United States Department of Energy, Washington, DC.)*

through the use of horizontal drilling followed by hydraulic fracturing, which is the induced fracturing of the rock formation, accomplished through the high pressure injection of sand, water, and chemicals (which can also include hydrochloric acid and ethylene glycol), in an attempt to release trapped oil and allow it to flow into the well (Chapter 5).

Other known tight formations (on a worldwide basis) include the R'Mah Formation (Syria), the Sargelu Formation (northern Persian Gulf region), the Athel Formation (Oman), the Bazhenov formation and Achimov Formation (West Siberia, Russia), the Coober Pedy formation (Australia), the Chicontepex formation (Mexico), and the Vaca Muerta field (Argentina) (US EIA, 2011, 2013). All are worthy of further investigation and possible development. However, tight oil formations are heterogeneous and vary considerably over relatively short distances. Thus, even in a single horizontal production well, the amount of oil recovered may vary and there may be variable recovery within a field or even variable recovery between adjacent wells. This makes evaluation of tight oil resources and decisions regarding the profitability of wells on a particular lease difficult. In addition, tight reservoirs which contain only crude oil (without natural gas as the pressurizing agent) may require a different economic treatment to make production and economic proposition (US EIA, 2011, 2013).

Thus while tight oil is found throughout the world, the United States, Canada, China, and Argentina are currently the only four countries in the world that are producing commercial volumes of either natural gas from shale formations (shale gas) or crude oil from tight formations (tight oil). Furthermore, the United States is by far the leading producer of both natural gas and crude oil from tight formations, while Canada is the only other country to produce *both* natural gas and crude oil from tight formations.

The challenges associated with the production of tight oils are a function of their compositional complexities and the varied geological formations where they are found. These oils are light, but they are very waxy and reside in oil-wet formations. These properties create some of the main difficulties associated with tight oil extraction. Such problems include scale formation, salt deposition, paraffin wax deposits, destabilized asphaltenes, corrosion, and bacteria growth. Multicomponent chemical additives are added to the stimulation fluid to control these problems.

In summary, common characteristics of the tight oils (Tables 8.1–8.3) can create unforeseen issues that require attention during refining although the sulfur content and other properties including distillate yields may compare favorably with similar properties for other light crude oils.

Table 8.1 Comparison of Properties of Selected Crude Oils to Illustrate the Relative Properties (API Gravity Sulfur Content, and Volatility) of Bakken and Eagle Ford Crude Oils

Crude Oil	API	Sulfur	Light Ends
	Gravity	% w/w	% v/v[a]
Bakken	40–43	0.1	7.2
Eagle Ford	48	0.1	8.3
West Texas Intermediate	37–42	0.4	6.1
Louisiana Light Sweet	36–40	0.4	3.0
Brent (North Sea)	37–39	0.4	5.3
Arabian Light	33	2.0	4.0
Arabian Heavy	28	3.0	3.0

[a]Light ends: low molecular weight organic constituents such as methane, ethane, propane, and butane which are included as components in the oil; in some crude assays, pentane and hexane may be included in the light ends fraction.

Table 8.2 Common Characteristics of Tight Oils

Advantages
Gravity ranges 40–65 degrees API
High yield of distillates
Low sulfur levels
Low levels of nitrogen
High paraffin content
Heavy metals (Ni & V) are low
Low yield of residuum
Disadvantages
Batch to batch variability
Unstable blends when mixed with some crude oils
The presence of hydrogen sulfide can be an issue
High paraffin content
Level of alkaline metals may be high
Other contaminants (Ba, Pb) may be present
Filterable solids: greater volume and smaller particle size
Presence of production chemicals or contaminants
Low yield of residuum and (therefore) low yield of asphalt

2 TIGHT OIL PROPERTIES

The crude oils found in reservoirs classed as tight reservoirs are typically light sweet crude oils (high API gravity) with a low sulfur content and a relatively high proportion of lower-molecular-weight volatile constituents (Table 8.1, Fig. 8.1). In fact, crude oils from tight formation are highly

Table 8.3 Comparison of Selected Properties of Crude Oils From Tight Formations (Eagle Fiord, Bakken) With Conventional Light Crude Oils (Louisiana Light Sweet Crude Oil) and Brent Crude Oil

	Eagle Ford	Bakken	Louisiana Light Sweet (LLS)	Brent (North Sea)
API	44–46	42–44	36–38	37–39
Sulfur, % w/w	0.2–0.3	0.05–0.10	0.35–0.45	0.35–0.45
N, ppm	200–400	300–500	900–1200	900–1100
TAN[a]	0.05–0.1	0.01–0.05	0.5–0.6	0.05–0.10
Light ends,[b] % v/v	13–14%	15–16	9–11	10–12
Naphtha, % v/v	22–24	25–27	19–21	19–21
Middle–distillates, % v/v	31–33%	31–32	33–34	29–31
Vacuum gas oil, % v/v	24–26%	22–24	28–29	28–30
Residuum, % v/v	4–6%	3–5	7–9	9–11

[a]TAN, total acid number.
[b]Light ends: low-molecular-weight organic constituents such as methane, ethane, propane, and butane which are included as components in the oil; in some crude assays, pentane and hexane may be included in the light ends fraction.

variable in their respective properties (Bryden et al., 2014). For example, density and other properties can show wide variation, even within the same field. However, as an illustration of all things not being equal, the light tight oils from the two different formations have different properties although sharing a common crude oil production technique.

American Petroleum Institute (API) gravity is a specific gravity scale developed by the oil industry to identify crude oils and other petroleum liquids based on the relative density. It is expressed in degrees (°API) and is an inverse measure of the relative density of a petroleum product and the density of water. Thus, the lower the API gravity, the denser (heavier) the oil. Arbitrary numbers have been assigned to identify the various crude oils, but these numbers should not be used to classify crude oils (Speight, 2014a, 2015a). For example, light crude oil has an API gravity higher than 30 degrees while a medium crude oil falls in the range APL medium crude between 20 and 30 degrees API and heavy crude oil is assigned an API gravity below 20 degrees API. Sweet crude oils contain less than 0.5% (w/w) sulfur (sour crudes contain more than that). Crude oils from tight formations are typically light and sweet (low sulfur) with an API gravity on the order of 40–50 degrees API.

Typical of the crude oil from tight formations (*tight oil*, *tight light oil*, and *tight shale oil* have been suggested as alternate terms, although the term *shale oil* is incorrect and adds confusion to the nomenclature) (Chapter 1) is the Bakken crude oil which is a light sweet highly volatile crude oil with an API gravity on the order of 40–43 degrees API and a sulfur content on the order of 0.2% (w/w), or less (Table 8.1, Fig. 8.1). The relatively high quality of Bakken crude oil high quality is an advantage insofar as these properties make the oil easier to refine into commercial products but is also a disadvantage insofar as unless the oil is stabilized by removal of the light ends, it is highly flammable when compared to many conventional crude oils. The *flash point*—the lowest temperature at which ignition can occur—is lower for Bakken crude oil than it is for many conventional crude oils, which should be (must be) interpreted as the Bakken crude oil is particularly flammable (in fact, it is highly flammable) and, moreover, when flammable gases (methane and the low-boiling hydrocarbons) are dissolved in oil, the oil should be stabilized (*degasified*) before transportation.

2.1 General Properties

Unlike most conventional light crude oils, tight oils are light sweet oils, with a high paraffin content and low acidity. They also have minimal asphaltenic content and varying contents of filterable solids, hydrogen sulfide (H_2S), and mercaptans. There are significant differences in the sulfur content and the filterable solids loading. In addition, the streams from a tight oil production region can have significant variability, with colors ranging from pale amber to black.

Tight oils are, in general, light crude oils, and the feed API derived from these oils are higher than API of feeds from conventional gas oils and typically have a lower amount of contaminants than conventional crude oils. However, the properties of tight oils are significantly different than typical crude oils and, as a result, a series of challenges exists that need to be solved to ensure uninterrupted transportation and refining of the tight oils to contend with and mitigate the potential disadvantages of the tight crude oils (Table 8.2).

On a comparative basis, tight oils typically have a higher API gravity than traditional crude oils, as well as much different properties. For example, Bakken crude oil and Eagle Ford crude oil have an API gravity on the order of 42–46 degrees, whereas Louisiana light sweet crude oil (a light sweet crude oil from the United States) and Brent crude oil (and international crude oil from the North Sea) have an API gravity on the order of 36–39 degrees. In keeping with the higher API gravity, Bakken crude oil and Eagle Ford crude oil give a higher yield of light ends and naphtha than Louisiana light sweet crude oil US, crude) or Brent crude oil as well as a lower yield of residuum (Table 8.3).

In addition, the crude oil assay reflects the yield pattern of distillates and is key information for determining the refinery products (Table 8.4). Thus, the tight formation crude oils (such as Bakken crude oil, Eagle Ford crude oil, and Utica crude oil) are typically lighter with higher API gravities and have predominantly higher gas constituents, naphtha with less heavy vacuum gas oil vacuum residue materials. Furthermore, the API gravity of crude oil can be used to approximate key properties, such as distillate yields, contaminants, and paraffin concentration—there is a general trend of increasing lower-boiling constituents with increasing API (Speight, 2014a, 2015a). Consequently, Eagle Ford oil with an API on the order of 48 degrees has (would be expected to have a higher yield of naphtha (boiling range: 50–200°C (120–390°F) than Bakken crude oil with an API on the order of 40–43 degrees. Contaminant levels also trend with gravity. For example, sulfur and nitrogen concentrations are generally lower for the light sweet crude oils produced from tight oil plays and higher for lower-gravity crude oils. However, higher-API gravity crudes typically have higher paraffin (wax) concentrations.

However, although tight oil is considered sweet (i.e., low sulfur content in the crude oil itself), hydrogen sulfide gas comes out of the ground with the crude oil. This gas is flammable (producing the noxious sulfur dioxide during combustion) and poisonous

$$H_2S + O_2 \rightarrow SO_2 + H_2O$$

Table 8.4 Yields of Distillate From Eagle Ford Crude Oil

Fraction	**Boiling Range**		**Yield**
	IBP	FBP	% v/v
C1–C4	<85	<85	1
Light naphtha	85	200	14
Heavy naphtha	200	350	23
Kerosene	350	450	12
Light gas oil	450	650	21
Vacuum gas oil	650	1050	24
Residuum	105+		5

Therefore, the presence of hydrogen sulfide must be monitored at the drilling site and during loading at the wellhead, transportation, and offloading of the crude oil at the refinery. Prior to transportation, amine-based scavengers should be added to the crude oil prior to transport to refineries. These scavengers react with the hydrogen sulfide to produce non-volatile products:

$$H_2S + RNH_2 \rightarrow RNH_3{}^{+}HS$$

However, mixing in the railcar due to movement, along with a change in temperature that raises the vapor pressure of the crude oil (an easily achievable event considering the volatility of the crude oil, Tables 8.1 and 8.3), can cause the release of hydrogen sulfide during offloading, thereby creating a safety issues, especially when the crude oil is handled in warmer climates.

2.1.1 Corrosivity

Corrosivity due to the presence of naphthenic acids in crude oil is a particular concern for refineries and had been raised as an issue in permitting the Keystone XL pipeline to transport Canadian oil sand derived crude oil. It is measured as the number of milligrams of potassium hydroxide (milligrams KOH per gram of crude oil, mg KOH/g/g) needed to neutralize the acids in one gram of oil and reported as *total acid number* (TAN) (Speight, 2014b). As a rule-of-thumb, crude oils with a TAN greater than 0.5 are considered potentially corrosive. Bakken crude oil and Eagle Ford crude oil each have a TAN that is less than 0.1 mg KOH per gram of crude oil.

In addition, some samples of Eagle Ford crude have been shown to contain olefin (>C=C<) and/or carbonyl (>C=O) compounds—both of which can act as fouling precursors (Speight, 2015b) that are not typically found in conventional crude oil.

2.1.2 Flammability

There is some risk with mixing any type of crude oil with air in the proper proportion which, in the presence of a source of ignition, can cause rapid combustion (flammable range) or, alternatively, an explosion (explosive range). The flammable range includes all concentrations of flammable vapor or gas in air, in which a flash will occur or a flame will travel if the mixture is ignited at or above a certain temperature (*flash point*). The *lower flammability limit* (lower flammable limit, LFL) is the minimum concentration of vapor or gas in air below which propagation of flame does not occur on contact with a source of ignition. The *upper flammability limit* (upper flammable limit, UFL) is the maximum proportion of vapor in air above which propagation of flame does not occur. The terms *lower explosive limit* and *upper explosive limit* are used interchangeably with LFL and UFL. Following from these definitions, liquids having a flash point at or above 100°F (37.8°C) are classed as *combustible* and liquids having a flash point below 100°F (37.8°C) are classed as *flammable*. However, crude oil may differ in terms of combustibility or flammability depending upon the proportion and properties of volatile and flammable constituents.

The flammability characteristics which include the flash point temperature, the auto ignition temperature, UFL, and LFL are some of important safety specifications that must be considered in assessing the overall flammability hazard potential of hydrocarbons, defined as the degree of susceptibility to ignition or release of energy under varying environmental conditions. Experimental values of these properties are available from data tables that list the properties of hydrocarbons. Typically, the *flash point* is used as the determinant for assessing the flammability of hydrocarbon fuels under a variety of conditions.

The *flash point* of a hydrocarbon fuel is the lowest temperature at which the constituents of the fuel can vaporize to form an ignitable mixture in air. The flash point is not to be confused with the autoignition temperature (Speight, 2015a), which does not require an ignition source, or the fire point, the temperature at which the vapor continues to burn after being ignited. On the other hand, the autoignition temperature of a hydrocarbon fuel is the lowest temperature at which it will spontaneously ignite without an external source of ignition, such as a flame or spark while The fire point of a hydrocarbon fuel is the temperature at which it will continue to burn for at least 5 s after ignition by an open flame (Speight, 2014a, 2015a).

A combustible gas-air mixture as illustrated by the lower boiling constituents of the Bakken crude oil can be burned over a wide range of compositions when subjected either to elevated temperatures or exposed to a catalytic surface at ordinary temperatures (Zabetakis, 1965). For example, methane and other volatile hydrocarbons emanating from a crude oil can be readily oxidized on a heated surface, and a flame will propagate from an ignition source at ambient temperatures and pressures.

2.1.3 Sulfur and Hydrogen Sulfide

Challenges in handling and processing crude oils form tight formations include hydrogen sulfide entrained in the crude and amine-based hydrogen sulfide scavengers added in the pipeline or railcars prior to transportation.

In addition, the amount of free sulfur in a crude oil is an indication of potential corrosivity from the formation of acidic sulfur compounds (such as the sulfur oxides, SO_x where x is 2 or 3, and the sulfur acids, H_2SO_3 and H_2SO_4). Sulfur oxides released into the air during combustion of refined petroleum products are also a major air pollutant. During the decomposition of organic matter that occurs with hydrocarbons in some geologic formations, sulfur may chemically combine with hydrogen to form hydrogen sulfide gas (H_2S), a highly corrosive, flammable, and toxic gas. Oil and gas reservoirs with high concentrations of hydrogen sulfide can be particularly problematic to produce. In addition, hydrogen sulfide causes sulfide-stress-corrosion cracking in the standard steel casing and valves used to construct oil wells and thus require a switch to costly stainless steel (Speight, 2015b).

The sulfur content is measured as an overall percentage by weight (% w/w) of free sulfur and sulfur compounds in a crude oil (Speight, 2014a, 2015a). The total sulfur content in crude oils generally ranges from below 0.05% to 5% (w/w) and crude oils with less than 0.5% free sulfur or other sulfur-containing compounds are typically (and arbitrarily) referred to as *sweet*, and above 0.5% sulfur as *sour*. Bakken crude oil and Eagle Ford crude oils are sweet, each with a low sulfur content (Tables 8.1 and 8.3), but hydrogen sulfide may occur at problematic levels.

Tight oils are typically low in sulfur (an advantage) but may, nonetheless, have high concentrations of hydrogen sulfide (a disadvantage). Triazine-based H_2S scavengers are commonly used upstream for environmental, health, and safety compliance. The triazine derivatives are a class of nitrogen-containing heterocyclic compounds that are based on the molecular formula $C_3H_3N_3$:

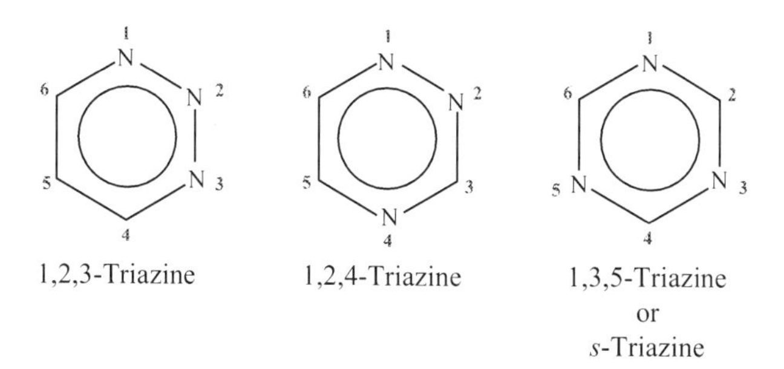

However, the triazine contaminates the crude oil and can contribute to disadvantageous consequence, such as (1) emulsion stabilization, (2) amine halide salt fouling, and under-deposit corrosion throughout the refinery.

Modification of the acidity in the desalting unit can be effective in removing the triazine derivatives from the crude oil.

Unfortunately, these amines (the triazine derivatives) contribute to corrosion problems that arise from the conversion of the triazine derivatives to monoethanolamine ($HOCH_2CH_2NH_2$, MEA) in the atmospheric distillation unit. Once formed, the MEA reacts rapidly with chlorine (from any residual brine associated with the crude oil to form chloride salts). These salts lose solubility in the hydrocarbon phase and become solids at the processing temperatures of the atmospheric distillation tower and form deposits on the trays or overhead system. The deposits are hygroscopic and, once water is absorbed, the deposits become very corrosive.

2.1.4 Volatility

Volatility refers to the evaporation characteristics or distillation characteristics of crude oil (and crude oil products) which includes the vapor pressure at 37.8°C (100°F) of petroleum products and crude oils with initial boiling point above 0°C (32°F) (ASTM D323). Vapor pressure is an important consideration for both crude oil producers and refiners in determining general handling and initial refinery treatment.

The liquids stream produced from the Bakken formation will include the crude oil, the low-boiling liquids, and gases that were not flared, along with the materials and by-products of the fracking process. These products are then mechanically separated into three streams: (1) produced salt water, often referred to as brine; (2) gases; and (3) petroleum liquids, which include condensates, natural gas liquids, and light oil. Depending on the effectiveness and appropriate calibration of the separation equipment which is controlled by the oil producers, varying quantities of gases remain dissolved and/or mixed in the liquids, and the whole is then transported from the separation equipment to the well-pad storage tanks, where emissions of volatile hydrocarbons have been detected as emanating from the oil.

The production of the crude oil also yields a significant amount of volatile gases (including propane and butane) and low-boiling liquids (such as pentane and natural gasoline), which are often referred to collectively as (low-boiling or light) naphtha. By definition, natural gasoline (sometime also referred to as *gas condensate*) is a mixture of low-boiling liquid hydrocarbons isolate from petroleum and natural gas wells suitable for blending with light naphtha and/or refinery gasoline (Mokhatab et al., 2006; Speight, 2007, 2014a). Because of the presence of low-boiling hydrocarbons, low-boiling naphtha (*light naphtha*) can become extremely explosive, even at relatively low ambient temperatures. Some of these gases may be burned off (flared) at the field wellhead, but others remain in the liquid products extracted from the well (Speight, 2014a).

2.1.5 Inorganic Constituents

Initially, the composition of the inorganic constituents of crude oil can be an issue—select samples of Bakken crude have contained salt concentrations as high as 500 ppm, as well as nonextractable salts. Salt composition is higher in terms of the concentration of calcium and magnesium salts (70–90%, w/w) versus typical crude oils which are sodium based (70–80%, w/w). The impact of the shift to calcium and magnesium salts is the potential for hydrolysis in the atmospheric tower fired heater. Hydrolysis is the conversion or decomposition of a salt to the ion and hydrogen chloride—sodium salts do not hydrolyze readily while magnesium and calcium do hydrolyze. The result is a constant total salt concentration the expectation is higher chloride levels in the atmospheric tower overhead leading to the potential for higher corrosion (Speight, 2014c, 2015b).

In addition, scale deposits of calcite ($CaCO_3$), other carbonate minerals (minerals containing the carbonate ion, CO_3^{2-}), and silicate minerals (minerals classified on the basis of the structure of the silicate group, which contains different ratios of silicon and oxygen) must be controlled during production or plugging problems arise. A wide range of scale additives is available which can be highly effective when selected appropriately. Depending on the nature of the well and the operational conditions, a specific chemistry is recommended or blends of products are used to address scale deposition.

2.2 Incompatibility

Briefly, the term *incompatibility* refers to the formation of a *precipitate* (or *sediment*) or *separate phase* when two or more liquids are mixed. The term *instability* is often used in reference the formation of color, sediment, or gum in the liquid over a period of time and is usually due to chemical reactions, such as oxidation, and is chemical rather than physical. This term may be used to contrast the formation of a precipitate in the near term (almost immediately). The phenomenon of *instability* is often referred to as *incompatibility*, and more commonly known as *sludge formation*, and *sediment formation*, or *deposit formation*. In petroleum and its products, *instability* often manifests itself in various ways (Table 8.5) (Stavinoha and Henry, 1981; Power and Mathys, 1992). Hence, there are different ways of defining each of these terms, but the terms are often used interchangeably. Both phenomena have extremely important

Table 8.5 Examples of Properties Related to Instability in Crude Oils and Crude Oil Products

Property	Comments
Asphaltene constituents	Influence oil-rock interactions
	Separates from oil when gases are dissolved
	Thermal alteration can cause phase separation
Heteroatom constituents	Provide polarity to oil
	Preferential reaction with oxygen
	Preferential thermal alteration
Aromatic constituents	May be incompatible with paraffinic medium
	Phase separation of paraffin constituents
Nonasphaltene constituents	Thermal alteration causes changes in polarity
	Phase separation of polar species
Wax constituents	Phase separation of high molecular weight paraffins
	Caused by changes in temperature or pressure
	Dependent upon solvent properties of medium

consequences for the production, transportation, and refining crude oils from tight formations.

The light tight crude oils from tight formation are predominantly paraffinic in character leading to the potential for incompatibility in the form of the deposition of the constituents of paraffin wax (Speight, 2014a). When this occurs, the deposited wax adheres to the walls of railcars, crude oil tank walls, and piping. In the refinery, the wax constituents can foul the preheat sections of crude heat exchangers (before they are removed in the crude desalter) (Speight, 2015b). On the other hand, crude oils from tight formations are characterized by low content of asphaltene constituents, low-sulfur content, and a significant molecular weight distribution of the paraffinic wax constituents. Paraffin carbon chains of C_{10}–C_{60} have been observed, with some tight oils containing carbon chains up to C_{72}. To control deposition and plugging in formations due to paraffins, the dispersants are commonly used. In upstream applications, these paraffin dispersants are applied as part of multifunctional additive packages where asphaltene stability and corrosion control are also addressed simultaneously.

In the refinery processing tight crude oil, each heat exchanger unit upstream of the crude desalter should be equipped with online temperature sensors to monitor the temperatures at the inlet and outlet of the heat exchanger to detect any changes in the rate of heat transfer as they occur. In addition, filterable solids also contribute to fouling in the crude preheat exchangers and since a tight crude oil can contain more filterable solids than a conventional crude oil, filterable solids can be a major issue when refining crude oil from tight formations. In order to mitigate plugging of filters, the filters at the entrance of the refinery require assiduous and careful (automated) monitoring because of the need to capture the solid matter in the feedstocks. In addition, wetting agents may be added to the desalter to help capture excess solids in the water, rather than allowing the undesired solids to travel further downstream into the refinery and thereby causing problems with the various refinery punts, especially in the preheater units to the distillation tower and even within the distillation tower itself (Speight, 2015b).

An additional aspect of incompatibility occurs when crude oil is blended with other crude oils to generate a composite refinery feedstock. It has become common refinery practice that, in order to make optimal use of the available crude oils, to blend several crude oil, which includes blending tight oils with conventional and even with heavier crude oils to produce a more consistent (in terms of composition) feedstock to the distillation unit to facilitate an optimal operation. If tight oil refinery feedstocks are blended with heavier crudes, the lighter oil can create a bottleneck in the crude overhead and naphtha processing units and limit production in bottom-of-the-barrel processing units such as the delayed coker.

The stability of asphaltene constituents in crude oil has always played a role in blending asphaltene-containing crudes; the high paraffin content of tight oils greatly increases the potential impact of asphaltene precipitation upon blending, and its negative impact on the refinery process. There are several established and developing test methods that can evaluate an oil, or a blend, for asphaltene stability. Typically, when blending asphaltene-containing crude oils, the initial blending of the oils may seemingly produce a stable homogeneous mixture but over time (which varies from minutes to days depending upon the properties of the crude oils or other liquids in the blend), the asphaltene constituents start to associate in such a manner that such that they form a separate detectable phase in the fluid. Finally, after significant association (leading to agglomeration) has occurred and the asphaltene constituents have formed larger particles, phase separation occurs. These phase-separated asphaltene constituents contribute to fouling and can also stabilize desalter emulsions.

When crude oils are incompatible, increased asphaltene precipitation accelerates fouling in the heat exchanger train downstream of the crude desalter (Speight, 2015b) which may necessitate an unscheduled shutdown for cleaning. In addition, high naphtha content of tight oils also creates favorable conditions for separation and precipitation of the asphaltene constituents, but it is not just missing the crude oils that can cause problems; it is the relative amounts of each crude oil in the blend. For example, pentane—a precipitating medium for asphaltene constituents—when mixed in specific proportions does not cause separation and precipitation of the asphaltene constituents (Mitchell and Speight, 1973; Speight, 2014a, 2015b). Manual monitoring of heat exchanger fouling often fails to detect crude blends that are incompatible, so it is beneficial to apply a series of test methods to determine the compatibility of the various crude oils in the refinery feedstock blend.

Another concern with tight oil refining is the potential for the phase separation of wax constituents. These waxes can contain very long-chain paraffins and isoparaffins (C_{35} and higher). The subsequent formation of waxy sludge can be problematic in transportation; in storage tanks, impacting storage capacity, and tank drains; as well as increasing the potential for fouling in the cold train. In addition to asphaltene constituents and wax constituents, tight crude oils may introduce other contaminants that may cause fouling which can adversely affect heat transfer throughout the refinery. Chemical treatment programs can effectively address many of the issues related to processing both Eagle Ford and Bakken blends, although the best treatment program for one blend may be very different from the best for the other.

Oil from tight formations is characterized by low-asphaltene content, low-sulfur content and a significant molecular weight distribution of the paraffinic wax content (Speight, 2014a, 2015a). High-molecular-weight paraffins having carbon chains of C_{10}–C_{60} have been found, with some tight oils containing carbon chains up to C_{72}. To control deposition and plugging in formations due to paraffins, the dispersants are commonly used. In upstream applications, these paraffin dispersants are applied as part of multifunctional additive packages where asphaltene stability and corrosion control are also addressed simultaneously (Speight, 2014a,b,c, 2015a).

Generally, the distillates containing high amounts of long-chain paraffins tend to exhibit poor cold flow properties because the paraffin constituents, which make these crude oils suitable for diesel production, are prone to wax production creating flocculation seen as high cloud points and as they crystallize with high pour points. The wax formation starts with a nucleation point for small crystals to collect (*cloud point*). As the crystal grows the larger crystals combine or agglomerate into larger collected masses until the fuel begins to gel (*pour point*). The thermodynamics are such that the heat of fusion can be measured by the differential temperature of the mixture. As a result, cold flow properties require improving via chemical additives, isomerization, or blending.

Crude oil blending operations and/or the product blending operations are the mainstays operations of many refineries, regardless of size or overall configuration. In the blending operations, refinery streams (crude oil or products) are blended in various proportions to produce the desired refinery feedstock mix or the desired finished refined products whose properties meet all applicable industry and government standards, at minimum cost. Production of each finished product requires multicomponent blending because: (1) refineries produce no single blend component in sufficient volume to meet demand for any of the primary blended products such as gasoline, jet fuel, and diesel fuel; (2) many blend components have properties that satisfy some but not all of the relevant standards for the refined product into which they must be blended; and (3) cost minimization dictates that refined products be blended to meet, rather than exceed, specifications to the extent possible. Typically, gasoline is a mixture of approximately 6–10 blend stocks, and diesel fuel is a mixture of approximately 4–6 blend stocks.

Gasoline blending is the most complex and highly automated blending operation. In modern refineries, automated systems meter and mix blend stocks and additives while on-line analyzers (supplemented by laboratory analyses of blend samples) continuously monitor the properties of the blend. Computer control and prebending mathematical models establish composition of the blends to produce the required (stable) product volumes and ensure that all blend specifications are met.

Wax deposition is particularly evident during recovery of crude oils from tight formations in addition to transportation, refining, and product (or crude oil) storage. The properties of the crude oils from tight formations are significantly different to the properties of typical conventional crude oils because of the high paraffin content. In addition, the varying content of filterable solids, hydrogen sulfide (H_2S), and mercaptan derivatives (RSH) also offer challenges during recovery, transportation, and storage. Solids content (and the potential for inorganic deposition) of samples from a single producing region can be highly variable and associated with the stage of fracturing and production from which the oil is produced.

The challenges associated with the production of crude oil from tight shale formation are a function of the wax content which leads to wax deposition. To control deposition and plugging in formations due to paraffins, dispersants are commonly used and, in upstream applications, the dispersants are applied as part of multifunctional additive packages where asphaltene stability and deposition control are also addressed simultaneously. Chemical and mechanical solutions are used to mitigate wax deposition, and these deposit problems and preventive deposition control programs have been developed to manage the wax deposition occurring in storage tanks. Use of the appropriate chemical treatment to control wax buildup during production and during storage can accommodate and transfer larger quantities of waxy crude oil without significant plugging issues.

The high paraffin content of the light tight crude oils is a main contributor that leads to the issue of wax deposition during blending. Mixing oil from tight shale with an asphaltene-containing oil leads to destabilization of the asphaltene constituents, separation, and deposition (asphaltene deposition). In addition, several crude oils form tight formations have a high content of hydrogen sulfide (H_2S) that can lead to the need for hydrogen sulfide scavengers prior to distillation. The scavengers are often amine-based products—such as methyl triazine—that are converted into monoethanolamine ($HOCH_2CH_2NH_2$, MEA) in the atmospheric distillation unit and which contribute to problems in the unit. Once MEA forms, it rapidly reacts with chlorine [from any associated formation water (brine)] to form chloride salts which lose solubility in the hydrocarbon phase and become solids at the processing temperatures of the atmospheric unit and form deposits on the trays or overhead system. The deposits are hygroscopic, and, once water is absorbed, the deposits become very corrosive.

While the basic approach toward developing a tight oil play is expected to be similar from area to area, the application of specific strategies, especially with respect to well completion and stimulation techniques, will almost certainly differ from play to play, and often even within a given play. The differences depend on the geology (which can be very heterogeneous, even within a play) and reflect the evolution of technologies over time with increased experience and availability.

In general, a modern refinery must continue to adapt to increasing variability in crude oil quality. The onset of tight oil refining combined with blending of tight oils into the standard conventional crude oil refinery slate, and any form of standardized refinery operations is difficult and has brought new issues for consideration into the refinery. Thus, processing difficult blends of crude oils, inclosing the light tight crude oils, can have a significant negative impact on overall refinery efficiency which affects product quality, unit reliability, on-stream time, and (above all) profitability.

Determining the means by which a new crude oil (specifically, a light tight crude oil) fits into a refinery operation requires a comprehensive understanding of the physical properties and unique characteristics of that crude and how it will interact with the rest of the typical crude slate

(Speight, 2014a, 2015a,b). Furthermore, the low yield of residuum (Table 8.3) is a disadvantage for refineries that configured for bottom-of-the-barrel upgrading and can limit the amount of tight oil that can be added to the blended cured oils that are the refinery feedstock. In order to balance the mix of products in the crude distillation tower to fit many refinery operations, blending tight oils with heavy asphaltic crude may be possible since the blend can result in a desirable distillation profile for the refinery, but the blend may also exhibit a tendency for instability with the deposition of an insoluble (incompatible) phase (Speight, 2014a, 2015b).

2.3 Wellhead Processing

Producing well sites are unmanned facilities which are visited only periodically by persons performing such functions as collecting crude oil or wastes, maintaining equipment and testing or otherwise operating wells. Since well site equipment, for the most part, is unattended, process and equipment reliability is crucial. However, there are situations in which the wellhead requires management and this is typically when the curd oil received preliminary process at the wellhead in the form of conditioning and/or stabilization. Thus, *wellhead processing* (also called *wellhead stabilization* and *wellhead conditioning*) is the processing of crude oil in proximity to the production well to remove impurities or stabilize the crude oil (by removing highly volatile constituents) prior to transport which has the potential to impact combustion-relevant crude oil properties.

It is recognized (and generally accepted) that no two wells are the same. Consequently, even though there are a limited number of basic types of equipment and configurations involved in crude oil conditioning, the specific configuration and operating conditions adopted for a particular well site depend upon the nature of the producing formation, the location and condition of the well, and ambient conditions at a particular point in time. Time is an important factor in well behavior for many tight oils because the depletion curve, which describes the change in petroleum production rate of a well over time, decreases over time.

The purpose of conditioning is to remove impurities from crude oil with the intent to eliminate compounds (typically nonhydrocarbon compounds) from the oil that lack fuel value or that unnecessarily elevate the hazard level of the petroleum, for example, presence of toxic hydrogen sulfide. The principal purpose of stabilization, in contrast to conditioning, is to remove hydrocarbon compounds which possess fuel value, but which also possess higher vapor pressures (i.e., have lower boiling points) and to reduce the volatility of the crude oil. It should be noted, however, that stabilization also can remove remaining traces of higher vapor pressure impurities remaining in the oil. Even though the two processes have similar effects, that is, removing components from crude oil, they are quite different in the effort and equipment required to perform the separation because, in a sense, water and inorganic gas impurities have inherent propensities to form separate phases from hydrocarbons in crude oil which aids separation, whereas lower-boiling (lower-molecular-weight) hydrocarbons which are the major contributor to crude oil volatility, have inherent affinity for the oil which impedes separation. The impact of this difference is that carefully designed fluid motion and step drops in pressure using gravity and available wellhead pressure, as well as possible chemical and heat addition, can be produced and introduced by equipment of relatively simple and reliable design with limited energy requirements to accomplish satisfactory conditioning.

Conditioning is the removal of water and solids from the crude oil. Equipment is a series of gravity-assisted, pressurized multiphase separators. The number, size, and design of these separators depend upon wellhead conditions, rate of production, relative amounts and phases of impurities, and the nature or composition of the petroleum. Different compositions of petroleum possess different affinities for impurities, such as water or sediment. Depending on the composition of the petroleum, for example, the crude oil and water might easily form two phases with little treatment or the two might form an emulsion that could require chemical additives, heat, quiet flow, centrifugation or other means to separate. After transit through separators, the conditioned petroleum as well as segregated wastewater and solids streams is retained on-site in tanks until collected and transported away from well site by truck or pipeline.

Depending on the quality of the produced water, the water may be recycled or treated and disposed of—often by deep-well injection. Any solids that were carried with the oil also require treatment and proper disposal. Minimally, conditioning that is performed at the well site yields a liquid product that can be safely and economically transported to a refinery or to another facility for further processing.

Stabilization is a process whereby higher vapor pressure components in the crude oil are removed and marketed separately from the stabilized crude oil. The gaseous products include hydrocarbon low-molecular-weight hydrocarbons (such as methane, ethane, propane, and butane) and inorganic gases that either made up a separate phase exiting the well or that were released from the crude oil during conditioning. The gaseous product has economic value, so it is generally collected in relatively low-pressure gathering lines which convey the gases to gas plants which, then, process the stream and ultimately sell it as natural gas and natural gas liquids or, in the absence of gathering lines, is flared at the well site.

On the other hand, stabilization which typically requires more complex processing and equipment, as well as more rigorous, energy-consuming vaporization and condensation of the crude oil. Additionally, while stabilization produces a lower vapor pressure crude oil, it also produces a vapor product that contains higher boiling constituents than those produced from its associated upstream conditioning system. Because of their tendency to condense in pipelines, their greater energy density in the vapor state and other reasons, natural gas pipeline operators have restricted how much of these constituents can be introduced into a natural gas pipeline. Consequently, the lighter-components stream produced by stabilization requires: (1) a process to remove components that are not permitted to be transported in natural gas pipelines from those that are permitted and (2) a transportation system which is separate from both natural gas and stabilized crude oil.

Another approach to crude oil stabilization involves a multistage flash process. Similar to the conditioning processes, multistage flash units can be installed at well sites. However, these systems differ from conventional stabilization systems by possessing more stages that make more efficient use of pressure and heat management. Consequently, multistage flash units require more detailed fabrication as well as more careful monitoring likely have more restricted operating regimes than conventional conditioning equipment.

Fractionation by means of a distillation column is the conventional means of performing crude oil stabilization. In this process, conditioned crude oil is fed into a vertical column containing carefully fabricated structures that enhance vapor and liquid contact and promote mixing of vapor and liquid such that more volatile components in the oil transfer to the vapor phase while the less volatile condense in the liquid phase. Heating the liquid in a reboiler at the bottom of the column promotes movement of more volatile components in the oil to vaporize and move upward in the column, exiting at the top as a vapor. The complexity of the design and equipment makes the process more expensive and less reliable and requires more attention than simpler conditioning processes. The associated equipment includes items such as a furnace to heat the reboiler heating medium, a source of cooling medium for the crude oil cooler and a cooler for the vapor product, as well as ancillary pumps.

3 TRANSPORTATION AND HANDLING

Another challenge encountered with crude oil from tight formations is the need for transportation—usually to a refinery—and the necessary transportation infrastructure (Andrews, 2014). Rapid distribution of the light sweet shale oils to the refineries is necessary to maintain consistent plant throughput. Some pipelines are in use, and additional pipelines need to be constructed to provide consistent supply to the refinery, remembering that the light sweet crude oil may be one of several crude oils blended to produce the refinery feedstock. During the interim, barges and railcars are being used, along with a significant expansion in trucking to bring the various oils to the refineries. Eagle Ford production has been estimated to increase by a factor of 6 over the past years from 350,000 bpd ultimately to approximately 2,000,000 bpd by 2017. Thus, a more reliable infrastructure is needed to distribute this oil to multiple locations. Similar expansion in oil production is estimated for the Bakken crude oil and other identified (and perhaps as yet unidentified) crude oils from tight formations.

Thus, the dramatic increase in crude oil production from tight formations in the United States, coupled with the increase in crude oil transport by rail, has raised questions about whether properties of these crude oils (such as the volatility of the unstabilized crude oil leading to flammability issues)—particularly Bakken crude oil from North Dakota—differ sufficiently from other crude oils to warrant any additional handling considerations. In fact, the United States Pipeline and Hazardous Materials Safety Administration (PHMSA) has already issued a Safety Alert to notify emergency responders, shippers, carriers, and the public that recent derailments crude oil-carrying railcars and the fires that result from such derailments indicate that the type of crude oil transported from the Bakken region of North Dakota may be more flammable than traditional light, medium, and heavy crude oils. The alert reminds emergency responders that light sweet crude oils, such as the crude oil being shipped from the Bakken region, pose significant fire risk if released from the package (tank car) in an accident. As a result, the Pipeline and Hazardous Materials Safety Administration has expanded the scope of lab testing to include other factors that affect proper characterization and classification of crude oil such as volatility, corrosivity, hydrogen sulfide content, and composition/concentration of the entrained gases (the low molecular weight hydrocarbons such as methane, ethane, propane, and butane) in the unstabilized crude oil.

However, it must be remembered that all crude oils are flammable, to a varying degree depending upon the composition, especially the presence of lower-boiling flammable hydrocarbons. The growing perception is that light sweet highly volatile crude oil, such as Bakken crude oil, is a root cause for catastrophic incidents and thus may be too hazardous to ship by rail. Furthermore, crude oils exhibit other potentially hazardous characteristics as well and equally hazardous and flammable liquids from other sources are routinely transported by rail, tanker truck, barge, and pipeline, though not without accident. A key question for state and federal legislators is whether or not the characteristics of Bakken crude oil make it particularly hazardous to ship by rail, or whether or not there other causes of transport incidents, such as: (1) the lack of the necessary prescribed tank cars that should be used to transport such volatile crude oils, (2) the lack of the correct maintenance practices, (3) the lack of adequate safety standards, or unfortunately (4) the occurrence of human error which is a collective grouping of the first three categories.

Crude oil shipments by rail have increased in recent years with the development of oil fields in North Dakota. With several derailments and releases of Bakken crude oil, regulators and others are concerned that it contains low molecular weight hydrocarbon constituents that exacerbate its volatility and thus its flammability. Absent new pipeline capacity, rail provides the primary takeaway capacity for Bakken producers. Unit train shipments of Bakken crude now supply refineries on both the East Coast and the West Coast.

However, the United States Gulf Coast region (Texas and Louisiana) makes up the most prolific region of domestic crude oil production, and the crude oils produced (such as West Texas intermediate crude oil, Eagle Ford crude oil, and Louisiana light sweet crude oil, among others) rival Bakken in the characteristics and, as a result, the alert posted by the United States Pipeline and Hazardous Materials Safety Administration has been called into scrutiny. The Gulf Coast does benefit from existing pipeline infrastructure but, however, producers are relying on rail to access new markets, as evidenced by Eagle Ford crude oil moving from East Texas to St. James, Louisiana by rail. A further policy question is whether Bakken crude oil significantly differs from other crude oils that the standard practices do not apply, and if this is the case, a series of policy steps should be put into place to mitigate the potential for accidents and to remedy safety concerns.

With some degree of justification, Bakken crude oil producers feel that they may have victimized as a result of the alert posted by the United States Pipeline and Hazardous Materials Safety Administration. True or not, serious efforts must be made and safety procedure put into place to mitigate the potential for such disastrous railcar derailments and the resulting fires (some of which have been explosive).

In addition, the paraffin content of the light sweet paraffin-rich crude oils is impacting all transportation systems. Deposits of paraffin wax have been found to coat the walls of railroad tank cars, barges, and trucks. Bakken tight oil is typically transported in railcar, although pipeline expansion projects are in progress to accommodate the long-term need. These railcars require regular steaming and cleaning for reuse. Similar deposits are being encountered in trucks being used for tight oil transportation. The wax deposits also create problems in transferring the tight oils to refinery tankage. The wax deposits that occur in pipelines (to the refinery or in the refinery) regularly require *pigging* to maintain full throughput.

Pigging in the context of pipelines refers to the practice of using devices known as *pigs* to perform various maintenance (cleaning, dewaxing) operations and can be accomplished without stopping the flow of the product in the pipeline. Cleaning (dewaxing) by pigging is accomplished by inserting the pig into a *pig launcher* (*launching station*), which is an oversized section in the pipeline, then reducing to the normal pipeline diameter. The launcher/launching station is then closed and the pressure-driven flow of the product in the pipeline is used to push it along down the pipe until it reaches the receiving trap (the *pig catcher* or *receiving station*). There are various types of pigs—typically, the pig is cylindrical or spherical, pigs sweep the line by scraping the sides of the pipeline and pushing debris ahead.

Multiple chemical and mechanical solutions are used to mitigate the problems cause by wax and other deposits in pipelines. A combination of chemical-additive treatment solutions involving paraffin dispersants and flow drag-reducer technologies has proven to be effective in pipeline applications. Wax dispersants and wash solvents have been used to clean transportation tanks and refinery storage vessels. In the case of pipeline fouling management, a combination of these technologies, coupled with frequent pigging, are the main means to mitigate wax deposition. Preventive fouling control programs have been developed to manage the wax deposition occurring in storage tanks. By injecting the proper chemical treatment to control wax buildup in storage tanks, the production field and refinery can handle and transfer larger quantities of oil without significant plugging issues.

One other problem encountered in storing and transporting tight oils is the concentrations of the light ends (volatile hydrocarbon gases) that accumulate in the vapor spaces, requiring increased safety and relief systems. Shipping Bakken crude via barges was challenged by the increased levels of volatile organic compounds and vapor-control systems ate necessary to ensure a safe environment.

Due to the paraffinic nature of tight oils and the absence of a substantial amount of high-boiling constituents that usually form the distillation residuum (Table 8.3), most refineries blend the light sweet crude with other crude oil feedstocks. Unfortunately, the light sweet crude oils have low aromatic content, so mixing with conventional crude oil often leads to asphaltene destabilization. If blended oils are transported, the deposits can consist of waxes and precipitated asphaltenes resulting in asphaltene deposition (incompatibility). Dispersants specifically designed for both hydrocarbon types can control deposit formation during transportation. Until safety issues in the transportation of the light sweet crude oil from tight formations are satisfied and the potential for accidents is diminished by the institution of a safe and reliable transportation infrastructure, significant issues relating to shipments of these crude oils and the potential for accidents remain distinct possibilities. In addition to this, refineries are already experiencing the impact of the variations in quality (composition) of light sweet crude oil from tight formations and the ensuing processing challenges.

Finally, prior to transportation, producers of crude oil are required to determine the appropriate hazard classification of their produced crude oil production at various stages in the process and for various purposes. As an example, the wellhead storage tanks should be marked with diamond shaped warning placards of different color and number to show the appropriate hazard classification of their contents.

Considering the volatility of crude oil from tight formations, such as Bakken crude oil, safety precautions are necessary prior to shipment. The most dangerous flammable liquids based on their flash point and initial boiling point. For example, naphtha (a precursor to gasoline) vaporizes easily at a relatively low temperature to form a flammable mixture with air. Bakken crude oil, which contains naphtha constituents, is also highly flammable and precautions must be addressed before the crude oil is shipped. In addition, during transportation, stratification of the crude oil can occur in which the lower-boiling lower density constituents can move to the top of the liquid. These constituents are the most dangerous in terms of low flash point and high flammability and caution is advised when opening the containers to transfer the liquid to storage vessels, especially of the transportation containers are top loaders and unloaders.

4 BEHAVIOR IN REFINERY PROCESSES

In a very general sense, petroleum refining can be traced back over 5000 years to the times when asphalt materials and oils were isolated from areas where natural seepage occurred. Any treatment of the asphalt (such as hardening in the air prior to use) or of the oil (such as allowing for more volatile components to escape prior to use in lamps) may be considered to be refining under the general definition of refining. However, petroleum refining as we know it is a very recent science and many innovations evolved during the 20th century.

Briefly, petroleum refining is the use of series on integrated unit processes to separate of petroleum into fractions and the subsequent treating of these fractions to yield marketable products (Speight, 2014a). In fact, a refinery is essentially a group of manufacturing plants which vary in number with the variety of products produced (Fig. 8.4). Refinery processes must be selected and products manufactured to give a balanced operation in which petroleum is converted into a variety of products in amounts that are in accord with the demand for each. For example, the manufacture of products from the lower-boiling portion of petroleum automatically produces a certain amount of higher-boiling components. If the latter cannot be sold as, say, heavy fuel oil, these products will accumulate until refinery storage facilities are full. To prevent the occurrence of such a situation, the refinery must be flexible and be able to change operations as needed. This usually means more processes: thermal processes to change an excess of heavy fuel oil into more gasoline with coke as the residual product, or a vacuum distillation process to separate the heavy oil into lubricating oil stocks and asphalt.

The first major step in refining is distilling or separating crude oil into different hydrocarbon streams by boiling point (Fig. 7). Light crude oil will have a high share of "light weight" material that boils at low temperatures, such as light and heavy naphtha, while heavy crude oils will have a smaller share of those light materials and a larger share of material that is heavy and boils at very high temperatures, like those present in atmospheric tower bottoms. As domestic refiners use more U.S. super light crude oils and less of other heavier crude types, they sometimes must alter the equipment that processes crude oil before it reaches the distillation unit and may even alter the crude distillation unit. Regardless, the use of more super light crude oils would change what comes out of the crude distillation unit. Using

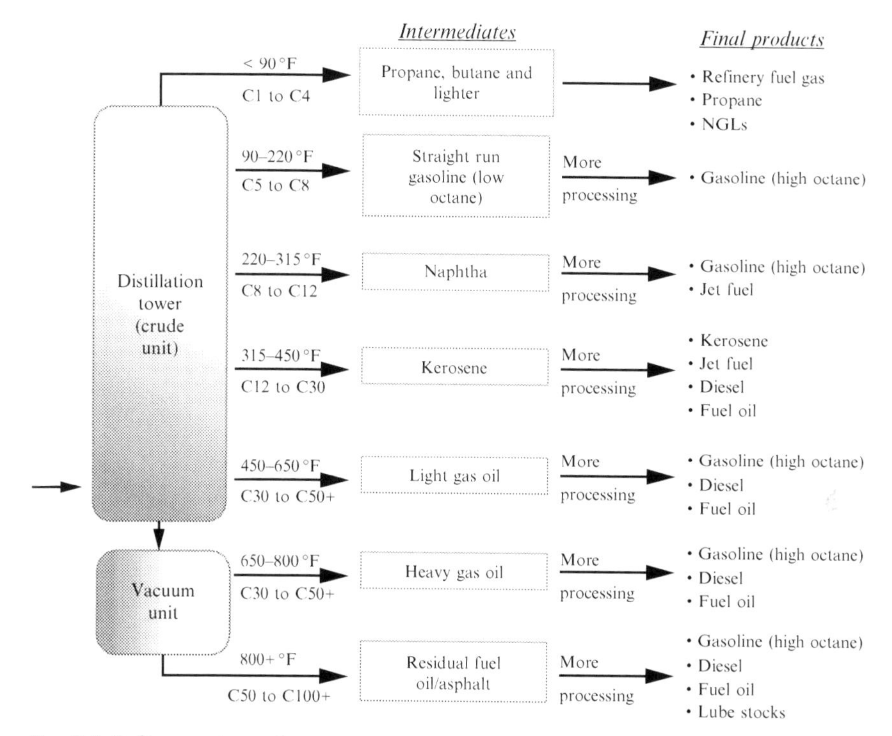

Fig. 8.4 Refinery schematic.

proportionally more light crude oil results in higher yields (i.e., volume shares) of light streams like naphtha, and lower yields of heavier streams.

In terms of refinery operations, the properties of crude oil from tight formations are highly variable (Bryden et al., 2014). Typically, the tight crude oils light, paraffinic, and sweet have a low amount of gas oil (feedstock to a fluid catalytic cracking unit) and while mot tight oils are generally easy to process, challenges arise when these crudes are the predominant feedstock in refineries designed for heavier crude oils. In addition, tight oils, like other light sweet crude oils, have a much higher ratio of atmospheric distillates (650°F^{-}) to atmospheric residuum (650°F^{+}) compared to conventional crude oils. For example, this distillation-to-residuum ratio for Bakken crude oil is on the order of 2:1 while typical conventional light oil has a ration closer to 1:1 (Bryden et al., 2014). Thus, a refinery accepting a light tight oil as feedstocks would need to accommodate the higher proportion of atmospheric distillates which could cause on overload on ancillary equipment, such as the naphtha processing units. At the same time, the light tight oil provides insufficient feedstock (gas oil) for the catalytic cracking unit and feedstock (residuum) for the coking unit.

Light oil is by no means new to the petroleum industry, but the recent expansion in the extraction of crude oil from the Bakken formation, the Eagle Ford formation, and similar tight formations does represent a new and unexpected development for the petroleum industry. The initial scientific and technical reaction to these types of crude oils is that such crude oils will be easy to refine and produce valuable low-boiling fuel products. However, while this may be true to a point, it is not necessarily the complete story and several issues exist related to the production, transportation, and the development of unconventional tight reservoirs producing light tight oil leading to the storage, transportation, and refining of these light sweet crude oils that has presented the petroleum industry with several issues that have been and are still being considered and resolved.

In fact, there are many areas of a refinery that can be impacted by the introduction of light sweet oils from tight formations, beginning at the tank farm. The presence of wax constituents, solids, and incompatibility as a result of blending can lead to unloading problems, wax sludge buildup, and tank drain plugging. Increased solids, salts, and other contaminants contribute to fouling in heat exchangers, furnaces, and atmospheric distillation columns (Speight, 2014a, 2015b).

Most refineries process are no longer limited to a single crude oil as the refinery feedstock and each crude oil can require different processing conditions in the major refinery units. Typically, a refinery is designed to process a feedstock with a particular composition and produce products with specified properties, with some flexibility based on the capabilities of equipment such as the reactors (and ancillary equipment such as pumps and heat exchangers) as well as the particular catalysts within the reactors. As a result, refiners try to match the crude oil composition to the configuration of the refinery, usually by blending several crude oils. However, over the past four decades, many refineries have made major investments to enable them to process heavier sour crude oils from sources such as Venezuela and Canada. These changes were made before the technological advancements that triggered the recent high interest in light sweet crude oils from tight formations. Moreover, because tight oil contains minor amounts of residuum and is low in sulfur content, there is often a mismatch between properties of the light sweet crude oil and the crude oil properties as required by the modified/upgraded refinery.

With the increase in supply of domestic unconventional oils, especially tight oils; for the first time in two decades, United States oil production has exceeded oil imports by the United States. Major Tight oil sources in the

United States come from the Bakken, Eagle Ford, and Permian Basin formations (Deepak et al., 2014), but current estimates of the total crude oil reserves in tight formations have been on the order of 50 billion barrels (50×10^9 bbls) of crude oil from tight formations in the United States (this number is continually being reevaluated and by the time of publication of this book could be considerably different), indicates that tight oil is here to stay, and will play an important role in current and future refinery activities and economics. Hence, fully understanding tight oil will be a crucial part of the current and future oil resource for refiners.

Common observations about light tight crude oils are they produce high value (high API) low sulfur products but require changes to the operation to adjust to the differences in crude oil properties. As a consequence of these variations in the properties of the crude oils from tight formations, it is increasingly more important for refiners to be able to identify, interpret, and respond quickly to changes in crude feed properties. Generally, these crude oils present numerous challenges, which may be reflected in the price when compared to crude oils such as Brent crude oil (a global benchmark crude oil) or West Texas intermediate crude oil (a US benchmark crude oil benchmark). The challenges typically imposed by the light crude oils from tight formations include: (1) difficulty in transportation due to high volatility; (2) the presence of hydrogen sulfide, which requires the addition of amine-based scavengers in the pipeline, truck, or railcar prior to transportation; (3) the presence of paraffin waxes that can cause fouling in piping, tank walls, and in crude preheat exchangers; (4) the presence of relatively high amounts of filterable solids; (5) the requirement of blending with other crude oils to balance the input to the atmospheric distillation unit; (6) the potential for incompatibility with other types of crude oils used in the blending operations; and (7) there may be cold-flow property deficiencies because of the presence of wax constituents that may require modifications to process catalysts. Other issues such as the requirement for energy balancing across the various preheat exchangers must also be taken into consideration.

The light, sweet crude oil characteristic of light sweet crude oils from tight oil plays is a distinct contrast to the heavy, sour Canadian tar sands bitumen (Speight, 2013; Wier et al., 2016). These light crude oils have very different yield patterns for refined products, as well as contaminant levels, impacting refiners' processing units and product slates. Tight oil crudes typically have higher light and heavy naphtha yields, presenting increasing challenges to the naphtha complex, typically consisting of naphtha hydrotreating, light naphtha isomerization, and catalyst regeneration platforming units.

More specifically, the light sweet crude oils from tight (low-permeability) formations have a composition that allows the yield of atmospheric distillate to approach 80% (v/v) of the crude oil, while the yield of vacuum residue may be as low as 5% (v/v) (Furimsky, 2015). The waxy behavior of these crude oil is a result of a high content of *n*-paraffin derivatives and, thus, blending of tight oils and/or feedstocks derived from tight oils requires a preblending series of test methods (Speight, 2014a, 2015a) to determine the potential for incompatibility. In addition, safety precautions must be taken during all stages of handling and transportation of these crude oil because of a high content of methine-to-butane (C_1–C_4) hydrocarbons.

Thus, many factors can affect the ability of refineries to process these crude oils, and the effects are not only specific to the source of the light crude oil but also are very specific to each refinery and to individual process unit within a refinery. The effects may be manifested in the atmospheric distillation unit or in the downstream processing of the volatile constituents of the crude oil into finished products. Examples of such effects are: (1) the atmospheric distillation unit does not have the necessary capacity for the more volatile constituents of the crude oil, (2) the heater or heat exchanger has insufficient heater flexibility or limited ability to cool and condense higher volume of light ends, (3) the saturated gas plant has insufficient capacity to process additional volume, and (4) the downstream processing capacity of the refinery limits the ability of the refinery to convert the intermediate products into finished (saleable) products.

While these *tight oils* have revolutionized crude supplies in the United States, they also are presenting refiners with new challenges. These crudes produce hydrocracker feeds with significantly fewer contaminants than oil sands derived feeds or conventionally produced gas oils, but there are reports that they may be contributing to crude blending compatibility issues, furnace and heat exchanger fouling, yield selectivity changes, and product quality issues. Only through technology advances such as new-generation catalysts and process designs can refiners match the need to optimize existing and new facilities to take advantage of expanding production of the light sweet crude oils.

4.1 Desalting

Before separation of petroleum into its various constituents can proceed, there is the need to clean the petroleum, which may also occur at the wellhead as well as inside the refinery. This is often referred to as desalting and dewatering in which the goal is to remove water and the constituents of the brine that accompany the crude oil from the reservoir to the wellhead during recovery operations. Field separation, which occurs at a field site near the recovery operation, is the first attempt to remove the gases, water, and dirt that accompany crude oil coming from the ground. The separator may be no more than a large vessel that gives a quieting zone for gravity separation into three phases: gases, crude oil, and water containing entrained dirt.

The desalting unit is the first line of defense for successful refinery crude unit corrosion and fouling control is optimal operation of the desalter (Dion, 2014). Most crude oils vary greatly in terms of quality and the processing challenges they represent. Additionally, crudes and their blends can be incompatible, precipitating asphaltene constituents and high-molecular-weight aliphatic compounds (wax constituents). This precipitation can increase the stability of emulsions and contribute to downstream fouling. Fluctuating crude quality and compatibility issues elevate the importance and challenge to effective desalter operation. Adopting an integrated approach to refinery operations when processing opportunity crudes can help anticipate and negate many of the negative impacts to downstream units, such as the waste water treatment plant, as well as provide an opportunity to improve overall plant reliability, such as with the use of low salting boiler amines to minimize crude unit overhead amine salt corrosion potential.

Tight oil is recovered from the reservoir mixed with a variety of substances: gases, water, and dirt (minerals) and, in that sense, is analogous to many opportunity crudes in regard to initial crude oil cleaning (Dion, 2014). Thus, refining actually commences with the production of fluids from the well or reservoir and is followed by pretreatment operations that are applied to the crude oil either at the refinery or prior to transportation. Pipeline operators, for instance, are insistent upon the quality of the fluids put into the pipelines; therefore, any crude oil to be shipped by pipeline or, for that matter, by any other form of transportation must meet rigid specifications in regard to water and salt content. In some instances, sulfur content, nitrogen content, and viscosity may also be specified.

Desalting is a water-washing operation performed at the production field and at the refinery site for additional crude oil cleanup. If the petroleum from the separators contains water and dirt, water washing can remove much of the water-soluble minerals and entrained solids. If these crude oil contaminants are not removed, they can cause operating problems during refinery processing, such as equipment plugging and corrosion as well as catalyst deactivation.

The usual practice is to blend crude oils of similar characteristics, although fluctuations in the properties of the individual crude oils may cause significant variations in the properties of the blend over a period of time. Blending several crude oils prior to refining can eliminate the frequent need to change the processing conditions that may be required to process each of the crude oils individually. However, simplification of the refining procedure is not always the end result. Incompatibility of different crude oils, which can occur if, for example, a paraffinic crude oil is blended with heavy asphaltic oil, can cause sediment formation in the unrefined feedstock or in the products, thereby complicating the refinery process (Mushrush and Speight, 1995).

Desalter performance has traditionally been measured by salt removal, oil dehydration, and chloride control efficiency. However, the recent influx of oils from tight formations has introduced tremendous variability in the quality of the crude blends, prompting many refiners to rethink the role of the desalter. Refiners are now often running this equipment more as an extraction vessel, removing many more contaminants than just salt. While the individual desalter challenges may not be particularly new, the combination of issues is new.

The high API gravity tends to improve desalting by creating a greater density difference between crude and water increasing the stokes settling velocity, but there can be difficulties in the desalting unit due to the formation of emulsions and the high amount of filterable solids adds to the desalter load and reduces efficiency. On the other hand, in keeping with the high API gravity, there is an increase in the yield of naphtha and other distillate fractions compared to conventional crude oils. However, wax formation in the cold crude and distillate hydrotreater preheat exchangers causes fouling (Speight, 2015b) which reduces heat transfer and has increased pressure drop requiring cleaning at shortened intervals. Furthermore, the presence of hydrogen sulfide requires use of hydrogen sulfide-scavengers that leads to the formation of amine salts and corrosion.

However, compatibility problems can result from blending highly paraffinic crudes with asphaltenic crudes, which lead to asphaltene destabilization that can stabilize emulsions, as well as accelerate preheat and furnace fouling. Tight oils can cause wax precipitation, which can degrade desalter temperatures and plug cold train exchangers. Variability, due to raw salt and BS&W, can stabilize emulsions in the desalter, as well as impact corrosion control in the overhead system. Increased solids loading may exceed the design capability of the desalter, resulting in emulsion control issues, accelerated fouling of the preheat train and furnace, as well as more difficult phase separation downstream in the primary wastewater and slop oil handling systems.

Wetting agent adjunct chemistry can also be very helpful when processing tight oils. The fracking process used to extract tight oils increases the amount of entrained solids. Compared to those found in traditional crudes, these solids are smaller and the volume is typically higher as well. The increased solids loading can easily overwhelm the ability of the desalter to remove solids. Loadings as high as 300 pounds per thousand barrels in the raw crude have been documented when processing certain tight oils. The higher loadings can cause stabilized emulsions, which can lead to water carry-over in the oil and oily effluent brine as well. Entrained oil in the effluent brine can cause problems in the wastewater treatment plant. The solids are inorganic particles that are coated in oil. Wetting agents help strip the oil layer from the particles and make it easier for them to be removed from the desalter.

In addition to the desalting and corrosion challenges associated with processing tight oils, fouling of equipment beyond the desalter can be a major concern. The processing of lighter crudes, with low asphaltenes, is not typically thought of as being particularly problematic. However, there are specific issues associated with these crudes that have been identified to cause issues in the refinery process. For example, when paraffinic tight oils are blended with asphaltenic crudes, the asphaltene constituents can destabilize and agglomerate, leading to emulsion stabilization, increased oil in the effluent, as well as preheat exchanger and furnace fouling. This is a lesser issue when refining tight crude oils because of the relative lack of asphaltene constituents. However, wax separation can be a major problem. Also, the cold train can experience wax precipitation, with the resultant loss of heat transfer causing low desalter temperatures in addition to increased pressure drop across the cold train heat exchangers. Also, increased preheat and furnace fouling potential can be experienced with these crudes due to asphaltene precipitation, metal-catalyzed polymerization, and/or solids deposition.

Thus, a critical understanding of the characteristics of the tight oils, as well as the blends being processed, is needed to properly address poor desalter performance, corrosion, and fouling (Speight, 2014c).

4.2 Fouling During Refining

One of the major causes of fouling during refining tight oils relates to the use of these oils as a refinery blend stock to produce the right blend for the utilization of existing downstream units. Having a more consistent feed to the crude unit allows for the opportunity to optimize operation. If light tight oil feeds are not blended, the lighter oil can create an overflow (bottleneck) in the atmospheric distillation unit and also in the naphtha processing units. In addition, when the light tight oil is incompatible with another constituent of the blend (or with other constituents of the blend), accelerated fouling occurs in the crude unit heat exchanger train due to asphaltene separation and precipitation (Speight, 2015b).

There are typically two types of fouling in the hot train and furnaces. Coke and inorganic solids are the primary culprits. The coke can result from asphaltene precipitation or polymerization by-products that fall out of the bulk fluid onto the tube surfaces and dehydrogenates. Metal-catalyzed polymerization is somewhat rare in crude oil but does occasionally occur due to sporadic spikes in the levels of reactive metals. Finally, high solids loading, common with these crudes, along with any carryover from the desalter can significantly contribute to fouling issues. Most refiners run years before fouling requires the furnace to be cleaned. Recently, some refiners have experienced as little as 3 months between turnarounds to clean the crude furnace.

In many crude oil, the separation of unreacted asphaltene constituents or reacted asphaltene constituents can cause solids deposition and fouling (Speight, 2015b). Refiners employ many performance management strategies to reduce or mitigate equipment fouling, including operational and mechanical adjustments as well as antifouling chemistries. Some of the common operational or mechanical approaches are: reducing solids and salts by optimizing desalter performance, increasing fluid velocities to minimize deposition potential, and modifying furnace flame patterns by cleaning or changing burner tips to maximize performance and minimize impingement that can cause coking.

Using proper characterization methods to understand the root cause of the fouling can help determine the most appropriate management strategies. When starting to process problematic crude blends or increasing a problematic crude type in a blend, effective baseline monitoring is extremely important to understand the status of the current system, as well as to anticipate what limitations may develop. Thus, the unique challenges associated with processing tight oils can be overcome with a combination of baseline and ongoing monitoring, defining and implementing new operating envelopes, and utilizing multifunctional chemical treatment programs that provide refiners with the tools and the flexibility to address the specific process problems as they arise.

4.3 Refining Options

To convert crude oil into desired products in an economically feasible and environmentally acceptable manner (Bryden et al., 2014). Refinery process for crude oil is generally divided into three categories: (1) separation processes, of which distillation is the prime example; (2) conversion processes, of which coking and catalytic cracking are prime examples; and (3) finishing processes, of which hydrotreating to remove sulfur is a prime example (Speight, 2014a, 2016).

The simplest refinery configuration (Table 8.6) is the *topping refinery* (*skimming refinery*) (Fig. 8.5) which is designed to prepare feedstocks for petrochemical manufacture or for production of industrial fuels in remote oil-production areas. A version of the topping refinery may be installed as a wellhead refining unit. The topping refinery consists of tankage, a distillation unit, recovery facilities for gases and light hydrocarbons, and the necessary utility systems (steam, power, and water-treatment plants). Topping refineries produce large quantities of unfinished oils and are highly

Table 8.6 Comparison of Various Refinery Types

Refinery Type	Processes	Alternate Type Name	Complexity	Comparative Range*
Topping	Distillation	Skimming	Low	1
Hydroskimming	Distillation Reforming Hydrotreating	Hydroskimming	Moderate	3
Conversion	Distillation Fluid catalytic cracking Hydrocracking Reforming Alkylation (etc.) Hydrotreating	Cracking	High	6
Deep conversion	Distillation Coking Fluid catalytic cracking Hydrocracking Reforming Alkylation (etc.) Hydrotreating	Coking	Very high	10

*Indicates complexity on an arbitrary numerical scale of 1–10.

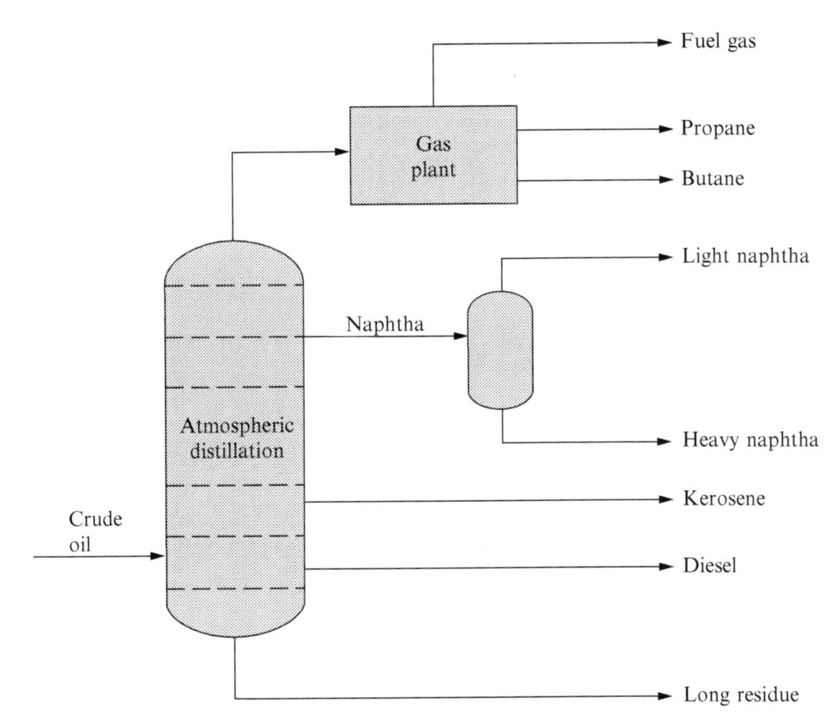

Fig. 8.5 A topping refinery.

dependent on local markets, but the addition of hydrotreating and reforming units to this basic configuration results in a more flexible *hydroskimming refinery* which can also produce desulfurized distillate fuels and high-octane gasoline. These refineries may produce up to half of their output as residual fuel oil, and they face increasing market loss as the demand for low-sulfur (even no-sulfur) high-sulfur fuel oil increases.

The most versatile refinery configuration is the *medium conversion refinery* (also called a cracking refinery) which incorporates all of the basic units found in both the topping and hydroskimming refineries, but it also features gas oil conversion plants such as catalytic cracking and hydrocracking units, olefin conversion plants such as alkylation or polymerization units but not always coking units (Fig. 8.6). Modern conversion refineries may produce two-thirds of their output as unleaded gasoline, with the balance distributed between liquefied petroleum gas, jet fuel, diesel fuel, and a small quantity of coke. Many such refineries also incorporate solvent extraction processes for manufacturing lubricants and petrochemical units with which to recover propylene, benzene, toluene, and xylenes for further processing into polymers.

The *high conversion refinery* (*coking refinery*) is, as the name implies, a special class of conversion refinery which includes not only catalytic cracking

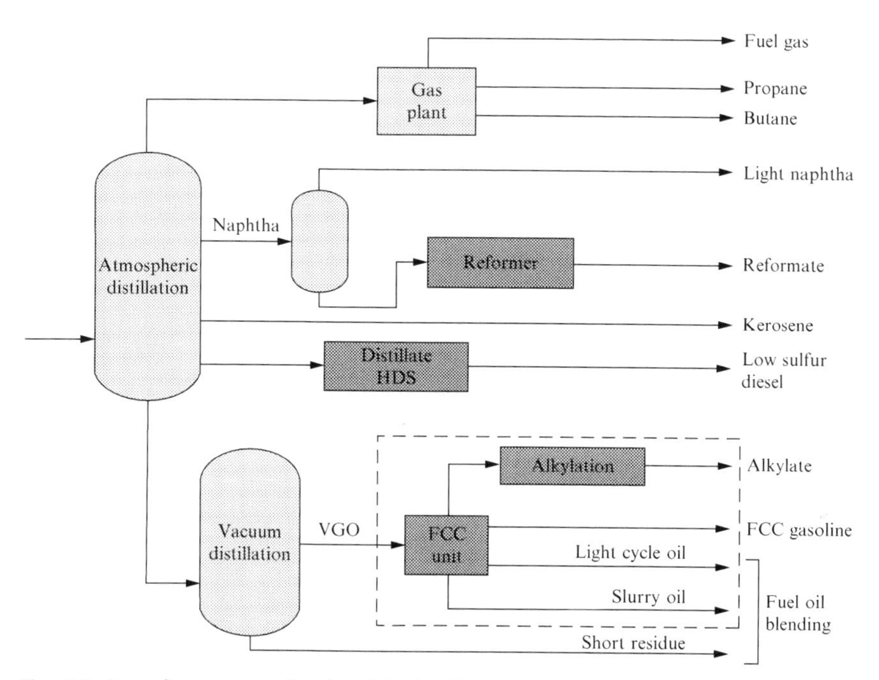

Fig. 8.6 A medium conversion (cracking) refinery.

and/or hydrocracking to convert gas oil fractions, but also a *coking unit* for reducing or eliminating the production of residual fuels and residua (Fig. 8.7). The function of the coking unit is to convert the highest boiling and least valuable crude oil fraction (*residua*) by converting it into lower-boiling product streams that serve as additional feedstocks to other conversion processes (such as the catalytic cracking unit) and to upgrading processes (such as the catalytic reforming unit) that produce the more valuable light products. A high conversion refinery with sufficient coking capacity will convert all of the residua in the crude oil feedstocks slates and produce lower boiling products or, alternatively, to leave a portion of the residuum for asphalt production. Almost all of the refineries on the United States are either *medium conversion refineries* or *high conversion refineries*, as are the newer refineries in Asia, the Middle East, South America, and other areas experiencing rapid growth in demand for lower-boiling products. By contrast, most refining capacity in Europe and Japan is in *hydroskimming refineries* and *medium conversion refineries*.

Finally, the yields and quality of refined petroleum products produced by any given refinery depends on the mixture of crude oil used as feedstock and the configuration of the refinery facilities. Light/sweet crude oil is generally

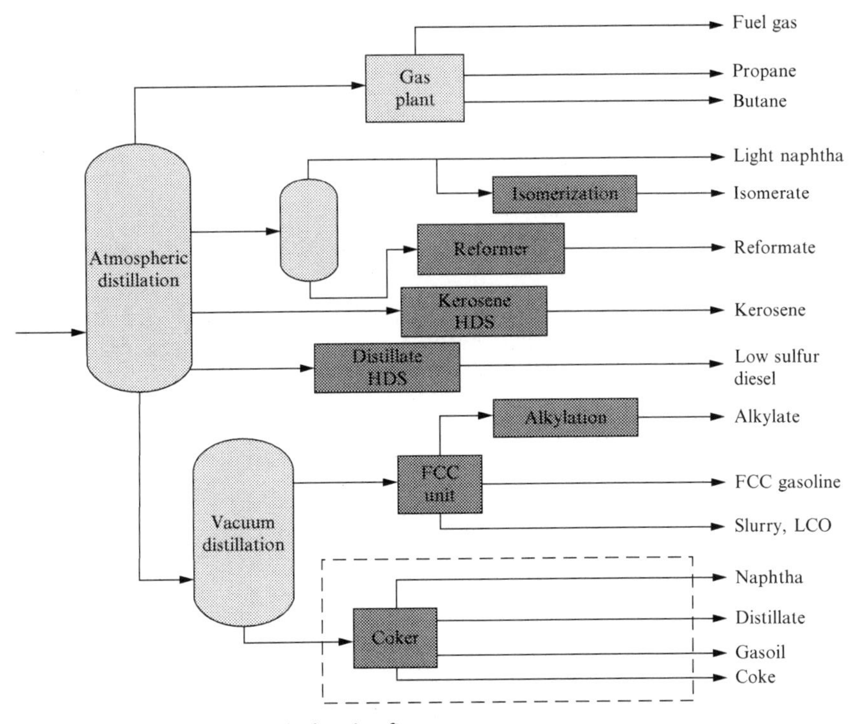

Fig. 8.7 A high conversion (coking) refinery.

more expensive and has inherent great yields of higher value low-boiling products such naphtha, gasoline, jet fuel, kerosene, and diesel fuel. Heavy sour crude oil is generally less expensive and produces greater yields of lower value higher boiling products that must be converted into lower boiling products. The configuration of refineries may vary from refinery to refinery. Some refineries may be more oriented toward the production of gasoline (large reforming and/or catalytic cracking), whereas the configuration of other refineries may be more oriented toward the production of middle distillates such as jet fuel and gas oil.

The means by which a refinery operates in terms of producing the relevant products depends not only on the nature of the petroleum feedstock but also on its configuration (i.e., the number of types of the processes that are employed to produce the desired product slate), and the refinery configuration is, therefore, influenced by the specific demands of a market. Therefore, refineries need to be constantly adapted and upgraded to remain viable and responsive to ever changing patterns of crude supply and product market demands. As a result, refineries have been introducing increasingly complex and expensive processes to gain higher yields of lower boiling products from the higher boiling fractions and the residua.

The yields and quality of refined petroleum products produced by any given oil refinery depends on the mixture of crude oil used as feedstock and the configuration of the refinery facilities. Light/sweet crude oil is generally more expensive and has inherent great yields of higher value low-boiling products such as naphtha, gasoline, jet fuel, kerosene, and diesel fuel. Heavy sour crude oil is generally less expensive and produces greater yields of lower value higher boiling products that must be converted into lower-boiling products (Speight, 2013, 2014a).

Since a refinery is a group of integrated manufacturing plants (Fig. 4.1) which are selected to give a balanced production of saleable products in amounts that are in accord with the demand for each, it is necessary to prevent the accumulation of nonsaleable products, the refinery must be flexible and be able to change operations as needed. The complexity of petroleum is emphasized insofar as the actual amounts of the products vary significantly from one crude oil to another (Speight, 2014a, 2016). In addition, the configuration of refineries may vary from refinery to refinery. Some refineries may be more oriented toward the production of gasoline (large reforming and/or catalytic cracking), whereas the configuration of other refineries may be more oriented toward the production of middle distillates such as jet fuel and gas oil.

Each refinery has a unique configuration, set of processing objectives, equipment limitations, and budget constraints. Therefore, there is no universal solution to maximize profitability. Solutions for refiners located in the Gulf Coast and Midwest regions that can access and sell products more readily may differ from refiners in other regions who incur higher product transportation costs.

4.4 Other Refining Issues

In general, tight crude oils have low nitrogen and high paraffin content (Table 8.1) which can offer advantages or disadvantages to refiners. Heavy metals, such as nickel and vanadium, are generally low, but alkaline metals (calcium, sodium, and magnesium) may be high. This is highly variable and, in addition, other contaminants such as barium and lead may be elevated. Filterable solids can be higher than conventional crude oils, with greater volume and smaller particle size.

Refining tight oil extracted through fracturing from fields such as Eagle Ford, Utica, and Bakken has become prevalent in many areas of the United

States. Although these oils are appealing as refinery feedstocks due to their availability and low cost, processing can be more difficult. The quality of the tight oils is highly variable. These oils can be high in solids with high melting point waxes. The light paraffinic nature of tight oils can lead to asphaltene destabilization when blended with heavier crudes. These compositional factors have resulted in cold preheat train fouling, desalter upsets, and fouling of hot preheat exchangers and furnaces. Problems in transportation and storage, finished-product quality, as well as refinery corrosion have a high potential.

In addition to catalyst selection, an equally critical component to minimizing risks and challenges associated with processing unconventional feeds is solid technical service support (Bryden et al., 2014). Understanding feed impacts earlier allows opportunity to optimize the operating parameters and catalyst management strategies, enabling a more stable and profitable operation.

Hydroprocessing of naphtha from tight oils is necessary to increase octane number and that of middle distillates to attain cold flow properties of diesel fuel. Also, hydroprocessing of atmospheric residue from tight oils yields additional naphtha and middle distillate feedstocks for other refinery operations. Because of a great variability in properties, the prehydroprocessing of tight oil feedstocks feeds may be necessary. Hydroisomerization is the principal reaction during the hydroprocessing of atmospheric distillates feedstocks, while hydrocracking and ring opening are the principal reactions when the atmospheric residuum from tight oil is hydroprocessed (Furimsky, 2015). In fact, one of the significant challenges for hydrocrackers processing significant amount of straight run material from tight oil crudes is a lower overall gas oil make. This lower gas oil yield provides less overall hydrocracker feed and has the potential to underutilize the unit capacity compared to processing conventional heavier crudes.

Refining the light tight crude oils is not an easy process and consideration must be given to desalter performance, corrosion, and fouling control. Furthermore, while tight oils have many physical properties in common, the characteristics that differentiate them from one another are, in many cases, the root cause of a variety of processing challenges. For example, the methods used to extract tight oil supplies often result in the oil containing more production chemicals and increased solids with smaller particle size than conventional crudes. When introduced to the refining process, tight oils can stabilize emulsions in the desalter, increase the potential for system corrosion and fouling, as well as negatively impact waste water treatment.

5 MITIGATING REFINERY IMPACT

Due to the variation in solids loading and their paraffinic nature, processing tight oils in refinery operations offers several challenges not the least of which are bottlenecks in the atmospheric distillation unit due to the higher amount of light ends in tight oil (EIA, 2015) as well as solids deposition leading to corrosion (Fig. 8.8) (Speight, 2014c. Problems can be found from the tank farm to the desalter, preheat exchangers and furnace, and increased corrosion in the atmospheric distillation unit. In the refinery tank farm, entrained solids can agglomerate and rapidly settle, adding to the sludge layer in the tank bottoms. Waxes crystallize and settle or coat the tank walls, thus reducing storage capacity. Waxes will stabilize emulsions and suspend solids in the storage tanks, leading to slugs of sludge entering the atmospheric distillation unit. Waxes will also coat the transfer piping, resulting in increased pressure drop and hydraulic restrictions. In addition, blending asphaltene-containing crude oil with paraffin-containing tight oils leads to asphaltene destabilization that contributes to stable emulsions and sludge formation through asphaltene deposition.

Solutions include using tank farm additives to control the formation of sludge layers, along with specially designed asphaltene dispersants and aggressive desalter treatments to ensure optimum operation. Pretreatment, coupled with high-performance desalter programs, has provided the best overall desalter performance and desalted crude quality; multiple treatment options for both areas can ensure maximum performance. A crude-oil tank treatment program was initiated that broke waxy emulsions in tankage, enabling improved water resolution of the raw crude oil and minimizing sludge and solids entering the desalter. This program provided significant improvement of solids released into the desalter brine water compared to previous operations. Prior to initiating the pretreatment program, solids in the brine averaged 29 PTB, and the emulsion band control was sporadic. After the tank pretreatment program started, the desalter emulsion band could be controlled with the emulsion breaker program, and solids removal to the brine water increased by a factor of 8 to an average of 218 PTB.

Desalter operations may suffer from issues related to the tight oil properties. Solids loading can be highly variable, leading to large shifts in solids removal performance. Sludge layers from the tank farm may cause severe upsets, including growth of stable emulsion bands and intermittent increases of oil in the brine water. Agglomerated asphaltenes can enter from storage

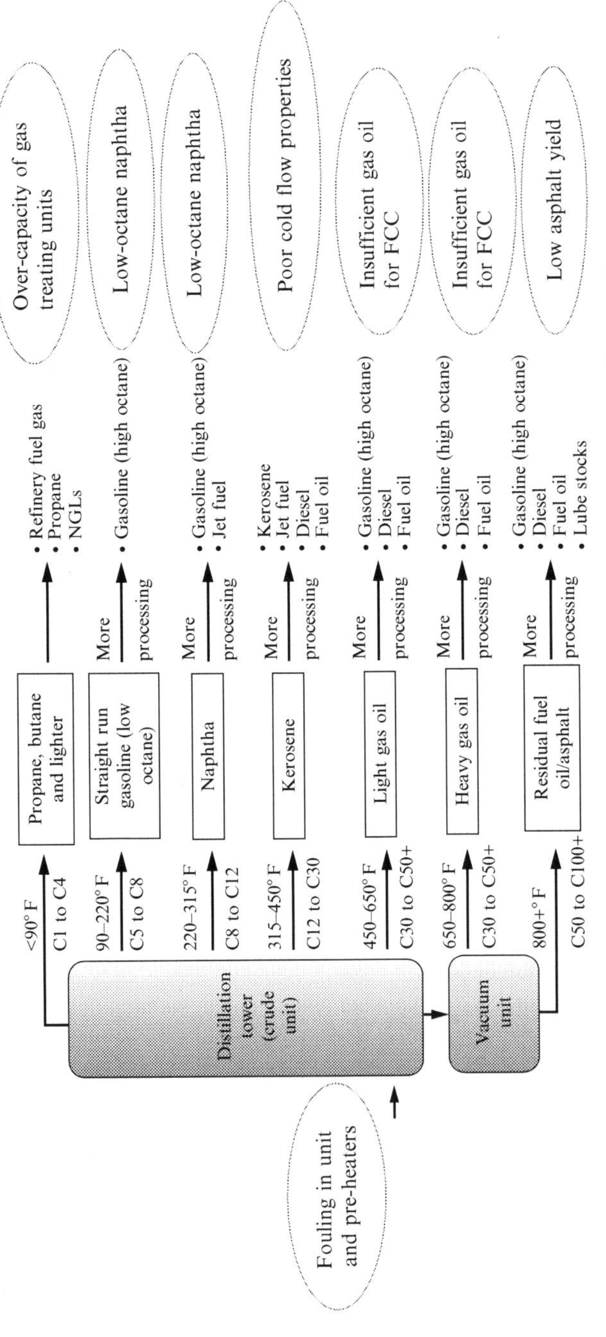

Fig. 8.8 Sites for potential issues during refining of crude oils from tight formations (compare with Fig. 8.4).

tanks or can flocculate in the desalter rag layer, leading to oil slugs in the effluent brine.

Preheat exchanger fouling has been observed in the cold train before the desalting units and in the hot train after the desalting operation. Cold train fouling results from the deposition of insoluble paraffinic hydrocarbons, coupled with agglomerated inorganic solids. Solutions to cold train exchanger fouling include the addition of wax dispersants and other oil management best practices to ensure consistent solids loading with minimum sludge processing. Crude oil management can include additives to stabilize asphaltenes and surfactants that resolve emulsions and improve water separation. These practices also include proactive asphaltene stability testing to ensure that the crude blends to be processed retain an acceptable compatibility level (Speight, 2014a, 2015a).

Hot train fouling occurs from destabilized asphaltenes that agglomerate and form deposits. These materials entrain inorganics, such as iron sulfide and sediments from production formations, into the deposit matrix. Some deposits, including high molecular-weight paraffins, can form complexes with the asphaltene aggregates. Blending tight light sweet crude oils with asphaltene-containing crude oil results in rapid asphaltene agglomeration. Rapid hot train exchanger fouling will occur in units running crude oil blends with asphaltene concentrations on the order of 1% or less. This hydrogen-to-carbon ratio is consistent with deposits of asphaltene constituents.

If the asphaltenes in the crude blend were not being rapidly destabilized, the asphaltene stability would have been well above 120. These data show that mixing certain crude oils with tight oil can result in rapid asphaltene deposition. New technology can provide the capability to rapidly perform asphaltene stability measurements on-site with a high degree of accuracy.

Hot-train exchanger fouling can be controlled through antifoulant additives designed to control the agglomeration and deposition of asphaltenes and entrained inorganic solids. Another fouling control strategy is to do regular analysis of the stability of the asphaltenes in the crude oil blend under consideration for processing. This information can guide operations to minimize fouling problems.

Tight oils contain high concentrations of hydrogen sulfide that require treatment with scavengers due to safety purposes. Amine-based scavengers often decompose as the crude oil is preheated through the hot preheat train and furnace, forming amine fragments. MEA, one of the most commonly used amines, readily forms an amine-chloride salt in the atmospheric tower which will deposit in the upper sections of the unit. Often, under-deposit corrosion is the major cause of failures in process systems because the atmospheric distillation tower under-salt corrosion rates can be 10–100 times faster than a general acidic attack. Mitigation strategies include controlling chloride to minimize the chloride traffic in the tower top and overhead, increasing the overhead operating temperature so that the salts move further downstream in the overhead system, and acidifying the desalter brine water to increase removal of amines into the water phase.

The quality of the finished fuels from refining tight oils has changed significantly. As the tight oils have higher light ends content, one benefit is increased production of naphtha for gasoline, and stable diesel and jet distillates which may be beneficial to the refinery but due to the properties (chemical nature) of these tight oil feedstocks, several challenges can be encountered. The streams are more paraffinic and, thus, they suffer from poor pour point and cloud point properties. In addition, tight oils are lower in sulfur content, so the need for lubricity additives is anticipated. Effective additives can be used to improve all distillate stream properties but to optimize any chemical treatment program, testing on specific product streams is required and suitable product selection should be customized.

Having a sufficiently advanced analysis capability is critical and, ideally, this analysis will be capable of rapidly discriminating between different types of crude, thereby providing information about distinguishing characteristics (Speight, 2014a, 2015a). Such data can then serve as the source for any plan to deal with incoming feedstock. But more important, the analysis must also capture data about crude properties in general. Often, the engineering and operations team at a refinery may not know exactly what will cause problems with heat exchangers, distillation columns, and other components of a crude processing unit.

Refineries operate in the real world and, consequently, there are some constraints on how they can be configured. However, the range of possible solutions can be expanded because there are two sides to this optimization puzzle. On the one hand, there is the refinery and its capabilities while, on the other hand, there is the feedstock. Any mismatch between the two will cause problems so, by carefully blending incoming feedstocks, it may be possible to create an input stream to the refinery that is a better match than any crude source may be on its own.

What would an ideal blending solution look like? It would be responsive and react in real-time to input from analyzers characterizing incoming crude. It would have algorithms built into it that would give the optimum ratio of different feedstocks for a given situation.

More importantly, the best crude oil blend would vary, depending upon refinery configuration and the desired fuel target to maximize during favorable market conditions. At one time, it might be gasoline. In another case, it might be either ultra-low sulfur diesel or jet aviation fuel, Jet A1. Lubricants and asphalts are another class of products that might be part of the decision matrix. Projects in Europe and Americas using an analyzer-driven blending system have shown that it is possible to determine such important properties of crude as its true boiling point in less than a minute. In turn, this enables blending adjustments that result in benefits running into millions of dollars. Scheduling in such a blending scheme will be improved by having visibility into incoming crude characteristics as soon as possible. That implies that analysis should be done as soon as feasible in the supply chain.

Speaking of the supply chain, disturbances happen. A crude shipment may be delayed, for instance, and a refinery will then run with what is in the tank because it costs too much for a facility to shut down and sit idle while waiting for a shipment to show up. Even in this case, though, knowing the characteristics of what is on hand can be used to minimize the disturbance of a crude switch. For instance, automatically making set-point changes in a fractionator via advanced process control may help smooth out any supply hiccups.

The last item brings up an important point: process control. Actual situations will necessarily have some unevenness in feedstock properties. Therefore, the third and final part of the ideal solution involves implementing intelligent refining solutions. For instance, it not only changes in feedstock that can impact the distillation process that is at the core of refinery operations. Changes in the weather can also push distillation column operation out of optimum.

One way to avoid going out of spec as a result of feedstock, weather or other changes is to over-purify, leading to wasted energy and lower product yield. A better solution is to use advanced process control technology to automatically adjust distillation column parameters to optimal targets without violating constraints. Doing so will reduce product quality variation and off-spec production while minimizing energy consumption per unit of feed. It will also increase recovery of more valuable products and column throughput. In the context of a world where feedstocks are more variable, this will mean that blending operations will have more leeway. This is because the blended feedstock can have a wider range of characteristics, yet the refinery will still be able to produce the required product blend.

This distillation process control also has to be dynamic. After all, it is not only the feedstocks that are changing over time. The output goal is also changing. The possible targets include reduced energy consumption or production at maximum capacity, just to name two. It may be necessary in the first case to minimize purification, while in the second the requirement may be to continuously operate a distillation train at the most limiting overall constraint. Thus, refinery process control should be nimble so as to cope with changes in operational targets and feedstocks.

The world of crude is changing due to the appearance of significant supplies of tight oil and other highly variable feedstock. In addition to becoming increasingly abundant as compared to traditional feedstock, these opportunity crudes may be discounted and so cannot be ignored. Because their properties are so variable, an old-style crude assay does not make sense. Hence, the need is for real-time analysis to provide the data for *ad-hoc* blending of crude supplies and intelligent process control. Together these can make a solution. Given the reality of the world refineries face today, it is important that such optimization solutions be implemented. It is also important that when this is done, it must be backed by deep expertise, as this is the best way to successfully handle a complex and changing situation.

Because of the influx of light sweet crude oil from tight formations, the nature of crude is changing and the refining industry must change with it. At one time, oil wells could be adequately characterized by a crude assay that would change only slowly with time and, thus, refineries could count on fairly stable feedstock. That is no longer the case, with the arrival of light sweet crude oil from tight formations has been the driver in a wider variability of refinery feedstocks. Dealing with this situation requires three components: (1) an ability to do accurate, on the fly, rapid analysis, (2) the capability to adjust feedstock blending as needed, and (3) smart process control to achieve the best possible processing.

REFERENCES

Andrews, A., 2014. Crude Oil Properties Relevant to Rail Transport Safety: In Brief. In: Report no. 7-5700. Prepared for Members and Committees of Congress. Congressional Research Service, Washington, DC.

ASTM D323, 2015. Standard Test Method for Vapor Pressure of Petroleum Products (Reid Method). Annual Book of Standards. ASTM International, West Conshohocken, Pennsylvania.

Bryden, K., Federspiel, M., Habib Jr., E.T., Schiller, R., 2014. Processing Tight Oils in FCC: Issues, Opportunities and Flexible Catalytic Solutions. Issue No. 114, Grace Catalysts Technologies, Catalagram, pp. 1–22. https://grace.com/catalysts-and-fuels/en-us/Documents/114-Processing%20Tight%20Oils%20in%20FCC.pdf. accessed February 18, 2016.

Deepak, R.D., Whitecotton, W., Goodman, M., Moreland, A., 2014. Challenges of Processing Feeds Derived from Tight Oil Crudes in the Hydrocracker. Paper np. AM-14-15, In: Proceedings of American Fuel & Petrochemical Manufacturers Meeting, Orlando, Florida, March 23-26. American Fuel & Petrochemical Manufacturers, Washington, DC.

Dion, M., 2014. Challenges and solutions for processing opportunity crudes. Paper No. AM-14-13, In: Proceedings of AFPM Annual Meeting, Orlando, Florida. March 23-25. American Fuel & Petrochemical Manufacturers, Washington, DC.
Furimsky, E., 2015. Properties of tight oils and selection of catalysts for hydroprocessing. Energy Fuels 29 (4), 2043–2058.
Mitchell, D.L., Speight, J.G., 1973. The solubility of asphaltenes in hydrocarbon solvents. Fuel 52, 149.
Mokhatab, S., Poe, W.A., Speight, J.G., 2006. Handbook of Natural Gas Transmission and Processing. Elsevier, Amsterdam, Netherlands.
Nordquist, J.W., 1953. Mississippian stratigraphy of Northern Montana. In: 4th Annual Field Conference Guidebook, Billings Geological Society, pp. 68–82.
Power, A.J., Mathys, G.I., 1992. Characterization of distillate fuel sediment molecules: functional group derivatization. Fuel 71, 903–908.
Speight, J.G., 2007. Natural Gas: A Basic Handbook. GPC Books, Gulf Publishing Company, Houston, Texas.
Speight, J.G., 2013. Heavy and Extra Heavy Oil Upgrading Technologies. Gulf Professional Publishing, Elsevier, Oxford, United Kingdom.
Speight, J.G., 2014a. The Chemistry and Technology of Petroleum, fifth ed. CRC Press, Taylor and Francis Group, Boca Raton, Florida.
Speight, J.G., 2014b. High Acid Crudes. Gulf Professional Publishing, Elsevier, Oxford, United Kingdom.
Speight, J.G., 2014c. Oil and Gas Corrosion Prevention. Gulf Professional Publishing, Elsevier, Oxford, United Kingdom.
Speight, J.G., 2015a. Handbook of Petroleum Product Analysis, second ed. John Wiley & Sons Inc., Hoboken, New Jersey
Speight, J.G., 2015b. Fouling in Refineries. Gulf Professional Publishing, Elsevier, Oxford, United Kingdom.
Speight, J.G., 2016. Handbook of Petroleum Refining. CRC Press, Taylor and Francis Group, Boca Raton, Florida.
Stavinoha, L.L., Henry, C.P. (Eds.), 1981. Distillate Fuel Stability and Cleanliness. Special Technical Publication No. 751. American Society for Testing and Materials, Philadelphia.
US EIA, 2011. Review of Emerging Resources. US Shale Gas and Shale Oil Plays. In: Energy Information Administration, United States Department of Energy, Washington, DC.
US EIA, 2013. Technically Recoverable Shale Oil and Shale Gas Resources: An Assessment of 137 Shale Formations in 41 Countries outside the United States. Energy Information Administration, United States Department of Energy, Washington, DC.
US EIA, 2015. Technical Options for Processing Light Tight Oil Volumes within the United States. Energy Information Administration, United States Department of Energy, Washington, DC.
USGS, 2008. Assessment of Undiscovered Oil Resources in the Devonian-Mississippian Bakken Formation, Williston Basin Province, Montana and North Dakota. Fact Sheet 2008-2031, United States Geological Survey, Reston, Virginia.
Wier, M.J., Sioui, D., and Metro, S. 2016. Catalysts optimize tight oil refining. American oil & gas reporter. April 11. http://www.aogr.com/web-exclusives/exclusive-story/catalysts-optimize-tight-oil-refining.
Zabetakis, M., 1965. Flammability Characteristics of Combustible Gases and Vapors. Bulletin No. 727, Bureau of Mines, United States Department of the Interior, Washington, DC.

CHAPTER NINE

Environmental Impact

1 INTRODUCTION

Most oil- or gas-bearing tight formation in the United States tend to be at least 4600 ft below the surface, whereas aquifers are generally no more than 1550 ft below the surface. Given the thickness of rock separating target tight formation formations from overlying aquifers, and the extremely low permeability of tight formation formations themselves, and also assuming the implementation of good oilfield practices (such as casing and cementing), it is considered by the industry that the risk of contamination of overlying aquifers as a result of hydraulic fracturing operations is remote. Instances where contamination of aquifers has been alleged are generally believed to have involved poor drilling practices, in particular poor casing and cementing of a well or poor construction of surface storage facilities.

Natural gas and crude oil production from hydrocarbon-rich (tight formation, sandstone, and carbonate) formations (Chapter 2) is one of the most rapidly expanding trends in domestic natural gas and crude natural gas and crude oil exploration and production. Moreover, this type of resource development is spreading to other countries and regions of the world (Chapter 3). In some areas, this has included bringing drilling and production to regions and countries that have seen little or no activity of these technologies in the past. New natural gas and crude oil developments bring change to the environmental and socioeconomic landscape, particularly in those areas where natural gas and crude oil development is a relatively new activity, especially in terms of the development of these unconventional natural gas and crude oil resources (Table 9.1) (Reig et al., 2014). With these changes have come questions about the nature of such natural gas and crude oil development, the potential environmental impacts, and the ability of the current regulatory structure to deal with this development. Regulators, policy makers, and the public need an objective (and not an emotional) source of information on which to base answers to these questions and decisions

Deep Shale Oil and Gas
http://dx.doi.org/10.1016/B978-0-12-803097-4.00009-7

Table 9.1 Environmental Impacts From Deep Tight Formation Development

Site preparation
Land clearing and infrastructure construction
Storm water flows
Habitat fragmentation and disruption
Effect on surface water quality
Drilling
Venting of methane—air quality
Casing and cementing—casing accidents and cementing accidents
Drilling fluids/cuttings
Fracturing fluids
Flowback and produced water
Effect on groundwater quality
Fracturing and completion
Freshwater withdrawals
Surface water and groundwater availability
Storage of fracturing fluids
Effect on surface water quality
Venting of methane and air quality
Storage/disposal of fracturing fluids and flowback water
Fracturing fluids, flowback water, and produced water
On-site pit/pond storage
Effect on surface water quality and groundwater quality
Water treatment by municipal wastewater treatment plants
Water treatment by industrial wastewater treatment plants

about how to manage the challenges that may accompany natural gas and crude oil development.

Extracting natural gas or crude oil from tight formations poses a number of risks to the environment and requires large quantities of nearby water (Muresan and Ivan, 2015). Much of this water is needed for fracturing the tight formation to allow the gas and/or oil to flow to the surface. Environmental impacts associated with natural gas and crude oil development occur at the global and local levels since resources in tight formations are not always located where water is abundant—for example, China, India, South Africa, and Mexico have large quantities of natural gas and crude oil but limited supplies of freshwater for use in recovery operations (Reig et al., 2014).

In fact, at least one-third of the area where resources in tight formations are located is arid or under significant water stress (water availability issues).

These factors pose significant social, environmental, and financial challenges to accessing water and could limit tight formation development. Thus, as countries escalate the exploration of gas and oil resources in tight formations, limited availability of freshwater could become a stumbling block. Extracting resources from tight formations requires large amounts of water for drilling and hydraulic fracturing (Chapters 5 and 6). In most cases, the demands for water are met by freshwater sources, making companies developing tight formation significant users and managers of water at local and regional levels often, unfortunately, in competition with ranches, farms, households, and other water-dependent industries.

A typical hydraulic fracturing fluid is more than 98% (v/v) water and sand. The other 2% consists of additives which may vary depending on the particular well and operator. Typically, additives include many substances that are commonly found in small measure in various household products. During a typical hydraulic fracturing process, the fracking fluid is transmitted down a cased wellbore to the target zones and then forced deep into the targeted natural gas and crude oil formations. In order to minimize the risk of any groundwater contamination, good drilling practice normally requires that one or more strings of steel casing are inserted into the well and cemented into place so as to ensure that the entire wellbore, other than the production zone, is completely isolated from the surrounding formations including aquifers. Thus, the focus of the environmental impact has tended to be on hydraulic fracturing fluids and, to some extent, the impact of the resource development on climate change (Shine, 2009; Schrag, 2012), local air quality, water availability, water quality, seismicity, and local communities (Brown, 2007; WHO, 2011; Clark et al., 2012). A number of suspected groundwater contamination due to hydraulic fracturing has been documented but, in some cases, the direct link between hydraulic fracturing and groundwater contamination has not been clearly established (Mall, 2011). In addition, there are also pertinent questions about the reliability of the sampling techniques and the analytic technique that have been used to implicate the effects of anthropogenic activities on climate change (Islam and Speight, 2016).

Rapid development of tight formation resources through horizontal drilling and hydraulic fracturing is significantly increasing the contribution of natural gas, natural gas liquids, and crude oil to the global energy supply mix (BP, 2015). Continued growth could transform the global energy market. While profitable production has yet to spread outside the United States and Canada, governments, investors, and companies have begun to explore the commercial potential of tight formation resources around the world

(Chapter 3). However, it is not sufficient to understand the potential benefits of resources in tight formations relative to other energy sources; it is also necessary to know if the natural gas or crude oil can actually be extracted from the formation. This depends on several factors, including but not limited to: (1) the amount of resource in the reservoir, (2) the mineralogy of the reservoir, (3) the pore size distribution in the reservoir rock, (4) processes that contribute to diagenesis, (5) the texture of the reservoir rock, as well as (6) the reservoir pressure and the reservoir temperature (Bustin et al., 2008).

Generally, the rapid expansion in natural gas and crude oil production from tight formations has given rise to concerns around the impact of operations in areas such as water, road, air quality, seismic, and greenhouse gas emissions (GHG) (Howarth et al., 2011a,b; Stephenson et al., 2011; O'Sullivan and Paltsev, 2012). However, the differences between the tight formations and the differences even within a tight formation add a degree of complicity to the development of such resources, especially when environmental management is taken into consideration. For example, the process of horizontal drilling and hydraulic fracturing (Chapter 5) to recover natural gas and crude oil tight reservoir requires significant volumes of water (which differs according to the reservoir character) and can cause carrying amounts of GHG compared to conventional gas wells (Spellman, 2013). There is already significant resistance to natural gas and crude oil development due to these water use and water disposal (Chapter 6) and emission concerns in many parts of the United States and Western Europe, with some countries imposing a nationwide moratorium on natural gas and crude oil production because of the environmental disturbances caused by the hydraulic fracturing technology. In fact, the regulation of natural gas and crude oil extraction from tight formations is an evolving issue because the rate of the rapid development of the industry that has often outpaced the rate of availability of information that regulators need to develop specific guidance and policies (Clark et al., 2012).

Nevertheless, there are justifiable environmental concerns because of the specialized techniques that are used to exploit natural gas and crude oil resources in tight formations (Arthur et al., 2008, 2009; GAO, 2012). There is potential for a heavy draw on freshwater resources because of the large quantities required for hydraulic fracturing fluids (Chapters 5 and 6). However, the land-use footprint of this type of natural gas and crude oil development is not expected to be much more than the environmental footprint of conventional natural gas and crude oil recovery operations, despite higher well densities, because advances in horizontal drilling technology allow for up to 10 or more wells to be drilled and produced from the same well site. Finally, there is, however, the potential for a high carbon footprint because of the emissions of carbon dioxide (CO_2), which is a natural impurity in some natural gas and crude oil reservoirs.

In fact, estimates of resource productivity and their implications for the environmental footprint of unconventional natural gas and crude oil development are, of necessity, ongoing. However, a more comprehensive approach is needed to understand how much energy will ultimately be extracted vis-à-vis the environmental costs. In this respect, important issues include: (1) the reservoir characteristics and fluid-transport mechanisms that govern resource storage and production in tight formation and other low-permeability formations, (2) the types of estimation techniques can provide the most accurate estimated ultimate recovery for the unconventional natural gas and crude oil resources, (3) the technical pathways toward improving recovery strategies and well completion to enhance short- and long-term well productivity, and (4) the effectiveness of stimulation methods, such as hydraulic fracturing, at enhancing well productivity and maximizing ultimate recovery of the resource.

Furthermore, public concerns about the environmental impacts of hydraulic fracturing have accompanied the rapid growth in energy production. These concerns include: (1) the potential for groundwater and surface water pollution, (2) the potential for a decrease in the air quality, (3) the potential for fugitive GHG, (4) the disturbance to the ecosystem, and (5) the potential for induced seismic events. Many of these issues are not unique to unconventional natural gas and crude oil production. However, the scale of hydraulic fracturing operations is much larger than for conventional exploration. Moreover, particularly worthy of note and giving rise to concern by local communities is the fact that extensive industrial development and high-density drilling are occurring in areas where there has been little or no previous natural gas and crude oil production activities.

In order to mitigate such concerns, it is vital that resource development companies assist local communities to understand these issues so that the residents can obtain a clearer picture of the productive capacity of unconventional resources and how intensive its development will be over the coming decades, particularly as unconventional natural gas and crude oil are promoted as a bridge the gap between current energy generation and use and the real, potential for lower-carbon-based energy future.

In terms of the development of unconventional natural gas and crude oil, if the correct approaches are not taken with regulatory agencies and with local communities, regulatory constraints related to environmental concerns could hinder (perhaps even terminate) developments, but if handled correctly the potential economic and energy security benefits from the development of unconventional natural gas and crude oil could result in significantly higher growth of the production natural gas and crude oil.

Thus, it is the purpose of this chapter to introduce the reader to the various aspects of natural gas and crude oil production that can cause pollution and to review that various regulations that apply to natural gas and crude oil production.

In summary, one tight reservoir is not necessarily the same as the next tight reservoir, and there is even variability within a specific reservoir. Furthermore, the economic viability of the resources, that is, the cost and feasibility of extraction, as well as on the on-site environmental and social considerations must also be given consideration. For natural gas and crude oil to be extracted successfully from tight reservoirs, governments and the developing companies must cooperate to overcome a series of technical, environmental, legal, and social challenges. Without good management, these challenges will, without doubt, impede development.

2 ENVIRONMENTAL REGULATIONS

The condition of the environment is important to floral and faunal life on Earth, and any disadvantageous disturbance of the environment can have serious consequences for continuation of this life.

The regulation of natural gas and crude oil production has traditionally occurred primarily at the state level with most states that produce natural gas and crude oil issuing more rigorous standards that take primacy over federal regulations, as well as additional regulations that control areas not covered at the federal level, such as hydraulic fracturing. Within states, regulation is carried out by a range of agencies.

The specific regulations vary considerably among states, such as different depths for well casing, levels of disclosure on drilling and fracturing fluids, or requirements for water storage. Currently, the majority of states that produce natural gas and crude oil have varying hydraulic fracturing regulations, specifically regulations related to (1) the disclosure of the components of hydraulic fracturing fluids, (2) the proper casing of wells to prevent aquifer contamination, and (3) management of wastewater from flowback and produced water (Chapters 4–6). The disposal of wastewater by underground injection (Chapter 6) has emerged as a concern for state regulators due to large interstate flows of wastewater to states with suitable geology for the water disposal as well as the potential for seismic activity near some well sites.

In terms of the development of unconventional natural gas and crude oil resources, a major environmental concern is the potential contamination of water courses. Apart from water, which makes up 99.5% (v/v) of the hydraulic fracturing fluid, the fluid also contains chemical additives to improve the process performance. The additives are varied and can include acid, friction reducer, surfactant, gelling agent, and scale inhibitor (Table 9.2) (API, 2010). The composition of the fracturing fluid is tailored to differing geologies and reservoir characteristics in order to address particular challenges including scale buildup, bacterial growth, and proppant transport.

Unfortunately, in the past many of the chemical compounds used in the hydraulic fracturing process lacked scientifically based maximum contaminant levels, making it more difficult to quantify their risk to the environment (Colborn et al., 2011). Moreover, uncertainty about the chemical makeup of fracturing fluids persists because of the limitations on required chemical disclosure (Centner, 2013; Centner and O'Connell, 2014; Maule et al., 2013).

The measures required by state and federal regulatory agencies in the exploration and production of natural gas and crude oil from deep tight formations have been very effective, for example, in protecting drinking water aquifers from contamination. In fact, a series of federal laws govern most environmental aspects of natural gas and crude oil development (Table 9.3). However, federal regulation may not always be the most effective way of assuring the desired level of environmental protection. Therefore, most of these federal laws have provisions for granting primacy to the state governments, which have usually developed their own sets of regulations. By statute, the different states may adopt these standards of their own, but they must be at least as protective as the federal principles they replace and, as a result, may be more protective in order to address local conditions.

State regulation of the environmental practices related to natural gas and crude oil development can more easily address the regional- and state-specific character of the activities, compared to a one-size-fits-all management by the federal level. Some of these factors include: geology, hydrology, climate, topography, industry characteristics, development history, state legal structures, population density, and local economics, and thus, the regulation of natural gas and crude oil production is a detailed monitoring of each stage of the development through the many controls at the state level. Each state has the necessary powers to regulate, permit, and enforce all activities—from drilling and fracturing of the well, to production operations, to managing

Table 9.2 Additives Used in the Hydraulic Fracturing Process

Water and sand: approximately 98% (v/v)			
Water	Expand the fracture and delivers sand	Some stays in the formation, while the remainder returns with natural formation water as produced water (actual amounts returned vary from well to well)	Landscaping and manufacturing
Sand (proppant)	Allows the fractures to remain open so that the oil and natural gas can escape	Stays in the formation, embedded in the fractures (used to "prop" fractures open)	Drinking water filtration, play sand, concrete, and brick mortar
Other additives: ~2%			
Acid	Helps dissolve minerals and initiate cracks in the rock	Reacts with the minerals present in the formation to create salts, water, and carbon dioxide (neutralized)	Swimming pool chemicals and cleaners
Antibacterial agent	Eliminates bacteria in the water that produces corrosive by-products	Reacts with microorganisms that may be present in the treatment fluid and formation; these microorganisms break down the product with a small amount returning to the surface in the produced water	Disinfectant; sterilizer for medical and dental equipment
Breaker	Allows a delayed breakdown of the gel	Reacts with the crosslinker and gel in the formation making it easier for the fluid to flow to the borehole; this reaction produces ammonia and sulfate salts, which are returned to the surface in the produced water	Hair colorings, as a disinfectant and in the manufacture of common household plastics
Clay stabilizer	Prevents formation clays from swelling	Reacts with clays in the formation through a sodium–potassium ion exchange; this reaction results in sodium chlorine (table salt), which is returned to the surface in the produced water	Low-sodium table salt substitutes, medicines, and IV fluids

Corrosion inhibitor	Prevents corrosion of the pipe	Bonds to the metal surfaces, such as pipe, downhole; any remaining product that is not bonded is broken down by microorganisms and consumed or returned to the surface in the produced water	Pharmaceuticals, acrylic fibers, and plastics
Crosslinker	Maintains fluid viscosity as temperature increases	Combines with the "breaker" in the formation to create salts that are returned to the surface in produced water	Laundry detergents, hand soaps, and cosmetics
Friction reducer	Minimizes friction	Remains in the formation where temperature and exposure to the breaker allows it to be broken down and consumed by naturally occurring microorganisms; a small amount returns to the surface with the produced water	Cosmetics including hair, makeup, nail, and skin products
Gelling agent	Thickens the water to suspend the sand	Combines with the breaker in the formation making it easier for the fluid to flow to the borehole and return to the surface in the produced water	Cosmetics, baked goods, ice cream, toothpastes, sauces, and salad dressings
Iron control	Prevents precipitation of metal in pipe	Reacts with minerals in the formation to create simple salts, carbon dioxide and water, all of which are returned to the surface in the produced water	Food additives, food and beverages, and lemon juice
Nonemulsifier	Breaks or separates oil/water mixtures (emulsions)	Generally, returns to the surface with produced water, but in some formations it may enter the gas stream and return to the surface in the produced oil and natural gas	Food and beverage processing, pharmaceuticals, and wastewater treatment
pH adjusting agent	Maintains the effectiveness of other components, such as crosslinkers	Reacts with acidic agents in the treatment fluid to maintain a neutral (nonacidic, nonalkaline) pH; this reaction results in mineral salts, water and carbon dioxide—a portion of each is returned to the surface in the produced water	Laundry detergents, soap, water softeners, and dish washer detergents

Table 9.3 Examples of Federal Laws (*Listed Alphabetically*) in the United States to Monitor Hydraulic Fracturing Projects

Act	Purpose
Clean Air Act	Limits air emissions from engines, gas processing equipment, and other sources associated with drilling and production
Clean Water Act	Regulation of surface discharges of water associated with natural gas and crude oil drilling and production, as well as storm water runoff from production sites
Energy Policy Act	Exempted hydraulic fracturing companies from some regulations; may disclose chemicals through a report submitted to the regulatory authority but in some instances, chemical information may be exempt from disclosure to the public as trade secrets
NEPA[a]	Requires that exploration and production on federal lands be thoroughly analyzed for environmental impacts
NPDES[b]	Requires tracking of any toxic chemicals used in fracturing fluids
Oil Pollution Act	Regulation of ground pollution risks relating to spills of materials or hydrocarbons into the water table; also regulated under the Hazardous Materials Transport Act
Safe Drinking Water Act	Directs the underground injection of fluids from natural gas and crude oil activities; disclosure of chemical content for underground injections; after 2005, see Energy Policy Act
TSCA[c]	Suggestion that this act be used to regulate the reporting of hydraulic fracturing fluid information

[a]National Environmental Policy Act.
[b]National Pollutant Discharge Elimination System.
[c]TSCA: Toxic Substance Control Act.
Notes: The Fracturing Responsibility and Awareness of Chemicals Act (the FRAC Act) was an attempt to define hydraulic fracturing as a federally regulated activity under the Safe Drinking Water Act; no significant moves or passage of this act at the time of writing (https://www.congress.gov/bill/114th-congress/senate-bill/785/text).

and disposing of wastes, and to abandoning and plugging the production well(s). These powers are a means of assuring that natural gas and crude oil operations do not have an adverse impact on the environment.

Moreover, because of the regulatory makeup of each state—which can vary from state to state—different states take different approaches to the regulation and enforcement of resource development, but the laws of each state generally give the state agency responsible for natural gas and crude oil development the discretion to require whatever is necessary to protect the environment, including human health. In addition, most have a general prohibition against pollution from natural gas and crude oil production.

A majority of the state requirements are written into rules or regulations, but some of the regulations may be added to permits on a case-by-case basis as a result of (1) environmental review, (2) on-site inspections, (3) commission hearings, and (4) public comments.

Finally, the organization of regulatory agencies within the different states where natural gas and crude oil is produced varies considerably. Some states have several agencies that oversee different aspects (with some inevitable overlap powers) of natural gas and crude oil operations, particularly the requirements that protect the environment. In different states, the various approaches have developed over time to create a structure that best serves its citizenry and all of the industries that it must oversee. The one constant is that in each state where natural gas and crude oil are produced, there is one agency that has the responsibility for issuing permits for gas and oil development projects. The permitting agencies work with other agencies in the regulatory process and often serve as a central organizing body and a useful source of information about activities related to natural gas and crude production (Arthur et al., 2008).

2.1 General Aspects

The productivity of an unconventional well is typically estimated using two factors: (1) the initial production (IP) rate after well completion and (2) the production decline curve. The IP rate quantifies the maximum production of natural gas or crude oil from a well, which is usually a calculated averaged of production over the first month. The decline curve describes the rate of decline of natural gas or crude oil production from which an estimate of the number of years that production can be expected from that well in addition to any potential environmental impacts will be. The variables that determine IP rate and decline curves are complex and include geological factors, such as: (1) the inorganic sedimentary composition of the formation, (2) the organic composition of the formation, (3) the burial history, (4) the natural fracturing patterns, (5) various petrophysical factors, such as formation porosity and formation permeability, and (5) other factors such as the level of induced fracturing during well completion (Lash and Engelder, 2009; Miller et al., 2011; Clarkson et al., 2012).

A number of federal laws direct natural gas and crude oil development (Table 9.3) (US EPA, 2012; Spellman, 2013). These regulations affect water management and disposal, as well as air quality (Gaudlip et al., 2008; Veil, 2010). In addition, current regulations require within 30 days of completing the last stage of a hydraulic fracturing operation, the responsible development company must disclose the type and amount of the chemicals used in the hydraulic fracturing process (i.e., in the fracturing fluid) for each well, as well as submittal of a report which details the handling and disposal of recovered fluids to the relevant regulation authority.

The disclosure of chemicals used in hydraulic fracturing operations is predominately a state issue. In some instances, chemical information may be exempt from disclosure to the public as trade secrets (Centner, 2013; Centner and O'Connell, 2014; Maule et al., 2013; Shonkoff et al., 2014). However, a company seeking such as an exemption will have to submit an affidavit and, if requested by the regularity authority, the chemical information for evaluation. If the regulatory authority determines that the information is not exempt from disclosure, the authority will notify the company prior to releasing any information about the chemical content of the fracturing fluid to allow the company to seek a court order preventing release of the information.

2.2 New Regulations

The hydraulic fracturing technique has evolved or the past 70 years to be carefully engineered and carefully monitored process for the generation of an extensive network of fractures within a tight formation and thereby produce from the formation a larger portion of the in-place natural gas or crude oil. This innovation has led not only to drilling many more natural gas and crude oil wells in tight (shale, sandstone, and carbonate) formations but also to increased attention on potential environmental effects.

Briefly and historically, hydraulic fracturing of gas wells began in 1949, but the process remained, for the most part, unregulated until significant unconventional gas production began (with the subsequent production of unconventional crude oil) at the beginning of the 21st century. As production of natural gas and crude oil from tight formations increased, there were reports of the contamination of drinking water sources which led the United States Environmental Protection Agency to commission a study into the risks of hydraulic fracturing to drinking water. In 2004, this study found that hydraulic fracturing of coal-bed methane posed minimal threat to underground sources of drinking water, which was a significant finding in support of the industry. Furthermore, in 2005, the federal Energy Policy Act granted hydraulic fracturing a specific exemption from the Safe Drinking Water Act (SDWA), which regulates all underground injection.

Since the Energy Policy Act passed in 2005, natural gas and crude oil production in the United States has grown significantly, and this rapid growth—along with continued reports of environmental effects—has led to renewed calls for the federal government to provide increased regulation or guidance. This pressure led to the introduction to Congress in 2009 of the Fracturing Responsibility and Awareness of Chemicals Act (the FRAC Act) to define hydraulic fracturing as a federally regulated activity under the SDWA (Table 9.3). The proposed requires the energy industry to disclose the chemical additives used in the hydraulic fracturing fluid. The Act went did not receive any action, was reintroduced in 2011, and appears to have been held in limbo since that time.

In the absence of new federal regulations, the various states that produce natural gas and crude oil by hydraulic fracturing have continued to use existing natural gas and crude oil and environmental regulations to manage natural gas and crude oil development, as well as introducing individual state regulations for hydraulic fracturing (Table 9.3). In fact, the current regulations are comprised of an overlapping collection of federal, state, and local regulations and permitting systems, and these regulations cover different aspects of the development and production of a natural gas and crude oil with the intention that the regulations combine to manage any potential impact on the surrounding environment, including any effects on water management. However, the hydraulic fracturing process that has not previously been regulated under current laws and, therefore, in terms of water management, emissions management, and site activity, the existing regulations are being (must be) reassessed for suitability for application to the hydraulic fracturing process. In the meantime, many states (including Wyoming, Arkansas, and Texas) have already implemented regulations requiring disclosure of the materials used in hydraulic fracturing fluids, and the United States Department of the Interior has indicated an interest in requiring similar disclosure for sites on federal lands.

The Bureau of Land Management's (BLM) of the Department of the Interior has proposed draft rules for natural gas and crude oil production on public lands, and these proposals would require disclosure of the chemical components used in hydraulic fracturing fluids. The proposed rule requires that an operations plan should be submit to the relevant authority—prior to initiation of the hydraulic fracturing project—that would allow the BLM to evaluate groundwater protection designs based on (1) a review of the local geology, (2) a review of any anticipated surface disturbance, and (3) a review of the proposed management and disposal of project-related fluids. In addition, the BLM would require submittal of the information necessary to confirm wellbore integrity before, during, and at the conclusion of the stimulation operation. Furthermore, before hydraulic fracturing commenced, the company would have to certify that the fluids comply with all applicable Federal, state, and local laws, rules, and regulations. After the conclusion of the hydraulic fracturing stage of the project, a follow-up report would be required in which the actual events that occurred during fracturing activities would be summarized, and this report would have to include the specific chemical makeup of the hydraulic fracturing fluid.

On Apr. 17, 2012, the United States Environmental Protection Agency released new performance standards and national emissions standards for hazardous air pollutants (HAPs) in the natural gas and crude oil industries. These rules include the first federal air standards for hydraulically fractured gas wells, along with requirements for other sources of pollution in the natural gas and crude oil industry that currently are not regulated at the federal level. These standards require either flaring or *green completion* on all natural gas wells developed prior to Jan. 1, 2015, with only green completions allowed for wells developed on and after that date.

Briefly, *green completion* requires natural gas companies capture the gas at the wellhead immediately after well completion instead of releasing it into the atmosphere or flaring the gas. Thus, the green completion systems are systems to reduce methane losses during well completion. After a new well completion or workover, the wellbore and formation must be cleaned of debris and fracturing fluid. Conventional methods for doing this include producing the well into an open pit or tank to collect sand, cuttings, and reservoir fluids for disposal. Typically, the natural gas that is produced is vented or flared, and the large volume of natural gas that is lost may not only affect regional air quality. When using green completion systems, gas and hydrocarbon liquids are physically separated from other fluids (a form of wellhead gas processing) (Chapter 7)—there is no venting or flaring of the gas—and delivered directly into equipment that holds or transports the hydrocarbons for productive use. Furthermore, by using portable equipment to process natural gas and natural gas condensate, the recovered gas can be directed to a pipeline as sales gas. The use of truck- or trailer-mounted portable systems can typically recover more than half of the total gas produced.

3 ENVIRONMENTAL IMPACT

It is important to recognize the ever-present environmental risks of natural gas and crude oil recovery from tight formations and the damage that can be caused (Muresan and Ivan, 2015). Particular attention should be given to those areas of the region or state which have not been accustomed to natural gas and crude oil development, and where all of the necessary regulatory and physical infrastructure may not yet be in place (Arthur et al., 2008).

The fracturing process (Chapter 5) entails pumping the fracturing fluid, primarily water with sand proppant, and chemical additives, at sufficiently high pressure to overcome the compressive stresses within the tight target formation for the duration of the fracturing procedure. In the process, formation pressure is increased to exceed the critical fracture pressure which creates narrow fractures in the formation. A proppant (typically sand) is then pumped into these fractures to maintain a permeable pathway for fluid flow after the fracture fluid is withdrawn and the operation is completed. While there are several environment-related issues that must receive attention, the major focus has been on the fracturing process, which poses risk to the shallow groundwater zones that may exist above or near to the gas-bearing formation or the oil-bearing formation.

As described previously (Chapters 5 and 6), multiple layers of cement and casing are put in place to protect the freshwater zones as the fracture fluid is pumped from the surface down into the tight formation. This protection is tested at high pressures before the fracturing fluids are pumped downhole. Once the fracturing process is underway, the large vertical separation between the tight formation sections being fractured, and the shallow zones prevent the growth of fractures from the tight formation into shallow groundwater zones. It should be noted here that only shallow zones contain potable water because the salinity of the groundwater increases with depth to the point where the water is too saline to be of any use.

During the process, gas escape or escape of volatile hydrocarbons can have an impact on air quality while contamination of groundwater aquifers with drilling fluids or natural gas while drilling and setting casing through the shallow zones can also occur. There is also the potential for on-site surface spills of drilling fluids, fracture fluids, and wastewater from fracture flowback.

After the hydraulic fracturing treatment, the water pressure in the well is reduced to allow the fracturing fluid to flow back out of the well followed by the natural gas and crude oil. As the fluid flows back to the surface, a process commonly referred to as flowback, the sand, and other proppants pumped into the formation are left behind to prop open the new and enlarged cracks. As flowback continues, the composition of the fluid carries higher and higher proportions of hydrocarbons. Within the first few weeks of flowback, some or most of the fracturing fluid returns to the surface as wastewater.

In North America, estimates of the volume of flowback vary between 10% and 75% (v/v) of the fracturing fluid originally injected. Because of its chemical content, this wastewater is recycled and treated for reuse, placed into disposal wells, or treated and discharged into surface waters. If not managed properly, flowback water and other wastewater from hydraulic fracturing operations can cause significant degradation to surface water and groundwater that could pose serious risks to the ecosystems and communities that depend on them.

Moreover, anywhere from 5% to 20% of the original volume of the fluid will return to the surface within the first 10 days as flowback water. An additional volume of water, equivalent to anywhere from 10% to almost 300% of the injected volume, will return to the surface as produced water over the life of the well. It should be noted that there is no clear distinction between the so-called flowback water and produced water, with the terms typically being defined by operators based upon the timing, flow rate, or sometimes composition of the water produced.

The rate at which water returns to the surface is highly dependent upon the geology of the formation. In the Marcellus play, operators recycle 95% of the flowback, whereas in the Barnett and Fayetteville plays, operators typically recycle 20% of the flowback. Water management and reuse are local issues and often depend upon the quality and quantity of water and the availability and affordability of management options (Veil, 2010). Over a 30-year life cycle, assuming a typical well is hydraulically fractured three times during that time period, construction and production of natural gas and crude oil typically consumes between 7,090,000 and 16,810,000 gallons of water per well.

Thus, the primary risks to the environment are: (1) air pollution, (2) water pollution, and (3) surface effects.

3.1 Air Pollution

Air pollutant emission sources from the development of tight formations can be grouped into two main (1) emissions from drilling, processing, well completions, servicing, and other gas-producing activities and (2) emissions from transportation of water, sand, chemicals, and equipment to and from the well pad. Thus, natural gas and crude oil production activities from tight formations can produce significant amounts of air pollution that can have an impact on air quality in areas surrounding the reservoir development program. In addition to GHG, fugitive emissions of natural gas can release volatile organic compounds (VOCs) and HAPs, such as benzene into the air.

Nitrogen oxides (NO_x) are also pollutants of concern since drilling, hydraulic fracturing, and compression equipment—typically powered by large internal combustion engines—produce these emissions.

However, uncertainty about the impacts of these emissions may exist, since air quality is highly dependent on local conditions. For example, in some areas emissions of VOCs will not be the primary source of organic emissions, and in such cases, there is the necessity to understand the impact of these emissions on local air quality. In addition, while elevated levels of benzene emissions have been found near production sites, concentrations are variable—some are below health-based screening levels—and with little data on how the emissions of HAPs behave in such circumstances, further examination is needed.

GHG and other air emissions from tight formation well sites are also a key environmental concern. These emissions are generated through the whole gamut of natural gas and crude oil operations from exploration through recovery, wellhead processing, transportation, and distribution. The United States Environmental Protection Agency has finalized GHG reporting regulations from many of these emissions sources under the Mandatory Reporting of Greenhouse Gases Rule (US EPA, 2015). Additional air emissions regulations on a state and federal level impact many of these operations as well.

Gaseous emissions such as hydrogen sulfide (H_2S), ammonia (NH_3), carbon monoxide (CO), sulfur dioxide (SO_2), NO_x, and trace metals are sources for air pollution. Such emissions are at least conceivable in oil tight formation processing operations and the lifetime of these species in the atmosphere is relatively short, and if they were distributed evenly, their harmful effects would be minimal. Unfortunately, these man-made effluents are usually concentrated in localized areas, and their dispersion is limited by both meteorological and topographical factors. Furthermore, synergistic effects mean that the pollutants interact with each other: in the presence of sunlight, CO, NO_x, and unburned hydrocarbons lead to photochemical smog, while when SO_2 concentrations become appreciable, sulfur oxide-based smog is formed.

It is worth remembering that even the nontoxic, but nonlife supporting and suffocating, CO_2 may have an important effect on the environment. The surface of the earth emits infrared radiation with a peak of energy distribution in the region where CO_2 is a strong absorber. This results in the situation, whereby this infrared radiation is trapped by the atmosphere and the temperature of the earth's surface is raised. As a result of the combustion of fossil fuels, the concentration of CO_2 in the atmosphere is increasing from its present level. Although many factors are involved, it does seem that an increase in the CO_2 concentration in the atmosphere would result in a temperature increase at the surface of the earth which could cause an appreciable reduction in the polar ice-caps and this in turn would result in further heating.

Acid gases (SO_x and NO_x) emitted into the atmosphere during tight formation processing provide the essential components in the formation of acid rain. Sulfur is present in oil tight formation as both an organic and inorganic compounds. On processing, most of the sulfur is converted to SO_2 with a small proportion remaining in the ash as sulfite:

$$S + O_2 \rightarrow SO_2$$

In the presence of excess air, some sulfur trioxide is also formed:

$$2SO_2 + O_2 \rightarrow 2SO_3$$

$$SO_3 + H_2O \rightarrow H_2SO_4$$

Only a small amount of sulfur trioxide can have an adverse effect as it brings about the condensation of sulfuric acid and causes severe corrosion. Also, nitrogen inherent in the kerogen can be converted to NO_x during processing which also produce acidic products thereby contributing to the acid rain:

$$2N + O_2 \rightarrow 2NO$$

$$NO + H_2O \rightarrow \underset{\text{Nitrous acid}}{HNO_3}$$

$$2NO + O_2 \rightarrow 2NO_2$$

$$NO_2 + H_2O \rightarrow \underset{\text{Nitric acid}}{HNO_3}$$

Most stack gas scrubbing processes are designed for SO_2 removal; NO_x are controlled as far as possible by modification of combustion design and flame temperature regulation (Mokhatab et al., 2006; Speight, 2007, 2008, 2013, 2014a). However, processes for the removal of SO_2 usually do remove some NO_x; particulate matter can be removed efficiently by commercially well-established electrostatic precipitators.

Most natural gas cleaning processes require removal of the traces of other hydrocarbons and impurities from the natural gas stream (Chapter 7). The recovery of natural gas liquids such as propane, butane, pentanes, and higher molecular weight hydrocarbons is a value-adding process throughout much of the gas processing industry (Mokhatab et al., 2006; Speight, 2007, 2014a,b,c).

Other trace products such as H_2S and CO_2 (the *acid gases*) must be removed from the gas stream to prevent corrosion of pipelines and equipment for safety reasons (Speight, 2014c).

Another local air pollutant of growing concern is crystalline silica dust, which can be generated from the sand proppant. Silica dust can be generated in the mining and transporting of sand to the well site and in the process of moving and mixing sand into the hydraulic fracturing fluid on the well pad. Crystalline silica dust within the respirable size range (<4 μm) is considered a HAPs and a carcinogen. In addition to an increased risk of lung cancer, exposure to crystalline silica can lead to a chronic, inflammatory lung disease called silicosis.

3.2 Water Pollution

In terms of waste water, a number of issues face the hydraulic fracturing industry, and the main driver behind these issues is (1) the presence of chemical additives that are added to the water before the fracturing process begins and (2) chemicals leached from the formation during the process and while the water returns to the surface. However, efforts by the industry are being continued to reduce and, if possible, eliminate completely any water contamination. In addition to monitoring organic compounds in the water, metals concentration must be continually monitored since heavy metals can be extremely hazardous when released to the environment.

For the most part, aqueous wastes are regulated under the SDWA which requires that the water be recycled through many stages during the purification process and passes through several treatment processes, including a wastewater treatment plant, before being released into surface waters. The wastes discharged into surface waters are subject to state discharge regulations and are regulated under the Clean Water Act. These discharge guidelines limit the amounts of sulfides, NH_3, suspended solids, and other compounds that may be present in the wastewater. Although these guidelines are in place, contamination from past discharges may remain in surface water bodies.

At the surface facilities of the fracturing operation, the options for water treatment are: (1) solid settling and removal; (2) the use of bacteria and aeration to enhance organic degradation; and (3) filtration through activated carbon, applications of ozone, and chlorination. Thus, toxic constituents in the water must be identified and plans must be developed to alleviate any problems. In addition, regulators have established and continue to establish water-quality standards for priority toxic pollutants.

In addition, for deep formations, contamination may occur due to defects in the wellbore. When the annulus between the well casing and surrounding geology is not adequately sealed during well installation, methane can migrate from the tight formation resource up the outside of the wellbore to shallow aquifers where it could dissolve in the drinking water. Another possible pathway for contamination is a defect in the casing at a shallow depth, allowing gas to flow from inside the wellbore to the aquifer. Faulty well construction appears to have caused one of the largest documented instances of water contamination. In addition to faulty well construction, uncased, abandoned wells may also provide pathways for methane migration to occur (Osborn et al., 2011). The most obvious, and perhaps most easily prevented, pathway for contamination is intentional dumping or accidental spilling of flowback water on the surface. A common cause of accidental spillage is overflows from retention ponds during major rain events.

Water for tight formation energy projects is used most intensely in the fracturing portion of the life cycle of a project. Under current practices, fracturing—typically a water-dependent activity—may require up to 10 million gallons of water for each horizontal well and the associated fractures. Production activities and management and treatment of the wastewater produced during natural gas or crude oil from tight formations (including flowback from fracturing and water produced from source formations) have raised concerns over the potential contamination of groundwater and surface water and also the potential for induced seismicity associated with wastewater injection wells.

In fact, constraints on the ability of the producer to use water resources may seriously effect the future development of the production of natural gas and crude oil from formations that require horizontal drilling and hydraulic fracturing. The potential for such projects to cause adverse environmental effects—particularly effects associated with fluids management (Chapter 6)—have prompted the various regulatory actions to introduce and enforce regulations to protect water supplies. Future actions may influence development of tight formations that hold natural gas and crude oil reserves through additional regulatory oversight or other policy actions. At the same time, innovations and advances in the extraction of natural gas and crude oil from tight formations as well as innovations and improvements in water use and wastewater management protocols may reduce some of the potentially adverse environmental development impacts (Tiemann et al., 2014). Furthermore, water management issues are relevant to the entire life cycle of energy development from tight formations because fluids will continue to be produces even after a well is drilled, hydraulically fractured, and produced natural gas and/or crude oil. Furthermore, it must not be forgotten that in the United States, where tight formation basins are found across most of the contiguous 48 states (Chapter 3) (US EIA, 2011a), the only commonality is the use of the umbrella term *tight formation* and, as is the case with processing the natural gas (Chapter 7) and crude oil (Chapter 8), each tight formation presents an individual set of unique challenges with respect to water contamination and water resource management.

Water consumption for hydraulic fracturing occurs during: (1) drilling, (2) extraction and processing of proppant sands, (3) testing natural gas transportation pipelines, and (4) gas processing plants. Typically, for most tight formation basins, water is acquired from local water supplies, including: (1) surface water bodies, such as rivers, lakes, and ponds; (2) groundwater aquifers; (3) municipal water supplies; (4) treated wastewater from municipal and industrial treatment facilities; and (5) produced and/or flowback water that is recovered, treated, and reused (Chapter 6). Water contamination can occur because of the presence of (1) suspended solids, (2) dissolved inorganic constituents, (3) dissolved organic constituents, as well as (4) a variety of other contaminants that will vary with the locale and the characteristics of the project, including the characteristics of the reservoir.

Suspended solids will occur primarily in water from the dust-control systems used in tight formation as well as in site drainage water. In above-ground wellhead treating systems, some fine particulate matter may be entrained in the produced natural gas as well as in the gas condensate. A wellhead cooling tower that may be associated with, and necessary for, natural gas and crude oil treatment before transportation may cause contamination of the water due to dust picked up from the atmosphere. Also, precipitated salts and biological matter may also be present in the cooling tower blowdown.

Dissolved inorganic constituents will occur in produced water, flowback water, and in site drainage water—these streams have a propensity to leach sodium, potassium, calcium, and magnesium ions from the tight formation. Anions such as sulfate (SO_4^{2-}), bicarbonate (HCO_3^-), and chloride (Cl^-) are associated with the cations (sodium, Na^+, potassium, K^+, calcium, Ca^{2+}, and magnesium, Mg^{2+}) may also be present depending upon the contact the water has made with the tight formation. The inorganic constituents that are leached from the reservoir (assuming that the wellbore and the fracturing operation have not allowed the water to have contact with other formations) include acidic materials, highly alkaline materials, and dilute concentrations of heavy metals. These materials can have an adverse effect on the indigenous flora and fauna by creating a hostile environment (often through poisoning the waterways) and, in some cases, cause the destruction of flora and fauna.

Dissolved organic constituents arise largely from the organic compounds in the natural gas, the gas condensate, or the crude oil. The types of organics in each condensate will probably depend on the solubility of the organic constituents and the reservoir temperature at which the gas and oil came into contact with the water. Water-soluble acidic compounds, that occur in some crude oils (Speight, 2014a, b), accumulate in water that has been in contact with the crude oil-containing tight formation. Constituents of natural gas, such as H_2S and thiols (RSH), also have a measurable solubility in water as do certain hydrocarbon species. In terms of soluble organic constituents, the general rule of thumb is to assume that soluble organic constituents will also have an adverse effect on the indigenous flora and fauna by creating a hostile environment (often through poisoning the waterways) and, in some cases, cause the destruction of flora and fauna.

In addition to constituents solubilized in the water while in the reservoir, water that originates from hydraulic fracturing (flowback water and produced water) often contains chemical additives to help carry the proppant and may become enriched in salts after being injected into tight formation formations (Chapters 5 and 6). Some of the chemicals used in hydraulic fracturing are common and generally considered to be harmless, and this is not true of all chemicals used in the process. In fact, consideration must be given to the injection of the so-called harmless chemicals into the formations and (since these chemicals are not indigenous to the formation) and the effect of these harmless but nonindigenous chemicals on the environment. Consideration must also be given to the injection of the so-called harmless chemicals into the formations and (even if the chemicals are identical to chemicals that are indigenous to the formation) and the effect of these indigenous chemicals on the environment. Within the local environment, these chemicals will be present in a measurable concentration, but the flora and fauna present in that ecosystem may be fatally susceptible to such chemicals when they are present in a concentration that is above the indigenous concentration of the chemicals. Thus, it is essential that companies participating in the hydraulic fracturing process must be willing to provide to regulatory authorities the name of the ingredients that are used in the hydraulic fracturing fluid, and there must also be a thorough understanding of the potential risks that these chemicals pose to the environment. Without knowing the identity of any proprietary components, regulators cannot test for their presence (in order to establish an ecosystem baseline), and this prevents government regulators from establishing baseline levels of the substances prior to a hydraulic fracturing project and changes in the levels of these chemicals cannot be documented, thereby making it more difficult to prove the effect (adverse or benign) that hydraulic fracturing has on the environment.

Until recently, the chemical composition of fracturing fluids was considered a trade secret and was not made public. This position has fallen increasingly out of step with public insistence that the community has the right to know what is being injected into the ground. Since 2010, voluntary disclosure has become the norm in most of the United States. The industry is also looking at ways to achieve the desired results without using potentially harmful chemicals. "Slick-water," made up of water, proppant, simple drag reducing polymers, and biocide, has become increasingly popular as a fracturing fluid in the United States, though it needs to be pumped at high rates and can carry only very fine proppant. Attention is also being focused on reducing accidental surface spills, which most experts regard as a more significant risk of contamination to groundwater.

Therefore, the water that is recovered during natural gas or crude oil production from tight formations must be either treated or disposed of in a safe manner—typically by injection into deep, highly saline formations through one or more wells drilled specifically for that purpose and by following clearly defined regulations (Chapter 6). There has been a tendency of late to investigate (and advocate) further the use of flowback water for reuse in hydraulic fracturing project. In addition to the presence of what-might-be-termed dangerous chemicals (to the flora and fauna of the ecosystem) in the water, there are also additional effects that can cause dissolution of more chemicals into the water—the potential for corrosion or scaling (Speight, 2014c), where the dissolved salts may precipitate out of the water and clog parts of the well or the formation may become a reality, thereby introducing more environmentally detrimental chemical species into the ecosystem.

In addition to the chemical additives in the hydraulic fracturing fluid, wastewater from natural gas and crude oil extraction from tight formations may contain high levels of total dissolved solids metals and naturally occurring radioactive materials. Furthermore, the amount of saline formation water produced from tight formation formations varies widely—up to several hundred barrels of water per day deepening upon the formation. The water comes from either (1) the tight reservoir formation or from (2) any adjacent formations that are connected through the fracture-induced fracture network. The water, like flowback water, is normally highly saline and must be treated and/or sent for disposal—typically by injection into deep saline formations, which are also protected and injection of nonindigenous fluids is also subject to clearly defined regulations. In fact, many states where natural gas and crude oil are produced have (though the various state regulatory agencies) implemented regulations regarding the disclosure of chemicals used in the process of hydraulic fracturing. There are, however, agreements that protect the states from disclosure of the name of proprietary chemicals, thereby protecting the intellectual property of the production company (Chapter 6).

The well casing does, in general, provide a protective barrier from potential contamination from hydraulic fracturing fluid, oil, and natural gas flowing from the well. Nevertheless, risks to water contamination (and thence, risks to water quality) occur from ground and surface spills. These spills include natural gas and crude oil drilling water contamination or other mishandling of wastewater, rather than from the hydraulic fracturing process itself (EIUT, 2012). Thus, the potential for contamination of the groundwater raises another environmental concern. The hydraulic fracturing process requires the use of copious quantities of water treated with chemicals that facilitate both the suspension of the proppant (such as sand) and the lubrication (for friction reduction) of the conveying medium (the hydraulic fracturing fluid) (Table 9.2). In the development of an entire field, the amount of water injected into a tight formation could be potential on the order of hundreds of millions of gallons. Although field operators retrieve most of the injected water upon completion of the hydraulic fracturing stimulation, a significant quantity of water and chemicals remains within the formation. As a result, there have been claims that the potential leakage of the chemicals used in the hydraulic fracturing process pose a risk to the flora and fauna of the local ecosystem (including a risk to human health and safety) and are calling for stricter regulation.

Thus, in order to protect water sources (including surface water and groundwater sources), existing regulations must (typically, the regulations do) require that natural gas and crude oil recovery projects must use cement well casings using cement of a certain quality and depth. Under any such rule, hydraulic fracturing operations will be required to follow any prevailing and new standards. In addition, the cement used for hydraulic fracturing operations must be adequately bonded and of a prescribed minimum depth. In certain instances, there may be the requirement for the project operator to submit cement evaluation data to demonstrate compliance to the regulations as well as test data to show the surface pressure of any casing or fracturing string prior to conducting fracking operations. In fact, some states have issued regulatory requirements for *responsible development* of tight formation formations, and these regulations include guidelines for the use and disposal of water, the protection of groundwater, and the use of chemicals. Furthermore, the regulatory requirements include: (1) review of each drilling application for environmental compliance, (2) complete environmental assessment of all proposed oil or gas well that is within a prescribed distance—usually a distance on the order of 2000 ft—of a municipal water well, (3) strict review of the well design to ensure groundwater protection, (4) on-site of inspection of drilling operations, and (5) enforcement of strict restoration rules when drilling ends. Improper discharge of produced water is an issue and is often best addressed by recycling, but one of the limitations to recycling and reusing water is that the amount of flowback returned to the surface varies between and within tight formations (Reig et al., 2014).

3.3 Aquifer Protection

Environmental concerns about natural gas and crude oil exploitation have received significant attention in the media. The issues raised are fresh water usage in competition with other uses such as farming, improper disposal of produced water, and contamination of fresh water aquifers. Thus, concerns over water quality focus on potential drinking water contamination by methane or fluids from hydraulic fracturing activities (WHO, 2011). The possible pathways for this contamination include underground leakage from the wellbore to drinking water aquifers and improper disposal or accidental leakage of hydraulic fracturing fluids to surface water bodies. Owing to the depth of most tight formation plays, it is unlikely that a credible pathway (independent of the wellbore) exists for fluids to flow from the fractures within the tight formation through thousands of feet of overlaying rock into a drinking water aquifer. Although natural gas and crude oil wells use up to 6 million gallons (6×10^6 gallons) of water per well, the water volume used per unit of energy produced is small compared to a number of alternatives. Although this usage is relatively low compared to alternatives, any usage of water may appear to be in competition with other uses, especially in draught years. To address this situation, salt water might be used in place of fresh water. Recent advances in fracturing permit this with small modifications to the needed chemicals.

There are two potential ways in which natural gas and crude oil operations could contaminate aquifers. One is through leakage of the chemicals used in fracturing. These then would be liquid contaminants. The second is the infiltration of aquifers by produced methane. It is a gaseous contaminant, albeit it gets dissolved in the water. If methane is present, a portion may be released as a gas. The distinction between potential liquid and gaseous contamination is important because the hazards are different, as are the remedies and safeguards. Also, because well water could not naturally have the liquid contaminants, their presence is evidence of a man-made source.

Therefore, simple testing of wells proximal to drilling operations is sufficient, with the only possible complication being the influence from some source other than drilling, such as agricultural runoff, for example. This is easily resolved because of the specificity in the chemicals used for fracturing.

Methane leakage can happen because of possible combination of not locating cement in the right places and of a poor cement job. Many wells will have intervals above the producing zone that are charged with gas, usually small quantities in coal bodies and the like. If these are not sealed off with cement, some gas will intrude into the wellbore. This will still be contained unless the cement up near the fresh water aquifers has poor integrity. In that case the gas will leak. Wells constructed to specification will not leak.

3.4 Induced Seismic Activity

Disposal of flowback water from hydraulic fracturing depends upon the availability of suitable injection wells. For example, the limited availability of suitable geology in Pennsylvania has led to hauling flowback water to Ohio for injection. The increased injection activity has been linked to *seismic events* or earthquakes. Additional studies have indicated that injection activities in Arkansas have been linked to nearby earthquakes (Horton, 2012).

A properly located injection well will not cause earthquakes. A number of factors must be present to induce seismic events at a disposal site. In order for earthquakes to occur, a fault must exist nearby and be in a near-failure state of stress. The injection well must have a path of communication to the fault, and the fluid flow rate in the well must be at a sufficient quantity and pressure for a long enough time to cause failure along the fault or system of faults. A recent National Research Council study concludes that the majority of disposal wells for hydraulic fracturing wastewater do not pose a hazard for induced seismicity. This report also concludes that the process of hydraulic fracturing itself does not pose a high risk for inducing felt seismic events (NRC, 2012).

Because it creates cracks in rocks deep beneath the surface, hydraulic fracturing always generates small seismic events; these are actually used by petroleum engineers to monitor the process. In general, such events are several orders of magnitude too small to be detected at the surface: special observation wells and very sensitive instruments need to be used to monitor the process. Larger seismic events can be generated when the well or the fractures happen to intersect, and reactivate, an existing fault.

Hydraulic fracturing is not the only anthropogenic process that has been claimed to trigger small earthquakes. Any activity that creates underground stresses carries such a risk. Examples linked to construction of large buildings, or dams, have been reported. Geothermal wells in which cold water is circulated underground have been known to create enough thermally induced stresses to generate earthquakes that can be sensed by humans (Cuenot et al., 2011) and the same applies to deep mining (Redmayne et al., 1998).

In order to circumvent any such issues arising from hydraulic fracturing, it is essential for unconventional gas development engineers to make a careful survey of the geology (with the geologists) of the area to assess whether deep faults or other geological features present an enhanced risk and to avoid such areas for fracturing. In any case, multidisciplinary monitoring is necessary so that operations can be suspended and corrective actions taken if there are signs of increased seismic activity.

4 REMEDIATION REQUIREMENTS AND OUTLOOK

The issue of natural gas and crude oil regulation is dominated by hydraulic fracturing, the key feature of natural gas, and crude oil that separates it from well-regulated conventional gas production. However, existing regulations to protect water resources during natural gas and crude oil development are also affected by the greater intensity of water, energy, and infrastructure used in natural gas and crude oil operations.

This consequence is driving significant uncertainty in the United States, which is still adapting to the new industry. The speed of industry growth has outpaced the availability of rigorous data on its potential impact, which has hindered the ability of government to adequately assess and regulate operations. To resolve this issue, there has been renewed focus by the US federal government on establishing better understanding of the potential impacts of natural gas and crude oil development, to most effectively regulate this critical new energy resource.

A large volume of water is needed for the development of natural gas and crude oil plays. Water is used for drilling, where it is mixed with clay minerals to form drilling mud. This mud is used to cool and lubricate the drill-bit, provide wellbore stability, and also carry rock cuttings to the surface.

The treatment of waste water is a critical issue for unconventional gas production—especially in the case of the large amounts of water customarily used for hydraulic fracturing. After being injected into the well, part of the fracturing fluid (which is often almost entirely water) is returned as flowback in the days and weeks that follow. The total amount of fluid returned depends on the geology; for tight formation, it can run from 20% to 50% of the input; the rest bound to the clays in the tight formation rock. Flowback water contains some of the chemicals used in the hydraulic fracturing process, together with metals, minerals, and hydrocarbons leached from the reservoir rock. High levels of salinity are quite common and, in some reservoirs, the leached minerals can be weakly radioactive, requiring specific precautions at the surface. Flowback returns (like waste water from drilling) require secure storage on site, preferably fully contained in stable, weather-proof storage facilities as they do pose a potential threat to the local environment unless handled properly.

Once separated out, there are different options available for dealing with waste water from hydraulic fracturing. The optimal solution is to recycle it for future use and technologies are available to do this, although they do not always provide water ready for reuse for hydraulic fracturing on a cost-effective basis. A second option is to treat waste water at local industrial waste facilities capable of extracting the water and bringing it to a sufficient standard to enable it to be either discharged into local rivers or used in agriculture. Alternatively, where suitable geology exists, waste water can be injected into deep rock layers.

Most of the fluids used in hydraulic fracturing is water and chemicals (typically 1%, v/v, of the water). The formulas for fracturing fluids vary, partly depending on the composition of the gas- or oil-bearing formations, remembering that all gas- and oil-bearing formations are not the same even when the formations are composed of the same minerals (tight formation, sandstone, or carbonate). In addition, some of the chemical additives can be hazardous if not handled carefully and caution is advised since the amount of the chemical(s) must not exceed the amount specified in regulatory requirements related to handling hazardous materials. Even if the chemical is one that is indigenous to the subsurface (and supposedly benign because it is found naturally), the amount used must not exceed the indigenous amount—in some case, exceeding the indigenous amount of a chemical can cause environmental problems.

Safe handling of all water and other fluids on the site, including any added chemicals, must be a high priority, and compliance with all regulations regarding containment, transport, and spill handling is essential. When it comes to disposal of the fracturing fluid, there are options. For example, the fluid, when possible without causing adverse effects to the environment, can be reused for additional wells in a single field—this reduces the overall use of fresh water and reduces the amount of recovered water and chemicals that must be sent for disposal. However, in such cases, recognition of the geological or mineralogical similarities or difference within a site must have been determined to assure minimal environmental damage. In addition, tanks (or *lined* storage pits) for the storage of recovered water are also a necessity until the water can be sent for disposal to a permitted saltwater injection disposal well or taken to a treatment plant for processing. The linings of such pits must be in accordance with local environmental regulations.

All injection wells must be designed to meet the regulations set by the national agency (e.g., the United States Environmental Protection Agency) or any local agency to protect the groundwater. In addition, production zones should have that multiple confining layers above the zone to keep the injected fluids within the target gas- or oil-bearing formation. In addition, multiple layers of well casing and cement (similar to production wells) should be used with periodic mechanical integrity tests to verify that the casing and cement are holding the liquids. The amount and pressure of the injected fluid (specified in each well permit) should be monitored to maintain the fluids in the target zone, and the pressure in the injection well and the spaces between the casing layers (also called the annuluses) should also be monitored check and verify the integrity of the injection well.

More particularly, the occurrence and production of natural gas from fractured, organic-rich Paleozoic and Mesozoic tight formation formations in the United States may be better understood by considering source rock, reservoir, seal, trap, and generation-migration processes within the framework of a petroleum system. The system concept must be modified, however, inasmuch as organic tight formation formations are both source and reservoir rocks and, at times, seals. Additional consideration must be given to the origin of the gas, whether biogenic or thermogenic, in defining the critical moment in the evolution of potentially producible hydrocarbons.

Remediation requirements become of greater importance as well as reach the end of their life cycles. More than half of the total production of a well is usually achieved in the first 10 years of operative well life. When a well can no longer produce natural gas and crude oil economically, it is plugged and abandoned according to the standards of each state. Disturbed areas, such as well sites and access roads, are reclaimed back to the native vegetation and contours, or to conditions specified by the landowner.

Improperly closed or abandoned natural gas and crude oil wells may create human health and safety risks, as well as air pollution and surface and groundwater contamination risks. Most states require operators to post a bond or some form of financial security not only to ensure compliance but also to ensure there are funds to properly plug the well once production ceases. However, the size of the bond may cover only a small fraction of the site reclamation costs.

The economics of natural gas and crude oil development encourages the transfer of assets from large entities to smaller ones. With the assets go the liabilities, but without a mechanism to prevent the new owners from assuming reclamation liabilities beyond their means, the economics favor default on well-plugging and site restoration obligations. In fact, a combination of improved technology and tight formation-specific experience has also led—and will continue to lead—to improvements in recovery factors and reductions in decline rates. It is now recognized that each gas tight formation resource requires a specific completion technique, which can be determined through careful analysis of rock properties. Continuing efforts to make the correct selection of well orientation, stimulation equipment, fracture size, and fracking fluids will serve to enhance the performance of a well and the overall recovery of natural gas or crude oil. Indeed, for developed tight formation formations in North America, the combined benefits of improved technology and increased experience will continue to provide enhanced production over time. Both the expected ultimate recovery per well and the peak production per well will continue to increase as developed natural gas and crude oil formations move to maturity.

Following on from the advances in the extraction of natural gas and crude oil from tight formations in the United States, a number of natural gas and crude oil companies will be willing to apply the techniques developed in North America in new geological basins and markets outside North America. A considerable number of regions around the world have been the focus of interest for their tight formation potential—in fact 48 major tight formation basins are identified in 32 countries around the world that are prospects for development (US EIA, 2011b).

These prospects include a number of tight formation formations across Europe where organic-rich tight formation sediments are present, including: (1) Lower Paleozoic tight formation formations, which extend from Eastern

Denmark and Southern Sweden to Northern and Eastern Poland, (2) Carboniferous tight formation formations, which extend from North-West England through Netherlands and North-West Germany to South-West Poland, and (3) Lower Jurassic bituminous tight formation formations, which extend from the South of England to the Paris Basin in France, the Netherlands, Northern Germany, and Switzerland. Poland and France are identified (US EIA, 2011b) as countries with some of the largest estimated natural gas and crude oil technically recoverable resources in Europe—both countries are currently highly dependent on imported gas to meet domestic demand.

Furthermore, horizontal wells with horizontal legs up to 1 mile or more in length are widely used to access the reservoir to the greatest extent possible. Multistage hydraulic fracturing, where the tight formation is cracked under high pressures at several places along the horizontal section of the well, is used to create conduits through which gas can flow. Microseismic imaging allows operators to visualize where this fracture growth is occurring in the reservoir.

Although fracture and matrix permeability, enhanced by application of appropriate well stimulation treatments, are key to achieving economical gas flow rates, sufficient amounts of organic matter (either for generation of thermogenic gas or as a microbial feedstock) must initially have been present to have generated the reservoir gas. Therefore, deciphering the thermal history of the organic matter within the tight formations and analyzing the rock mechanics response of the tight formation matrix and organic matter to local and regional stresses are critical steps in establishing their complex relationship to gas producibility. The poor quality of one factor (e.g., low adsorbed gas) may be compensated for by another factor (e.g., increased reservoir thickness); however, tight formation-gas production cannot always be achieved even where optimum combinations of geological and geochemical factors apparently are present.

However, as a technology-driven resource, the rate of development of natural gas and crude oil may become limited by the availability of required resources, such as fresh water, fracture proppant, or drilling rigs capable of drilling wells miles in length. Thus, to important challenges for developing the Natural gas and crude oil resources are: (1) the significant depth and (2) the lack of information for many of the resources.

In areas where the resources are present, companies must continue to focus on the careful environmental develop and companies before setting their sights on a deeper target with an uncertain payoff. On the other hand, in areas where the natural gas and crude oil development has already occurred and new resources are discovered and opened up to development, there may be an infrastructure advantage. Drilling pads, roadways, pipelines, gathering systems, surveying work, permit preparation data, and landowner relationships might still be useful for developing future tight formation resources.

There are some environmental concerns with the specialized techniques used to exploit natural gas and crude oil that need to be continually addressed. There is potential for a heavy draw on freshwater resources because of the large quantities required for hydraulic fracturing fluid. The land-use footprint of natural gas and crude oil development is not expected to be much more than the footprint of conventional operations, despite higher well densities, because advances in horizontal drilling technology allow for up to 10 or more wells to be drilled and produced from the same well site.

Finally, there is potential for a high carbon footprint through emissions of CO_2, a natural impurity in some natural gas and crude oil.

REFERENCES

API, 2010. Water Management Associated With Hydraulic Fracturing, API Guidance Document Hf2. American Petroleum Institute, Washington, DC.

Arthur, J.D., Langhus, B., Alleman, D., 2008. An Overview of Modern Natural Gas and Crude Oil Development in the United States. ALL Consulting, Tulsa, Oklahoma. http://www.all-llc.com/publicdownloads/ALLTightformationOverviewFINAL.pdf.

Arthur, J.D., Bohm, B., Cornue, D., 2009. Environmental considerations of modern tight formation developments. In: Proceedings of the SPE Annual Technical Meeting, New Orleans, Louisiana, Oct. 4–7. Paper No. SPE 122931.

BP, 2015. Statistical Review of World Energy 2015. BP plc, London, United Kingdom.

Brown, V.J., 2007. Industry Issues: Putting the Heat on Gas. Environmental Health Perspectives. United States National Institute of Environmental Health Sciences, Washington. DC Report No. 115–2.

Bustin, A.M.M., Bustin, R.M., Cui, X., 2008. Importance of fabric on the production of gas shales. In: Proceedings of the Unconventional Gas Conference, Keystone, Colorado, Feb. 10–12. SPE Paper No. 114167.

Centner, T.J., 2013. Oversight of shale gas production in the United States and the disclosure of toxic substances. Resour. Policy 38, 233–240.

Centner, T.J., O'Connell, L.K., 2014. Unfinished business in the regulation of shale gas production in the United States. Sci. Total Environ. 476–477, 359–367.

Clark, C., Burnham, A., Harto, C., Horner, R., 2012. Hydraulic Fracturing and Natural Gas and Crude Oil Production: Technology, Impacts, and Policy. Argonne National Laboratory, Argonne, Illinois.

Clarkson, C.R., Nobakht, M., Kavaini, D., Ertekin, T., 2012. Production analysis of tight-gas and tight formation-gas reservoirs using the dynamic-slippage concept. SPE J. 17, 230–242.

Colborn, T., Kwiatkowski, C., Schultz, K., Bachran, M., 2011. Natural gas operations from a public health perspective. Hum. Ecol. Risk. Assess. 17, 1039–1056.

Cuenot, N., Frogneux, M., Dorbath, C., Calo, M., 2011. Induced microseismic activity during recent circulation tests at the EGS site of Soultz-sous-Forêts (France). In: Proceedings of the 36th Workshop on Geothermal Reservoir Engineering, Stanford, California.
EIUT, 2012. Fact-Based Regulation for Environmental Protection in Shale Gas Development Summary of Findings. The Energy Institute, University of Texas at Austin, Texas, Austin. http://energy.utexas.edu.
GAO, 2012. Information on Tight Formation Resources, Development, and Environmental and Public Health Risks. United States Government Accountability Office, Washington, DC. Report No. GAO-12-732. Report to Congressional Requesters.
Gaudlip, A.W., Paugh, L.O., Hayes, T.D., 2008. Marcellus water management challenges in Pennsylvania. In: Proceedings of the Natural Gas and Crude Oil Production Conference, Nov. 16–18. Ft. Worth, Texas. Paper No. SPE 119898.
Horton, S., 2012. Disposal of hydrofracking waste fluid by injection into subsurface aquifers triggers earthquake swarm in central arkansas with potential for damaging earthquake. Seismol. Res. Lett. 83 (2), 250–260.
Howarth, R.W., Santoro, R., Ingraffea, A., 2011a. Methane and the greenhouse-gas imprint of natural gas from tight formations. Clim. Chang. 106 (4), 1–12.
Howarth, R., Santoro, R., Ingraffea, A., 2011b. Methane and the greenhouse-gas footprint of natural gas from tight formations. Clim. Change 106, 679–690.
Islam, M.R., Speight, J.G., 2016. Peak Energy—Myth or Reality? Scrivener Publishing, Beverly, Massachusetts.
Lash, G.C., Engelder, T., 2009. Tracking the burial and tectonic history of Devonian tight formation of the Appalachian Basin by analysis of joint intersection style. Geol. Soc. Am. Bull. 121, 265–272.
Mall, A., 2011. Incidents Where Hydraulic Fracturing is a Suspected Cause of Drinking Water Contamination. Switchboard: Natural Resources Defense Council, Washington, DC.
Maule, A.L., Makey, C.M., Benson, E.B., Burrows, I.J., Scammell, M.K., 2013. Disclosure of hydraulic fracturing fluid chemical additives: analysis of regulations. New Solution 23, 167–187.
Miller, C., Waters, G., Rylande, E., 2011. Evaluation of production log data from horizontal wells drilled in organic tight formations. In: Proceedings of the SPE North American Unconventional Gas Conference, Jun. 14–16. The Woodlands, Society of Petroleum Engineers, Richardson, Texas. Paper No. SPE-144326.
Mokhatab, S., Poe, W.A., Speight, J.G., 2006. Handbook of Natural Gas Transmission and Processing. Elsevier, Amsterdam, The Netherlands.
Muresan, J.D., Ivan, M.V., 2015. Controversies regarding costs, uncertainties and benefits specific to shale gas development. Sustainability 7, 2473–2489.
NRC, 2012. Induced Seismicity Potential in Energy Technologies. National Research Council, The National Academies Press, Washington, DC.
O'Sullivan, F., Paltsev, S., 2012. Natural gas and crude oil production: potential versus actual greenhouse gas emissions. Environ. Res. Lett. 7, 1–6.
Osborn, S.G., Vengosh, A., Warner, N.R., Jackson, R.B., 2011. Methane contamination of drinking water accompanying gas-well drilling and hydraulic fracturing. Proc. Natl. Acad. Sci. U. S. A. 108 (20), 8172–8176.
Redmayne, D.W., Richards, J.A., Wild, P.W., 1998. Mining-induced earthquakes monitored during pit closure in the Midlothian coalfield. Q. J. Eng. Geol. Hydrogeol. 31 (1), 21–36.
Reig, P., Luo, T., Proctor, J.N., 2014. Global Natural Gas Development: Water Availability and Business Risks. World Resources Institute, Washington, DC.
Schrag, D.P., 2012. Is natural gas and crude oil good for climate change? Dædalus 141 (2), 72–80.

Shine, K.P., 2009. The global warming potential—the need for an interdisciplinary retrial. Clim. Chang. 96 (4), 467–472.

Shonkoff, S.B.C., Hays, J., Finke, M.L., 2014. Environmental public health dimensions of shale and tight gas development. Environ. Health Perspect. 122 (8), 787–795.

Speight, J.G., 2007. Natural Gas: A Basic Handbook. GPC Books, Gulf Publishing Company, Houston, Texas.

Speight, J.G., 2008. Synthetic Fuels Handbook: Properties, Processes, and Performance. McGraw-Hill, New York.

Speight, J.G., 2013. The Chemistry and Technology of Coal, third ed. CRC-Taylor and Francis Group, Boca Raton, Florida.

Speight, J.G., 2014a. The Chemistry and Technology of Petroleum, fifth ed. CRC-Taylor and Francis Group, Boca Raton, Florida.

Speight, J.G., 2014b. High Acid Crudes. Gulf Professional Publishing, Elsevier, Oxford, United Kingdom.

Speight, J.G., 2014c. Oil and Gas Corrosion Prevention. Gulf Professional Publishing, Elsevier, Oxford, United Kingdom.

Spellman, F.R., 2013. Environmental Impacts of Hydraulic Fracturing. CRC Press, Taylor & Francis Group, Boca Raton, Florida.

Stephenson, T., Valle, J.E., Riera-Palou, X., 2011. Modeling the relative GHG emissions of conventional and natural gas and crude oil production. Environ. Sci. Technol. 45, 10757–10764.

Tiemann, M., Folger, P., Carter, N.T., 2014. Tight formation energy technology assessment: current and emerging water practices. In: Prepared for Members and Committees of Congress, Jul. 14. Congressional Research Service, Washington, DC. CRS Report R43635.

US EIA, 2011a. Natural Gas and Crude Oil and Tight Formation Oil Plays. Energy Information Administration, United States Department of Energy, Washington, DC. www.eia.gov.

US EIA, 2011b. World Natural Gas and Crude Oil Resources: An Initial Assessment of 14 Regions Outside the United States. Energy Information Administration, United States Department of Energy, Washington, DC. www.eia.gov.

US EPA, 2012. Regulation of Hydraulic Fracturing Under the Safe Drinking Water Act. United States Environmental Protection Agency, Washington, DC.

US EPA, 2015. Greenhouse Gas Reporting Program (GHGRP): GHGRP and the Oil and Gas Industry. United States Environmental Protection Agency, Washington, DC. https://www.epa.gov/ghgreporting/ghgrp-and-oil-and-gas-industry (accessed 28.03.16.).

Veil, J.A., 2010. Water management technologies used by Marcellus natural gas and crude oil producers. Argonne National Laboratory, Argonne, Illinois. Report No. ANL/EVR/R-10/3.

WHO, 2011. Guidelines for Drinking Water Quality, fourth ed. World Health Organization, Geneva, Switzerland.

CONVERSION FACTORS

1 acre = 43,560 square feet (43,560 ft^2)
1 acre-foot = 7758.0 bbl
1 atmosphere = 760 mmHg = 14.696 psia = 29.91 in. Hg
1 atmosphere = 1.0133 bars = 33.899 ft H_2O
1 barrel (oil) = 42 gal = 5.6146 ft^3
1 barrel (water) = 350 lb at 60°F
1 barrel per day = 1.84 cm^3/s
1 Btu = 778.26 foot-pound (ft-lb)
1 cubic foot = 28,317 cm^3 = 7.4805 gal
Density of water at 60°F = 0.999 g/cm^3
1 gallon = 231 cubic inch = 3785.4 cm^3 = 0.13368 ft^3
1 horsepower-hour—0.7457 kWh = 2544.5 Btu
1 horsepower = 550 ft-lb/s = 745.7 W
1 in. = 2.54 cm
1 m = 100 cm = 1000 mm = 10 μm = 10 Å
1 ounce = 28.35 g
1 pound = 453.59 g = 7000 grains
1 square mile = 640 acres
1 tonne of oil = 7.3 barrels of oil equivalent
1 tonne of condensate = 8.0 barrels of oil equivalent
1 billion cubic meters of gas = 6.6 million barrels of oil equivalent
1 billion cubic meters of gas = 35.3 billion cubic feet of gas
1 billion cubic meters of gas = 0.9 million tonnes of oil equivalent

NATURAL GAS CONVERSION TABLE[a]

1 cubic foot (ft^3) = 1000 Btu
1 CCF = 100 cubic feet = 1 Therm = 100,000 Btu
1 MCF = 1000 cubic feet = 1 MM Btu
1 MMCF = 1 million cubic feet = 10,000 MMBtu
1 BCF = 1 billion cubic feet = 1×10^9 ft^3 = 1 million MMBtu
1 TCF = 1 trillion cubic feet–1×10^{12} ft^3

[a] Based upon an approximate natural gas heating value of 1000 Btu per cubic foot.

Ft^3	CCF	MCF	MMCF	Therm	Dekatherm	Btu	MMBtu	kJ	kWh
1	0.01	0.001	0.000001	0.01	0.001	1000	0.001	1054	0.293
100	1	0.1	0.0001	1	0.1	100000	0.1	105461.5	29
1000	10	1	0.001	10	1	1000000	1	1054615	293
1000000	10000	1000	1	10000	1000	1.00E+09	1000	1054615000	293071
100	1	0.1	0.0001	1	0.1	100000	0.1	105500	29
1000	10	1	0.001	10	1	1000000	1	1054615	293
0.001	0.00001	0.000001	0.000000001	0.00001	0.0001	1	0.000001	1.055	0
1000	10	1	0.001	10	1	1000000	1	1054615	293
0.0009	0.00001	0.000001	0	0.00001	0.0001	0.9482	0.000001	1	0
3.345	0.033	0.003	0.000003	0.034	0.003	3412	0.003	3600	1

GLOSSARY

Abandonment pressure A direct function of the economic premises, the static bottom pressure at which the revenues obtained from the sales of the hydrocarbons produced are equal to the well's operation costs.

Abiogenic gas Gas formed by inorganic means.

Absolute permeability Ability of a rock to conduct a fluid when only one fluid is present in the pores of the rock.

Absolute pressure The total pressure equal to gauge pressure plus 14.7 lbs./sq. in. at sea level.

Absorbed gas Natural gas that has been dissolved into the rock and requires hydraulic fracturing to be released.

Absorber See Absorption tower.

Absorption The process by which the gas is distributed throughout an absorbent (liquid); depends only on physical solubility and may include chemical reactions in the liquid phase (*chemisorption*).

Absorption oil Oil used to separate the heavier components from a vapor mixture by absorption of the heavier components during intimate contacting of the oil and vapor; used to recover natural gasoline from wet gas.

Absorption plant A plant for recovering the condensable portion of natural or refinery gas, by absorbing the higher boiling hydrocarbons in an absorption oil, followed by separation and fractionation of the absorbed material.

Absorption tower A tower or column which promotes contact between a rising gas and a falling liquid so that part of the gas may be dissolved in the liquid.

Accumulation A pool of petroleum locally confined by subsurface geologic features.

Acid deposition (acid rain) Occurs when sulfur dioxide (SO_2) and, to a lesser extent, NO_x emissions are transformed in the atmosphere and return to the Earth as dry deposition or in rain, fog, or snow.

Acid gas Carbon dioxide and hydrogen sulfide; see also Sour gas.

Acid gas loading The amount of acid gas, on a molar or volumetric basis, which will be picked up by a solvent.

Acidity The presence of acid-type constituents whose concentration is usually defined in terms of neutralization number; the constituents vary in nature and may or may not markedly influence the behavior of the oil; see Neutralization number.

Acidizing A technique used in tight gas and tight oil extraction and involves pumping acids into a well to dissolve the limestone ($CaCO_3$) and dolomite ($CaCO_3 \cdot MgCO_3$) cement between the sediment grains of the reservoir rocks.

Acid rain See Acid deposition

Acoustic log See Sonic log.

Adsorbed gas Natural gas that has accumulated on the surface of a solid forming a thin film.

Adsorption Transfer of a substance from a solution to the surface of a solid resulting in relatively high concentration of the substance at the place of contact; molecular bonding of a gas to the surface of a solid. In the case of shale, natural gas is adsorbed or bonded to the organic material in the shale.

Air pollution The discharge of toxic gases and particulate matter introduced into the atmosphere, principally as a result of human activity.

Air quality A measure of the amount of pollutants emitted into the atmosphere and the dispersion potential of an area to dilute those pollutants.

Alkylation A process for manufacturing high octane blending components used in unleaded petrol or gasoline.

American Society for Testing and Materials (ASTM) The official organization in the United States for designing standard tests for petroleum and other industrial products.

Amine washing A method of gas cleaning whereby acidic impurities such as hydrogen sulfide and carbon dioxide are removed from the gas stream by washing with an amine (usually an alkanolamine).

Aniline point The aniline point of a petroleum product is the minimum equilibrium solution temperature with an equal volume of freshly distilled aniline.

Annulus The space between two concentric objects, such as the space between the casing and the wellbore or surrounding rock.

Anticline An area of the Earth's crust where folding has made a dome-like shape in the once flat rock layers. Anticlines often provide an environment where natural gas can become trapped beneath the Earth's surface, and extracted.

API American Petroleum Institute.

API gravity A measure of the lightness or heaviness of petroleum that is related to density and specific gravity.

Aquifer An underground layer of water-bearing rock or gravel, sand or silt; the subsurface layer of rock or unconsolidated material that allows water to flow within it; aquifers can act as sources for groundwater, both usable fresh water and unusable salty water.

Aquifer storage and recovery (ASR) A process in which clean water is injected into a shallow aquifer, stored for some period of time, and then withdrawn for later use.

Aromatics A range of hydrocarbons which have a distinctive sweet smell and include benzene and toluene, occur naturally in petroleum and are also extracted as a petrochemical feedstock, as well as for use as solvents.

Ash The inorganic residue remaining after ignition of combustible substances determined by definite prescribed methods; see ASTM.

Asphaltene fraction The brown to black powdery material produced by treatment of petroleum, heavy oil, bitumen, or residuum with a low-boiling liquid hydrocarbon.

Assessment unit An assessment unit is a volume of rock within the total petroleum system that encompasses fields, discovered and undiscovered, sufficiently homogeneous in terms of geology, exploration strategy and risk characteristics to constitute a single population of field characteristics with respect to criteria used for resource assessment.

Associated gas Natural gas that is in contact with and/or dissolved in the crude oil of the reservoir. It may be classified as gas cap (free gas) or gas in solution (dissolved gas).

Associated gas in solution (or dissolved gas) Natural gas dissolved in the crude oil of the reservoir, under the prevailing pressure and temperature conditions.

ASTM American Society for Testing and Materials; the quality specifications for petroleum products are determined by ASTM test methods.

Atmospheric distillation The process used to separate the desalted crude into specific fractions such as naphtha, middles distillates, and light gas oil.

Atomization characteristics The ability of an oil to be broken up into a fine spray by some mechanical means.

Autoignition temperature Minimum temperature at which a fuel-air mixture ignites.

BACT Best available control technology

Baghouse A filter system for the removal of particulate matter from gas streams; so called because of the similarity of the filters to coal bags.

Barrel (bbl) The unit of measure used by the petroleum industry; equivalent to approximately 42 US gallons or approximately 34 (33.6) Imperial gallons or 159 L; 7.2 barrels are equivalent to one tonne of oil (metric).

Barrel of oil equivalent (boe) The amount of energy contained in a barrel of crude oil, that is, approximately 6.1 GJ (5.8 million Btu), equivalent to 1700 kWh.

Base gas The quantity of gas needed to maintain adequate reservoir pressures and deliverability rates throughout the withdrawal season; base gas usually is not withdrawn and remains in the reservoir and all gas native to a depleted reservoir is included in the base gas volume.

Baseline water stress The ratio of total water withdrawals from municipal, industrial, and agricultural users relative to the available renewable surface water. Higher values may indicate more competition among users and greater depletion of water resources.

Basin A geological receptacle in which a sedimentary column is deposited that shares a common tectonic history at various stratigraphic levels; a closed geologic structure in which the beds dip toward a center location; the youngest rocks are at the center of a basin and are partly or completely ringed by progressively older rocks; see also Province.

Basin-centered gas Regionally pervasive gas accumulations that are abnormally pressured, commonly lack a down-dip water contact and have low-permeability reservoirs; in the deeper parts of basins that are actively generating gas there can be hundreds (even thousands) of feet of stacked reservoirs of different lithology with gas in tight sandstone formations, siltstone formations, shale formations, and coal seams.

Bbl See Barrel.

Bcf (billion cubic feet) Gas measurement approximately equal to one trillion (1,000,000,000,000) British thermal units.

Billion 1×10^9.

Biocide An additive used in hydraulic fracturing fluids (and often drilling muds) to kill bacteria that could otherwise reduce permeability and fluid flow.

Biogenic Material derived from bacterial or vegetation sources.

Biogenic gas Natural gas produced by living organisms or biological processes.

Biomass Biological organic matter.

Bitumen The extractable material in oil shale; also, on occasion, referred to as native asphalt, and extra-heavy oil; more correctly—a naturally occurring hydrocarbonaceous material that is immobile under reservoir conditions and which cannot be recovered through a well by conventional oil well production methods including currently used enhanced recovery techniques; current methods involve mining for bitumen recovery.

Black shale Thinly bedded shale that is rich in organic matter, formed by anaerobic decomposition of the precursors; black shale formations occur in thin beds in many areas at various depths and are of interest both historically and economically.

Blender A device for mixing two or more crude oils fuel oils to achieve an acceptable refinery feedstock; also used for mixing petroleum products, such as fuel oils, to achieve a less viscous and more uniform fuel.

Blowdown The difference between the set pressure and the reseating pressure of a safety valve expressed in percent of the set pressure or in psi, bar, or kPa.

Blowout An uncontrolled release of fluids during the drilling, completion, or production of crude oil and natural gas; in former times, the blow out may have been referred to as a gusher.

Boiling liquid expanding vapor explosion (BLEVE) An explosion resulting from the failure of a vessel containing a liquid at a temperature significantly above its boiling point at normal atmospheric pressure.

Borehole A generalized term for a shaft bored into the ground.

Bottom simulating reflector (BSR) A seismic reflection at the sediment to clathrate stability zone interface caused by the different density between normal sediments and sediments laced with clathrates.

Brackish water Water that is generally saltier than freshwater, but not as salty as seawater.

Breccia pipe A mass of breccia (a rock composed of broken fragments of minerals or rock cemented together by a fine-grained matrix) often in an irregular and cylindrical shape; rock fragments usually consist of fragments of the host rock (the rock layer they are contained in) cemented together by silica; these formations are often hosts for ore deposition especially in copper and uranium mining districts.

British thermal unit (Btu) A non-metric unit of heat, still widely used by engineers; One Btu is the heat energy needed to raise the temperature of one pound of water from 60 to 61°F at one atmosphere pressure. 1 Btu = 1055 joules (1.055 kJ).

Brittleness index (BI) A measure of the ability of rock to fracture.

BTU See British thermal unit.

Bubble point pressure (BPP) The pressure at which gas bubbles first form within a liquid at a known temperature. This is also a special case of the true vapor pressure where the volume ratio of gas to liquid is zero ($V/L = 0$), and any incremental increase in temperature or decrease in pressure will incur gas formation and $V/L > 0$. The definition implies that the liquid is contained at a high pressure that is gradually decreased to a point where bubbles start to form, analogous to the situation where production fluids are depressurized and at some discrete pressure and temperature condition, separate into liquid and gas phases.

Bunker fuel oil Heavy, residual fuel oil used in ships.

Burning velocity Velocity at which a fuel-air mixture issuing from a burner burns back to the burner.

C_1, C_2, C_3, C_4, C_5 fractions A common way of representing natural gas fractions containing a preponderance of hydrocarbons having 1, 2, 3, 4, or 5 carbon atoms, respectively, and without reference to hydrocarbon type.

CAA Clean Air Act; this act is the foundation of air regulations in the United States.

Calorie The amount of heat required to raise the temperature of 1 gram of water by 1°C, at or near maximum density.

Calorific value The amount of heat produced by the complete combustion of a unit weight of fuel; usually expressed in calories per gram or BTU per pound, the latter being numerically 1.8 times the former.

Carbonate rock A rock consisting primarily of a carbonate mineral such as calcite or dolomite, the chief minerals in limestone and dolostone, respectively.

Carbonate washing A process using a mild alkali (e.g., potassium carbonate) for emission control by the removal of acid gases from gas streams.

Carbon dioxide (CO_2) A product of combustion that acts as a greenhouse gas in the Earth's atmosphere, trapping heat and contributing to climate change.

Carbon monoxide (CO) A lethal gas produced by incomplete combustion of carbon-containing fuels in internal combustion engines. It is colorless, odorless, and tasteless. (As in flavorless, we mean, though it's also been known to tell a bad joke or two.)

Casing Steel pipe inserted into a wellbore and cemented into place; also used to protect freshwater aquifers or otherwise isolate a zone and serves to isolate fluids, such as water, gas, and oil, from the surrounding geologic formations; used to line the walls of a gas well to prevent collapse of the well, and also to protect the surrounding Earth and rock layers from being contaminated by petroleum or the drilling fluids.

Catalyst A substance that accelerates a chemical reaction without itself being affected. In refining, catalysts are used in the cracking process to produce blending components for fuels.

Catalyst fines Hard, abrasive crystalline particles of alumina, silica, and/or alumina silica that can be carried over from the fluidic catalytic cracking process of residual fuel stocks; the particle size can range from submicron to >60 μm in size.

Cat cracker A large refinery vessel for processing reduced crudes or other feedstocks in the presence of a catalyst, as opposed to the older method of thermal cracking, which employs heat and pressure only; the catalytic cracking process is generally preferred since it produces less gas and other highly volatile byproducts such as naphtha of higher octane than the thermal process.

CFR Code of Federal Regulations; Title 40 (40 CFR) contains the regulations for protection of the environment.

Christmas tree The series of pipes and valves that sits on top of a producing gas well; used in place of a pump to extract the gas from the well.

Class II injection well A well that injects fluids into a formation rather than produces fluids. A Class II injection well is a well associated with oil or natural gas production. Such wells include enhanced recovery wells, disposal wells, and hydrocarbon storage wells.

Clarifier A machine used for a liquid-sludge separation in which the particles with a higher specific gravity are separated from the lower specific gravity of the liquid; a clarifier bowl has one outlet for the light phase oil; the heavier phase particles are retained on the bowl wall.

Clastic Composed of pieces of preexisting rock.

Clathrate A solid compound in which molecules of one substance are trapped in the crystal lattice of another; also known as hydrate; see Gas hydrate.

Clay Silicate minerals that also usually contain aluminum and have particle sizes <0.002 μm; used in separation methods as an adsorbent and in refining as a catalyst.

Clean Air Act Amendments of 1990 Legislation to improve the quality of the atmosphere and curb acid rain promotes the use of cleaner fuels in vehicles and stationary sources.

Cleat The vertical cleavage of coal seams; the main set of joints along which coal breaks when mined.

Cloud point The temperature at which wax begins to crystallize from crude oil or from a distillate fuel.

Coal A readily combustible black or brownish-black rock whose composition, including inherent moisture, consists of more than 50% *w/w* and more than 70% *v/v* of carbonaceous material; formed from plant remains that have been compacted, hardened, chemically altered, and metamorphosed by heat and pressure over geologic time.

Coal basin A region in which coal deposits of known or possible economic value occur within a basin-type structure.

Coalbed methane (coal bed methane) Methane from coal seams; released or produced from the seams when the water pressure within the seam is reduced by pumping from either vertical or inclined to horizontal surface holes; see also Biogenic coal bed methane and Thermogenic coal bed methane.

Coalescer A mechanical process vessel with wettable, high-surface area packing on which liquid droplets consolidate for gravity separation from a second phase (e.g., gas or immiscible liquid).

Coal quality The character or nature of the amount of impurities (ash and trace elements) in coal; coal quality parameters of greatest interest include ash, moisture, sulfur, and energy value (heat content, Btu/lb).

Code of Federal Regulations (CFR) For example, Title 40 (40 CFR) contains the regulations for protection of the environment.

Coking A thermal method used in refineries for the conversion of bitumen and residua to volatile products and coke (see Delayed coking and Fluid coking).

Cold stacked drilling rig (sometimes called mothballed drilling rig) Refers to a drilling rig that has been shut down and stored in a harbor, shipyard, or designated offshore area; also referred to as mothballing, cold stacking is a cost reduction step taken when the contracting prospects for a drilling rig are minimal or the available contract terms do not justify an adequate return on the investment needed to make the unit work ready (e.g., repairs or refurbishment).

Combined-cycle gas turbine A gas-powered electricity generator in which the exhaust heat from a gas turbine is used to drive a steam turbine, producing electricity with efficiency levels of up to 60%.

Completion Includes the steps required to drill and assemble casing, tubes, and equipment to efficiently produce oil or gas from a well. For shale gas wells, this includes hydraulic fracturing activities.

Composition The make-up of a gaseous stream.

Compressed natural gas (CNG) Natural gas compressed to a pressure at or above 2900–3600 psi and stored in high-pressure containers; used as a fuel for natural gas-powered vehicles.

Compression The means of reduction is the volume of natural gas as for transportation and storage.

Condensate A mixture of light hydrocarbon liquids obtained by condensation of hydrocarbon vapors: predominately butane, propane, and pentane with some heavier hydrocarbons and relatively little methane or ethane; see also Natural gas liquids.

Conditioning Processing of crude oil to remove impurities prior to transport; impurities include gases, water, and solids that were coproduced with the crude oil.

Conductor casing Prevents collapse of the loose soil near the surface of a borehole.

Contingent resource The amounts of hydrocarbons estimated at a given date that are potentially recoverable from known accumulations but are not considered commercially recoverable under the economic evaluation conditions corresponding to that date.

Continuous oil accumulation An oil resource that is dispersed throughout a geologic formation rather than existing as discrete, localized occurrences, such as those in conventional accumulations; unconventional resources often require special technical drilling and recovery methods.

Conventional crude oil (conventional petroleum) Crude oil that is pumped from the ground and recovered using the energy inherent in the reservoir; also recoverable by application of secondary recovery techniques.

Conventional gas Natural gas that is extracted from underground reservoirs using traditional exploration and production methods.

Conventional limit The reservoir limit established according to the degree of knowledge of (or research into) the geological, geophysical, or engineering data available.

Conventional oil (conventional crude oil) Crude oil that is produced by conventional primary or secondary recovery methods in which a well is drilled into a geologic formation in which the reservoir and fluid characteristics permit the oil to readily flow to the wellbore.

Conventional natural gas (conventional gas) Natural gas oil that is produced by recovery methods in which a well is drilled into a geologic formation in which the reservoir and fluid characteristics permit the oil to readily flow to the wellbore.

Conventional oil and gas accumulations Discrete accumulations with well-defined hydrocarbon-water contacts, where the hydrocarbons are buoyant on a column of water; conventional accumulations commonly have relatively high matrix permeability, have obvious seals and traps, and have relatively high recovery factors.

Core A cylindrical rock sample taken from a formation when drilling, used to determine the rock's permeability, porosity, hydrocarbon saturation, and other productivity-associated properties.

Corrosion The detrimental change in the size or characteristics of material under conditions of exposure or use; usually results from chemical action either regularly and slowly, as in rusting (oxidation), or rapidly, as in metal pickling.

Corrosion inhibitor A chemical compound that decreases the corrosion rate of a metal or an alloy.

Cracking A secondary refining process that uses heat and/or a catalyst to break down high molecular weight chemical components into lower molecular weight products which can be used as blending components for fuels.

Cricondenbar The maximum pressure at which two phases can coexist; see Cricondentherm temperature, Dew point.

Cricondentherm temperature The highest dew point temperature seen on a liquid-vapor curve for a specific gas composition over a range of pressure, for example, 200–1400 psia; see Cricondenbar, Dew point.

Cryogenic plant A processing plant capable of producing liquid natural gas products, including ethane, at very low operating temperatures.

Cryogenic process A process involving low temperatures.

Cubic foot A unit of measurement for volume; an area one foot long, by one foot wide, by one foot deep.

Cutter stock A petroleum product which is used to reduce the viscosity of a heavier residual product by dilution.

Cutting A piece of rock or dirt that is brought to the surface of a drilling site as debris from the bottom of well; often used to obtain data for logging.

Cyclone A device for extracting dust from industrial waste gases. It is in the form of an inverted cone into which the contaminated gas enters tangential from the top; the gas is propelled down a helical pathway, and the dust particles are deposited by means of centrifugal force onto the wall of the scrubber.

Darcy's law An equation that describes the flow of liquid through a porous medium.

Debutanization Distillation to separate butane and lighter components from higher boiling components.

Decline rate The rate at which the production rate of a well decreases.

Deethanization Distillation to separate ethane and lighter components from propane and higher boiling components; also called deethanation.

Deflagration Classification of an explosion; burning of a fuel-air mixture where the flame travels at subsonic velocities.

Degrees API (141.5/sp gr @ 60°F)–131.5.

Dehydration Water removal from natural gas streams.

Delayed coking A coking process in which the thermal reactions are allowed to proceed to completion to produce gaseous, liquid, and solid (coke) products.

Demethanization The process of distillation in which methane is separated from the higher boiling components; also called demethanation.

Demulsibility The resistance of an oil to emulsification, or the ability of an oil to separate from any water with which it is mixed; the better the demulsibility rating, the more quickly the oil separates from water.

Density The mass (or weight) of a unit volume of any substance at a specified temperature; see also Specific gravity.

Depentanizer A fractionating column for the removal of pentane and lighter fractions from a mixture of hydrocarbons.

Depleted reservoirs Reservoirs that have already been tapped of all their recoverable natural gas.

Depleted storage field A subsurface natural geological reservoir, usually a depleted gas or oil field, used for storing natural gas.

Deposit Mineral deposit or ore deposit is used to designate a natural occurrence of a useful mineral, or an ore, in sufficient extent and degree of concentration to invite exploitation.

Depropanization Distillation in which lighter components are separated from butanes and higher boiling material; also called depropanation.

Desalter A unit which mixes a crude oil stream with a small amount of fresh water (e.g., 10% by volume) forming a water-in-oil emulsion after which the emulsion is subjected to an electric field wherein the water is coalesced as an under flow from the upper flow of a relatively water-free, continuous hydrocarbon phase; the desalted stream has a very small residual salt content and the performance of this unit can be improved with a demulsifier.

Desalting The process used, prior to distillation, to remove corrosive salts as well as metals and other suspended solids from crude oil.

Desorption The reverse process of adsorption whereby adsorbed matter is removed from the adsorbent; also used as the reverse of absorption (*q.v.*).

Desulfurization The removal of sulfur or sulfur compounds from a feedstock.

Detonation Classification of an explosion; burning of a fuel-air mixture where the flame travels at supersonic velocities; a violent explosion involving high-velocity pressure waves; in a gasoline engine, the spontaneous combustion of part of the compresses charge after spark occurs. Detonation usually produces a characteristic metallic sound or knock.

Developed proved reserves Reserves that are expected to be recovered in existing wells, including reserves behind pipe, which may be recovered with the current infrastructure through additional work and with moderate investment costs. Reserves associated with secondary and/or enhanced recovery processes will be considered as developed when the infrastructure required for the process has been installed or when the costs required for such are lower. This category includes reserves in completed intervals that have been opened at the time when the estimation is made but that have not started flowing due to market conditions, connection problems, or mechanical problems and whose rehabilitation cost is relatively low.

Development Activity that increases or decreases reserves by means of drilling exploitation wells.

Development well A well drilled in a proved area in order to produce hydrocarbons.

Dewatering A process used in coalbed methane extraction whereby water is pumped from a coal seam to lower the pressure, which causes the gas to desorb and flow via cleats into the well.

Dew point (dew point temperature) The temperature below which the water vapor in a volume of humid gas at a given constant barometric pressure will condense into liquid water at the same rate at which it evaporates; the temperature to which a given volume of gas must be cooled, at constant barometric pressure, for vapor to condense into liquid; the saturation point; see Cricondenbar. Cricondentherm temperature, Hydrocarbon dew point.

Diesel fuel A distillate of fuel oil that has been historically derived from petroleum for use in internal combustion engines; also derived from plant and animal sources.

Diesel index The product of the API gravity and the aniline point (in degrees Fahrenheit) of a diesel fuel, divided by 100; an indication of the ignition quality of the fuel.

Diluent Typically, a hydrocarbon fluid that is used to dilute heavy, extra-heavy crude oil, or tar sand bitumen in order to reduce its viscosity for easier transportation; generally, a distillate is used for heavy oil dilution and transportation; the added diluent may be recovered at the destination using distillation and the diluent may be subsequently pumped back for blending.

Directional drilling The technique of drilling at an angle from a surface location to reach a target formation not located directly underneath the well pad; see Horizontal drilling.

Discovered resource The volume of hydrocarbons tested through wells drilled.

Discovery The incorporation of reserves attributable to drilling exploratory wells that test hydrocarbon-producing formations.

Disposal well A well which injects produced water into a regulated and approved deep underground formation for disposal.

Distillate Any petroleum product produced by boiling crude oil and collecting the vapors produced as a condensate in a separate vessel, for example gasoline (light distillate), gas oil (middle distillate), or fuel oil (heavy distillate).

Distillation The primary distillation process which uses high temperature to separate crude oil into vapor and fluids which can then be fed into a distillation or fractionating tower.

Doctor test A qualitative method of detecting undesirable sulfur compounds in petroleum distillates, that is, of determining whether oil is *sour* (high sulfur) or *sweet* (low sulfur).

Dominant water user The sector (agricultural, municipal, or industrial) with the largest annual water withdrawals.

Drilling mud A fluid used to aid the drilling of boreholes; a mixture of clay, water, and other ingredients that are pumped downhole through the drill pipe and drill bit that enable the removal of the drill cuttings from the well bore and also stabilize the penetrated rock formations before casing is installed in the borehole.

Drilling rig The machine that creates the holes in the ground—typically large standing structures.

Drilling water Water that is used, often in conjunction with other chemicals, to cool and lubricate the drill bit and carry out drill cuttings during the drilling of the borehole.

Drought severity The average length of droughts multiplied by the dryness of the droughts from 1901 to 2008; higher values indicate areas subject to periods of more severe drought.

Dry gas Natural gas containing negligible amounts of hydrocarbons heavier than methane; natural gas which remains after: (1) the liquefiable hydrocarbon portion has been removed from the gas stream (i.e., gas after lease, field, and/or plant separation); and (2) any volumes of nonhydrocarbon gases have been removed where they occur in sufficient quantity to render the gas unmarketable; also known as consumer-grade natural gas; the parameters for measurement are cubic feet at 60°F and 14.73 pounds per square inch absolute.

Dry hole A well that cannot produce oil or gas in sufficient quantities to warrant completion.

Economic limit The point at which the revenues obtained from the sale of hydrocarbons match the costs incurred in its exploitation.

Economic reserves The accumulated production that is obtained from a production forecast in which economic criteria are applied.

Effective permeability A relative measure of the conductivity of a porous medium for a fluid when the medium is saturated with more than one fluid; this implies that the effective permeability is a property associated with each reservoir flow, for example, gas, oil, and water; a fundamental principle is that the total of the effective permeability is less than or equal to the absolute permeability.

Effective porosity A fraction that is obtained by dividing the total volume of communicated pores by the total rock volume.

Effluent The liquid or gas discharged from a process or chemical reactor, usually containing residues from that process.

Emissions Waste substances discharged into the air, land, or water, usually specified by mass per unit time.

Emission control The use gas cleaning processes to reduce emissions.

Emission standard The maximum amount of a specific pollutant permitted to be discharged from a particular source in a given environment.

Emulsion A liquid mixture of two or more liquid substances not normally dissolved in one another, one liquid held in suspension in the other; water-in-oil emulsions have water as the internal phase and oil as the external, while oil-in-water have oil as the internal phase and water as the external phase.

End-of-pipe emission control The use of specific emission control processes to clean gases after production of the gases.

Energy-efficiency ratio A number representing the energy stored in a fuel as compared to the energy required to produce, process, transport, and distribute that fuel.

Energy security A term used to refer to the reliability of future energy supply; dependant on a number of different factors, including the availability of energy supplies, affordability, and the capacity to extract them in an environmentally sustainable manner.

EPA Environmental Protection Agency.

EPACT (Energy Policy Act of 1992) Comprehensive energy legislation designed to expand natural gas use by allowing wholesale electric transmission access and providing incentives to developers of clean fuel vehicles.

EPCRA Emergency Planning and Community Right-to-Know Act.

Equation of state (EOS) Mathematical model relating state variables pressure, density, and temperature; an EOS is useful for interpreting experimental phase behavior data and also predicting P-V-T behavior in process simulations and for conditions where experimental data are not available.

Estimated additional amount in place The volume additional to the proved amount in place that is of foreseeable economic interest. Speculative amounts are not included.

Estimated additional reserves recoverable The volume within the estimated additional amount in place which geological and engineering information indicates with reasonable certainty might be recovered in the future.

Exploration The process of identifying a potential subsurface geologic target formation and the active drilling of a borehole designed to assess the natural gas or oil.

Exploratory well A well that is drilled without detailed knowledge of the underlying rock structure in order to find hydrocarbons whose exploitation is economically profitable.

Extraction process The process used for drawing out resources from the ground; for unconventional gas and unconventional crude oil, this is mainly hydraulic fracturing.

Extra-heavy oil A hydrocarbonaceous material similar to tar sand bitumen that occurs in the solid or near-solid state and is generally mobile under reservoir (or deposit) conditions.

Fabric filters Filters made from fabric materials and used for removing particulate matter from gas streams (see Baghouse).

Facies One or more layers of rock that differs from other layers in composition, age, or content.

Fault A fractured surface of geological strata along which there has been differential movement; a fracture surface in rocks along which movement of rock on one side has occurred relative to rock on the other side.

Final boiling point The highest temperature indicated on the thermometer inserted in the flask during a standard laboratory distillation. This is generally the temperature at which no more vapor can be driven over into the condensing apparatus.

Fireball Partially premixed diffusion flames which rapidly combust due to enhanced turbulent mixing and atomization.

Fire point The lowest temperature to which gas must be heated under prescribed conditions of the method to burn continuously when the mixture of vapor and air is ignited by a specified flame.

Flammability limits Range of vapor concentration in air that will support combustion termed lower flammability limit (LFL) and upper flammability limit (UFL).

Flares The burning of fuel vapors at the source of a release.

Flash fire The burning of a fuel vapor cloud which was ignited at a location away from the release point.

Flash point The temperature to which gas must be heated under specified conditions to give of sufficient vapor to form a mixture with air that can be ignited momentarily by a specified flame; dependant on the composition of the gas and the presence of other hydrocarbon constituents.

Flowback (flow-back) water The water that returns to the surface from the wellbore within the first few weeks after hydraulic fracturing; it is composed of fracturing fluids, sand, and water from the formation, which may contain hydrocarbons, salts, minerals, naturally occurring radioactive materials; often mixed with water found in the geological formation; the amount and quality (often poor) of flowback water returning to the surface varies depending on local geologic conditions and hydraulic fracturing fluids utilized.

Flow line A small-diameter pipeline that generally connects a well to the initial processing facility.

Flow rate The rate that expresses the volume of fluid or gas passing through a given surface per unit of time (e.g., cubic feet per minute).

Fluid coking A continuous fluidized solids process that cracks feed thermally over heated coke particles in a reactor vessel to gas, liquid products, and coke.

Fluidized-bed boiler A large, refractory-lined vessel with an air distribution member or plate in the bottom, a hot gas outlet in or near the top, and some provisions for introducing fuel; the fluidized bed is formed by blowing air up through a layer of inert particles (such as sand or limestone) at a rate that causes the particles to go into suspension and continuous motion.

Formation (geologic) A rock body distinguishable from other rock bodies and useful for mapping or description; formations may be combined into Groups or subdivided into members.

Fossil fuel Solid, liquid, or gaseous fuels formed in the ground after millions of years by chemical and physical changes in plant and animal residues under high temperature and pressure. Oil, natural gas, and coal are fossil fuels.

Fracking The process of using high-pressured fluid containing water, sand, and chemicals into subsurface rock formations—the fluid fractures the rocks, improving the flow of natural gas into the well bore.

Frac tank This is where the water or the proppant is held while a well is being fractured.

Fracture A natural or man-made crack in a reservoir rock.

Fracturing The breaking apart of reservoir rock by applying very high fluid pressure at the rock face; see Fracking.

Free associated gas Natural gas that overlies and is in contact with the crude oil of the reservoir—it may be gas cap.

Free gas Natural gas contained in pores in rock.

Friction reducer An additive that reduced the friction of a fluid as it flows through small spaces.

Fuel oil A heavy residue, black in color, used to generate power or heat by burning in furnaces; the heavy distillates from the oil refining process; used as fuel for power stations, marine boilers.

Gas condensate See Condensate.

Gaseous pollutants Gases released into the atmosphere that act as primary or secondary pollutants.

Gas hydrate A molecule consisting of an ice lattice or cage in which low molecular weight hydrocarbon molecules, such as methane, are embedded; gas hydrates can form under conditions of low temperature and high pressure in pipelines and block them; they are generally removed by reducing pressure, heating, or using chemicals, such as methanol, to dissolve them; methane hydrates also occur in great quantities in permafrost regions and subsea sediments and are a potentially vast energy resource; see Clathrate.

Gas-in-place (GIP) The hypothetical amount of gas contained in a formation or rock unit; gas-in-place always represents a value that is more than what is economically recoverable and refers to the total resources that are possible.

Gas-liquid separator A vertical or horizontal separator in which gas and liquid are separated by means of gravity settling with or without a mist eliminating device.

Gas-oil ratio (GOR) The volume ratio of gas to liquid evolved from an oil that is depressurized to known P, T conditions; volume units are in standard cubic feet of gas per standard barrel of liquid (scf/bbl); standard conditions for reported gas standard cubic feet per industry standards are $P=1$ atmosphere and $T=60°F$.

Gas processing The preparation of gas for consumer use by removal of the non-methane constituents; synonymous with gas refining.

Gas refining See Gas processing.

Geological province A region of large dimensions characterized by similar geological history and development history.

Geological survey The exploration for natural gas that involves a geological examination of the surface structure of the Earth to determine the areas where there is a high probability that a reservoir exists.

Geophones Equipment used to detect the reflection of seismic waves during a seismic survey.

Geothermal Pertaining to the heat of the interior of the Earth.

Global warming An environmental issue that deals with the potential for global climate change due to increased levels of atmospheric greenhouse gases.

Glycol-amine gas treating A continuous, regenerative process to simultaneously dehydrate and remove acid gases from natural gas or refinery gas.

Greenhouse effect The phenomenon where by thermal radiation from the Earth's surface is absorbed by atmospheric greenhouse gases and then reradiated, causing the average surface temperature to rise; see Global warming.

Greenhouse gas Any of the atmospheric gases that contribute to the greenhouse effect by absorbing infrared radiation produced by the solar warming of the surface of the Earth; greenhouse gases include carbon dioxide, methane, nitrous oxide, and water vapor, any of which can be naturally occurring or produced by anthropogenic activities.

Greenhouse gases Gases that trap the heat of the sun in the Earth's atmosphere, producing the greenhouse effect. The two major greenhouse gases are water vapor and carbon dioxide. Other greenhouse gases include methane, ozone, chlorofluorocarbons, and nitrous oxide.

Groundwater Water located beneath the surface of the Earth: subsurface water that is in the zone of saturation and is the source of water for wells, seepage, and springs; the top surface of the groundwater is the *water table*.

Groundwater stress The ratio of groundwater withdrawal to its recharge rate over a given aquifer; values above 1.0 indicate where unsustainable groundwater consumption could impact groundwater availability and groundwater-dependent ecosystems.

GWPC Ground Water Protection Counsel.

Halo oil Oil that exists in the fringe regions halos, surrounding the areas of historical production; traditional technology cannot produce economically because the reservoir permeability is low; see Tight oil.

HAP(s) Hazardous air pollutant(s).

HCPV Hydrocarbon pore volume.

Heat of combustion (energy content) The amount of energy that is obtained from burning natural gas; measured in British thermal units (Btu).

Heat of combustion (gross) The total heat evolved during complete combustion of unit weight of a substance, usually expresses in BTU per pound.

Heat of combustion (net) The gross heat of combustion minus the latent heat of condensation of any water produced.

Heat value The amount of heat released per unit of mass or per unit of volume when a substance is completely burned. The heat power of solid and liquid fuels is expressed in calories per gram or in Btu per pound. For gases, this parameter is generally expressed in kilocalories per cubic meter or in Btu per cubic foot.

Heavy oil (heavy crude, heavy crude oil) Oil that is more viscous that conventional crude oil, has a lower mobility in the reservoir but can be recovered through a well from the reservoir by the application of a secondary or enhanced recovery methods.

Heteroatom compounds Chemical compounds that contain nitrogen and/or oxygen and/or sulfur and/or metals bound within their molecular structure(s).

Heterogeneity Lack of uniformity in reservoir properties such as permeability.

HHV (*gross energy value, upper heating value, gross calorific value, higher calorific value*) The same value as the thermodynamic heat of combustion since the enthalpy change for the reaction assumes a common temperature of the compounds before and after combustion, in which case the water produced by combustion is liquid.

Homogenizer A mechanical device which is used to create a stable, uniform dispersion of an insoluble phase (asphaltene constituents) within a liquid phase (fuel oil).

Horizontal drilling A drilling procedure in which the well bore is drilled vertically to a kick-off depth above the target formation and then angled through a wide 90-degree arc such that the producing portion of the well extends horizontally through the target formation; see Directional drilling.

Horsehead (balanced conventional beam, sucker rod) pump A common type of cable rod lifting equipment for recovery of oil and gas; so-called because of the shape of the counter weight at the end of the beam.

Hydraulic fracturing (fracking, fracing, fraccing) A stimulation technique performed on low-permeability reservoirs such as shale to increase oil and/or gas flow from the formation and improve productivity. Fluids and proppant are injected at high pressure and flow rate into a reservoir to create fractures perpendicular to the wellbore according to the natural stresses of the formation and maintain those openings during production.

Hydrocarbon An organic compound containing only carbon and hydrogen. Hydrocarbons often occur in petroleum products, natural gas, and coals.

Hydrocarbonaceous material A material such as bitumen that is composed of carbon and hydrogen with other elements (heteroelements) such as nitrogen, oxygen, sulfur, and metals chemically combined within the structures of the constituents; even though carbon and hydrogen may be the predominant elements, there may be very few true hydrocarbons.

Hydrocarbon compounds Chemical compounds containing only carbon and hydrogen.

Hydrocarbon dew point The temperature (at a given pressure) at which the hydrocarbon constituents of any hydrocarbon-rich gas mixture, such as natural gas, will start to condense out of the gaseous phase; the maximum temperature at which such condensation takes place is called the *cricondentherm temperature*; the hydrocarbon dew point is a function of the gas composition as well as the pressure; the hydrocarbon dew point of a gas is a different concept from the water dew point, the latter being the temperature (at a given pressure) at which water vapor present in a gas mixture will condense out of the gas; see Cricondenbar, Cricondentherm temperature; Dew Point.

Hydrocarbon resource Resources such as petroleum and natural gas which can produce naturally occurring hydrocarbons without the application of conversion processes.

Hydrodesulfurization The removal of sulfur by hydrotreating.

Hydrofracking See Hydraulic fracturing.
Hydrofracturing See Hydraulic fracturing.
Hydrology The study of water.
Hydrometer An instrument for determining the specific gravity of a liquid.
Hydroprocesses Refinery processes designed to add hydrogen to various products of refining.
Hydrostatic pressure The pressure exerted by a fluid at rest due to its inherent physical properties and the amount of pressure being exerted on it from outside forces.
Hydrotreating The removal of heteroatomic (nitrogen, oxygen, and sulfur) species by treatment of a feedstock or product at relatively low temperatures in the presence of hydrogen.
Ideal gas A gas in which all collisions between atoms or molecules are perfectly elastic and in which there are no intermolecular attractive forces.
Impure natural gas Natural gas as delivered from the well and before processing (refining).
Inclined grate A type of furnace in which fuel enters at the top part of a grate in a continuous ribbon, passes over the upper drying section where moisture is removed, and descends into the lower burning section. Ash is removed at the lower part of the grate.
Independent producer A nonintegrated company which receives nearly all of its revenues from production at the wellhead; by the IRS definition, a firm is an independent if the refining capacity is less than 50,000 barrels per day in any given day or their retail sales are less than $5 million for the year.
Inhibitor injection techniques The injection of substances such as methanol to assist with the extraction of natural gas hydrates; such an injection will dissolve the methane from the hydrate so the gas is released.
Initial boiling point (IBP) The temperature at which a liquid begins to boil during a distillation process, typically at atmospheric pressure; the result for a liquid containing a wide range of boiling components is highly dependent on method.
Injection well Any bored, drilled, or a driven shaft or a dug hole, where the depth is greater than the largest surface dimension that is used to inject fluids underground; Class II injection wells are used to inject produced water into the ground, inject other fluids underground to increase the recovery of hydrocarbons, or to store hydrocarbons underground.
Innage Space occupied in a product container.
In-place resources The quantities of crude oil and natural gas that are estimated, as of a given date, to be contained in known accumulations prior to production; the quantity which can be commercially produced or mined, may be significantly less than the volumes estimated to be in place.
In situ Geological term meaning in the original location or position, such as an outcrop that has not been upset by faults.
Intermediate casing Casing used on longer drilling intervals—set after the surface casing and before the production casing and prevents caving of weak or abnormally pressured formations.
IOGCC Interstate Oil and Gas Commission
Isopach A line on a map designating points of equal formation thickness.
Kerogen A complex carbonaceous (organic) material that occurs in sedimentary rock and shale; generally insoluble in common organic solvents.
Kerosene A light middle distillate that in various forms is used as aviation turbine fuel or for burning in heating boilers or as a solvent, such as white spirit.
Kinematic viscosity The ratio of the absolute viscosity of a liquid to its specific gravity at the temperature at which the viscosity is measured; expressed in stokes or centistokes.

Kitchen The underground deposit of organic debris that is eventually converted to petroleum and natural gas.

Kriging A technique used in reservoir description for interpolation of reservoir parameters between wells based on random field theory.

Latent heat Heat required to change the state of a unit weight of a substance from solid to liquid or from liquid to vapor without change of temperature.

Lean gas Natural gas in which methane is the major constituent.

Lease A legal document that conveys to an operator the right to drill for oil and gas. Also, the tract of land on which a lease has been obtained and where producing wells and production equipment are located.

Light crude Crude oil with a low specific gravity and high API gravity due to the presence of a high proportion of light hydrocarbon fractions and low metallic compound.

Light ends The more volatile products of petroleum refining, such as butane, propane, naphtha.

LHV Lower heating values; see HHV.

Liquefied natural gas The liquid form of natural gas; natural gas takes up 1/600 of its gaseous volume making it easier and cheaper to transport if pipelines are not available; the gas is liquefied by cooling to a temperature of −162°C (−260°F).

Liquefied petroleum gas (LPG) Hydrocarbons, primarily composed of propane and butane, obtained during processing of crude oil, which are liquefied at low temperatures and moderate pressure. It is similar to NGL but originates from crude oil sources.

Lithology The geological characteristics of the reservoir rock; the study of rocks; important for exploration and drilling crews to have an understanding of lithology as it relates to the production of gas and oil.

LLS Louisiana light sweet crude oil.

Logging Lowering of different types of measuring instruments into the wellbore and gathering and recording data on porosity, permeability, and types of fluids present near the current well after which the data are used to construct subsurface maps of a region to aid in further exploration.

Long ton An avoirdupois weight measure equaling 2240 pounds.

LPG Liquefied petroleum gas.

LTO Light tight oil.

MACT Maximum achievable control technology. Applies to major sources of hazardous air pollutants.

Magnetometer A device to measure small changes in the Earth's magnetic field at the surface, which indicates what kind of rock formations might be present underground.

Maintenance water Water required to continue production over the life of a well; some wells may require flushing with freshwater to prevent salt accumulation in pipelines.

Manufactured gas A gas obtained by destructive distillation of coal or by the thermal decomposition of oil, or by the reaction of steam passing through a bed of heated coal or coke. Examples are coal gases, coke oven gases, producer gas, blast furnace gas, blue (water) gas, carbureted water gas; the Btu content varies widely.

Marcellus Shale A rock formation that extends from the base of the Catskills in New York and extends southwest to West Virginia, Kentucky, and Ohio.

Marginal resources A crude oil or natural gas resource for which the economics of the field are barely able to cover the costs of production.

Marine diesel oil A middle distillate fuel oil which can contain traces of 10% or more residual fuel oil from transportation contamination and/or heavy fuel oil blending. The MDO does not require heated storage.

Mcf (thousand cubic feet) One thousand cubic feet; a unit of measure that is more commonly used in the low volume sectors of the gas industry.

MCL Maximum contaminant level as dictated by regulations.

Membrane technology Gas separation processes utilizing membranes that permit different components of a gas to diffuse through the membrane at significantly different rates.

MER See Most efficient recovery rate.

Mercaptan A hydrocarbon group (usually a methane, ethane, or propane) with a sulfur group (−SH) substituted on a terminal carbon atom.

Metamorphic rocks Rocks resulting from the transformation that commonly takes place at great depths due to pressure and temperature. The original rocks may be sedimentary, igneous, or metamorphic.

Methane (CH_4) Commonly (often incorrectly) known as natural gas; colorless and naturally odorless, and burns efficiently without many byproducts.

Methanogens Methane-producing microorganisms

Methanol A fuel typically derived from natural gas, but which can be produced from the fermentation of sugars in biomass.

Metric ton A weight measure equal to 1000 kg, 2204.62 pounds, and 0.9842 long tons.

Micron A unit of length. One millionth of a meter or one thousandth of a millimeter; one micron equals 0.00004 of an inch.

Microseismic The process of using seismic recording devices to measure the location of fractures that are created during the hydraulic fracturing process; mapping of these microseismic events allows the extent of fracture development to be determined.

Microseismic fracture mapping A technique that provides images of fractures by detecting microseisms or microearthquakes that are set off by shear slippage on bedding planes or natural fractures adjacent to hydraulic fractures.

Middle distillate A term applied to hydrocarbons in the so-called middle range of refinery distillation; examples: heating oil, diesel fuels, and kerosene.

Migration (primary) The movement of hydrocarbons (oil and natural gas) from mature, organic-rich source rocks to a point where the oil and gas can collect as droplets or as a continuous phase of liquid hydrocarbon.

Migration (secondary) The movement of the hydrocarbons as a single, continuous fluid phase through water-saturated rocks, fractures, or faults followed by accumulation of the oil and gas in sediments (traps, *q.v.*) from which further migration is prevented.

MilliDarcy A Darcy (or Darcy unit) and milliDarcys (mD) are units of permeability.

Million 1×10^6.

Mineral rights The rights of the owner of the property to mine or produce any resources below the surface of the property.

Minimum ignition energy Minimum energy required to ignite a flammable fuel-air mixture.

Mist extractor A device installed in the top of scrubbers, separators, tray, or packed vessels to remove liquid droplets entrained in a flowing gas stream.

Moisture content The weight of the water contained in a fossil fuel, usually expressed as a percentage of weight, either oven-dry or as received.

Most efficient recovery rate (MER) The rate at which the greatest amount of natural gas may be extracted without harming the formation itself.

Motor gasoline A blended refinery product produced by blending several refinery streams; a complex mixture of relatively volatile hydrocarbons with or without small quantities of additives, that have been blended to form a fuel suitable for use in spark-ignition engines.

MSDS Material Safety Data Sheet. A document that provides pertinent information and a profile of a particular hazardous substance or mixture; the MSDS is normally developed by the manufacturer or formulator of the hazardous substance or mixture; the MSDS is required to be made available to employees and operators whenever there is the likelihood of the hazardous substance or mixture being introduced into the workplace; some manufacturers prepare the MSDS documents for products that are not considered to be hazardous to show that the product or substance is not hazardous.

Muds Used in drilling to lubricate the drilling bit in rotary drilling rigs.

Multistage fracturing The process of undertaking multiple hydraulic fracture stimulations in the reservoir section where parts of the reservoir are isolated and separate hydraulic fracturing.

Multilateral drilling A drilling technique that is similar to stacked drilling in that it involves the drilling of two or more horizontal wells from the same vertical well bore and the horizontal wells access different areas of the shale at the same depth but in different directions.

Multiple completions The result of drilling several different depths from a single well to increase the rate of production or the amount of recoverable gas.

NAAQS National Ambient Air Quality Standards; standards exist for the pollutants known as the criteria air pollutants: nitrogen oxides (NO_x), sulfur oxides (SO_x), lead, ozone, particulate matter, <10 μm in diameter, and carbon monoxide (CO).

Naphtha A volatile, colorless product of petroleum distillation. Used primarily as paint solvent, cleaning fluid, and blend stock in gasoline production, to produce motor gasoline by blending with straight-run gasoline.

Naphthenes One of three basic hydrocarbon classifications found naturally in crude oil; naphthenes are widely used as petrochemical feedstock; examples are cyclopentane; methyl-ethyl cyclopentane, and propyl cyclopentane.

Native gas Gas in place at the time that a reservoir was converted for use as an underground storage reservoir in contrast to injected gas volumes.

Natural gas Natural gas is a naturally occurring gas mixture consisting of methane and other hydrocarbons; used as an energy source to heat buildings, generate electricity and recently, to power motor vehicles.

Natural Gas Act Passed in 1938 and give the Federal Power Commission (now the Federal Energy Regulatory Commission or FERC) jurisdiction over companies engaged in interstate sale or transportation of natural gas.

Natural gas basin A depressed area in the Earth's crust, of tectonic origin, in which sediments have accumulated and natural gas has been generated and/or accumulated, and/or migrated.

Natural gas field A region or area that possesses or is characterized by natural gas.

Natural gas field facility A field facility designed to process natural gas produced from more than one lease for the purpose of recovering condensate from a stream of natural gas; however, some field facilities are designed to recover propane, normal butane, pentanes plus and to control the quality of natural gas to be marketed.

Natural gas hydrates Solid, crystalline, wax-like substances composed of water, methane, and usually a small amount of other gases, with the gases being trapped in the interstices of a water-ice lattice; they form beneath permafrost and on the ocean floor under conditions of moderately high pressure and at temperatures near the freezing point of water.

Natural gas liquids (NGL) Hydrocarbons, typically composed of propane, butane, pentane, hexane, and heptane, obtained from natural gas production or processing which are liquefied at low temperatures and moderate pressure; may be similar to liquefied petroleum gas (LPG) but originate from natural gas sources.

Natural gasoline A mixture of liquid hydrocarbons extracted from natural gas (*q.v.*) suitable for blending with refinery gasoline.

Natural gas plant liquids Those hydrocarbons in natural gas that are separated as liquids at natural gas processing plants, fractionating and cycling plants, and in some instances, field facilities; lease condensate is excluded; products obtained include liquefied petroleum gases (ethane, propane, and butanes), pentanes plus, and iso-pentane; the component products may be fractionated or mixed.

Natural gas play The active exploration or leasing of land for natural gas.

Natural Gas Policy Act of 1978 One of the first efforts to deregulate the gas industry and to determine the price of natural gas as dictated by market forces, rather than regulation.

Natural gas processing plant Facilities designed to recover natural gas liquids from a stream of natural gas that may or may not have passed through lease separators and/or field separation facilities; these facilities control the quality of the natural gas to be marketed; cycling plants are often classified as gas processing plants.

Natural Gas Resource Base An estimate of the amount of natural gas available, based on the combination of proved reserves, and those additional volumes that have not yet been discovered, but are estimated to be "discoverable" given current technology and economics.

NDPC North Dakota Petroleum Council

NES (National Energy Strategy) A 1991 federal proposal that focused on national security, conservation, and regulatory reform, with options that encourage natural gas use.

NESHAP National Emissions Standards for Hazardous Air Pollutants; emission standards for specific source categories that emit or have the potential to emit one or more hazardous air pollutants; the standards are modeled on the best practices and most effective emission reduction methodologies in use at the affected facilities.

Net thickness The thickness resulting from subtracting the portions of the reservoir that have no possibilities of producing hydrocarbon from the total thickness.

Neutralization number The weight in milligrams of an alkali needed to neutralize the acidic material in one gram of oil; the neutralization number of an oil is an indication of its acidity; see Acidity.

NGLs Natural gas liquids.

Nitrogen oxides (NOx) Products of combustion that contribute to the formation of smog and ozone.

Nonassociated gas Natural gas found in reservoirs that do not contain crude oil at the original pressure and temperature conditions; sometimes called *gas well gas*; gas produced from geological formations that typically do not contain much, if any, crude oil, or higher boiling hydrocarbons (*gas liquids*) than methane; can contain nonhydrocarbon gases such as carbon dioxide and hydrogen sulfide.

Nonhydrocarbon gases Typical nonhydrocarbon gases that may be present in reservoir natural gas, such as carbon dioxide, helium, hydrogen sulfide, and nitrogen.

Non-proved reserves Volumes of hydrocarbons and associated substances, evaluated at atmospheric conditions, resulting from the extrapolation of the characteristics and parameters of the reservoir beyond the limits of reasonable certainty or from the assumption of oil and gas forecasts with technical and economic scenarios other than those in operation or with a project in view.

NORM (naturally occurring radioactive material) Some hydrocarbon bearing formations contain naturally occurring radionuclides: crude oil, natural gas, or produced water may contain small quantities of NORM as a result of being in contact with the formation rock for many years; typically, the concentrations of NORM in natural gas, crude oil, or water are not high enough to cause concern but NORM can accumulate in pipe scale or tank sludge.

Normal fault The result of the downward displacement of one of the strata from the horizontal. The angle is generally between 25 and 60 degrees and it is recognized by the absence of part of the stratigraphic column.

NPDES permit National Pollutant Discharge Elimination System permit is the regulatory agency document issued by either a federal or state agency which is designated to control all discharges of pollutants from point sources into US waterways; the permits regulate discharges into navigable waters from all point sources of pollution, including industries, municipal wastewater treatment plants, sanitary landfills, large agricultural feed lots, and return irrigation flows.

Observation wells Wells that are completed and equipped to measure reservoir conditions and/or sample reservoir fluids, rather than to inject or produce reservoir fluids.

Oil and gas basin A region in which oil and gas of known or possible economic value occurs within a basin-type structure.

Oil from tar sand Synthetic crude oil (*q.v.*).

Oil mining Application of a mining method to the recovery of bitumen.

Oil shale A fine-grained impervious sedimentary rock which contains an organic material called kerogen.

Olamine process A process that used an amine derivative (an olamine) to remove acid gas from natural gas streams.

Olamines Compounds that are widely used in gas processing; examples are ethanolamine (monoethanolamine, MEA), diethanolamine (DEA), triethanolamine (TEA), methyldiethanolamine (MDEA), diisopropanolamine (DIPA), and diglycolamine (DGA).

Olefins A class of unsaturated paraffin hydrocarbons recovered from petroleum; typical examples include butene, ethylene, and propylene.

Organic sedimentary rocks Rocks containing organic material such as residues of plant and animal remains/decay.

Original gas volume in place The amount of gas that is estimated to exist initially in the reservoir and that is confined by geologic and fluid boundaries, which may be expressed at reservoir or atmospheric conditions.

Original pressure The pressure prevailing in a reservoir that has never been produced. It is the pressure measured by a discovery well in a producing structure.

Original reserve The volume of hydrocarbons at atmospheric conditions that are expected to be recovered economically by using the exploitation methods and systems applicable at a specific date. It is a fraction of the discovered and economic reserve that may be obtained at the end of the reservoir exploitation.

OSHA The Occupational Safety and Health Act of 1970 (OSHA) is a law designed to protect the health and safety of industrial workers and treatment plant operators which regulates the design, construction, operation, and maintenance of industrial plants and wastewater treatment plants; the Act does not apply directly to municipalities, except in those states that have approved plans and have asserted jurisdiction under

Section 18 of the OSHA Act; wastewater treatment plants have come under stricter regulation in all phases of activity as a result of OSHA standards; OSHA also refers to the federal and state agencies which administer the OSHA regulations.

Oxidation Combining elemental compounds with oxygen to form a new compound.

Oxidizing agent Any substance such as oxygen and chlorine that can accept electrons; when oxygen or chlorine is added to wastewater, organic substances are oxidized; the oxidized organic substances are more stable and less likely to give off odors or to contain disease bacteria.

Ozonation The application of ozone to water, wastewater, or air, generally for the purposes of disinfection or odor control.

Pad drilling A technique in which a drilling company uses a single drill pad to develop as large an area of the subsurface as possible; as many as six to eight horizontal wells can originate from the same pad.

PAH Polycyclic aromatic hydrocarbons.

Particulate A small, discrete mass of solid or liquid matter that remains individually dispersed in gas or liquid emissions.

Particulate emissions Particles of a solid or liquid suspended in a gas, or the fine particles of carbonaceous soot and other organic molecules discharged into the air during combustion.

Particulate matter (particulates) Particles in the atmosphere or on a gas stream that may be organic or inorganic and originate from a wide variety of sources and processes.

Pay zone thickness The depth of an oil shale deposit from which shale oil can be produced and recovered.

PCB Polychlorinated biphenyls; polychlorobiphenyls.

Perforation A hole in the casing, often generated by means of explosive charges, which enables fluid and gas flow between the wellbore and the reservoir.

Permeability A measure of the ability of a material to allow fluids to pass through it; it is dependent upon the size and shape of pores and interconnecting pore throats; a rock may have significant porosity (many microscopic pores) but have low permeability if the pores are not interconnected; permeability may also exist or be enhanced through fractures that connect the pores.

Petrochemical An intermediate chemical derived from petroleum, hydrocarbon liquids, or natural gas, such as ethylene, propylene, benzene, toluene, and xylene.

Petroleum (crude oil) A naturally occurring mixture of gaseous, liquid, and solid hydrocarbon compounds usually found trapped deep underground beneath impermeable cap rock and above a lower dome of sedimentary rock such as shale; most petroleum reservoirs occur in sedimentary rocks of marine, deltaic, or estuarine origin.

pH An expression of the intensity of the basic or acidic condition of a liquid; mathematically, pH is the logarithm (base 10) of the reciprocal of the hydrogen ion concentration; the pH may range from 0 to 14, where 0 is most acidic, 14 most basic, and 7 is neutral.

Physical limit The limit of the reservoir defined by any geological structures (faults, unconformities, change of facies, crests, and bases of formations, etc.), caused by contact between fluids or by the reduction to critical porosity of permeability limits or by the compound effect of these parameters.

Play A group of fields sharing geological similarities where the reservoir and the trap control the distribution of oil and gas; a geologic area where hydrocarbon accumulations occur—also called a *resource*; for shale gas, examples include the Barnett and Marcellus plays.

Pollutant A chemical (or chemicals) introduced into the land water and air systems of that is (are) not indigenous to these systems; also an indigenous chemical (or chemicals) introduced into the land water and air systems in amounts greater than the natural abundance.

Pollution The introduction into the land water and air systems of a chemical or chemicals that are not indigenous to these systems or the introduction into the land water and air systems of indigenous chemicals in greater-than-natural amounts.

Pooling or unitization A provision that allows landowners to combine land to form a drilling unit.

Population density The average number of people per square kilometer.

Pore space A small hole in reservoir rock that contains fluid or fluids; a 4-in. cube of reservoir rock may contain millions of interconnected pore spaces.

Pore volume The total volume of all pores and fractures in a reservoir or part of a reservoir.

Porosity The percentage of void space in a rock that may or may not contain oil or gas.

Possible reserves Reserves where there is an even greater degree of uncertainty but about which there is some information.

Potential reserves Reserves based upon geological information about the types of sediments where such resources are likely to occur and they are considered to represent an educated guess.

Pour point The lowest temperature at which an oil will pour or flow under certain prescribed conditions.

Pressure cores Cores cut into a special coring barrel that maintains reservoir pressure when brought to the surface; this prevents the loss of reservoir fluids that usually accompanies a drop in pressure from reservoir to atmospheric conditions.

Primary term The length of a lease in years.

Probable reserves Mineral reserves that are nearly certain but about which a slight doubt exists.

Process heat Heat used in an industrial process rather than for space heating or other housekeeping purposes.

Produced water The water that is brought to the surface during the production of oil and gas; typically consists of water already existing in the formation, but may be mixed with fracturing fluid if hydraulic fracturing was used to stimulate the well.

Producer The company generally involved in exploration, drilling, and refining of natural gas.

Producibility The rate at which oil or gas can be produced from a reservoir through a wellbore.

Producing well A well in an oil field or gas field used for removing fluids from a reservoir.

Production casing The final interval in a well and the smallest casing which forms the outer boundary of the annulus.

Production rate The rate of production of oil and/or gas from a well; usually given in barrels per day (bbls/day) for oil or standard cubic feet ($scft^3$/day) for gas.

Proppant Refers to particles (the propping agent) mixed with fracturing fluid to maintain fracture openings after hydraulic fracturing; these typically include sand grains, but they may also include engineered proppants; silica sand or other particles pumped into a formation during a hydraulic fracturing operation to keep fractures open and retain the induced permeability.

Proppant flowback Refers to the loss of proppant from a fracture which is a leading cause of production decline; proppant flowback can lead to damage of production equipment; can result in significant nonproductive time and may require well intervention thereby causing significant production loss.

Prospective resource The amount of hydrocarbons evaluated at a given date of accumulations not yet discovered but which have been inferred and are estimated as recoverable.

Proved amount in place The volume originally occurring in known natural reservoirs which has been carefully measured and assessed as exploitable under present and expected local economic conditions with existing available technology.

Proved area The known part of the reservoir corresponding to the proved volume.

Proved recoverable reserves The volume within the proved amount in place that can be recovered in the future under present and expected local economic conditions with existing available technology.

Proved reserves (proven reserves) Mineral reserves that have been positively identified as recoverable with current technology.

Proved resources Part of the resource base that includes the working inventory of natural gas; volumes that have already been discovered and are readily available for production and delivery.

Province A group of fields often found in a single geologic environment; see also Basin.

Psi Pounds per square inch.

Psia Pounds per square inch absolute.

Psig Pounds per square inch gauge.

Pyrolysis The thermal decomposition of organic materials at high temperatures (typically >350°C or 640°F) in the absence of air; the end product of pyrolysis is a mixture of solids (char), liquids (oxygenated oils), and gases (methane, carbon monoxide, and carbon dioxide) with proportions determined by operating temperature, pressure, oxygen content, and other conditions.

QA/QC Quality assurance/quality control.

Quad An abbreviation for a quadrillion (1,000,000,000,000,000) Btu; approximately equivalent to one trillion (1,000,000,000,000) cubic feet, or 1 Tcf. See also Bcf, Mcf, Tcf.

Quadrillion 1×10^{15}.

RACT Reasonably Available Control Technology standards; implemented in areas of nonattainment to reduce emissions of volatile organic compounds and nitrogen oxides.

Raw natural gas Impure natural gas as delivered from the well and before processing (refining).

Reclamation The act of restoring a site to a state suitable for use after removal of any contaminants.

Recovery factor The ratio between the original volume of oil or gas at atmospheric conditions and the original reserves of the reservoir.

Recovery rate The rate at which natural gas or crude oil is removed from a reservoir.

Recycled water Water utilized a second time in hydraulic fracturing operations after undergoing treatment for contaminants.

Reduced crude Crude oil that has undergone at least one distillation process to separate some of the lighter hydrocarbons; reducing crude lowers its API gravity, but increases the handling safety by raising the flash point.

Reduced emission completion (REC or green completion) An alternative practice that captures and separates natural gas during well completion and workovers activities instead of allowing it to vent into the atmosphere.

Reducing agent Any substance, such as the base metal (iron) or the sulfide ion that will readily donate (give up) electrons; the opposite of an oxidizing agent.

Refinery gas Noncondensate gas collected in petroleum refineries.

Reforming The process of converting methane to hydrogen using a catalyst and steam.

Refractory lining A lining, usually of ceramic, capable of resisting and maintaining high temperatures.

Reid vapor pressure (RVP) Measurement of petroleum product vapor pressure per ASTM D323; measures the equilibrium pressure of an air saturated petroleum product with an air filled vapor chamber with four times the volume of the attached liquid filled chamber at a temperature of 100°F—the equilibrium pressure is reported as the Reid vapor pressure.

Relative permeability The permeability of rock to gas, oil, or water, when any two or more are present, expressed as a fraction of the sir phase permeability of the rock.

Remaining reserves The volume of hydrocarbons measured at atmospheric conditions that are still to be commercially recoverable from a reservoir at a given date, using the applicable exploitation techniques. It is the difference between the original reserve and the cumulative hydrocarbon production at a given date.

Reserve additions Volumes of the resource base that are continuously moved from the resource category to the proved resources category.

Reserve depth interval The range of depths of the prospective shale area; deeper formations generally require more water for drilling.

Reserve growth The increase in estimated volumes of oil and natural gas that can be recovered from existing fields and reservoirs through time; most reserve growth results from delineation of new reservoirs, field extensions, or improved recovery techniques thereby improving efficiency, and recalculation of reserves due to changing economic and operating conditions.

Reserve-production ratio The result of dividing the remaining reserve at a given date by the production in a period. This indicator assumes constant production, hydrocarbon prices, and extraction costs, without variation over time, in addition to the nonexistence of new discoveries in the future.

Reserve replacement rate A rate that indicates the amount of hydrocarbons replaced or incorporated by new discoveries compared with what has been produced in a given period. It is the coefficient that arises from dividing the new discoveries by production during the period of analysis. It is generally referred to in annual terms and is expressed as a percentage.

Reserves Well-identified resources that can be profitably extracted and utilized with existing technology; the estimated volume of gas economically recoverable from single or multiple reservoirs. Reserve estimates are based on strict site-specific engineering criteria.

Reservoir An area that contains a resource. In fracking, well operators are seeking to tap into natural gas reservoirs deep underground.

Reservoir energy The underground pressure in a reservoir that will push the petroleum and natural gas up the wellbore to the surface.

Reservoir simulation Analysis and prediction of reservoir performance with a computer model.

Reservoir sweet spot The most productive area of a reservoir; particularly applicable terminology for shale reservoirs and tight reservoirs.

Residual fuel oil Heavy fuel oil (viscous fuel oil) produced from the nonvolatile residue from the fractional distillation process; heavy oil that is leftover from various refining processes; heavy black oil that is used in marine boilers and in heating plants.

Residue gas Natural gas from which the higher molecular weight hydrocarbons have been extracted; mostly methane.

Residuum (pl. residua, also known as resid or resids) The nonvolatile portion of petroleum that remains as residue after refinery distillation; hence, atmospheric residuum, vacuum residuum.

Resource The total amount of a commodity (usually a mineral but can include nonminerals such as water and petroleum) that has been estimated to be ultimately available; also called a *play*.

Resource assessment The process by which one estimates the location, amounts, and production of a resource; the USGS assesses resources in terms of (1) undiscovered resources and (2) technically recoverable resources.

Reused water Water utilized a second time in hydraulic fracturing operations with minimal treatment requirements.

Reverse fault The result of compression forces where one of the strata is displaced upwards from the horizontal.

Revision The reserve resulting from comparing the previous year's evaluation with the new one, in which new geological, geophysical, operation, and reservoir performance information is considered, in addition to variations in hydrocarbon prices and extraction costs. It does not include well drilling.

Rich gas A gaseous stream is traditionally very rich in natural gas liquids (NGLs); see Natural gas liquids.

Rock matrix The granular structure of a rock or porous medium.

Royalty A payment received by the lessor from the oil or gas company, based on the production of the well and market prices.

R/P (reserves/production) ratio Calculated by dividing proved recoverable reserves by production (gross less reinjected) in a given year.

RVP Reid vapor pressure.

Salinity A measure of salt content in a water sample; can also be expressed as total dissolved solids or electrical conductivity; high salinity waters cannot be put to beneficial reuse with first treating the water to remove some of the salt.

Sand A course granular mineral mainly comprising quartz grains that is derived from the chemical and physical weathering of rocks rich in quartz, notably sandstone and granite.

Sandstone A sedimentary rock formed by compaction and cementation of sand grains; can be classified according to the mineral composition of the sand and cement.

Scrubbing Purifying a gas by washing with water or chemical; also but less frequently used to describe the removal of entrained materials.

Seasonal variability The variation in water supply between months of the year. Higher values indicate more variation in water supply within a given year, leading to situations of temporary depletion or excess of water.

Secondary pollutants A pollutant (chemical species) produced by interaction of a primary pollutant with another chemical or by dissociation of a primary pollutant or by other effects within a particular ecosystem.

Secondary term The length of a lease after a well is drilled.

Sedimentary Formed by or from deposits of sediments, especially from sand grains or silts transported from their source and deposited in water, as sandstone and shale; or from calcareous remains of organisms, as limestone.

Sedimentary basin Any geological feature exhibiting subsidence and consequent infilling by sediments.

Sedimentary strata Typically consist of mixtures of clay, silt, sand, organic matter, and various minerals; formed by or from deposits of sediments, especially from sand grains or silts transported from their source and deposited in water, such as sandstone and shale; or from calcareous remains of organisms, such as limestone.

Seismic event An earthquake—induced seismicity is an earthquake caused by human activities.

Seismic section A seismic profile that uses the reflection of seismic waves to determine the geological subsurface.

Seismic survey A survey that consists of emitting waves through the subsoil and recording their return using groups of sensors; one of the basic and essential methods used in oil and gas exploration to generate information concerning the shape of the underground strata in the explored region.

Seismograph An instrument used to detect and record earthquakes, is able to pick up and record the vibrations of the Earth that occur during an earthquake; when seismology is applied to the search for natural gas, seismic waves, emitted from a source, are sent into the Earth and the seismic waves interact differently with the underground formation (underground layers), each with its own properties.

Seismology The study of the movement of energy, in the form of seismic waves, through the Earth's crust.

Shale A fine-grained sedimentary rock that is formed from compacted mud—black shale sometimes breaks down to form natural gas or oil.

Shale basin A large shale formation defined by similar geologic characteristics.

Shale gas Natural gas deposits found in shale reservoirs; see Unconventional gas.

Shale play The prospective areas of a shale basin where gas and oil could potentially be commercially extracted.

Shale resource Hydrocarbon resources found in shale plays, such as natural gas, natural gas liquids, and tight oil.

Shut-in royalty A payment to the lessor in lieu of a production royalty. This is received when a well cannot produce due to production problems.

SimDist Simulated distillation

Skin factor A factor used to refer to whether the reservoir is already stimulated or, perhaps, damaged; the most commonly used measure of formation damage in a well is the skin factor, *S*, which is a dimensionless pressure drop caused by a flow restriction in the near-wellbore region; typical values for the skin factor range from −6 for an infinite-conductivity massive hydraulic fracture to more than 100 for a poorly executed gravel pack.

Slagging Formation of hard deposits on boiler tubes and/or piston crowns, usually due to the presence of sodium, vanadium, and sulfur.

Slant drilling See Horizontal drilling.

Slick water fracturing (Slickwater fracturing) A method of hydraulic fracturing that adds chemicals to water to increase the fluid flow and increases the speed at which the pressurized fluid can be pumped into the wellbore.

Sludge Deposits in fuel tanks and caused by the presence of wax, sand, scale, asphaltene constituents, tars, and water; the sludge formed in a #6 fuel oil storage tank is mostly composed of heavy hydrocarbons; the sludge formed in diesel storage tanks is a combination of water with fungus and bacteria, which grow on the unevenly mixed water/fuel interface.

Sonic log A well log based on the time required for sound to travel through rock, useful in determining porosity.

Sour crude oil Crude oil containing a relatively high mass% sulfur; standards vary according to context; API (2011) identifies sour oils as having more than 1% *w/w* sulfur whereas the United States Strategic Petroleum Reserve identifies sour oils as having between 0.5% and 2.0% *w/w* sulfur; the SPR currently does not accept oils with 2.0% *w/w* sulfur or higher.

Sour gas Natural gas that contains hydrogen sulfide.

Spacing The optimum distance between hydrocarbon-producing wells in a field or reservoir.

Specifications A term referring to the properties of a given crude oil or petroleum product, which are specified since they often vary widely even within the same grade of product; in the normal process of negotiation, seller will guarantee buyer that product or crude to be sold will meet certain specified limits and will agree to have such limits certified in writing.

Specific gravity The mass (or weight) of a unit volume of any substance at a specified temperature compared to the mass of an equal volume of pure water at a standard temperature.

Specific heat The quantity of heat required to raise the temperature of a unit weight of a substance by 1 degree; usually expresses as calories/gram/C or BTU/lb/F.

S-shaped wells Wells drilled vertically several thousand feet and then extend in arc-shapes beneath the surface of the Earth; usually drilled from single pad to minimize surface disturbance.

Stabilization Removing higher vapor pressure components from a crude oil to lower the volatility of the oil to make it more acceptable for sale and further transport to a refinery or other manufacturing plant.

Stabilize To convert to a form that resists change. Organic material is stabilized by bacteria which convert the material to gases and other relatively inert substances. Stabilized organic material generally will not give off obnoxious odors.

Stacked drilling A technique used to drill several wells at different levels giving the appearance of a stack of wells, one on top of the other.

Stacked wells The drilling of horizontal where shale is sufficiently thick or multiple shale rock strata are found layered on top of each other; one vertical well bore can be used to produce gas from horizontal wells at different depths.

Standard conditions The reference amounts for pressure and temperature—in the English system, they are 14.73 pounds per square inch for the pressure and 60°F for temperature.

Stimulation Any of several processes used to enhance near reservoir permeability.

STP Standard temperature (25°C) and pressure (300 mmHg)

Straight-run Refers to a petroleum product produced by the primary distillation of crude oil, free of cracked components.

Stranded resource (stranded reserve) A resource (reserve) that is not economical to recover and transport to an existing market.

Strata Layers including the solid iron-rich inner core, molten outer core, mantle, and crust of the Earth.

Stratigraphy The subdiscipline of geology that studies the origin, composition, distribution, and succession of rock strata.

Stripper wells Natural gas wells that produce less than 60,000 cubic feet of gas per day.

Surface-active agent The active agent in detergents that possesses a high cleaning ability; used in a spray solution to improve its sticking and wetting properties when applied to plants, algae, or petroleum.

Surface casing A pipe that protects fresh-water aquifers and it also provides structural strength so that other casings may be used.

Surfactant A compound that lowers the surface tension of a liquid.

Sweet crude oil Crude oil containing a relatively low mass% sulfur. Standards vary according to context; API (2011) identifies sweet oils as having less than 1% *w/w* sulfur; the United States Strategic Petroleum Reserve identifies sweet oils as having less than 0.5% *w/w* sulfur.

Sweetening process A process for the removal of hydrogen sulfide and other sulfur compounds from natural gas.

Sweet gas Natural gas that contains very little, if any, hydrogen sulfide.

Sweet spot The most productive area of a reservoir; particularly applicable terminology for shale reservoirs and tight reservoirs: the area of a gas or oil play with the best potential for commercial exploitation due to its higher porosity and permeability.

Synthetic crude oil (syncrude) A hydrocarbon product produced by the conversion of coal, oil shale, or tar sand bitumen that resembles conventional crude oil; can be refined in a petroleum refinery.

Synthesis (syngas) A mixture of carbon monoxide and hydrogen; produced by reacting steam, or steam and oxygen, with a heated carbon-containing material such as natural gas.

TAN (total acid number) A measure of acidity that is determined by the amount of potassium hydroxide in milligrams that is needed to neutralize the acids in one gram of crude oil; a measure of the quality of crude oil.

Tar sand (bituminous sand) A formation in which the bituminous material (bitumen) is found as a filling in veins and fissures in fractured rocks or impregnating relatively shallow sand, sandstone, and limestone strata; a sandstone reservoir that is impregnated with a heavy, extremely viscous, black hydrocarbonaceous, petroleum-like material that cannot be retrieved through a well by conventional or enhanced oil recovery techniques; the several rock types that contain an extremely viscous hydrocarbon which is not recoverable in its natural state by conventional oil well production methods including currently used enhanced recovery techniques.

Tar sand bitumen A black hydrocarbonaceous material that occurs in the solid or near-solid state and is immobile under deposit conditions; see Bitumen.

TBN Total base number (ASTM D2896) which is measured in milligrams. KOH needed to neutralize an acidic solution through a reverse titration; TBN is the ability of the product to neutralize acid; in a motor oil, this is a property which allows the oil to neutralize acids from combustion that would otherwise degrade the oil; see Acidity, Neutralization number.

Tcf (trillion cubic feet) Gas measurement approximately equal to one quadrillion (1,000,000,000,000,000) British thermal units.

Technically recoverable reserves Natural gas reserves or crude oil reserves that are known to exist; the technology is available to drill, complete, stimulate, and produce this gas but the gas cannot be booked as reserves until the wells are drilled and the reservoirs are developed.

Technical reserves The accumulative production derived from a production forecast in which economic criteria are not applied.

Termination The end of a lease.

Thermal maturity The amount of heat to which a rock has been subjected; a thermally immature rock has not been subjected to sufficient heat to start the process of converting organic material into oil or gas.

Thermogenic Generated or formed by heat.

Thermogenic gas Gas formed by pressure effects and temperature effects on organic debris.

Three-dimensional (3D) seismic survey Allows producers to see into the crust of the Earth to find promising formations for retrieval of gas.

Tight formation A formation consisting of extraordinarily impermeable hard rock, typically sandstone or carbonate; the formations are relatively low permeability, non-shale, sedimentary formations; when a significant amount of organic matter has been deposited with the sediments, the shale rock can contain organic solid material (kerogen).

Tight gas Natural gas trapped in fine-grained sedimentary rocks with extremely low permeability, such as shale, sandstone, or carbonate.

Tight oil Oil trapped in fine-grained sedimentary rocks with extremely low permeability, such as shale, sandstone, or carbonate.

Tight reservoir A formation that contains natural gas and/or crude oil and gas; when a significant amount of organic matter has been deposited with the sediments, the shale rock can contain organic solid material (kerogen).

Time-lapse logging The repeated use of calibrated well logs to quantitatively observe changes in measurable reservoir properties over time.

Ton (long) An avoirdupois weight measure equal to 2240 pounds; see Tonne.

Ton (metric) A weight measure equal to 1000 kg, 2204.62 pounds, and 0.9842 long tons; see Tonne.

Ton (US ton) 2000 pounds

Tonne (Imperial ton, long ton, shipping ton) A weight measure equal to 2204.62 pounds; see Ton (metric).

Topped crude oil Oil from which the light ends (low-boiling constituents) have been removed by a simple refining process.

Total dissolved solids (TDS) The amount of inorganic compounds dissolved in a specific volume of sample; can easily be correlated to salinity; often used as a regulatory limit.

Total existent sediment The combination of inorganic and hydrocarbon sediments existing in a fuel as delivered.

Total organic carbon (TOC) The concentration of carbon material derived from decaying vegetation, bacterial growth, and metabolic activities of living organisms or chemicals found in source rocks.

Total petroleum system (TPS) The essential elements (source, reservoir, seal, and overburden rocks) and processes (generation-migration-accumulation and trap formation) as well as all genetically related petroleum that occurs in seeps, shows, and accumulations (discovered and undiscovered) whose provenance is a pod or closely related pods of active source rock.

Total thickness The thickness from the top of the formation of interest down to a vertical boundary determined by a water level or by a change of formation.

Toxicity The relative degree of being poisonous or toxic; a condition which may exist in wastes and will inhibit or destroy the growth or function of certain organisms.

Tracer test A technique for determining fluid flow paths in a reservoir by adding small quantities of easily detected material (often radioactive) to the flowing fluid and monitoring their appearance at production wells.

Transgression A geological term used to define the immersion of one part of the continent under sea level as a result of a descent of the continent or an elevation of the sea level.

Transmissibility (transmissivity) An index of producibility of a reservoir or zone, the product of permeability and layer thickness.

Traps A generic term for an area of the Earth's crust that has developed in such a way as to *trap* gas beneath the surface.

Traveling grate A type of furnace in which assembled links of grates are joined together in a perpetual belt arrangement. Fuel is fed in at one end and ash is discharged at the other.

TRI Toxics Release Inventory.

Triaxial borehole seismic survey A technique for detecting the orientation of hydraulically induced fractures, wherein a tool holding three mutually seismic detectors is clamped in the borehole during fracturing; fracture orientation is deduced through analysis of the detected microseismic perpendicular events that are generated by the fracturing process.

Trillion 1×10^{12}

Trillion cubic feet A volume measurement of natural gas. Approximately equivalent to one quad.

True vapor pressure (TVP) Pressure exerted by a gas that is in thermodynamic equilibrium with a liquid phase; the true vapor pressure for pure liquids is typically a known material property and constant at a given temperature; the true vapor pressure for mixtures such as crude oil is a function of the ratio of gas to liquid volume (V/L) and temperature.

Ullage The amount which a tank or vessel lacks of being full.

Ultimate recovery The cumulative quantity of oil and/or that will be recovered when revenues from further production no longer justify the costs of the additional production.

Unconformity A surface of erosion that separates younger strata from older rocks.

Unconventional gas Gas that occurs in tight sandstones, siltstones, sandy carbonates, limestone, dolomite, and chalk; a collective term for shale gas, tight gas, and coal bed methane; the gas is produced by methods that do not meet the criteria for conventional production; these methods include horizontal drilling, hydraulic fracturing, surface mining, and in situ processes. See also Shale gas.

Unconventional oil Crude oil that occurs in tight sandstones, siltstones, sandy carbonates, limestone, dolomite, and chalk; the oil is produced by methods that do not meet the criteria for conventional production; these methods include horizontal drilling, hydraulic fracturing, surface mining, and in situ processes.

Unconventional resource An umbrella term for oil and natural gas produced by means that do not fit the criteria for conventional production; the term is currently used to reference oil and gas resources whose porosity, permeability, fluid trapping mechanism, or other characteristics differ from conventional sandstone and carbonate reservoirs.

Undeveloped proved area The plant projection of the extension drained by the future producing wells of a producing reservoir and located within the undeveloped proved reserve.

Undeveloped proved reserves The volume of hydrocarbons that is expected to be recovered through wells without current facilities for production or transportation and future wells. This category may include the estimated reserve of enhanced recovery projects, with pilot testing or with the recovery mechanism proposed in operation that has been predicted with a high degree of certainty in reservoirs that benefit from this kind of exploitation.

Undiscovered resource The volume of hydrocarbons with uncertainty but whose existence is inferred in geological basins through favorable factors resulting from the geological, geophysical, and geochemical interpretation. They are known as prospective resources when considered commercially recoverable.

Unsaturated zone A zone where the soil and the rock contains air as well as water in its pores and which is above the groundwater table. The unsaturated zone does not contain readily available water, but it does provide water and nutrients to the biosphere.

Utica Shale A natural gas containing rock formation below the Marcellus Shale. The Utica Shale formation extends from eastern Ohio through much of Pennsylvania to western New York.

Vacuum distillation A secondary distillation process which uses a partial vacuum to lower the boiling point of residues from primary distillation and extract further blending components.

Vadose zone The layer of Earth between the land's surface and the position of groundwater is at atmospheric pressure.

Vanadium inhibitor An organic and/or inorganic metal bearing chemical intended to chemically and/or physically combine with the compounds formed during combustion of heavy fuel oil to improve the surface properties of the treated ash compounds.

Vapor density The density of any gas compared to the density of air with the density of air equal to unity.

Vapor-liquid equilibrium (VLE) A condition achieved when the chemical potential of each component in a liquid is equalized across the phase boundary, and there is no net transfer of vapor to liquid or liquid to vapor.

Vertical seismic profiling A method of conducting seismic surveys in the bore hole for detailed subsurface information.

Viscosity A measure of the ability of a liquid to flow or a measure of its resistance to flow; the force required to move a plane surface of area 1 m^2 over another parallel plane surface 1 m away at a rate of 1 m/s when both surfaces are immersed in the fluid; the higher the viscosity, the slower the liquid flows.

VOC (VOCs) Volatile organic compound(s); volatile organic compounds are regulated because they are precursors to ozone; carbon-containing gases and vapors from incomplete gasoline combustion and from the evaporation of solvents.

Volatile A volatile substance is one that is capable of being evaporated or changed to a vapor at a relatively low temperature; volatile substances also can be partially removed by air stripping.

Volatile organic compounds See VOC (VOCs)

VSP Vertical seismic profiling, a method of conducting seismic surveys in the borehole for detailed subsurface information.

Vug A small cavern or cavity within a carbonate rock.

Warm stacked drilling rig (also called *hot stacked rig* or *ready stacked rig*) A term that refers to a drilling rig that is deployable (warm) but idle (stacked); warm stacked rigs are typically mostly crewed, actively marketed, and standing by ready for work if a contract can be obtained; during the idle time, routine rig maintenance is continued, and daily costs may be modestly reduced but are typically similar to levels incurred in drilling mode; rigs are generally held in a ready stacked state if a contract is expected to be obtained relatively quickly.

Waste streams Unused solid or liquid byproducts of a process.

Water consumption The volume of freshwater that is taken from surface or groundwater resources and is not returned.

Water rights The legal authorization to withdraw and/or use natural water resources from both surface water and groundwater; usually administered through state laws and each state may have its own unique way of allocating and sharing water.

Water risk The ways in which water-related issues potentially undermine business viability.

Water scarcity The volumetric abundance, or lack thereof, of freshwater supply and increasingly accounts for water flow required to maintain the ecological health of rivers and streams.

Water stress A measures of the total annual water withdrawals (municipal, industrial, and agricultural) expressed as a percentage of water available.

Water withdrawals The volume of freshwater that is taken from surface or groundwater resources.

Well abandonment The final activity in the operation of a well when it is permanently closed under safety and environment preservation conditions.

Wellbore (well bore) The hole in the Earth comprising a well—this includes the inside diameter of the drilled hole bounded by the rock face; a channel created by drilling.

Well casing A series of metal tubes installed in the freshly drilled hole; serves to strengthen the sides of the well hole, ensure that no oil or natural gas seeps out of the well hole as it is brought to the surface, and to keep other fluids or gases from seeping into the formation through the well.

Well completion the process for completion of a well to allow for the flow of petroleum or natural gas out of the formation and up to the surface; includes strengthening the well hole with casing, evaluating the pressure and temperature of the formation, and then installing the proper equipment to ensure an efficient flow of natural gas out of the well; the complete outfitting of an oil well for either oil production or fluid injection; also the technique used to control fluid communication with the reservoir.

Well head (wellhead) The pieces of equipment mounted at the opening of the well to regulate and monitor the extraction of hydrocarbons from the underground formation; prevents leaking of oil or natural gas out of the well, and prevents blowouts due to high-pressure formations; the structure on the well at ground level that provides a means for installing and hanging casing, production tubing, flow control equipment, and other equipment for production.

Well logging A method used for recording rock (formation) and fluid properties to find gas and oil containing zones in subterranean formations.

Well logs The information concerning subsurface formations obtained by means of electric, acoustic, and radioactive tools inserted in the wells. The log also includes information about drilling and the analysis of mud and cuts, cores, and formation tests.

Wet gas Gas containing a relatively high proportion of hydrocarbons which are recoverable as liquids; see also Lean gas; natural gas that contains considerable amounts of higher molecular weight hydrocarbons other than methane.

Wet scrubbers Devices in which a countercurrent spray liquid is used to remove impurities and particulate matter from a gas stream.

Wobbe number (Wobbe Index) The calorific value of a gas divided by the specific gravity.

Workover The repair or refracturing of an existing oil or gas well to enhance or prolong production.

WTI West Texas intermediate crude oil

INDEX

Note: Page numbers followed by *f* indicate figures, and *t* indicate tables.

A

Absorption, 323
Absorption process, 341–342
Acid gases, 414
Acid-gas removal, 318
Acidizing, 217–218
Acid mine drainage, 279–280
Acid stage, 224–225
Acid treatment, 194
Additives, 225, 226–228*t*
Adsorption, 323–324
Air pollution
  acid gases, 414
  benzene emissions, 413
  CO_2 concentration, 413–414
  crystalline silica dust, 415
  fugitive emissions of natural gas, 412–413
  gaseous emissions, 413
  GHG, 413
  nitrogen oxides, 412–413
  sources, 412–413
Amount and distribution, 167–168
Annual salinity, 266
Antrim and new albany shale formations, 127
Antrim Shale, 130–132, 329–330
  black shale formations, 131
  composition, 131
  depths for, 131
  gas storage, 132
  intracratonic Michigan basin, 130–131
  mineralogy, 131
  natural fractures, 132
  organic matter, 131–132
  shallow wells, 132
  stratigraphy, 131
  vitrinite reflectance, 131–132
API gravity, 2–4, 3–4*t*, 349, 350*f*, 354–355
API gravity and sulfur content, 64–65*t*
Aquifer protection, 421–422
Argillaceous shale, 94
Asphaltene constituents, 21, 363–364
Asphaltene destabilization, 373
Asphalt treatment, 374
Associated gas, 314–315, 322
Atmospheric distillates, 375
Avalon and Bone Springs, 132–133

B

Bakken formation, 351–352
Bakken region of North Dakota, 370
Bakken shale, 133–134
Bakken tight oil, 22
  characteristics, 353*t*
Bakken tight oil corrosivity, 357
Bakken *vs.* Eagle Ford, 353*t*
Barnett shale, 99, 130, 134–136, 181–182, 330–331
Barnett shale formation, 330–331
Basins, 129*f*, 351*f*
Baxter shale, 136–137
Bed, 33–34
Benzene emissions, 413
Big Sandy shale gas, 137
Biofuels, 270
Biogenic and thermogenic shale gas, 14–15, 79
Biogenic gas, 14–15
Black shale formations, 131
Blending equipment, 230
Blending operations, 365
Borehole collapsing, 194
Brackish water, 279–280
Brittleness index of rocks, 113–114

C

Caney shale gas, 137–138
Capillary pressure, 222–223
Carbonaceous shale, 94
Carbonate, 92–93

Carbon dioxide concentration, 413–414
Carboniferous period, 69
Cardium formation, 158–159
Casing and cementing, 397
Casinghead gas, 8–9
Casting defect, 416
Cementation, 106
Cementitious materials, 73–74
Cement-pumping and acid-pumping equipment, 229–230
Chattanooga shale gas, 138
Chemical additives, 283, 415
Chemical-additive treatment solutions, 372
Chemical and physical processes, 317
Chemical compaction, 106
Chemicals dissolution, 419
Chemical treatment programs, 364
Clastic sedimentary rocks, 33
Clastic sources, 19
Clay genesis, 106–107
Clay minerals, 67*t*, 309, 315–316
Clay-size mineral grains, 93–94
Clean Air Act, 404–405*t*
Clean Water Act, 404–405*t*
Cleats, 11–12
Cloth filters, 324–325
Cloud point and pour point, 364
Coalbed methane, 280
- cleats, 11–12
- *vs.* conventional natural gas, 11
- dewatering, 11
- extraction, 12–13
- horizontal and multilateral drilling, 13
- hydraulic fracturing, 12
- permeability, 12, 28
- reservoirs, 27–28
- ventilation air methane, 11–12

Coking refinery, 384–385, 386*f*
Colorado group, 159
Composite refinery feedstock, 363
Compressive strengths, 47
Conasauga shale gas, 139
Condensate and crude oil removal, 336–338
Condensate blockage, 10
Conglomerate formations, 70
Conventional gas, 8, 11
Conventional petroleum extraction, 209
Conventional reservoir, 78–79, 168
Core analyses, 113
Cores, handling and testing, 110–111
Cracking refinery, 384, 385*f*
Cretaceous period, 69
Critical period of shale production, 217
Critical reservoir parameters, 216
Cross-bedding, 71–72
Crude oil(s)
- API gravity, 2–4, 3–4*t*, 4*t*, 349, 350*f*
- Bakken formation, 351–352
- Bakken *vs.* Eagle Ford, 353*t*
- in Canada, 352
- characteristics, 353*t*
- extraction problems, 352
- fractionating sequence, 6*f*
- geologic techniques, 5–6
- geothermal gradient, 4–5
- hydrocarbons, 5
- iron and copper, 2–4
- metal-containing constituents, 2–4
- nonhydrocarbon constituents, 5, 6*f*
- refinery processes
  - asphalt treatment, 374
  - atmospheric distillates, 375
  - categories, 383
  - challenges, 377
  - coking refinery, 384–385, 386*f*
  - cracking refinery, 384, 385*f*
  - desalting, 378–381
  - distillation, 374–375
  - domestic unconventional oils, 376–377
  - factors affecting, 378
  - fouling, 381–382
  - higher-boiling components, 374
  - hydroskimming, 383–384
  - issues, 387–388
  - light oil, 376
  - manufacturing plants, 387
  - mitigation strategies, 389–394
  - preheat exchanger fouling, 391
  - refinery feedstock, 376
  - skimming refinery, 383–384
  - waxy behavior, 378
  - yields and quality, 385–387
- reserves, 6

reservoir types, 349, 350f
sandstone and siltstone, 4–5
sedimentary basin, 4–5
sulfur content, 3–4t, 4t, 350f
thermal recovery methods, 5–6
transportation and handling
asphaltene destabilization, 373
Bakken region of North Dakota, 370
chemical-additive treatment solutions, 372
domestic crude oil production, 371
hazard classification, 373
lower-boiling flammable hydrocarbons, 371
pigging, 372
safety precautions, 373
sweet shale oil distribution, 370
in United States, 352
vanadium and nickel, 2–4
Crude oil blending, 25–26
Crude oil extraction, 21
Crude oil production, 168
Crude oil quality, 366
Cryogenic process, 342–343
Crystalline silica dust, 415

D

Deep formations, 416
Deep shale oil and gas projects, 269
Deep underground injection, 287
Dehydration, 271
Depositional rock types, 77–78
Depositional system, 180
Desalting, 378–381
Design considerations and procedures, 253
Development and production, 107–109
Devonian Duvernay shale, 160
Devonian period, 69
Devonian shale wells, 314–315
Dewatering, 11
Diagenesis, 92, 105
Direct far-field techniques, 246
Directional drilling technology, 212–213
Direct near-wellbore techniques, 247
Dissolved gas, 8–9
Dissolved inorganic constituents, 417–418
Dissolved organic constituents, 418
Distillate yields, 356t
Distillation, 319, 374–375
Dolomite, 70–71
Domestic crude oil production, 371
Domestic unconventional oils, 376–377
Downhole tiltmeter mapping technology, 245
Drainage area size, 80, 179–180
Drilling, 216, 230
formation damage and drilling speed, 183
horizontal drilling, 183–184, 186–190
multilateral, 190–193
pad drilling, 186–190
recovery technology, 185–186
stacked, 190–193
tight formations, 186
TOC content, 185
vertical well, 183–184
Drilling mud, 423–424
Drilling phase, 266
Drilling records, 126
Drinking water source contamination, 408
Dry natural gas, 8
Ductile failure, 248
Ductile reservoir, 216
Duvernay shale, 160

E

Eagle Ford shale gas, 140
Economically recoverable resources (ERR), 55
Economic and energy security benefits, 401–402
Economic assessment of reservoir, 193
Economic natural gas production, 200–201
Economic production, 176, 180–181, 194, 427
Effective stimulation volume, 242–243
Energy Policy Act, 404–405t, 408–409
Energy security, 51–53
Enhanced oil recovery, 269
Environmental impact
air pollution, 412–415

Environmental impact *(Continued)*
casing and cementing, 397
deep tight formation development, 398*t*
estimation techniques, 401
greenhouse gas emissions, 400
high carbon footprint, 400–401
hydraulic fracturing process, 399
public concerns, 401
remediation requirements
drilling mud, 423–424
economical gas flow rates, 427
economics, 426
flowback returns, 424
fracturing fluids formula, 424
human health and safety risks, 426
injection well checking, 425
microseismic imaging, 427
multistage hydraulic fracturing, 427
organic-rich tight formation sediments, 426–427
technology-driven resource, 427
waste water treatment, 424
water handling, 425
reservoir characteristics and fluid-transport mechanisms, 401
stimulation methods, 401
technical pathways, 401
tight formation resources, 399–400
unconventional natural gas and crude oil
aquifer protection, 421–422
economic and energy security benefits, 401–402
federal regulation, 403
hydraulic fracturing regulations, 402–403
induced seismic activity, 422–423
initial production (IP) rate, 407
local communities, 401
organization of regulatory agencies, 407
regulation and enforcement, 406–407
state regulation, 403–406
wastewater disposal, 402–403
water contamination, 403
water availability issues, 398–399
water pollution, 415–421
Environmental implications, 212
Environmentally sensitive headwater areas, 267
Environmental regulations, 225
EPA analytical methods, 302*t*
Evaporite minerals, 70–71
Extraction, 12–13, 19
Extraction problems, 352

F

Far-field techniques, 246
Faulty well construction, 416
Fayetteville shale, 331
Fayetteville shale gas, 140–141
Federal regulations and laws, 402, 406*t*
Field separator, 322, 323*f*
Filterable solids, 25, 362–363
Fischer assay test method, 45–46
Flammability, 357–358
Flash point, 357–358
Flowback fluid, 272–273
Flowback returns, 424
Flowback water and produced water, 195–196, 231, 272, 283, 287, 300
Flowback water recycling and reuse, 298–300
Floyd shale gas, 141–142
Fluid composition, 289–290
Fluid handling and storage, 290–291
Fluid properties, 20*t*
Fluid requirements, 288–289
Fluids management
fluids analysis
EPA analytical methods, 302*t*
flowback water, 300
method detection limits (MDLs), 301
organic compounds detection, 302*t*
produced water, 300
quality assurance, 304–305
quality control, 304–305
scaling, 301
waste fluids, 286–293
water contamination
coalbed methane, 280
groundwater contamination, 280
natural fractures, 280
scale remediation techniques, 281–282
scaling tendency of source water, 281

sulfate-reducing bacteria, 280–281
water disposal issues, 282–283
water requirements, 264–283
Flushing stage, 224–225
Fluvial systems, 180
Forced convection coolers, 325
Formation damage and drilling speed, 183
Formation evaluation, 218
Formation integrity test (FIT), 220–221
Fossils, 72
Fouling, 381–382
FRAC Act. *See* Fracturing Responsibility and Awareness of Chemicals (FRAC) Act
Fractionating sequence, 6*f*
Fractionation, 313–314, 343
Fracture fluid selection, 253–254
Fracture geometry and orientation, 215, 241
Fracture length, 214–215
Fracture networks, 195
Fracture patterns, 237–240
Fracture stimulation drilling technology, 18–19
Fracture treatment optimization, 252
Fracturing, 101
Fracturing fluid composition, 216, 419
Fracturing fluids formula, 424
Fracturing pressure, 201
Fracturing Responsibility and Awareness of Chemicals (FRAC) Act, 408–409
Fugitive emissions, 412–413

G

Gammon shale gas, 142
Gas and oil storage properties, 36
Gas-bearing shale formations, 328
Gas condensate, 307–308
Gas-condensate fluids, 96
Gas-condensate reservoirs, 7
Gaseous components, 310–311
Gaseous emissions, 413
Gas generation process, 313–314
Gas hydrates, 14
methane clathrates, 14
methane origin, 13–14
natural gas component, 13
physical appearance, 13
stability, 13
Gasoline blending, 365
Gas processing
absorption, 323
acid-gas removal, 318
adsorption, 323–324
associated gas, 322
chemical and physical process, 317
cloth filters, 324–325
dehydration, 319
distillation, 319
equipment, 320
field separator, 322, 323*f*
forced convection coolers, 325
impurities, 318
multistage gas-oil separation process, 324
natural convection coolers, 325
nonassociated gas, 322
oil/gas separator, 323
once-through wash operation, 317
pipeline-quality natural gas, 320, 322
process selectivity, 319–320
schematic flow for, 320–322, 321*f*
Selexol process, 318
sequence, 316
sulfur and carbon dioxide removal, 316–317
variables, 318
water coolers, 325
wet purifiers, 324
Gas production, 124–125
Gas recovery, 35–36
Gas storage, 132, 312–313
Geography, 270
Geologic age, 69, 69*t*, 91
Geological characteristics, 99
Geologic formations, 68
Geologic parameters, 79, 179, 219
Geologic techniques, 5–6
Geologic timescale, 124*t*
Geopressurized zones, 38–39
Geotechnical characterization, 219–220
Geothermal gradient, 4–5
GHG. *See* Greenhouse gas (GHG)
Glycol dehydration process, 339–340
Grade, 45–46

Graded bedding, 71–72
Grains, 72–73
Grain size analysis, 220
Gray shale formations, 121–124
Greenhouse gas emissions, 400
Greenhouse gas (GHG), 413
Groundwater, 276–277
Groundwater and surface water contamination, 416
Groundwater contamination, 280, 416

H

Handling and testing cores, 110–111
Harmless chemicals, 418–419
Haynesville/Bossier shale, 130
Haynesville shale, 331
Haynesville shale gas, 142–145
Hazard classification, 373
Hermosa shale gas, 145–146
Heterogeneity, 82–83
High carbon footprint, 400–401
Higher-boiling components, 374
High-molecular-weight paraffins, 364
Horizontal and directional drilling, 102
Horizontal and multilateral drilling, 13
Horizontal drilling, 26, 182–184, 186–190, 213
Horizontal drilling and hydraulic fracturing, 99
Horizontal fracture, 241
Horizontal hydraulic fracture, 231–232
Horizontal production well, 352
Horizontal variation, 314
Horizontal wells and multistage fracture, 127–128
Horn River basin, 161–162
Horton bluff group, 162
Human health and safety risks, 426
Hydraulic communication, 240–241
Hydraulic fracture parameters, 181
Hydraulic fracturing, 10, 12, 22–23, 26–27, 194, 204, 399
  acid stage, 224–225
  additives, 225, 226–228*t*
  conventional petroleum extraction, 209
  development of, 210*t*
  direct far-field techniques, 246
  directional drilling technology, 212–213
  direct near-wellbore techniques, 247
  environmental implications, 212
  environmental regulations, 225
  equipment
    blending equipment, 230
    cement-pumping and acid-pumping equipment, 229–230
    drilling, 230
    horizontal hydraulic fracture, 231–232
    pneumatic methods, 232–233
    Poisson ratio, 232
    produced water and flowback water, 231
  fluids management (*see* Fluids management)
  flushing stage, 224–225
  fracture monitoring
    microseismic monitoring, 244–245
    tiltmeters, 245
  fracture optimization
    effective stimulation volume, 242–243
    fracture geometry, 241
    horizontal fracture, 241
    hydraulic communication, 240–241
    nucleation and propagation, 242
    secondary porosity, 243
    shear fractures, 242
    transverse fractures, 243
  fracture patterns, 237–240
  horizontal drilling, 213
  hydrocarbon saturation, 229
  indirect fracture techniques, 247
  induced porosity, 214
  injection wells, 225–229
  local water planning agencies, 264–265
  modern oil-field production operations, 214
  pad stage, 224–225
  perforating gun, 213
  proppant type, 211*t*
  propping agent, 270
  prop sequence stage, 224–225
  pump pressure, 212
  reservoir evaluation
    blowouts, 218–219
    capillary pressure, 222–223
    critical period of shale production, 217

critical reservoir parameters, 216
drilling, 216
ductile reservoir, 216
formation evaluation, 218
formation integrity test, 220–221
fracture geometry and orientation, 215
fracture length, 214–215
fracturing fluid composition, 216
geologic parameters, 219
geotechnical characterization, 219–220
grain size analysis, 220
land seismic techniques, 217
mechanical properties, 223–224
moisture content, 220
permeability and porosity, 221–222
permeability estimation, 220
permeability levels, 214–215
residual fluid saturation, 222–223
shale matrix porosity and permeability, 216–217
slickwater fracturing, 216
unconfined compressive strength, 220
vertical wells, 218
well acidizing, 217–218
reservoir resource, 209–210
reservoir volume, 229
shale gas development, 211
slickwater, 213
tight reservoirs
brittle failure, 248
design considerations and procedures, 253
ductile failure, 248
fracture fluid selection, 253–254
fracture-propagation pressure, 247–248
fracture treatment optimization, 252
postfracture reservoir evaluation, 255
practices and procedures, 249–251t
production data, 255
reservoir selection, 248–252
stress-induced natural fractures, 248
tight sands, 213
water consumption estimation, 271
water pollution
chemical additives, 415
chemicals dissolution, 419
fracturing fluids composition, 419
groundwater and surface water contamination, 416
harmless chemicals, 418–419
metal concentration, 415
regulatory requirements, 420–421
saline formation water, 419–420
slick-water, 419
waste discharge, 415
water consumption, 417
water treatment options, 415
well casing, 420
watershed, 265
well development and completion
drilling equipments, 233–234
minimum compressive stress, 235
plug and perf method, 235
rotary drilling process, 234
sliding sleeve method, 235
Young's modulus, 235
Hydraulic fracturing operation, 196
Hydraulic fracturing process
additives, 403, 404–405t
chemical information, 408
disclosure of the materials, 409
drinking water source contamination, 408
Energy Policy Act, 408–409
federal laws, 406t
FRAC Act, 408–409
national emissions standards, 410
new performance standards, 410
Hydraulic fracturing process additives, 403, 404–405t
Hydraulic fracturing regulations, 402–403
Hydrocarbon(s), 5, 35, 76
Hydrocarbon constituents, 7
Hydrocarbon potential, 91
Hydrocarbon saturation, 231
Hydrogen sulfide, 328
Hydrogen sulfide gas, 25, 356
Hydrogen sulfide level, 23
Hydroskimming, 383–384
Hyperbolic production profile, 168

I

Indirect fracture techniques, 247
Induced fracturing, 37
Induced porosity, 214

Induced seismic activity, 422–423
Industrial waste treatment facilities, 297
Initial gas production, 326–327
Initial production (IP) rate, 407
Injection well(s), 225–229, 296
Injection well checking, 425
Inorganic constituents, 327, 361
Interstitial spaces, 100

J

Jurassic shale formations, 327

K

Kerogen oil, 20, 50–51, 177

L

Laminations, 34
Land seismic techniques, 204
Layers, 92
Leak-off tests (LOTs), 220–221
Lean gas, 8–9, 307–308
Lenticular/compartmentalized tight gas reservoir, 179–180
Lewis shale gas, 146
Light oil, 376
Limestone, 70–71
Lithology, 34
Local water planning agencies, 264–265
Lower-boiling flammable hydrocarbons, 371
Lower flammable limit (LFL), 357–358
Low-permeability reservoir(s), 104–105, 200
Low-permeability reservoir flow enhancement, 194

M

Mancos shale gas, 146–148
Manufacturing plants, 387
Marcellus shale, 96–97, 130, 293–294, 331
Marcellus shale gas, 148–149
Material deposition, 68
Matrix component, 73
Maturation processes, 314
Meandering formation, 178–179
Mechanical compaction, 106
Mechanical properties, 97, 223–224
Mercury, 310–311
Metal concentration, 415
Metal-containing constituents, 2–4
Methane, 310–311
Methane clathrates, 14
Methane origin, 13–14
Method detection limits (MDLs), 301
Michigan basin, 130–131
Microseismic imaging, 427
Microseismic monitoring, 244–245
Migration path, 76
Mineral constituents, 43–45, 72
Mineral dissolution, 106
Mineralogy, 131
Minimum compressive stress, 235
Mitigation strategies, 389–394, 390*f*
Modern oil-field production operations, 214
Moisture content, 219–220
Monterey/Santos, 149
Monterey/Santos shale, 149
Montney shale, 162–164
Morphology, 90–91
Mud deposition, 33
Mudstones, 93
Multilateral, 190–193
Multilateral drilling, 190–193, 197
Multistage flash process, 369
Multistage gas-oil separation process, 324
Multistage hydraulic fracturing, 427
Municipal waste water treatment plants, 296–297
Municipal water suppliers, 277

N

Naphtha, 23
National emissions standards, 410
Natural convection coolers, 325
Natural fractures, 132, 280
Natural gas
  associated gas, 8–9
  casinghead gas, 8–9
  composition, 7–8, 9*t*
  condensate, 10
  condensate blockage, 10
  constituents, 6–7, 7*t*
  conventional gas, 8
  dissolved gas, 8–9

dry natural gas, 8
gas-condensate reservoirs, 7
hydraulic fracturing, 10
hydrocarbon constituents, 7
lean gas, 8–9
nonhydrocarbon constituents, 7
residue gas, 8–9
sour gas, 8–10
sweet gas, 8–9
unconventional gas, 8
wet gas, 8–9
Natural gas and crude oil, 1, 63, 121, 175, 209, 262, 349, 397–398
Natural gas component, 13–14
Natural gas/crude oil extraction casing and cementing, 397
Natural gas imports, 166
Natural gas liquids (NGLs), 99, 310–311
Natural gas resources, 314–315
Neal shale, 149–151
Near-wellbore techniques, 247
NEPA, 404–405*t*, 406*t*
New Albany shale, 332
New Albany shale gas, 151–152
New performance standards, 410
Niobrara shale, 152
Nitrogen oxides, 412–413
Nonassociated gas, 322
Nonhydrocarbon constituents, 5, 6*f*, 7
Nonshale sedimentary tight formation, 121
North American shale gas, 201
NPDES, 404–405*t*, 406*t*

O

Odoriferous wax, 15
Ohio shale, 152–153
Oil-containing formation, 179
Oil/gas separator, 323
Oil Pollution Act, 404–405*t*
Oil shale
compressive strengths, 47
general properties, 42–43
grade, 45–46
kerogen, 50–51
mineral constituents, 43–45
permeability, 47
porosity, 46–47
thermal conductivity, 47–48
thermal decomposition, 48–50
as three-phase material, 39–40
Oil shale and tar sands, 269
Old drilling records, 126
Once-through wash operation, 317
Organic compounds detection, 302*t*
Organic content, 97–98
Organic matter, 131–132, 312–314
Organic-rich shale formations, 27, 327–328
Organic-rich tight formation sediments, 426–427
Original oil/gas in place, 54
Overbalanced drilling, 201–202

P

Pad drilling, 186–190
Pad stage, 224–225
Paraffin wax, 15–16, 25, 63
Paraffin wax deposition, 352
Pearsall shale, 153
Perforating gun, 213
Perforations, 195
Permeability and porosity, 12, 18, 18*f*, 28, 37, 47, 77–78, 83–90, 127–128, 221–222
Permeability differences, 176, 177*f*
Permeability estimation, 220
Permeability levels, 214–215
Permian period, 69
Petrographic rock types, 78
Physical and chemical processes, 78
Pierre shale, 154
Pigging, 372
Pipeline-quality natural gas, 320, 322
Plug and perf method, 234–235
Pneumatic methods, 232–233
Poisson ratio, 232
Pore spaces, 97
Porosity and permeability, 12, 18, 18*f*, 28, 37, 47, 77–78, 83–90, 127–128, 221–222
Porous rock, 178
Possible reserves, 55
Postfracture reservoir evaluation, 255
Potable water sources, 271
Potential reserves, 55

Preheat exchanger fouling, 391
Preproject comprehensive evaluation, 268
Pressure methane isotherms, 327
Pretreated seawater, 279–280
Probable reserves, 55
Process selectivity, 319–320, 334
Produced shale gas, 328
Produced water, 272–273
Produced water and flowback water, 231, 272–273, 300
Production data, 255
Production phase, 199–200
Production potential, 126–127
Productivity test data analysis, 178–179
Proper well construction, 288
Properties, 42–43
Proppant type, 211, 211*t*
Propping agent, 270
Prop sequence stage, 224–225
Proved reserves, 55–56
Public concerns, 401
Pump pressure, 212

Q

Quality assurance (QA), 304–305
Quality control (QC), 304–305

R

Rate of resource production, 168
Recovery operations, 287
Recovery technology, 185–186
Recycle operations, 284
Refinery feedstock, 376
Refinery process gas, 333
Regulation and enforcement, 406–407
Regulatory requirements, 420–421
Reserves, 6
Reservoir(s), 27–28, 266–267
Reservoir characteristics and fluid- transport mechanisms, 401
Reservoir development, 104–105
Reservoir drainage, 80
Reservoir evaluation blowouts, 218–219
Reservoir geology, 80
Reservoir parameters, 129–130
Reservoir pressure, 36
Reservoir properties, 166–167, 326
Reservoir resource, 209–210
Reservoir rock, 67–68, 75–76
Reservoirs and reservoir fluid
  core analyses, 110–114
  heterogeneity, 82–83
  hydrocarbons, 76
  migration path, 76
  morphology, 90–91
  natural gas and crude oil
    API gravity and sulfur content, 64–65*t*
    clay minerals, 67*t*
    constituents, 64*t*
    geologic formations, 68
    reservoir rock, 67–68
    source rock, 67–68
  paraffin wax, 63
  porosity, 77
  porosity and permeability, 83–90
  reservoir rocks, 75–76
  rock types
    biogenic and thermogenic shale gas, 79
    conventional reservoirs, 78–79
    depositional rock types, 77–78
    drainage area size, 80
    geologic parameters, 79
    hydraulic rock types, 78
    permeability-porosity relationships, 78
    petrographic rock types, 78
    physical and chemical processes, 78
    reservoir drainage, 80
    reservoir geology, 80–81
    unconventional reservoirs, 78–79
    well size, 80
  sediments (*see* Sedimentary rocks)
  source rocks, 75–76
  structural types, 81–82
  tight formations
    development and production, 107–109
    diagenesis, 92
    geologic age, 91
    hydrocarbon potential, 91
    layers, 92
    sandstone and carbonate formations, 102–107
    shale formations, 92–102
    unconventional reservoirs, 91

Reservoir selection, 248–252
Reservoir types, 349, 350*f*
permeability and production methods, 308*f*
Reservoir volume, 229
Residual fluid saturation, 222–223
Residue gas, 8–9, 307–308
Resource, 19, 97, 166–167
Resources and reserves
economically recoverable resources (ERR), 55
original oil/gas in place, 54
possible reserves, 55
potential reserves, 55
probable reserves, 55
proved reserves, 55–56
stranded resource, 56
subdivision, 54*f*
technically recoverable resources (TRR), 55
ultimately recoverable resources (URR), 54–55
Riparian zones, 284
Rock types, 77–81, 312
Rotary drilling process, 234
Routine core analyses, 111–112

S

Safe Drinking Water Act, 404–405*t*
Safety precautions, 373
Saline formation water, 283, 419–420
Sandstone, 69–70, 92–93
Sandstone and carbonate formations, 102–107
Sandstone and siltstone, 4–5
Sandstone sediments, 179
Scale deposition, 21
Scale remediation techniques, 281–282
Scaling, 301
Scaling tendency of source water, 281
Schematic flow for, gas processing, 320–322, 321*f*
Secondary porosity, 243
Sedimentary basin, 4–5, 98–99
Sedimentary rocks
bedding structure, 68
biochemical, 71
Carboniferous period, 69
cementitious materials, 73–74
chemical and organic, 70
composition, 72–74
conglomerate formations, 70
Cretaceous period, 69
cross-bedding, 71–72
Devonian period, 69
dolomite, 70–71
evaporite minerals, 70–71
fossils, 72
geologic age, 69, 69*t*
graded bedding, 71–72
grains, 72–73
limestone, 70–71
material deposition, 68
matrix component, 73
minerals, 72
Permian period, 69
sandstone, 69–70
stratification, 71–72
structure, 74–75
texture, 74
Sedimentary rocks bedding structure, 68
Sediments deposition, 94
Sediments source rocks, 75–76
Seismic and petrophysical properties, 94
Selexol process, 318–319
Sequence, 316
Shale basins, 268
Shale economics, 309
Shale formations, 91, 326
Barnett shale, 99
bed, 33–34
carbonate, 92–93
characteristics, 33
clastic sedimentary rocks, 33
clay-size mineral grains, 93–94
definition, 33–34
fissile, 33–34
fractures, 33–34
fracturing, 101
gas and oil storage properties, 36
gas-condensate fluids, 96
gas recovery, 35–36
geological characteristics, 99
horizontal and directional drilling, 102

Shale formations *(Continued)*
horizontal drilling and hydraulic fracturing, 99
hydrocarbons, 35
induced fracturing, 37
interstitial spaces, 100
laminations, 34
lithology, 34
Marcellus Shale, 96–97
mechanical properties, 97
mud deposition, 33
mudstones, 93
natural gas liquids, 99
organic content, 97–98
permeability, 37
pore spaces, 97
properties of, 102
reservoir pressure, 36
resource play, 97
sandstone, 92–93
sedimentary basins, 98–99
sediments deposition, 94
seismic and petrophysical properties, 94
shale reservoir, 93
shale thickness, 37–38
sweet spots, 35
technology-driven resource, 100
terrigenous, 33–34
thermal maturity, 35–36
total organic carbon, 37
total organic content, 35–36
types, 95–96
wellbore, 101–102
Shale gas and tight oil formations, 121, 122–123*t*
amount and distribution, 167–168
Antrim and New Albany Shale formations, 127
black color, 121–124
Canada, 128–129, 128*f*
Cardium formation, 158–159
Colorado group, 159
Devonian Duvernay shale, 160
Horn River Basin, 161–162
Horton Bluff Group, 162
Montney shale, 162–164
Utica shale, 164
conventional reservoir, 168
crude oil production, 168
economic recoverability, 167
estimated true size, 125–126
gas production, 124–125
geologic timescale, 124*t*
gray shale formations, 121–124
hyperbolic production profile, 168
largest shale gas resources, 126–127
natural gas imports, 166
new shale resource, 126–127
old drilling records, 126
production potential, 126–127
rate of resource production, 168
reservoir properties, 166–167
resource estimates, 166
storage properties, 167
tight well spacing, 168
United States
Antrim Shale, 130–132
Avalon and Bone Springs, 132–133
Bakken shale, 133–134
Barnett, 130, 134–136
Baxter Shale, 136–137
Big Sandy shale gas, 137
Caney shale gas, 137–138
Chattanooga shale gas, 138
Conasauga shale gas, 139
Eagle Ford shale gas, 140
Fayetteville shale gas, 140–141
Floyd shale gas, 141–142
Gammon shale gas, 142
Haynesville/Bossier, 130
Haynesville shale gas, 142–145
Hermosa shale gas, 145–146
horizontal wells and multistage fracture, 127–128
Lewis shale gas, 146
Mancos shale gas, 146–148
Marcellus, 130
Marcellus shale gas, 148–149
Monterey/Santos, 149
Neal shale, 149–151
New Albany shale gas, 151–152
Niobrara shale, 152
Ohio shale, 152–153
Pearsall shale, 153
permeability, 127
Pierre shale, 154
reservoir parameters, 129–130
tight sandstone, 127–128

unconventional resource accumulations, 127–128
Utah shale, 154–155
Utica shale, 155–156
Woodford shale, 157
United States and Canada, 122–123*t*
Shale gas development, 211
Shale gas production expansion, 270
Shale gas reservoirs, 93, 308*t*, 313
Shale gas resources, 126–127, 308
Shale matrix porosity and permeability, 216–217
Shale play, 349
Shale thickness, 37–38
Shallow wells, 132
Shear fractures, 242
Silt laminations, 107
Skimming refinery, 383–384
Slickwater, 213, 419
Slickwater fracturing, 216
Sliding sleeve method, 234–235
Solid-desiccant dehydration process, 340–341
Source rock, 67–68
Sour gas, 8–10, 307–308
Steel casing and cement, 197
Stimulation methods, 401
Storage properties, 167
Stranded resource, 56
Stratification, 71–72
Stratigraphy, 131
Stress-induced natural fractures, 248
Sulfate-reducing bacteria, 280–281
Sulfur and carbon dioxide removal, 316–317, 343–346, 345*t*
Sulfur and hydrogen sulfide, 358–360
Sulfur content, 3–4*t*, 4*t*, 24–25, 350*f*, 355
Surface and groundwater quality, 295–296
Surface water, 274–276
Surface water contamination, 416
Suspended solids, 417–418
Sweet gas, 8–9
Sweet shale oil distribution, 353–354, 370
Sweet spots, 35

T

Technically recoverable resources (TRR), 55
Technology-driven resource, 100, 427
Terrigenous, 33–34
Texture, 74
Thermal conductivity, 47–48
Thermal decomposition, 48–50
Thermal maturity, 19, 35–36, 313–314
Thermal recovery methods, 5–6
Thermogenic and biogenic pathways, 312–313
Three Forks formation, 22
Tight formation(s), 38, 186, 398*t*
basins, 129*f*, 351*f*
core analyses
brittleness index of rocks, 113–114
handling and testing cores, 110–111
routine core analyses, 111–112
development and production, 107–109
diagenesis, 92
gas and oil resources (*see* Shale gas and tight oil formations)
geologic age, 91
horizontal production well, 352
hydrocarbon potential, 91
layers, 92
natural gas/crude oil extraction
casing and cementing, 397
deep tight formation development, 398*t*
environmental impact (*see* Environmental impact)
nonshale sedimentary tight formation, 121
in North America, 349–351
sandstone and carbonate formations, 102–107
cementation, 106
chemical compaction, 106
clay genesis, 106–107
diagenesis, 105
low-permeability reservoirs, 104–105
mechanical compaction, 106
mineral dissolution, 106
reservoir development, 104–105
silt laminations, 107
shale formations, 92–102
tight oil (*see* Tight oil)
unconventional gas production, 125
unconventional reservoirs, 91
Tight formation basin, 293
Tight formation development, 269

Tight formation resources, 399–400
Tight gas
energy security, 51–53
permeability, 18, 18*f*
reservoirs, 28
shale gas
characteristics, 18
clastic sources, 19
extraction, 19
fracture stimulation drilling technology, 18–19
horizontal drilling, 26
hydraulic-fracturing operation, 26–27
organic-rich shale formations, 27
resource, 19
thermal maturity, 19
total organic content, 19
Tight gas processing
clay minerals, 315–316
Devonian shale wells, 314–315
differences in, 315
fluids, 311*t*
fractionation, 313–314
gaseous components, 310–311
gas generation process, 313–314
gas plant operations
glycol dehydration process, 339–340
solid-desiccant dehydration process, 340–341
water removal, 338–341
gas storage, 312–313
horizontal variation, 314
maturation processes, 314
mercury, 310–311
methane, 310–311
natural gas liquids, 310–311
natural gas liquids separation
absorption process, 341–342
cryogenic process, 342–343
fractionation, 343
natural gas resources, 314–315
organic matter, 313–314
process selectivity, 334
properties
Antrim shale, 329–330
Barnett shale formation, 330–331
Fayetteville shale, 331
gas-bearing shale formations, 328
Haynesville shale, 331
hydrogen sulfide, 328
initial gas production, 326–327
inorganic constituents, 327
Jurassic shale formations, 327
Marcellus shale, 331
New Albany shale, 332
organic-rich shale formations, 327–328
pressure methane isotherms, 327
produced shale gas, 328
reservoir properties, 326
reservoir properties and fracture parameters, 327
shale formations, 326
refinery process gas, 333
rock types, 312
shale gas reservoirs, 313
sulfur and carbon dioxide removal, 343–346
thermal maturity, 313–314
thermogenic and biogenic pathways, 312–313
transportation pipelines, 332
wellhead processing
condensate and crude oil removal, 336–338
water removal, 334–336
Tight oil
API gravity, 354–356
asphaltene constituents, 21
Bakken tight oil, 22
characteristics, 353*t*
corrosivity, 357
crude oil blend(s), 26
crude oil blending, 25
crude oil extraction, 21 (*see* Crude oils)
distillate yields, 356*t*
energy security, 51–53
filterable solids, 25
flammability, 357–358
fluid properties, 20*t*
hydraulic-fracturing process, 22–23
hydrogen sulfide gas, 25, 356–357
hydrogen sulfide level, 23
incompatibility
asphaltene constituents, 363–364

blending operations, 365
chemical treatment programs, 364
cloud point and pour point, 364
composite refinery feedstock, 363
crude oil quality, 366
filterable solids, 362–363
gasoline blending, 365
high-molecular-weight paraffins, 364
paraffin wax deposition, 362
wax deposition, 365–366
well completion and stimulation technique, 366
inorganic constituents, 361
kerogen oil, 20
lack of transportation infrastructure, 24
naphtha, 23
in North America, 21
paraffin carbon chains, 23–24
paraffin waxes, 25
scale deposition, 21
sulfur and hydrogen sulfide, 357–358
sulfur content, 24–25, 355
Three Forks Formation, 22
volatility, 360–361
wellhead processing
conditioning, 368
configuration and operating conditions, 367–368
fractionation, 369
multistage flash process, 369
stabilization, 368–369
Tight reservoirs and conventional reservoirs
abandonment and reclamation, 199–200
Barnett shale, 181–182
depositional system, 180
drainage area size, 179–180
drilling, 183–193
economic production, 180–181
economic viability, 181
fluvial systems, 180
geologically shale, 177
geologic parameters, 179
hydraulic fracture parameters, 181
lenticular/compartmentalized tight gas reservoir, 179–180
meandering formation, 178–179
oil-containing formation, 179
permeability differences, 177*f*
porous rock, 178
production phase, 199–200
production trends
economic natural gas production, 200–201
fracturing pressure, 201
hydraulic fracturing, 204
low-permeability reservoirs, 200
North American shale gas, 201
overbalanced drilling, 201–202
underbalanced drilling, 201–202
vertical wells drilling, 203
productivity test data analysis, 178–179
sandstone sediments, 179
spatial distribution, 178–179
ultralow permeability, 181–182
well completion, 193–198
well integrity, 198–199
well shape, 179–180
Tight reservoirs brittle failure, 248
Tight sands, 213
Tight sandstone, 127–128
Tight well spacing, 168
Tiltmeters, 245
Total organic carbon, 185
Total organic content, 19, 35–36
Transportation, 291–293
Transportation pipelines, 332
Transverse fractures, 243
TSCA, 404–405*t*

U

Ultimately recoverable resources (URR), 54–55
Ultralow permeability, 181–182
Unconfined compressive strength, 219–220
Unconventional gas, 8
Unconventional gas production, 125
Unconventional gas reservoirs, 178
Unconventional natural gas and crude oil, 198, 397–398, 401
Unconventional reservoirs, 78–79, 91
Unconventional resource accumulations, 127–128
Underbalanced drilling, 201–202
Upper flammable limit (UFL), 357–358

Utah shale, 154–155
Utica shale, 155–156, 164

V

Vanadium and nickel, 2–4
Vertical well, 183–184, 198–199
Vertical well drilling, 203
Volatility, 360–361

W

Waste discharge, 415
Waste fluids
  deep underground injection, 287
  flowback water, 287
  fracturing
    fluid composition, 289–290
    fluid handling and storage, 290–291
    fluid requirements, 288–289
    transportation, 291–293
  proper well construction, 288
  recovery operations, 287
  water flow control, 287–288
Wastewater and power plant cooling water, 277–278
Wastewater disposal, 402–403
Wastewater treatment, 424
Water acquisition, use, and management, 294*t*
Water availability, 398–399
Water consumption, 271, 284–285, 417
Water contamination
  coalbed methane, 280
  groundwater contamination, 280
  natural fractures, 280
  scale remediation techniques, 281–282
  scaling tendency of source water, 281
  sulfate-reducing bacteria, 280–281
  water disposal issues, 280–283, 403
Water disposal, 267, 282–286, 294–295
Water flow control, 287–288
Water handling, 425
Water management and disposal
  chemical additives, 283
  flowback water, 283
  flowback water recycling and reuse, 298–300
  industrial waste treatment facilities, 297
  injection wells, 296
  Marcellus tight formation, 293–294
  municipal wastewater treatment plants, 296–297
  recycle operations, 284
  Riparian zones, 284
  saline water formation, 283
  seasonal flow differences, 283–284
  surface and groundwater quality, 295–296
  tight formation basin, 293
  water acquisition, use, and management, 294*t*
  water consumption, 284–285
  water disposal options, 295
Water pollution, 415–421
  accidental spilling, 416
  casting defect, 416
  deep formations, 416
  dissolved inorganic constituents, 417–418
  dissolved organic constituents, 418
  faulty well construction, 416
  hydraulic fracturing
    chemical additives, 415
    chemicals dissolution, 419
    fracturing fluids composition, 419
    groundwater and surface water contamination, 416
    harmless chemicals, 418–419
    metal concentration, 415
    regulatory requirements, 420–421
    saline formation water, 419–420
    slick-water, 419
    waste discharge, 415
    water consumption, 417
    water treatment options, 415
    well casing, 420
  suspended solids, 417
Water quality, 266
Water recycling and reuse, 272–273
Water removal, 334–336, 338–341
Water requirements
  annual salinity, 266
  biofuels, 270
  challenges, 266–267
  deep shale oil and gas projects, 269
  drilling phase, 266
  enhanced oil recovery, 269

environmentally sensitive headwater areas, 267
geography, 270
oil shale and tar sands, 269
preproject comprehensive evaluation, 268
reservoirs, 266–267
shale basins, 268
shale gas production expansion, 270
tight formation development, 269
water disposal, 267
water quality, 266
water sources, 266
acid mine drainage, 279–280
brackish water, 279
chemicals, 273
contaminants, 274
flowback fluid, 272–273
groundwater, 276–277
municipal water suppliers, 277
potable water sources, 271
pretreated seawater, 279–280
produced water, 272
recycling and reuse, 272–273
reservoir water and recycled flowback water, 278–279
selection of, 273–274
surface water, 274–276
wastewater and power plant cooling water, 277–278
water stress, 267–268
Watershed, 264–265
Water sources, 266
Water stress, 267–268
Water treatment options, 415
Wax, 15–16
Wax deposition, 365–366
Waxy behavior, 378
Wellbore, 101–102
Well casing, 420–421
Well completion
acid treatment, 194
borehole collapsing, 194
economic assessment of reservoir, 193
economic viability, 194
flowback water, 195–196
fracture networks, 195
hydraulic fracturing, 194
hydraulic fracturing operation, 196
low-permeability reservoirs flow enhancement, 194
multilateral drilling, 197
perforations, 195
steel casing and cement, 197
Well completion and stimulation technique, 233–235, 366
Wellhead processing, 367–369
Well integrity, 198–199
Well shape, 179–180
Well size, 80
Wet gas, 8–9, 307–308
Wet purifiers, 324
Woodford shale, 157

Printed in the United States
By Bookmasters